D0850119

Benchmark Papers in Inorganic Chemistry

Series Editor: Harry H. Sisler
University of Florida

Other volumes in the series

Compounds Containing Phosphorous-Phosphorous Bonds, Alan H. Cowley
Chloramination Reactions, Stephen E. Frazier
Hard and Soft Acids and Bases, Ralph G. Pearson
Symmetry in Chemical Theory, John P. Fackler, Jr.
Ligand Field Theory, R. Carl Stoufer
Compounds Containing Arsenic-Nitrogen and Antimony-Nitrogen Bonds, Larry K. Krannich

Benchmark Papers in Inorganic Chemistry

METAL-AMMONIA SOLUTIONS

Edited by
WILLIAM L. JOLLY
University of California, Berkeley

Library of Congress Catalog Card Number: 72-80134
ISBN: 0-87933-008-2

Manufactured in the United States of America.

Exclusive distributor outside the United States and Canada:
JOHN WILEY & SONS, INC.

Acknowledgements and Permissions

ACKNOWLEDGEMENTS

Johann Ambrosius Barth, Leipzig—ANNALEN DER PHYSIK
"Über Metallammonium–Verbindungen"
The American Chemical Society—JOURNAL OF PHYSICAL CHEMISTRY
"The Electrolysis and Electrolytic Conductivity of Certain Substances Dissolved in Liquid Ammonia"
W. A. Benjamin, New York—SOLUTIONS MÉTAL-AMMONIAC; PROPRIÉTÉS PHYSICOCHIMIQUES edited by G. Lepoutre and M. J. Sienko
"On the Coexistence of Liquid Phases in Metal–Ammonia Systems and Some Surface Tension Studies on These Solutions above Their Consolute Points"

PERMISSIONS

The following papers have been reprinted with the permission of the authors, publishers and the present copyright owners.

Johann Ambrosius Barth, Leipzig—ANNALEN DER PHYSIK
"Über die Lösungen von Natrium in Flüssigem Ammoniak: Magnetismus; Thermische Ausdehnung; Zustand des Gelösten Natriums"

Gauthier-Villars—COMPTES RENDUS HEBDOMADAIRES DES SEANCES DE L'ACADEMIE DES SCIENCES
"Action du Phosphure d'Hydrogène sur le Potassammonium et Sodammonium"
"Sur le Baryum–Ammonium et l'Amidure de Baryum"
"Constantes d'Association dans les Solutions Diluées Métal–Ammoniac"

American Chemical Society—JOURNAL OF THE AMERICAN CHEMICAL SOCIETY
"Solution of Metals in Non-Metallic Solvents; On the Formation of Compounds between Metals and Ammonia"
"A Study of Sodium-Lead Compounds in Liquid Ammonia"
"The Constitution of Metallic Substances"
"Polarography in Liquid Ammonia. The Electron Electrode"
"The Heats of Reaction of Lithium, Sodium, Potassium and Cesium with the Ammonium Ion in Liquid Ammonia at −33°"
"The Oxidation of Lithium and the Alkaline Earth Metals in Liquid Ammonia"
"Liquid Ammonia Chemistry of the Methyl Phosphines"
"Chemical Generation of the Ammoniated Electron *via* Ytterbium"
"Solutions of Metals in Non-Metallic Solvents. The Electromotive Force of Concentration Cells of Solutions of Sodium in Liquid Ammonia and the Relative Speed of the Ions in These Solutions"
"Solutions of Metals in Non-Metallic Solvents. The Conductance of the Alkali Metals in Liquid Ammonia"
"Electronic Processes in Liquid Dielectric Media. The Properties of Metal-Ammonia Solutions"
"The Thermoelectric Properties of Metal-Ammonia Solutions. Theory and Interpretation of Results"
"The Activity Coefficient of Sodium in Liquid Ammonia"
"A Revised Model for Ammonia Solutions of Alkali Metals"
"Low-Temperature X-Ray Study of the Compound Tetraamminelithium (0)"
"Solubilization of Alkali Metals in Tetrahydrofuran and Diethyl Ether by Use of a Cyclic Polyether"

continued

American Chemical Society—INORGANIC CHEMISTRY
"The Reversibility of the Reaction of Alkali Metals with Liquid Ammonia"

American Chemical Society—JOURNAL OF PHYSICAL CHEMISTRY
"A Kinetic Study of the Reaction of Water and *t*-Butyl Alcohol with Sodium in Liquid Ammonia"
"Rate of the Reaction of the Ammoniated Electron with the Ammonium Ion at −35°"
"Solutions of Europium and Ytterbium Metals in Liquid Ammonia"
"Temperature Dependence of the Knight Shift of the Sodium–Ammonia System"
"The Conductance of Dilute Solutions of Lithium in Liquid Ammonia at −71°"
"Liquid–Liquid Phase Separation in Alkali Metal–Ammonia Solutions. Sodium and Potassium"
"Metal–Ammonia Solutions. Spectroscopy of Quaternary Ammonium Radicals"
"Metal–Ammonia Solutions. Optical Properties of Solutions in the Intermediate Concentration Region"

Akademische Verlagsgesellschaft—ZEITSCHRIFT FUR PHYSIKALISCHE CHEMIE
"Salzartige Verbindungen und Intermetallische Phasen des Natriums in Flüssigem Ammoniak"

The Electrochemical Society, Inc.—JOURNAL OF THE ELECTROCHEMICAL SOCIETY
"Potentiometric Titrations Involving Solutions of Metals in Liquid Ammonia"

Verlag der Zeitschrift für Naturforschung, Tübingen—ZEITSCHRIFT FÜR NATURFORSCHUNG
"Neuartige Reaktionen der Metallcarbonyle mit Alkalimetallen"

The Chemical Society, London—QUARTERLY REVIEW
"Reduction by Metal–Amine Solutions; Applications in Synthesis and Determination of Structure"

The Chemical Society, London—CHEMICAL COMMUNICATIONS
"The Mechanisms of Metal–Ammonia Reduction of Alkyl Halides"
"Some Competitive Rate Studies of the Ammoniated Electron with Metal Ions: Evidence for Transient Intermediates"

The Chemical Society, London—JOURNAL OF THE CHEMICAL SOCIETY
"The Reaction of Tetra-alkylammonium Halides with Potassium in Liquid Ammonia"
"Alkali-Metal-Amine Solutions. Spectroscopic and Magnetic Studies"
"Unstable Intermediates. Solvated Electrons: Solutions of Europium in Ammonia"

The American Physical Society—PHYSICAL REVIEW
"The Absorption Spectra of the Blue Solutions of Sodium and Magnesium in Liquid Ammonia"
"Bose-Einstein Condensation of Trapped Electron Pairs. Phase Separation and Superconductivity of Metal–Ammonia Solutions"
"Incoherent Scattering and Electron Transport in Metal–Ammonia Solutions"

The American Physical Society—REVIEW OF MODERN PHYSICS
"Paramagnetic Resonance Absorption in Solutions of K in Liquid NH_3"

National Research Council of Canada—JOURNAL OF CANADIAN CHEMISTRY
"Absorption Spectra of Metals in Solution"

The American Institute of Physics—JOURNAL OF CHEMICAL PHYSICS
"Model for Metal–Ammonia Solutions"
"Heats of Solution of Alkali Metals in Liquid Ammonia at 25°"
"Hyperfine Interactions in Solutions of Ca and Rb in Methylamine"
"Electronic Processes in Solutions of Alkali Metals in Liquid Ammonia. A Model Including Two Diamagnetic Species"
"Knight Shifts and Relaxation Times of Alkali–Metal and Nitrogen Nuclei in Metal–Ammonia Solutions"
"Spin Densities in Alkali–Metal–Ammonia Solutions"
"Phase Changes and Electrical Conductivity of Concentrated Lithium–Ammonia Solutions and the Solid Eutectic"
"Absence of a Volume Change upon Dilution of Dilute Sodium–Ammonia Solutions at −45° and Potassium–Ammonia Solutions at −34°"
"Thermal Properties of Solid Lithium Tetraamine"
"Origin of the 600 mμ Band in the Spectra of Alkali–Metal–Amine Solutions"
"Hall Effect in Lithium–Ammonia Solutions"
"Excess Electrons in Polar Solvents"
"Solvation Number of the Electron in Liquid Ammonia. A Nitrogen Magnetic Relaxation Study"

Van Nostrand Reinhold Co.—THE NITROGEN SYSTEM OF COMPOUNDS by E. C. Franklin
Excerpt from the Introduction

Butterworth and Company, Publishers, Ltd.—International Union of Pure and Applied Chemistry—SUPPLEMENT TO PURE AND APPLIED CHEMISTRY—Proceedings of an International Conference on the Nature of Metal–Ammonia Solutions held at Cornell University, Ithaca, New York, U.S.A., June 1969, Published, 1970.
"Pressure Effects in Metal–Ammonia Solutions"

Series Editor's Preface

This is the first volume in the series Benchmark Papers in Inorganic Chemistry in which outstanding papers in the development of various fields of inorganic chemistry research are collected by a major expert in the field and published along with explanatory notes which put these papers into context. This volume deals with the development of our knowledge of solutions of metals in liquid ammonia. Because of the unique and unusual character of metal-ammonia solutions, this topic is an appropriate one for this series. Professor William Jolly, an outstanding expert in liquid ammonia chemistry, is eminently qualified to prepare this important volume.

Harry H. Sisler

Preface

Electropositive metals such as sodium and calcium dissolve in liquid ammonia to form solutions which are blue when dilute and bronze-colored when concentrated. The behavior of these remarkable solutions has fascinated chemists and physicists for more than a century, and, although at least 2000 articles on this subject have been published, there is no decline in the rate of such publication. It seems that the more we learn about these solutions, the more intriguing they become. Work on metal-ammonia solutions can be roughly categorized into studies of the chemical properties and studies of the physical properties. Therefore the reprints in this book have been separated into these same categories. The first section, on chemical properties, contains some information on the use of the solutions in both inorganic and organic synthesis. The second section, on physical properties, contains many discussions of models used to interpret the physical properties. Obviously some readers will be more interested in one section than another; however, in many cases the categorization of articles was quite arbitrary, and therefore the serious student of this field should not ignore either section.

I have tried to choose articles which were important in the development of our present understanding of metal solutions. As editor, I have greatly profited from the reading (sometimes for the first time) of these and the many other articles screened when making the selections. Many important and first-rate articles were omitted because of the space limitations in this volume. However, I believe the included articles permit the reader to trace the development of the field satisfactorily. In many cases, the footnotes serve as useful bibliographies; the author index of this volume includes all the authors whose names appear in footnotes.

W.L.J.

Contents

II. PHYSICAL CHEMICAL STUDIES AND MODELS

Author List

Chemical Properties of Metal–Ammonia Solutions

I

The Discovery of Metal–Ammonia Solutions

Ammonium amalgams (solutions of NH_4 in mercury) were first prepared in 1808 by Seebeck,[1] and during the following fifty years or so these fascinating substances were sufficiently well characterized to establish that the dissolved species was NH_4. Therefore Weyl was not particularly eccentric when he proposed, in the *Annalen der Physik* of 1864, the possible existence of "metal–ammoniums," that is, compounds in which one or more of the hydrogen atoms in NH_4 were replaced with metal atoms. He believed that it would be possible to prepare a metal–ammonium of the type MNH_4 by direct combination of ammonia and a metal if the latter were highly electropositive. Weyl's paper was important, not so much because of his metal–ammonium theory, as because it contains the first description of what we now recognize as the preparation of a solution of a metal in liquid ammonia. A reproduction of the experimental portion of his paper is given in the following pages. A translation of two very pertinent paragraphs, beginning at the bottom of the first reproduced page, follows.

> For this purpose I let dry ammonia gas act on potassium, because this extremely electropositive metal seemed to offer the greatest prospect for success in the experiment. Since ammonia at ordinary pressure does not react with potassium, and since at high temperature displacement of hydrogen by potassium to form potassium amide occurs, I was left with

[1]Seebeck, *Ann. Chim.*, **66**, 191 (1808).

the option (in order to obtain an ammonium by the direct combination of its constituents) of allowing potassium and ammonia to act on one another under high pressure. A Faraday's tube of the form shown (on p. 5) was used.

First of all potassium, as free of oxide as possible, was transferred through the open end *c* to the leg *b*; then the front leg *a* was slightly narrowed by drawing it out in order to facilitate the later sealing of the tube at *a* in the flame. After this came the introduction of the silver chloride, which was saturated with dry ammonia gas; the end *c* was now sealed in the flame as quickly as possible, and the leg *a* of the now completely closed tube was immersed in a calcium chloride solution. Upon gradually raising the temperature of the solution to the boiling point, the potassium balls in leg *b* underwent a marked swelling. At the same time little silver-white balls appeared to break through the oxide layer onto the surface of the balls near the silver chloride; these gradually covered the entire surface, then obtained a brass-yellow appearance, and finally, with complete liquifaction of the balls, took on a copper-red color with strongly metallic luster.

1

Über Metallammonium–Verbindungen

W. WEYL

... Nichtsdestoweniger führt eine sorgfältigere Betrachtung der Bildungs- und Zersetzungsweise, namentlich der übrigen Metallammoniumverbindungen, zu einer Auffassungsweise der Constitution des Ammoniums, welche dem vierten Aequiv. Wasserstoff eine von den übrigen wesentlich verschiedene Stelle in der Ammoniumgruppe anweist. Besonders spricht dafür, dafs viele der Metallammoniumverbindungen beim Erhitzen entweder ihr Ammoniak vollständig, oder doch zum gröfsten Theil, abgeben. So habe ich selbst bei dem Chlortetramercur-Ammonium allen Grund zu vermuthen, dafs es in NHg_3 und $HgCl$ zerfällt, was ich in einer späteren Arbeit noch ausführlicher begründen werde. Ferner bildet sich der Salmiak durch directes Zusammentreten von Ammoniak und Salzsäure, und zerfällt, wie neuerdings Pebal nachgewiesen, bei hoher Temperatur wieder in beide, ja giebt selbst bei gewöhnlicher Temperatur an der Luft Ammoniak ab. Einen analogen Procefs habe ich bei einer Bildungsweise des weifsen schmelzbaren Präcipitats beobachtet. Bringt man nämlich Ammoniak und Quecksilberchlorid in alkoholischer Lösung zusammen, so fällt die Verbindung sofort als weifses Pulver nieder, und kann, mit Alkohol ausgewaschen, vollkommen rein und unzersetzt erhalten werden. Dafs ferner dem Ammoniumoxyd, welches in feuchtem Ammoniakgas doch als vorhanden angenommen werden mufs, das vierte Aequivalent Wasserstoff an Sauerstoff gebunden als Wasser verhältnifsmäfsig leicht entzogen werden kann, führte mich mit den erwähnten Thatsachen zu dem Schlufs, dafs das eventuell existirende Metallammonium als durch die nähern Bestandtheile Ammoniak und Wasserstoff gebildet, aufzufassen sey, und veranlafsten mich zu dem Versuche, ein Ammonium aus Ammoniak und einem dem vierten Aequivalent Wasserstoff correspondirenden, elektro-positiven Metall darzustellen.

Zu dem Zwecke liefs ich trockenes Ammoniakgas auf Kalium einwirken, da diefs Metall als das elektro-positivste die meisten Aussichten für ein Gelingen des Versuchs zu bieten schien. Da Ammoniak unter gewöhnlichem Druck auf Kalium nicht einwirkt, bei erhöhter Temperatur aber eine andere Art von Verbindungen durch Deplaciren des Wasserstoffs durch Kalium, Amide sich bilden, so stand mir noch, um ein Ammonium durch directes Zusammentreten der von mir supponirten Constituenten zu erhalten, der Versuch offen, Kalium und Ammoniak unter erhöhtem Druck aufeinander einwirken zu lassen. Es wurde eine Faraday'sche Röhre von beistehender Form angewandt.

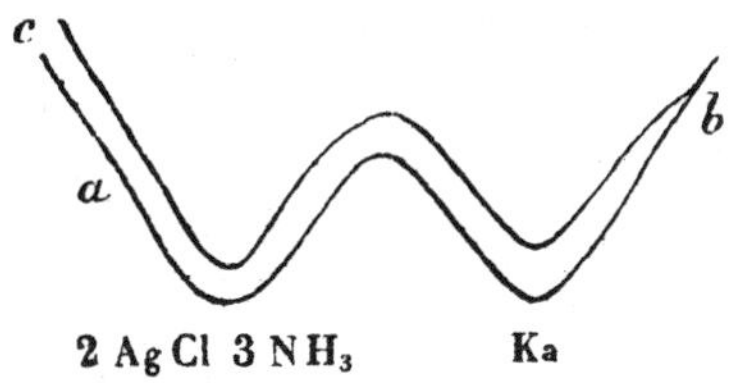

Zunächst wurde durch das offene Ende *c* möglichst oxydfreies Kalium nach dem Schenkel *b* eingefüllt, alsdann der vordere Schenkel *a* durch Ausziehen wenig verengt, um das spätere Schliefsen der Röhre bei *a* vor dem Gebläse zu erleichtern; hierauf erfolgte das Einfüllen des Chlorsilbers, welches mit trockenem Ammoniakgase gesättigt war; das Ende *c* wurde nun möglichst schnell vor dem Gebläse zugeschmolzen, und die nun so vollkommen geschlossene Röhre mit dem das Chlorsilber enthaltenden Schenkel *a* in eine Chlorcalcium-Lösung eingesenkt. Steigert man nun allmählich die Temperatur der Lösung bis zum Siedepunkt, so tritt nach Verlauf von spätestens einer halben Stunde ein merkliches Anschwellen der im Schenkel *b* enthaltenen Kaliumkugeln ein; zugleich zeigen sich auf der Oberfläche der dem Chlorsilber zunächst gelegenen Kugeln von Innen heraustretende, die Oxydschicht gleichsam durchbrechende, kleine silberweifse Kügelchen, die nach und nach die ganze Oberfläche bedecken, dann ein schwach messinggelbes Ansehen gewinnen, und zuletzt unter vollständiger Verflüssigung der Kugeln eine kupferrothe Farbe mit starkem Metallglanze annehmen.

Dieser Procefs schreitet im Verlauf einer Stunde durch die ganze Kaliummasse fort, und zeigt sich hierbei ein durch die Hand leicht merkliche Temperaturerhöhung des ganzen Schenkels. Zum Gelingen des Versuches ist erforderlich, während der ganzen Dauer der Einwirkung des Ammoniaks das im vordern Schenkel enthaltene Chlorsilber-Ammoniak stets geschmolzen zu erhalten, und den Kaliumschenkel der Röhre gegen die von der Chlorcalciumschale ausgehende strahlende Wärme, etwa durch Eintauchen in Wasser, zu schützen; widrigenfalls durch den übermäfsig gesteigerten Druck ein Springen der Röhre fast unvermeidlich ist. Der so erhaltene Körper überkleidet theilweise die Innenfläche des Schenkels, theilweise sammelt er sich im untern Theile desselben als leicht bewegliches, flüssiges Metall; nimmt daselbst gegen die von ihm benetzten Glaswandungen eine concave Oberfläche an, die beim Fliefsen stark convex wird. Bei senkrecht auffallendem Lichte zeigt die vollkommen undurchsichtige Flüssigkeit, wie schon oben bemerkt, Metallglanz mit intensiv kupferrother Farbe, die unter Umständen purpurfarben erscheint; bei sehr schief auffallendem Lichte jedoch messinggelb, ins grünliche spielend, erscheint. Sehr dünne an den Glaswandungen der Röhre adhärirende Schichten erscheinen bei durchfallendem vollem Tageslichte intensiv blau; doch mufs ich bis jetzt noch unentschieden lassen, ob letztere Eigenschaft dem beschriebenen Körper angehört, oder einer Verbindung, die sich bildet durch die gleichzeitige Einwirkung des Ammoniaks auf das kaum zu vermeidende Kaliumoxyd. Läfst man nach beendeter Bildung des bisher beschriebenen Körpers den Chlorsilberschenkel wieder erkalten, so tritt, in dem Maafse als das in der Röhre frei enthaltene Ammoniak absorbirt wird, ein Zerfallen der in dem Kaliumschenkel enthaltenen Verbindung in ihre ursprüngliche Componenten ein. Die Farbe des vorher kupferrothen Körpers wird hierbei stets heller, und geht zuletzt durch ein schwaches Gelb in die vollkommen silberweifse Farbe des zurückbleibenden Kaliums über, wodurch die Innenfläche des Schenkels wie mit einem Silberspiegel überzogen erscheint. Diese Rückzersetzung beendet sich im Laufe eines Tages, und läfst sich alsdann durch wiederholtes Erwärmen des Chlorsilberschenkels die Verbindung nach Belieben wieder herstellen. Oeffnet man die

Röhre, so zeigt sich, dafs das Kalium schwammig aufgetrieben ist, und noch geringe Mengen Ammoniak enthält, die jedoch bald entweichen, im Uebrigen seine charakteristischen Eigenschaften beibehalten hat. Wenn man die Röhre öffnet, bevor die Verbindung sich vollständig zerlegt hat, so findet ein plötzliches Zerfallen in Ammoniak und Kalium statt; ersteres entweicht explosionsartig, letzteres überkleidet die Innenfläche der Röhre bis zur Austrittsöffnung.

Aus den bisher beschriebenen Eigenschaften folgt unzweifelhaft, dafs der beobachtete Körper nur als durch directe Aneinanderlegung von Kalium und Ammoniak gebildet aufgefafst werden kann, insofern ein Entstehen einer amidartigen Verbindung in Folge theilweiser Substitution des Wasserstoffs durch Ammoniak die nachherige Regeneration unmöglich gemacht hätte. Diefs, sowie die für Metalle so charakteristisch physikalischen Eigenschaften, liefsen vermuthen, dafs hier das gesuchte Ammonium vorliege, und war es jetzt die nächste Aufgabe, durch die Analyse die Zusammensetzung dieses Ammoniums zu ermitteln. Da die Natur des in Rede stehenden Körpers nur die Analyse auf synthetischem Wege gestattete, so wurde im Wesentlichen der bei seiner Darstellung eingeschlagene Weg beibehalten.

Die zu diesem Versuche angewandte Röhre wog 143,404 Grm., nach dem Einfüllen des Kaliums 146,456 Grm., woraus sich die Menge des Kaliums ergiebt zu 3,052 Grm.

Chlorsilber absorbirt nach H. Rose, und, wie ich es versucht, bei gewöhnlichem Druck und gewöhnlicher Temperatur auf zwei Aequiv. Chlorsilber drei Aequiv. Ammoniak. Eine dem Kalium aequivalente Menge Ammoniak erforderte sonach 7,488 Grm. Chlorsilber, die mit sorgfältig getrocknetem Ammoniak unter den angeführten Bedingungen gesättigt in den vorderen Schenkel der Röhre eingefüllt wurden. Da beim Schliefsen der Röhre vor der Gebläselampe ein Ammoniakverlust eintrat, so konnte dieser Versuch nur zeigen, ob auch weniger als eine dem Kalium aequivalente Gewichtsmenge Ammoniak zur Bildung des beobachteten Körpers genügte; es zeigte sich jedoch, dafs die vorhandene Ammoniakmenge zu gering war, insofern zwar sämmtliche Kugeln an Volumen bedeutend zugenommen und einen leichten messinggelben metallischen Glanz zeigten, ohne jedoch die schon beschriebenen charakteristischen Eigenschaften bei weiterem Einwirken des Ammoniaks anzunehmen. Bei dem zweiten Versuch wurden verwandt 1,947 Grm. Kalium und die entsprechende Menge ungesättigten Chlorsilbers im Wasserstoffstrome in die an beiden Enden offene Röhre eingefüllt; alsdann durch Ueberleiten trockenen Ammoniakgases das Chlorsilber gesättigt, und nach der Sättigung die Röhre vor der Lampe geschlossen. Der Versuch ergab, dafs hier, wo ein Aequiv. Ammoniak auf ein Aequiv. Kalium einwirken konnte, die Verbindung sich vollkommen bildete, und folgt aus diesem für die quantitative Zusammensetzung des vorliegenden Körpers entscheidenden Resultate, dafs derselbe aufgefafst werden mufs es als ein Ammonium von der Formel $N \left\{ \begin{matrix} H \\ H \\ H \end{matrix} \right. \atop Ka$, in dem das vierte Aequivalent Wasserstoff durch Kalium vertreten ist. In wie weit es nun möglich seyn wird, die diesem Kalium-Ammonium entsprechenden Oxyde und Salze darzustellen, behalte ich einer weiteren Untersuchung vor, ebenso die Ermittelung der Function, welche der in dem vorliegenden Ammonium enthaltenen Ammoniakgruppe zuzuschreiben ist; d. h., ob das Kalium aufzufassen ist als das in dem hypothetischen Ammonium NH_4 enthaltene vierte Aequiv. Wasserstoff vertretend, oder ob, wie es in obiger Formel angedeutet, die Ammoniakgruppe NH_3 an Stelle des einen Atoms Kalium im Molecül $\left\{ \begin{matrix} Ka \\ Ka \end{matrix} \right.$ getreten ist.

In gleicher Weise wie auf das Kalium wurde auch die Einwirkung des Ammoniaks auf Natrium untersucht, und traten bei ihm ganz analoge Erscheinungen ein; die Einwirkung des Ammoniaks geht mit gleicher Leichtigkeit vor sich; auch hier bildet sich ein flüssiges, leicht bewegliches

Metall von kupferrother, etwas ins gelbliche spielender Farbe; auch das Natrium-Ammonium zersetzt sich nach dem Erkalten des Chlorsilberschenkels in Ammoniak und Natrium, und zeigt das zurückbleibende Natrium eine matte Oberfläche, während das Kalium sich durch seine silberweifse Farbe auszeichnet.

Zur Bestätigung der metallischen Natur der beschriebenen Verbindungen von Ammoniak mit Natrium und Kalium schien es mir wichtig, ihre Amalgamation mit Quecksilber zu versuchen. Ich liefs auf die oben beschriebene Weise Ammoniak auf ein Amalgam einwirken, das erhalten war aus nahe gleichen Theilen Natrium und Quecksilber. Das vorher pulverförmige Amalgam wurde nach etwa zweistündiger Einwirkung des Ammoniaks fest und zusammenhängend; unter der Loupe erschien es als eine homogene Masse, und zeigte, wo es an den Wandungen der Röhre fest anlag, metallischen Glanz und die röthliche Farbe einer kupferreichen Bronce, während die Innenfläche der Metallmasse matt und ziegelroth erschien. Diefs Amalgam zeigte ebenso wenig Beständigkeit, wie das Ammonium selbst, und zerfiel wenige Stunden nach dem Erkalten in Ammoniak, ein an Quecksilber ärmeres Natrium-Amalgam und Quecksilber.

Berlin, den 18. Februar 1864.

The First Recognition of the "Nonchemical" Solvent Action of Ammonia on Metals

In 1870 Seely stated that, on the basis of his own experiments,

> . . . liquid anhydrous ammonia is a solvent, without definite chemical action, of the alkali metals. I mean that these metals dissolve in the ammonia as salt dissolves in water—the solid disappears in the liquid, and on evaporating the liquid, the solid reappears in its original form and character. There is no definite atomic action in any such cases; the components of the solution are not changed in their chemical relations to other substances.

These clear and forthright statements seem to have been largely ignored, as shown by the fact that metal–ammonia solutions were usually referred to as metal–ammoniums well into the twentieth century. It was not until the classic experiments of Kraus that the reversible solubility of electropositive metals in ammonia was generally recognized. In the following pages we have reprinted the complete text of Seely's paper.

Mechanics, Physics, and Chemistry.

ON AMMONIUM AND THE SOLUBILITY OF METALS WITHOUT CHEMICAL ACTION.

By Chas. A. Seely.

In April and May last it came in my way to make a pretty careful study of the constitution of the so-called ammonium amalgam, and on May 9th, I reported to the Lyceum of Natural History of New York what I had done in the matter up to that time. Subsequently, I continued the study, and was led to the discovery of the solubility of the alkali metals in anhydrous ammonia, an account of which formed a paper read at the Troy meeting of the American Association for the Advancement of Science. Neither of these communications have been printed; the first was oral, and the latter was lost. The present paper is intended to be a concise and authoritative statement of the contents of the two papers and as a substitute for them.

In the Lyceum communication I proved that the so-called ammonium amalgam is in fact no amalgam at all, but a metallic froth of which the liquid part is mercury and the gaseous part a mixture of hydrogen and ammonia. This conception of the nature of the substance was at that time not altogether novel, but the considerations through which the conclusion was reached, were believed to be new and of such moment that taken together they have the effect of a demonstration. The considerations were these:

1st. In the formation of ammonium amalgam the mercury is increased in bulk tenfold or more, while in weight it is increased only one or two thousandths. This fact would be easily predicted by the froth theory while otherwise it is anomolous and utterly inexplicable.

2d. Ammonium amalgam has less of the mirror surface than mercury; it approaches in appearance the whiteness and lack of lustre of mat silver. This fact also finds a satisfactory explanation in the froth theory.

3d. Ammonium amalgam is readily compressible; when confined in a fire syringe it steadily contracts on forcing down the piston till the original bulk of the mercury is almost reached, and on relieving the pressure the original volume and appearance are resumed. It is only a froth which behaves in this manner.

Either (1) or (3) will be conclusive to many minds; (3) is put forward as a fair *experimentum crucis.* The condensation by pressure is easy to execute, and an elegant experiment for the lecture-room. The condensation is somewhat greater than in the case of atmospheric air for the reason that some of the ammonia gas goes into the liquid state.

Having concluded that the ammonium amalgam is a mercuric froth, a very important question presents itself: Why and how do metal and gases become mingled? A froth is a mechanical mixture, and in all ordinary cases the manner of mixing and the forces engaged are quite evident. In the formation of the mercuric froth, the mercury covered by a watery solution rises up spontaneously into the comparatively light water, and the still lighter gases descend into the mercury and mingle with it until they become diffused through its whole mass. What are the forces which thus overpower the action of gravity? Moreover, a froth implies a certain viscosity or plasticity of the liquid which seems not to be possessed by the mercury in its normal condition. Do hydrogen and ammonia in their nascent state have the power to modify the physical properties of mercury? The production of mercuric froth as a fact of physics simply may seem anomolous and paradoxical. Is it necessary after all to admit the existence of an ammonium amalgam, not of course as a constituent of the mercuric froth, but as a condition precedent of the formation of the latter,—in other words, of another ammonium amalgam to satisfy a new theory?

I was making satisfactory progress in the investigation of the theory of mercuric froth when, early in May, my attention was diverted to the researches of Weyl on ammonium and compound ammoniums as reported in *Watt's Dictionary* (article, Sodium), and in the last edition of *Fownes' Chemistry;* subsequently, I compared Weyl's original memoirs in *Poggendorff's Annalen* of 1864. Weyl's statements and conclusions were quite at variance with my own views; his alleged facts were inconsistent with my own experience. Weyl, it appeared, had actually isolated ammonium; if he represented the facts correctly all my work needed revision. I therefore at once devoted my whole care to a verification of his results. I repeated his experiments, some of them many times, and with such precautions, modifications and additions as the case seemed to require. The result of all is that I find that Weyl has had only an imperfect apprehension of the fundamental facts, and that thus building on a weak foundation the superstructure of reasoning and conclusion cannot stand. For the purpose of presenting in a clearer light what I suppose to be of more interest to the chemical world, I dismiss, for the time, the consideration of Weyl's views, and I proceed to give what in my judgment are the proved physical and chemical facts.

This discussion involves the relation of ammonia to the alkali metals, and the results which grow out of that relation. Now, the key to the whole subject is the fact that liquid anhydrous ammonia is a solvent, without definite chemical action, of the alkali metals. I mean that these metals dissolve in the ammonia as salt dissolves in water,—the solid disappears in the liquid, and on evaporating the liquid, the solid reappears in its original form and character There is no definite atomic action in any such cases; the components of the solution are not changed in their chemical relations to other substances.

I began my experiments by bringing together ammonia and the metals in sealed inverted U-shaped glass tubes. Chloride of silver saturated with ammonia was placed in one leg of the tube, and the metal under test in the other; the chloride of silver leg being immersed in boiling water, and the other in ice-cold water, the ammonia was liberated and passing over was gradually condensed on the metal. Such an arrangement, although sufficient to demonstrate the fact of the solution of the metal, proved inconvenient in several respects; the tube is liable to burst, the experiment can be made only on a pretty small scale, the chloride of silver does not absorb a desirable amount of ammonia, and by reason of its fusing or agglomerating, the rapidity of its absorption varies greatly. I therefore constructed an ammonia generator or retort of iron, to the horizontal neck of which turned downward at its end, I attached by means of an iron screw coupling the stout glass tube containing the metal or other substance to be subjected to the ammonia. In the place of the chloride of-silver I substituted anhydrous chloride of calcium saturated at 32° Fahr. with anhydrous ammonia. With this arrangement the experiments were continued with comparative ease and expedition.

When sodium is subjected in this apparatus to the condensing ammonia, before any ammonia is visibly condensed to the liquid

state, it gradually loses its lustre, becomes of a dark hue, and increases in bulk. The solid then appears to become pasty, and at last we have only a homogeneous mobile liquid. During the liquifaction, and for a little time after, the mass is of a lustrous, copper-red hue; the condensation of the ammonia and its mingling with the liquid steadily goes on, the liquid is progressively diluted, and passing through a variety of tints by reflected light at last it becomes plainly transparent and of a lively blue as well by reflected as by transmitted light; the liquid now closely resembles a solution of aniline blue or other pure blue dye-stuff. On reversing the process by cooling the ammonia generator, the ammonia gradually evaporates out of the liquid, and the changes observed during the condensation reappear in the reverse order, till at last the sodium is restored to its original bright metallic state. If the evaporation be conducted slowly and quietly the sodium is left in crystals of the forms seen in snow. The formation of the transparent blue liquid, and the restoration of the sodium are steadily progressive, and the repeated and closest scrutiny of the process has failed to reveal the slightest break or irregularity in its continuity. The inevitable conclusion from such facts is that the blue liquid is a simple solution of sodium in ammonia, not at all complicated or modified by any definite chemical action.

If crystallized sodium amalgam be made to replace the sodium in the experiment it remains unchanged although it is probable enough that at a higher temperature or with a larger excess of sodium, some of the sodium would be removed. This interesting fact is consistent with the simple solubility of sodium in ammonia, and indeed is some confirmation of it. The force, whatever it be, which keeps together sodium and mercury is certainly weaker than that exhibited between sodium and ammonia. If sodium amalgam is not a definite chemical compound, then surely the blue liquid cannot be. I have not yet made the experiment to determine whether mercury will remove sodium from its ammoniacal solution, but it seems extremely probable that it would so act.

The brilliant and varied colors exhibited in the experiment may seem anomalous to some, when, in fact, a closer scrutiny of the case will show that they might have been predicted. Sodium appears white to the eye, but with the white light reflected from its surface to the eye, there are always mingled red rays. If most of the incident white light were normally decomposed, sodium would appear as a brilliant red metal. Ammonia favors such a decomposition probably by reducing the density and opacity of the surface, and thus the concentrated solution of sodium is lustrous copper-red by reflected light. What should be the color by transmitted light? Not red, for the red rays do not penetrate the substance; the color must be looked for in that which is complementary to red; it must be blue or yellow or combinations of these. A continuance of the argument will bring the conclusion that the color by transmitted light will be blue. Intense blue tinctorial substance, like aniline blue, indigo and Prussian blue, all illustrate the phenomena of color of the sodium solution; they are metallic red when concentrated, and if the solvent be applied in vapor as in the sodium dissolving experiment, there will be the same modification of color exhibited. Sodium has a remarkable tinctorial power which seems not to be surpassed by that of any of the aniline colors.*

If a salt of a metal electrically negative to sodium be subjected with the sodium to the ammonia, in the apparatus, the sodium will replace the negative metal, and the latter will be reduced to the simple state. This reduction seems always to be accompanied with the evolution of heat, but does not commence till a visible quantity of the sodium solution is formed. It is possible that there may be exceptions to these generalizations. I need only say that they seem extremely reasonable, and are confirmed by a good many experiments. The reaction furnishes a very simple and elegant process for the reduction of rare metals. I anticipate that the process will prove of great value.

I have heretofore spoken of the solution of sodium only, for the reason that it is a typical case, and because demonstrations can be more satisfactorily made with it. Potassium behaves towards ammonia in almost precisely the same way. It is readily dissolved, the strong solution being copper-red and the diluted solution blue. Lithium gives also a blue solution, but is not nearly so soluble. The only other alkali metal I have tried is rubidium, and although the experiment was prematurely ended, it had progressed far enough to make it almost certain that rubidium is soluble like potassium, and I intend, as soon as I can procure material, to repeat the rubidium experiment.

Besides the metals named, I have subjected many others to the ammonia, and have found none outside of the alkali metals which are in the least affected. I have directly tried aluminium, magnesium, thallium, indium, mercury and copper.

* For a further elucidation of the matter of this paragraph I would refer the reader to my paper on the "Colors of Metals," printed in the proceedings of the Lyceum of Natural History, of New York, of October, 1870.

Chemical Laboratory, 26 Pine Street, N. Y.

The Reaction of Metal–Ammonia Solutions with Phosphine

In the period between 1889 and 1906, Joannis extensively investigated the chemical reactions of the "metal–ammoniums." In the reprinted paper which follows, first published in 1894, he reported his studies of the reaction of PH_3 with solutions of potassium and sodium in ammonia to form KPH_2 and $NaPH_2$. Joannis' experimental genius is evidenced by the facts that his analytical data are at least as accurate as those generally obtained in similar studies today and that his results and conclusions have stood the test of time. This work corresponds to one of the first applications of metal–ammonia solutions to the deprotonation of extremely weak acids.

3

CHIMIE. — *Action du phosphure d'hydrogène sur le potassammonium et le sodammonium.* Note de M. **A. Joannis.**

« *Action du phosphure d'hydrogène sur le potassammonium.* — Quand on fait arriver du phosphure d'hydrogène dans le potassammonium dissous dans de l'ammoniac liquéfié, on constate qu'une réaction se produit; le phosphure d'hydrogène disparaît peu à peu, en même temps que de l'hydrogène se dégage. Dans le tube où se fait l'expérience, on voit se former un liquide qui ne se mêle pas à la solution ammoniacale d'ammonium alcalin, mais qui dissout cependant une petite quantité de ce corps. Lorsque la réaction est presque terminée, le potassammonium restant nage, sous forme de gouttelettes qui paraissent huileuses, au-dessus de l'autre liquide. Puis, lorsque la réaction est terminée, ces gouttelettes ont disparu et l'on a un liquide très réfringent, rappelant, à ce point de vue, le sulfure de carbone; si on laisse alors partir l'ammoniac en excès, il se dépose de fines aiguilles d'un corps blanc, qui a pour composition PhH^2K : c'est un composé analogue à l'amidure de potassium AzH^2K et que l'on peut appeler par analogie *phosphidure de potassium*. Ce composé, qui n'avait pas encore été obtenu bien que l'amidure soit connu depuis Gay-Lussac, se trouve ainsi préparé à un grand degré de pureté ([1]).

» *Action du phosphure d'hydrogène sur le sodammonium.* — Lorsqu'on fait agir du phosphure d'hydrogène sur une solution de sodammonium faite dans l'ammoniac liquéfié, on observe d'abord les mêmes phénomènes qu'avec le potassammonium : disparition du phosphure d'hydrogène, mise en liberté d'une quantité d'hydrogène qui correspond à la formation d'un phosphidure de formule PhH^2Na. On voit aussi, vers la fin de l'expérience, des gouttes de sodammonium, avec leur couleur mordorée, flotter à la surface d'un liquide incolore, réfringent, et ayant, à cause de cela,

---

([1]) C'est ce que montrent les deux analyses suivantes :

	Calculé PhH^2K.	Trouvé. I.	Trouvé. II.
K	54,16	53,29	54,27
Ph	43,06	43,00	44,54
H	2,78	2,80	2,89
	100,00	99,09	101,70

l'apparence de sulfure de carbone; ce liquide se prend en masse quand on le refroidit fortement. Si l'on maintient, au contraire, le tube à 0°, la masse reste liquide et, si on laisse partir tout l'ammoniac qui peut s'échapper à cette température, on n'obtient pas, comme avec le potassium, de masse solide cristallisée. La matière est encore liquide; elle contient à la fois du phosphidure de sodium et de l'ammoniac. Le tube, ayant dégagé tout l'ammoniac qu'il pouvait perdre à 0° sous la pression atmosphérique, a été pesé après un séjour de quarante-huit heures dans la glace; puis on l'a laissé revenir à la température ambiante (13°); il a perdu une nouvelle quantité d'ammoniac, que l'on a déterminée par une nouvelle pesée. On l'a alors chauffé à 65°; la matière a cristallisé et a perdu une nouvelle quantité de gaz ammoniac, que l'on a encore déterminée par la pesée. Entre 65° et 69°, un peu de phosphure d'hydrogène ($0^{cc},3$) a commencé à se dégager; on a cessé de chauffer. Il est peu probable, d'après la façon dont l'ammoniac s'est dégagé, que le liquide, stable à 0°, soit une combinaison définie; pour une molécule de phosphidure de sodium, on a trouvé qu'il s'était dégagé $2^{mol},87$ d'ammoniac. On peut aussi enlever tout l'ammoniac en faisant le vide dans le tube; on obtient ainsi un corps solide, blanc, toujours souillé d'un peu de phosphure jaune $PhNa^3$ [1].

» La chaleur détruit ces composés, par une réaction analogue à celle qui transforme l'amidure en azoture

$$3PhH^2K = 2PhH^3 + PhK^3.$$

» L'eau les décompose, en mettant aussi en liberté du phosphure d'hydrogène.

» J'ai fait réagir, sur ces composés dissous dans l'ammoniac liquéfié, du protoxyde d'azote; l'action est bien différente de celle que j'avais obtenue avec les amidures alcalins dans les mêmes conditions (*Comptes rendus*, t. CXVIII, p. 715). Tandis qu'avec ces derniers composés il se forme de l'azoture de sodium Az^3Na, sel de l'acide azothydrique, le protoxyde d'azote étant absorbé sans mise en liberté d'azote, ici, au contraire, avec les phosphidures, il se dégage un volume d'azote égal au volume du protoxyde employé. J'étudie en ce moment les autres produits de la réaction. »

(1)

	Calculé PhH^2Na.	I.	II.	III.
Na..........	41,07	40,34	40,62	42,09
Ph..........	55,36	56,05	55,59	54,62
H...........	3,57	3,61	3,79	3,29

An Early Study of a Metal Hexaammine

We now know that calcium, strontium, and barium (as well as europium and ytterbium) do react with ammonia to form solid compounds of composition $M(NH_3/_6$. Apparently Mentrel was the first to recognize the existence of such a metal hexaammine. In the following reprinted paper, he reported the formation of barium hexaammine.

4

CHIMIE MINÉRALE. — *Sur le baryum-ammonium et l'amidure de baryum.* Note de M. MENTREL, présentée par M. Haller.

« M. Guntz a montré ([1]) que le baryum et le strontium métalliques se dissolvent dans l'ammoniac liquide pour donner des composés mordorés semblables aux autres ammoniums préparés par M. Joannis ([2]) et par M. Moissan ([3]).

» Nous avons étudié les conditions de formation du baryum-ammonium et ses propriétés.

» Lorsqu'on fait passer du gaz ammoniac sur le baryum, on constate que ce métal ne s'attaque pas au-dessus de + 28°. Au-dessous de cette température il se forme un produit solide rouge mordoré se transformant en un liquide bleu lorsque la température baisse au-dessous de — 23°. Vers — 50° il se sépare un liquide huileux bleu foncé, peu soluble dans l'ammoniac liquide qu'il colore en bleu pâle.

» Au-dessous de — 23° ces composés sont stables; à partir de — 15°, ils se transforment en amidure d'autant plus rapidement que la température est plus élevée.

» Voici les tensions de dissociation du baryum-ammonium que nous avons observées en opérant toujours en présence d'un excès de baryum :

Température.	Tensions en millimètres de mercure.
—63	19
—31	38
—19	59
0	158
+19	507
+28	785

» Nous avons analysé ce composé par la méthode de M. Joannis, en cherchant à diverses températures la composition du produit limité qui, en perdant une trace d'ammoniac, donne du baryum libre. On trouve ainsi : à 0°, Ba + 6,1 AzH^3 ; à —23°, Ba + 6,3 AzH^3 ; à —50°, Ba + 6,97 AzH^3, les tensions de ces composés étant, à température égale, les mêmes que les tensions de dissociation indiquées précédemment.

([1]) GUNTZ, *Bull. Soc. des Sciences de Nancy*, 1902.

([2]) JOANNIS, *Comptes rendus*, t. CIX, p. 900, 965; t. CXII, p. 392; t. CXIII, p. 795; t. CXV, p. 820.

([3]) MOISSAN, *Comptes rendus*, t. CXXVII, p. 685.

» A basse température, le baryum-ammonium renferme donc un léger excès d'ammoniac provenant de la dissolution de ce gaz dans le composé solide dont la formule semble être $Ba(AzH^3)^6$. M. Moissan avait trouvé pour le composé analogue du calcium la formule $Ca(AzH^3)^4$.

» Il semble donc que la proportion du gaz ammoniac combiné avec les métaux de cette famille augmente avec le poids atomique; pour le vérifier, nous nous proposons de déterminer la formule du strontium-ammonium.

» Les propriétés du baryum-ammonium sont semblables à celles des autres ammoniums; il prend feu au contact de l'air, se décompose très vivement par l'eau.

» L'oxygène à basse température est absorbé en donnant un mélange de bioxyde de baryum et de baryte.

» Avec le bioxyde d'azote, nous avons obtenu l'hypoazotite de baryum, solide blanc $Ba(AzO)^2$.

» L'action de l'oxyde de carbone sur la solution ammoniacale de baryum-ammonium nous a permis de préparer un compose nouveau, le baryum-carbonyle $Ba(CO)^2$, corps solide, jaune, se décomposant sans explosion au contact de l'air et par la chaleur, soluble dans l'eau avec décomposition.

» En faisant passer du gaz ammoniac sur le baryum chauffé dans une nacelle en fer, on constate que l'attaque a lieu à 280°. Il se forme un liquide gris devenant vert, puis rouge lorsque la température augmente. Il se forme de l'amidure de baryum :

$$Ba + 2AzH^3 = Ba(AzH^2)^2 + H^2.$$

» A 460°, l'amidure fondu bout en dégageant un mélange d'azote et d'hydrogène dans le rapport $\frac{H}{Az} = 3$.

» A 650°, il se forme un produit solide, jaune orangé, fusible seulement à 1000°. En abaissant la température et en opérant toujours dans un courant d'ammoniac, les phénomènes inverses se produisent; le composé redevient liquide vers 450°, puis se solidifie à 280°.

» Ces changements curieux sont dus à la transformation, par la chaleur, de l'amidure $Ba(AzH^2)^2$ en azoture Ba^3Az^2, et, par refroidissement, de l'azoture en amidure, comme les analyses nous l'ont montré, ces réactions étant accompagnées d'une décomposition illimitée de l'ammoniac en ses éléments.

» En opérant dans le vide, nous avons obtenu de l'azoture de baryum pur et exempt de fer. Il se produit donc, à chaque température, un équilibre entre Ba^3Az^2 et $Ba(AzH^2)^2$, d'après la réaction

$$3Ba(AzH^2)^2 \rightleftarrows Ba^3Az^2 + 4AzH^3.$$

» Nous avons vérifié que l'amidure de lithium donne nettement une transformation analogue, qui, probablement, se produit aussi pour l'amidure de sodium, mais en très faible proportion, aux températures où l'on peut opérer dans le vide sans dissocier totalement $NaAzH^2$. »

A Systematic Study of Compound Formation between Metals and Ammonia

Kraus was the first to systematically study the equilibrium vapor pressure of ammonia in determinations of the compositions of phases in various metal–ammonia systems. He reported, in the following reprinted paper, that lithium, sodium, and potassium do not form compounds, but that calcium (like strontium and barium) does form a hexaammine. We now know that lithium forms a tetraammine, $Li(NH_3)_4$, under conditions not investigated by Kraus, namely at low temperatures. (See the papers by Mammano and Coulter and Mammano and Sienko.)

VOL. XXX. MAY, 1908. No. 5.

THE JOURNAL

OF THE

American Chemical Society

5

[CONTRIBUTIONS FROM THE RESEARCH LABORATORY OF PHYSICAL CHEMISTRY OF THE MASSACHUSETTS INSTITUTE OF TECHNOLOGY No. 24.]

SOLUTION OF METALS IN NON-METALLIC SOLVENTS; II.[1] ON THE FORMATION OF COMPOUNDS BETWEEN METALS AND AMMONIA.

BY CHARLES A. KRAUS.

Received February 14, 1908.

Solutions of metals in liquid ammonia were first obtained by Weyl,[2] who brought together sodium and potassium with gaseous ammonia under pressure. However, he mistook the solutions which are formed under these conditions for simple compounds and assigned to them the formulae $NaNH_3$ and KNH_3, respectively. These compounds he supposed to be structurally analogous to the hypothetical free ammonium group, being derived therefrom by substitution of an atom of hydrogen by one of the metal in question. Taking account of this relationship in his nomenclature, he introduced the terms sodammonium and potassammonium, respectively, which nomenclature has been largely adopted by subsequent investigators and commentators.

The study of solutions of metals in ammonia was materially advanced by Seely,[3] who showed conclusively that solutions result in the action of ammonia on the alkali metals. From a consideration of the optical properties of these solutions, he concluded that a compound was not formed between the two components. Neither Weyl nor Seely, however, was able to adduce quantitative data in support of his contention.

These investigations on the action of ammonia on the alkali metals

[1] For the first paper of this series, "I. General Properties of Solutions of Metals in Liquid Ammonia," see THIS JOURNAL, 29, 1557–1571 (1907).

[2] *Ann. Physik*, 121, 601 (1864).

[3] *Chem. News*, 23, 169 (1871).

seem to have excited little active interest for, excepting an isolated observation by Gore,[1] we find no further investigations recorded until 1889, when Joannis[2] undertook an extended series of investigations in this field. To him belongs the credit of bringing quantitative data to bear on the problem of the compounds formed by sodium and potassium with ammonia. He devised a means of isolating and analyzing these compounds, to which he assigned the composition $NaNH_3$ and KNH_3 respectively, and to which, like Weyl, he ascribed an ammonium structure. Employing the method devised by Joannis, Moissan obtained the compounds $LiNH_3$, $Ca(NH_3)_4$,[3] and $LiCH_3NH_2$,[4] while Mentrel obtained the compound $Ba(NH_3)_6$[5] and Roederer the compound $Sr(NH_3)_6$.[6]

In the preceding paper[7] attention was called to the fact that the concentrated solutions of metals in ammonia exhibit metallic reflection and are consequently opaque. It is plain, therefore, that the formation of a compound cannot be ascertained by visual observations nor can separation of the different phases in these concentrated solutions be carried out by the simple means usually employed in the preparation of a pure substance. The method adopted by Joannis in preparing and identifying the compounds in question is therefore an indirect one, as will be seen from the description given below. Objections have been raised from time to time to the results obtained by Joannis as well as to those obtained by other chemists employing the same method.

It is the purpose of the present paper to determine, if possible, whether or not compounds are formed. To this end I shall first examine such evidence as is already at hand. The solutions of sodium and potassium in ammonia have been studied extensively and, as will be seen below, the available data are sufficient to enable us to draw the conclusion that solid compounds are not formed. In the case of other metals it has been found necessary to adduce new experimental evidence. It will thus be shown that lithium, like sodium and potassium, does not form a solid compound, while calcium forms the compound $Ca(NH_3)_6$ with ammonia. Finally the questions of constitution and nomenclature will be considered and the physical properties of the compound $Ca(NH_3)_6$ will be discussed briefly.

Criterion for the Appearance of New Phases.

In determining whether compounds are formed between a solvent and a dissolved substance, it is of primary importance to possess a clear

[1] *Phil. Mag.*, 44, 315 (1873).
[2] *Compt. rend.*, 109, 900 (1889).
[3] *Ibid.*, 127, 685 (1898).
[4] *Ibid.*, 128, 26 (1899).
[5] *Ibid.*, 135, 790 (1902); *Bull. soc. chim.*, 29, 493 (1903).
[6] *Compt. rend.*, 140, 1252 (1905).
[7] THIS JOURNAL, 29, 1570.

knowledge of the phase relations before attempting to identify any of these phases as compounds by means of chemical analysis. Since the solutions of metals in ammonia constitute two component systems, it follows that if ammonia is withdrawn from the system, the pressure must become constant as soon as a third phase appears. If, on continuing the withdrawal of ammonia, one phase disappears the pressure will again vary when ammonia is withdrawn. If, however, one of the three phases is substituted by a new third phase, the pressure changes abruptly to a new constant value. One of the phases present is always gaseous ammonia; the other two phases may either be both solid or liquid, or one may be liqiud and the other solid. It is an easy matter to determine how many new phases make their appearance in withdrawing ammonia from a dilute solution of a metal until the free metal and gaseous ammonia are left behind. The study of the vapor pressure of a system of metal and ammonia is therefore a necessary preliminary in determining whether compounds are formed. The nature of the phases present at any time may, in general, at once be determined by visual-examination, and by proper means it is always possible to transform the entire system into any desired phase, when its composition may be determined by analysis.

Non-existence of the Compounds $NaNH_3$ and KNH_3.

The pressure of the systems sodium-ammonia and potassium-ammonia have been carefully investigated by Joannis.[1] He finds that in the case of both these systems, if ammonia is withdrawn from a solution of the metal, the vapor pressure falls until a solid phase makes its appearance, after which the pressure of the system remains constant until only free metal and gaseous ammonia are left. This would seem to show that the solid phase which initially separates from solution is free metal. Joannis, however, believes that such is not the case. He observed that the solid, which initially precipitates, appears to possess the same color of metallic reflection as does the solution itself, while, as is well known, the free metal possesses white metallic reflection. He therefore considers that this solid may be a compound of sodium with ammonia. We shall simply call this substance "solid compound" in order to avoid circumlocution. This substance is evidently a new phase; the question only remains to show whether it is a compound, as Joannis believes it to be, or, otherwise, free metal and saturated solution as the vapor pressure relations indicate. According to Joannis, free metal does not make its appearance until the saturated solution is completely converted into the solid compound, after which, on further withdrawal of ammonia, the free metal appears, without being accompanied by any change in

[1] *Loc. cit.*: For details of the method described in this paragraph v. *Ann. chim. phys.*, 7, 13–36 (1906).

the pressure. He assumes, therefore, that the dissociation pressure of the "solid compound" is exactly equal to the vapor pressure of its saturated solution. Analysis of the "solid compound" gave him a composition corresponding to the simple formula $MeNH_3$, where Me may be either Na or K.

According to the phase rule, the system, vapor, saturated solution, "solid compound," and free metal must be an invariant one and obtainable at a single temperature only, since two components and four phases are present. Roozeboom[1] first called attention to these facts and suggested that Joannis had carried out his experiments at the temperature of this invariant point. Joannis, however, showed that at a series of temperatures no change occurs in the pressure from the moment that a solid phase begins to separate until only free metal and gaseous ammonia are left. This proves conclusively that the solid phase initially separating out of the solution is identical with the solid phase which is finally left behind, namely, free metal, and that a solid compound is not formed between the metal and ammonia. The "solid compound" can consist only of free metal and saturated solution of the same in ammonia. Ruff and Geisel[2] have recently expressed this view and in addition they have adduced some evidence to show that the solid compound consists of a free metal covered with a film of solution, which adheres through the action of strong surface forces. Joannis,[3] replying to this paper of Ruff and Geisel, throws some doubt on the correctness of the conclusion which they have drawn from their experimental results. In the section dealing with lithium, whose behavior is in every way similar to that of sodium and potassium, I shall adduce independent evidence showing conclusively that, in accordance with the view of Ruff and Geisel, the so-called compound $LiNH_3$ contains solution and free metal. This evidence was not obtained primarily to show that compounds do not exist, for as to this point the thermodynamic evidence would seem to be sufficient in itself. Since, however, Joannis[4] believes that his hypothesis may be reconciled with the phase rule, it may be as well to give evidence which is quite independent of any theoretical considerations.

This reconciliation of his hypothesis with the phase rule Joannis[5] believes to have been effected by Moutier.[6] A careful examination of Moutier's paper fails to show, however, that such a reconciliation is possible. A consideration of the free energy of a system can lead to no results other than those obtainable by other thermodynamic methods.

[1] *Compt. rend.*, **110**, 134 (1890).
[2] *Ber.*, **39**, 831 (1906).
[3] *Ann. chim. phys.*, **11**, 101 (1907).
[4] *Loc. cit.*
[5] *Loc. cit.*, 103; *Ann. chim. phys.*, **7**, 34 (1906).
[6] *Compt. rend.*, **110**, 518 (1890).

The conclusions reached by Moutier are in fact applicable to a single temperature only and not to a series of temperatures, as be believes them to be.

Vapor Pressure of the System $LiNH_3$.

According to Moissan,[1] the compound $LiNH_3$ is formed. Since he employed the same method as did Joannis in the case of sodium and potassium and since confirmatory vapor pressure measurements were lacking, it seemed advisable to determine the pressure relations of the system $Li–NH_3$.

The apparatus employed need not be described in detail. It consisted essentially of a tube permanently attached to a manometer and containing the solution. Provisions were made for connecting this tube with a source of pure ammonia vapor or a vacuum pump as desired. After introducing the metal into the containing tube, it was placed in a bath at —10° and ammonia vapor was introduced under a pressure slightly greater than one atmosphere. The process of condensation and solution takes place with great facility in the case of lithium, even at ordinary temperatures. When the process of solution was complete the container was placed in a bath at 20° and the excess of ammonia was allowed to escape under a pressure of about 100 cm. of mercury. On withdrawing ammonia by means of the pump, the pressure fell until it reached a value of about 10 centimeters, after which it remained constant. At this point a solid phase began to separate out and on continuing the withdrawal of ammonia the amount of solid increased. The solid phase appeared to have the same color as the solution and the solution apparently disappeared long before all the ammonia had been completely withdrawn. The pressure, however, remained constant until only free metallic lithium was left behind. When but very little ammonia was present the color distributed appeared uniformly over the entire surface of metal, giving it a slight tinge of color. On adding more ammonia the color became more pronounced while, at the same time, the crystals of metal, which previously were very sharply defined, lost their sharp outline, their edges becoming indistinct. The phenomenon is such as we should expect if the crystals were covered by a film of highly colored metallic solution.

The equilibrium pressure of the saturated solution of lithium in ammonia was measured at a number of temperatures with the following results:

Temperature.	Pressure.	Mean.	Temperature.	Pressure.	Mean.
19.3°	9.48 cm. 9.58 "	9.53 cm.	9.75°	5.68 cm. 5.72 "	5.70 cm.
0°	3.31 " 3.31 "	3.31 "	20.3°	9.91 " 10.06 "	9.98 "
0°	3.31 " 3.32 "	3.31 "	20.3°	9.90 " 10.06 "	9.98 "

[1] *Compt. rend.*, 127, 685 (1898).

The pressures first recorded were obtained after abstracting, the second after adding ammonia.

From the pressure data the heat evolved when one gram-molecule of ammonia vapor combines with metal under equilibrium conditions to form a saturated solution may be calculated from the equation:

$$Q = 4.6 \frac{T_1 T_2}{T_2 - T_1} \log. \frac{p_2}{p_1}.$$

Using the values of p at 0° and 20.3°, respectively, we thus obtain Q = 8698 calories. The value of P at 9.75° calculated from this value of Q is 5.699 cm. while the value 5.70 was found.

Independent Evidence Showing that the Solid Compound $LiNH_3$ is Not Formed.

The pressure relations recorded above show conclusively that a compound is not formed between lithium and ammonia. It seemed worth while to examine the behavior of lithium solutions somewhat further in order to determine, if possible, the sources of the error in the result of Moissan. In common with other investigators who have adopted the method of Joannis, he finds that so long as the amount of ammonia present exceeds that corresponding to the formula $LiNH_3$, that is, when saturated solution is present, any free bit of metal at once absorbs ammonia from the saturated solution to form the compound. On the other hand, when the amount of ammonia is less than that which corresponds to the formula $LiNH_3$, a free surface of metal remains permanently free from ammonia. This is, in fact, the criterion which these investigators employ in determining when the saturated solution disappears and the supposedly "solid compound" begins to dissociate. Since no change occurs in the pressure when the composition passes through the point corresponding to the composition $LiNH_3$, it is difficult to understand why there should be any difference in the behavior of the system on one side and the other of this point. The following experiments were undertaken for the purpose of obtaining more light on this question.

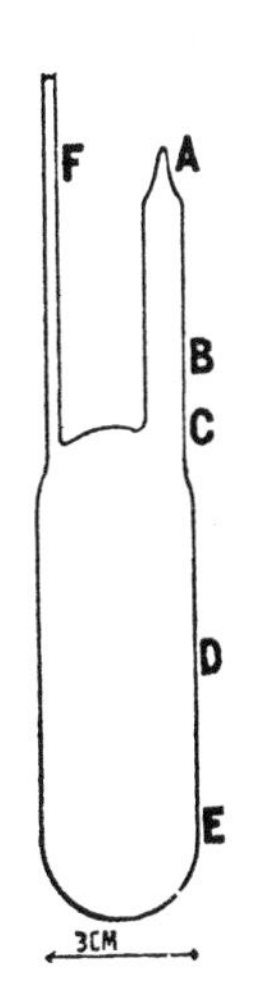

Fig. 1.—Apparatus employed in studying lithium solutions.

A quantity of lithium was introduced into a tube of the form outlined in Fig. 1. Ammonia was condensed at 0° until the metal had all dissolved, after which ammonia was withdrawn by means of the pump until a portion of the metal had been precipitated. A portion of the solution spattered along the walls of the tube EC, while a few isolated drops of solution could be seen clinging to the walls of the smaller tube AC.

On immersing the tube as far as E in a bath at 15°, 10° below room temperature, the solvent at once evaporated from the isolated drops in AC, leaving metal behind, while along the walls of the tube EDC the metal retained ammonia. This portion of the tube was then warmed with the hand, whereupon the ammonia evaporated. As soon as this heating process was discontinued, the solution from the bottom could be seen creeping up over the walls of the tube. The solution passed from crystal to crystal, tracing out characteristic figures, such as would naturally result if a liquid were to creep over a surface covered with irregularly spaced particles. Isolated particles of metal like those in the tube AC underwent no change whatever. On plunging the entire tube in the bath the particles in AC remained unaffected.

After repeating the above operation a number of times, it was observed that the amount of metal in the space DC had greatly increased. This is evidently due to the fact that the metal on the walls of the tube acts as a wick by means of which solution is drawn from the colder portion of the tube in the bottom to the warmer portions above. Here the solvent evaporates to condense again in the bottom, while the metal is left behind in the warmer portion of the tube. To test this further, the tube was left immersed as far as E in a bath at 12°. At the end of 45 minutes the metal had crept up the tube AC over a distance of more than a centimeter, forming a very heavy deposit. The deposit of metal in CD had now become so heavy that it could not be freed from ammonia by warming with the hand. The tube was left immersed as far as D in a bath at 0° for some time, after which all the ammonia was withdrawn. On examination it was found that no metal was left in the bottom, it having collected in the warmer portion of the tube DC.

A small quantity of ammonia was now introduced, sufficient to form only a thin film on the surface of the metal. The tube was placed in ice-water as far as C, while the metal deposit extended about 1.5 cm. above C in AC. On warming the tube at B with the hand, the metal deposit crept up the tube with visible speed, and at the end of about ten minutes the entire surface of the tube AC was coated with metal. The tube AC was kept warm for some time, after which the ammonia was again withdrawn. Nearly all the metal now appeared collected in the tube AC.

It is remarkable that this process should take place so rapidly against a temperature difference of nearly 30°. No better illustration could be given of the strength of the surface forces coming into play between the metal and its solution. It is to be remembered that the amount of ammonia was but a small fraction of that required to form a compound of the composition $LiNH_3$. This experiment, therefore, shows conclusively that liquid is present in the system even when, according to Moissan, only the compound $LiNH_3$ should be present. These experiments show, moreover, that the behavior of the system is the same irrespective of the amount of ammonia present so long as this is less than that necessary to form a saturated solution of all the metal present. Those phenomena which Moissan employed in isolating the supposed compound $LiNH_3$, I have been quite unable to reproduce. It is clear, then, that in accordance with the vapor pressure relations, the properties of lithium in the presence of small quantities of ammonia, are such as to preclude the possibility of a compound being formed at the temperatures of the present experiments.

The behavior of other alkali metals is doubtless similar to that of lithium. In the case of sodium I have observed this experimentally. Caesium and rubidium have not been investigated as regards their vapor pressure relations. Moissan[1] obtained what he believed to be the compounds $CsNH_3$ and $RbNH_3$. It seems not improbable, however, that the same errors underlie the results obtained with these metals as has been shown to underlie those obtained with the remaining alkali metals.

The Vapor Pressure of the System Ca-NH_3.

It having been shown that the alkali metals do not form compounds with ammonia, it seemed important to examine at least one member of the group of the alkaline earths. For this purpose calcium was selected. According to Moissan, the compound $Ca(NH_3)_4$ is formed, the method employed being that of Joannis, which has been shown to be untrustworthy. Pressure data were not obtained by Moissan, but he states that above 0° calcium combines with ammonia without liquefaction, while at lower temperatures liquefaction takes place. This indicates that a solid compound is formed whose dissociation pressure lies below one atmosphere, while the vapor pressure of its saturated solution lies above one atmosphere at temperatures in the neighborhood of 0°. That this is correct, follows from the experiments about to be described.

Observations on the phase relations in the more dilute solutions of calcium in ammonia were made in connection with some conductivity experiments. Solutions of calcium, like those of sodium, separate into two liquid phases. The point of complete miscibility of these two phases lies at much higher temperatures in the case of calcium than it does in that of sodium. Even at room temperatures, the concentration of the two phases differs very widely. At —33° the concentration of the dilute phase does not exceed 1/10 gram-atom per liter, while that of the concentrated phase is such that pronounced metallic reflection results. Owing to the tendency of the concentrated phase to cling to the walls of the containing tube and also, perhaps, because of the small difference in the specific gravity of the two phases, it was not possible to separate them into two layers, one above the other. The fact that the concentra-

[1] *Loc. cit.*

ted solution clings to the walls of the container made it very difficult to determine the nature of the system. However, on using very large quantities of solvent and small quantities of metal the concentrated phase could be plainly seen adhering at intervals to the walls of the tube or floating about in the dilute solution.

In studying the concentrated solutions of calcium, a tube of the form outlined in Fig. 2 was employed. The tube G serves as a receptacle for the metal or its solution. At F it is joined to the manometer system C by means of a bit of rubber tubing. The stop-cocks D and E are provided so that at any time G may be detached and weighed, the cocks being closed beforehand. When G is again attached, the air is exhausted from the tube ED before opening E. The cocks A and B make connection with a source of pure ammonia and a vacuum pump, respectively.

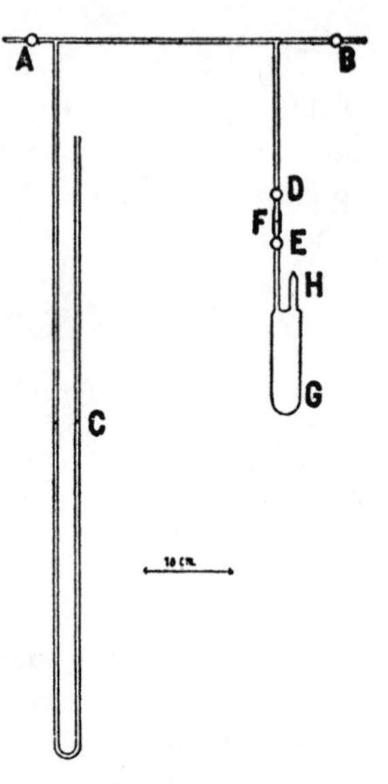

Fig. 2.—Apparatus employed in analyzing calcium compounds.

A piece of metal, freshly cleaned, is introduced into G through H, which is immediately sealed off, after which G is exhausted. The tube G having been weighed beforehand, the weight of metal is obtained by detaching G and weighing. The amount of ammonia present in G at any time may be obtained in a similar manner.

After having introduced a quantity of metal according to the method described above, G was placed in a bath at 20° and ammonia vapor was introduced under a pressure slightly greater than one atmosphere. At the end of 10 minutes appreciable absorption of ammonia had not taken place. Even at 0°, absorption was inappreciable at the end of 5 minutes. The containing tube was then placed in a bath at —33° and ammonia was condensed. It was noticeable that calcium dissolves at a much slower rate than do the alkali metals. When the metal was all in solution, the tube was allowed to warm up, the excess of ammonia being allowed to escape under a pressure of 1 1/3 atmospheres. In the neighborhood of 0°, the liquid disappeared, leaving behind what appeared to be a solid metallic substance identical in color with the solution from which it was precipitated and possessing the mechanical properties of a solid.

The tube was now placed in a bath at 22° and ammonia was withdrawn by means of the pump. The pressure fell rapidly and reached a constant value at 10 centimeters. Pressure readings were carried out at different temperatures with the following results:

Temperature.	Pressure.
10.8°	4.60
0°	2.28
21.7°	9.07
21.7°	9.07
43.7°	30.67

The last two determinations were made at different times. The equilibrium pressure could be reached from one side only. This was due to the extreme slowness with which ammonia combines with calcium. Such combination does take place, however, as will be seen below. In the last experiment a slow increase in pressure was noted, due, without doubt, to the formation of amide and hydrogen from the two constituents.

Calculating, as above, the value of Q from the pressures at 0° and 21.7° we find Q = 10,230 calories for the heat evolved when one gram-molecule of ammonia combines with calcium to form the compound in question. This value of Q gives the pressures 4.66 cm. and 30.20 cm. at 10.8° and 43.7°, respectively. The corresponding values found are 4.60 cm. and 30.67 cm.

The tube was now cooled to —33°, and ammonia was again condensed until the metal was dissolved. Leaving the tube G in its bath of boiling ammonia, solvent was withdrawn with the pump. The pressure soon reached a constant value at about 50 centimeters and a solid substance apparently crystallized from solution. Evidently a compound precipitates out of solution at this pressure, which in turn loses ammonia at a much lower pressure.

A fresh piece of calcium weighing 0.4189 gram was introduced into G, and ammonia was condensed until the process of solution was complete. Leaving the tube in its bath, ammonia was withdrawn until the pressure became constant and solid began to precipitate. The following pressure observations were then made:

Temperature.	Pressure.	Approached from
—32.5°	47.42 cm.	higher pressure
—32.5°	46.98 "	lower "
—32.5°	47.28 "	higher "
—50°	19.28 "	higher "
—50°	19.28 "	lower "
—32.5°	47.18 "	higher "
—32.5°	47.08 "	lower "
—32.5°	47.42 "	higher "

As may be seen, the pressure reading differs slightly according as the equilibrium is approached from higher or from lower pressures. This is due to the slowness with which equilibrium establishes itself. Taking the mean of these observations, we obtain the following values for the pressure of the saturated solution, namely: $t = 50°$, $p = 19.28$ cm., $t = -32.5°$, $p = 47.18$ cm.

For the heat of solution, we obtain from these data, by calculation, the value Q = 5458 calories when one gram-molecule of gaseous ammonia dissolves the compound under equilibrium conditions to form a saturated solution

The pressure-composition curve for calcium and ammonia at —32.5° is represented in Fig. 3, where the ordinates represent pressures in centimeters and the abscissae composition in mols of calcium per 100 mols of calcium and ammonia. Portions of the curve are exaggerated in order to bring out certain points. The correct pressures, however, appear on the margin. Along AB we have the change in pressure of a dilute solution with concentration. At B, a second liquid phase of concentration

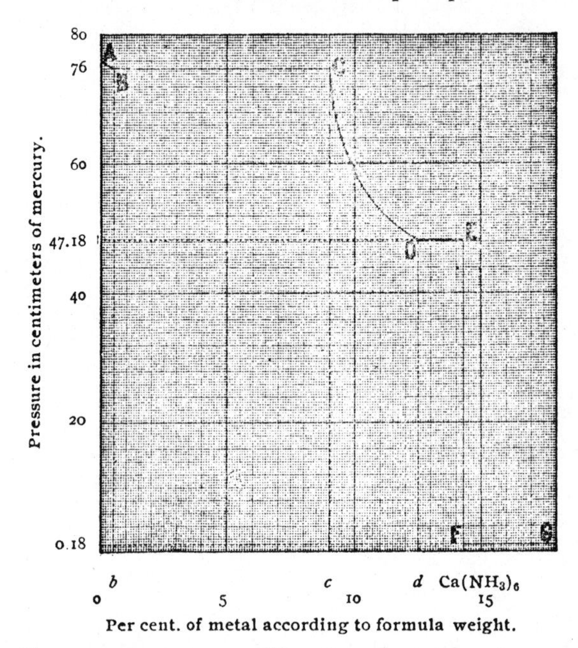

Fig. 3.—Vapor pressure—composition curve for calcium in ammonia.

c appears, and the pressure remains constant until the dilute phase of concentration *b* disappears. This pressure is within a millimeter of the atmospheric pressure, since *b* is very small. Along CD the pressure falls to 47.18 cm., when the solid compound of composition $Ca(NH_3)_6$ (see below) appears. The pressure now remains constant until the saturated solution of composition *d* disappears, when the pressure falls abruptly to 1.8 mm. (calculated) and metallic calcium appears. This pressure is maintained as long as the solid compound remains. That the solid compound does not lose its ammonia in two steps instead of one was shown by measuring the pressure: first when the compound had lost but a little ammonia, and second when only a small amount of ammonia was present in the system. This result will be referred to below in connection with the possible formation of the compound $Ca(NH_3)_4$.

Moissan[1] states that calcium does not absorb ammonia above 20°. As already stated, calcium in the massive form absorbs ammonia very slowly even at 0°. This is to be expected, since a compound is formed which does not liquefy in the presence of ammonia. By employing very finely divided calcium, as it may be obtained by completely withdrawing the ammonia from a solution of the same, it was found that at temperatures above 20° ammonia combines with calcium, although the process is a very slow one.

Composition of the Compound of Calcium and Ammonia.

Experiment 1.—In the preceding experiment in which the vapor pressure of a saturated solution of the compound of calcium in ammonia was determined, 0.4198 gram of metal was employed. At the end of the pressure experiments, which occupied in all about 8 hours, the solution was placed in a bath at 0° and ammonia was withdrawn until a pressure of about 10 centimeters was reached. The tube was then weighed. The contents of the tube weighed 0.9873 gram in excess of that of the metal present. Assuming that this excess in weight is due to ammonia combined with calcium, we may calculate n the number of molecules of ammonia per atom of calcium. We thus have

$$n = \frac{0.9873 \times 40.1}{0.4198 \times 17.06} = 5.529.$$

The correct value is probably either 5 or 6. In view of the fact that the solution was prepared eight hours before analysis was made, it is not improbable that a portion of the metal reacted with the solvent according to the equation: $Ca + 2NH_3 = Ca(NH_2)_2 + H_2$.

Experiment 2.—To avoid errors due to possible formation of amide the following experiments were carried out as rapidly as possible. In this experiment 0.2489 gram of metal was employed. Ammonia was withdrawn at 0° until the pressure nearly reached the dissociation pressure of the compound. There was found a gain of 0.6218 gram, corresponding to $n = 5.874$.

Experiment 3.—The calcium introduced weighed 0.2075 gram. After pumping off the excess solvent at 20°, there was found a gain of 0.5142 gram, giving $n = 5.825$.

Experiment 4.—In this experiment 0.5142 gram of metal was employed. The excess solvent was removed at 0° as in Experiment 2.

[1] *Loc. cit.*

There was found a gain of 1.2830 grams, from which n may be calculated to be 5.864.

After weighing, the tube was again attached to the pump and the ammonia was completely eliminated. The contents of the tube now weighed 0.0116 gram in excess of the weight of metal initially present. This indicates that amide is formed during the experiment. It was accordingly decided to carry out several experiments in which the solvent should be withdrawn at lower temperatures. It is to be mentioned, however, that at lower temperatures a longer time is required in removing the excess of solvent.

Experiment 5.—Employing 0.3888 gram of calcium from which the excess solvent was withdrawn at —33°, a gain of 0.9772 gram was obtained. This gives $n = 5.909$.

Experiment 6.—This experiment is a duplicate of No. 5. There was employed 0.4491 gram of metal and found a gain of 1.1255 grams. From these results n is found to be 5.891 molecules of ammonia per atom of calcium.

Collecting the results of the last five experiments we have:

Experiment No.	n.	Temperature.
2	5.874	0°
3	5.825	20°
4	5.864	0°
5	5.909	—33°
6	5.891	—33°

The temperatures here given are those at which the excess of ammonia was withdrawn. It is plain that at lower temperatures the value of n is consistently larger than at higher ones, while at the same temperature the results are in good accord. That a portion of the metal reacts with ammonia is indicated by the fact that a slow but steady increase in pressure may be observed, particularly at higher temperatures. We may conclude, therefore, that the composition of the compounds of calcium is represented by the formula $Ca(NH_3)_6$.

As already stated, Moissan describes a compound $Ca(NH_3)_4$. It might be though that the compound $Ca(NH_3)_6$ breaks down in two steps, in which case $Ca(NH_3)_4$ should appear as an intermediate product. The experiments described above show conclusively, however, that the compound dissociates according to the equation

$$Ca(NH_3)_6 \rightleftarrows Ca + 6NH_3 .$$

Compounds of Barium and Strontium.

A compound of barium, whose composition is $Ba(NH_3)_6$, has been described by Mentrel.[1] Systematic pressure determinations were not carried out for the purpose of determining the phase relations, but it is stated

[1] *Loc. cit.*

that above certain temperatures a solid compound is obtained which does not liquefy in excess of ammonia. It can scarcely be doubted, therefore, that a compound is formed. Employing the method of Joannis, he finds the decomposition decreasing from 6.97 molecules of ammonia per atom of metal at —50° to 6.10 at 0°. He believes the compound $Ba(NH_3)_6$ to be formed, the excess ammonia being present as a solid solution in this compound. There seems to be need of further evidence on this point.

Roederer[1] has investigated the action of ammonia on strontium. His results are in every way similar to those of Mentrel in the case of barium. He believes the compound $Sr(NH_3)_6$ to be formed. Here also the composition is a function of the temperature and a solid solution is suggested by way of explanation.

Nature of the Compound $Ca(NH_3)_6$.

It was stated at the beginning of this paper that Weyl, on discovering the solutions of sodium and potassium in ammonia, concluded that compounds resulted, to which he assigned an ammonium structure. Joannis, believing that he had isolated compounds of sodium and ammonia, retained these conceptions of Weyl as to the constitution, and other investigators have, for the most part, accepted this view. In the case of compounds containing a considerable number of ammonia molecules, the structural formulae advanced are rather complex to say the least, and they lack for support a single physical property or a single reaction that would indicate a structure such as has been proposed. How little this theory of constitution was based upon facts is well illustrated by the fact that those compounds which first led to the ammonium theory have now been shown to be non-existent. In the case of the metals of the alkaline earths where six molecules of ammonia are present per atom of metal and where all are given off at the same pressure, it is no longer necessary to take the ammonium theory into consideration. The term metal-ammonium should therefore be dropped from the literature as Ruff and Geisel have suggested.[1]

Before discussing further the question of the constitution of the compound $Ca(NH_3)_6$ it will be necessary to consider, briefly, the properties of this substance. It has already been mentioned that the compound is identical in appearance with the solution from which it is precipitated, *i. e.*, that it possessed the same optical properties. To whatever molecular condition these optical properties may be due, it is plain that they obtain in both solid and solution and are therefore not dependent on the physical state of the system. Now the compound and its solution not only possess the same optical properties but they likewise possess the same electrical properties, for both the solid and its solution exhibit

[1] *Loc. cit.*

metallic conduction. It seems possible, therefore, that the same factors which govern the optical properties of these two substances also govern their electrical properties. This suggests that if the factors governing the electrical and optical properties may be determined they will lead to some knowledge as to the state of the compound $Ca(NH_3)_6$ in its solid state. A further discussion of this point cannot be undertaken, however, until the data relating to the properties of the solutions in ammonia have been pesented.

That the compound $Ca(NH_3)_6$ is capable of existence is an important fact. We have here for the first time a compound in which a metal appears combined with a solvent without at the same time being combined with a strongly electronegative element or group of elements, as is commonly the case in solvated salts. The nature of the forces coming into play when calcium combines with ammonia must be quite different from those involved in the combination of a metal with a negative element to form a salt, for, while in the latter case all metallic properties are lost, in the former the metallic properties persist. The salts are usually considered to be valence compounds, and their formation is supposed to involve forces of an electric nature. The compound $Ca(NH_3)_6$ belongs to the class of compounds which, like the solvates, are commonly grouped under the head of molecular compounds. That electrical forces play only a minor part in the compound $(CaNH_3)_6$ is indicated by the persistence of the metallic properties.

It seems probable that the ammonia present in the calcium compound is combined with the metal in the same manner as in the case of ammoniated salts or solvated ions. Indeed the thought lies near that $Ca(NH_3)_6$ is simply a free positive ion which is present to some extent when a calcium salt is dissolved in ammonia. It is interesting, also, to note that this compound corresponds with both Abegg's theory of contravalences and Werner's co-ordination number.

Since the compound $Ca(NH_3)_6$ appears to be a solvate of the metal calcium, the name calcium, hexammoniate may be suggested as a consistent nomenclature.

Incidentally, attention may be called to the fact that the presence of a large number of nitrogen and hydrogen atoms in the compound does not interfere with its properties as a metal. The presence of nonmetallic elements in a compound does not necessarily preclude the existence of metallic properties. Further examples of compounds of this character will be discussed later.

Summary.

The existing experimental data relating to the supposed formation of the compounds $NaNH_3$ and KNH_3 are examined in the light of the phase rule and the conclusion is reached that these compounds do not exist.

In a similar manner the non-existence of the compound $LiNH_3$ is established. In this case independent evidence is given which shows conclusively that in a system containing lithium and a small molecular per cent. of ammonia a saturated solution of the metal in ammonia is formed. This result is in agreement with the phase relationships existing in the system and demonstrates the inapplicability of the method employed by Moissan in obtaining and identifying the supposed compound of lithium and ammonia.

It is shown that calcium forms a solid compound with ammonia whose composition is represented by the formula $Ca(NH_3)_6$. The optical properties of this compound are apparently identical with those of its saturated solution in ammonia, and like its solution, the compound exhibits metallic conduction.

The vapor pressures of saturated solutions of lithium and of $Ca(NH_3)_6$ in ammonia have been determined, as well as the dissociation pressures of the compound itself. The heats of formation of the corresponding solutions and of the compound from metal and gaseous ammonia are calculated to be 8700, 10230, and 5460 calories per gram-molecule of ammonia, respectively.

The constitution of the compound $Ca(NH_3)_6$ is discussed. It appears that this compound is of the nature of a solvate, corresponding, perhaps, to an ammoniated calcium ion. It is suggested that the compound be called calcium hexammoniate in order to take account of these relations in the nomenclature.

BOSTON, February 6th, 1908.

Reactions of Sodium with Metals and Nonmetals in Liquid Ammonia

Unusual polyanionic species can be formed by the reactions of metal–ammonia solutions with certain elements. Excerpts from three papers describing studies of such reactions are presented in the following pages. The first paper is by Smyth, who determined the composition of the sodium–lead reaction product; we have reproduced the introduction and summary from his paper. From the second paper, by Kraus, we have reproduced a brief discussion of the properties of metallic compounds of this type. The third paper, by Zintl, Goubeau, and Dullenkopf, describes a quantitative investigation of the reaction of sodium with various metals and nonmetals; for this paper we merely provide a translation of their summary.

VOL. XXXIX. JULY, 1917. No. 7.

THE JOURNAL

OF THE

American Chemical Society

with which has been incorporated the

American Chemical Journal

(Founded by Ira Remsen)

6

[CONTRIBUTION FROM THE RESEARCH LABORATORY OF PHYSICAL CHEMISTRY OF THE MASSACHUSETTS INSTITUTE OF TECHNOLOGY, No. 107.]

A STUDY OF SODIUM-LEAD COMPOUNDS IN LIQUID AMMONIA SOLUTION.

BY F. HASTINGS SMYTH.

Received May 7, 1917.

Introduction.—For some time it has been known that solutions of sodium and potassium in liquid ammonia are capable of dissolving certain other metals such as lead, antimony and bismuth, and since these metals are themselves insoluble in pure ammonia, it follows that their solution is attended with the formation of intermetallic compounds of the sodium or potassium.

Two compounds of lead and sodium have been isolated by Joannis,[1] one having the composition corresponding to the formula NaPb, which is formed by the agitation of a large volume of sodium solution with metallic lead, and in addition another compound having the composition $NaPb_4$, formed from the former compound by adding an excess of lead to the ammonia solution. This latter compound separates from its solution according to Joannis, with two molecules of ammonia of crystallization. Both compounds are only sparingly soluble in liquid ammonia.

[1] *Compt. rend.*, **113**, 797 (1890); **119**, 586 (1892).

Mathewson,[1] using Tammann's[2] method of thermal analysis, has established the existence of the series of compounds Na_4Pb, Na_3Pb, NaPb, Na_2Pb_5. These compounds were formed by the direct union of the elements and only one of them corresponds to the compound isolated by Joannis.

Kraus[3] first showed that solutions of the compounds of lead and sodium formed by the ammonia solution method are electrolytic in nature, and his work indicated that the ions Na^+ and Pb_2^- are present in them.

In an attempt to verify this result, Dr. Posnjak, working in this laboratory, undertook to electrolyze solutions of lead in sodium ammonia solutions. The investigation was of a preliminary nature and the method of procedure and results of his work will be briefly reviewed.

The solutions for the electrolysis were prepared in Dewar tubes, the ammonia being allowed to flow directly from a commercial ammonia cylinder into the tubes. The top of the Dewar flask was closed with a tight-fitting rubber stopper, through which passed three small glass tubes. Through one of the tubes small pieces of freshly cut sodium could be dropped into the ammonia in the flask, and through the other tubes a platinum gauze anode and a lead cathode could be lowered into or withdrawn from the solution. The anode was fitted with a pulley so that it could be rotated by a small motor. The solution was always allowed to stand for several hours in contact with an excess of metallic lead in the bottom of the Dewar tube before adjusting the previously weighed cathode and starting an electrolysis. It was thus hoped to make sure that all lead dissolved from the cathode should be the result of the electrolytic process only. With this arrangement air was probably excluded, since ammonia vapor passed continually out of the container. The concentration of the solution could not, however, be kept constant, neither could sufficient time be allowed before beginning an electrolysis to make certain that the sodium had completely reacted with the lead. Since commercial ammonia was used, dissolved air and water were also present. Even under the most favorable conditions some oxide was always introduced with the sodium and the initial concentration was only roughly estimated. The current used was measured with an ammeter, and the amount of electricity passed determined by taking ammeter readings at short intervals over a definite period of time.

After the completion of an experiment the electrodes were withdrawn from the solution by removing the rubber stopper and then plunged into a previously prepared Dewar tube containing pure ammonia.. This method of washing is not altogether satisfactory. As will be seen later, the elec-

[1] *Z. anorg. Chem.*, 50, 172 (1906).

[2] *Ibid.*, 37, 303 (1903).

[3] THIS JOURNAL, 29, 1571 (1907).

trodes are difficult to clean under the best conditions. Exposing them to the air even for a moment, moistened with electrolyte at a temperature of —33°, results in the condensation of water and consequent oxidation of sodium and lead. The preliminary results thus obtained are given in Table I.

TABLE I.
Electrolysis of Sodium-Lead Solutions.

Time. Minutes.	Current. Ampere.	Anode gain. G.	Cathode loss. G.	G. atoms lead per faraday.	
				Anode.	Cathode.
40	0.0225	0.2505	0.2533	2.164	2.188
40	0.0255	0.2232	0.2550	1.928	2.203
44	0.0215	0.2750	0.2686	2.260	2.207
41	0.0215	0.2487	0.2465	2.193	2.174
40	0.0215	0.1823	0.2455	1.648[1]	2.168
40	0.0215	0.2814	0.2444	2.602[1]	2.214
40	0.0215	0.2383	0.2410	2.154	2.178
				Av., 2.14	2.19

The table is self-explanatory. In the first column is given the time of passing the current through the solutions, and in Col. 2 the value of the current passed as read on an ammeter. The weights of lead deposited at the anode and dissolved at the cathode, respectively, are tabulated in Cols. 3 and 4. Obviously these values should agree. The gram-atoms of lead deposited and dissolved per faraday are given in the last two columns.

Although these results do not permit of more than a qualitative discussion, they point, nevertheless, clearly to the fact that there are not exactly 2 g. atoms of lead deposited per faraday when the sodium-lead solutions are electrolyzed.

It was the purpose of the renewed investigation to electrolyze under carefully regulated conditions the solutions obtained when metallic sodium solutions in liquid ammonia are agitated with an excess of metallic lead. It was hoped to obviate the sources of error in the preliminary work mentioned above, and to establish more accurately the amount of lead transferred from cathode to anode per faraday, and to determine whether this amount is dependent upon such conditions as the concentration of the solution or the magnitude of the current used in the electrolysis. It was proposed also to verify these results by a direct determination of the solubility of lead in sodium-ammonia solutions. . . .

Summary.—The compounds formed when solutions of metallic sodium in liquid ammonia are agitated with excess of metallic lead at —33° have been investigated. These compounds are very soluble in liquid ammonia and separate without ammonia of crystallization.

[1] Omitted in taking the average.

It has been found by direct solubility determinations that one atom of sodium carries into solution at least 2.20 atoms of lead, but that the true value is somewhat higher than this, since sodium is lost through interaction with the solvent.

Electrolysis of the sodium lead solutions shows that 2.26 g. atoms of lead are deposited at the anode and dissolved at the cathode for each faraday passed. This value is independent of the concentration of the solution up to a concentration of 0.1 molal sodium, and also of the electrolyzing current.

This unusual result has been explained by assuming an equilibrium to exist between a Pb_2^- ion and another negative ion of the form Pb_3^-, in the presence of metallic lead. On this basis the equilibrium ratio

$$\frac{(Pb_3^-)}{(Pb_2^-)} = \text{Const.} = \frac{0.26}{0.74} = 0.36,$$

has been calculated.

In conclusion, it is a pleasure to acknowledge the help and encouragement received from Professor C. A. Kraus, formerly of this laboratory, whose active interest has contributed much to the completion of this research.

CAMBRIDGE, MASS.

7

The Constitution of Metallic Substances

C. A. KRAUS

V. Reaction of Metallic Compounds in Liquid Ammonia Solution

The behavior of solutions of metallic compounds in liquid ammonia clearly indicates that these substances are true electrolytes. They exist in solution in equilibrium with other electrolytes as do ordinary salts and they react with other electrolytes in a similar manner. They undergo metathetical reactions the course of which is determined by the solubility of the compounds formed. With ordinary salts they form metallic compounds which are precipitated from solution when these are insoluble, while they remain in solution when soluble. Thus the lead compound reacts with a cadmium salt according to the equation, $Na_2Pb + Cd(NO_3)_2 = 2NaNO_3 + CdPb_x$, the compound $CdPb_x$ being precipitated. Whether this compound will remain as such after precipitation depends upon its stability and on other factors. This point has not, as yet, been fully cleared up. On the other hand, with metallic calcium, a reaction takes place according to the reaction equation, $K_2Pb_x + Ca = CaPb_x + 2K$. Here calcium plumbide is precipitated while metallic potassium remains in solution.[11] With an ammonium salt the following reaction occurs: $NaPb_x + NH_4I = NaI + NH_4Pb_x$. The resulting plumbide, however, is unstable, as might be expected, and decomposition occurs at once on precipitation with the formation of ammonia and the evolution of hydrogen, while pure lead is left behind.

Very interesting is the case in which an electronegative ion of a metal is precipitated by means of an electropositive ion of the same metal. For example, sodium plumbide, on the addition of lead nitrate, precipitates lead according to the equation, $Na_4Pb_9 + 2Pb(NO_3)_2 = Pb_2Pb_9 + 4NaNO_3$. The precipitated lead may here be looked upon as a lead plumbide. Whether or not the various lead atoms in the resulting precipitate are equivalent to one another or whether they differ remains undetermined. It is conceivable that, in the case of metals which are fairly electropositive and which, at the same time, exhibit a pronounced electronegative valence, the metal in the pure state may exist in the form of a compound, a portion of the atoms being positively charged and another negatively charged. Many facts are in harmony with such an hypothesis. It is well known that the more electronegative elements, such as selenium,[16] tellurium,[17] arsenic[18] and antimony,[17,18] are complex in the vapor state. It is not unreasonable to assume that in these complexes some of the atoms function electropositively and others electronegatively. Furthermore, if these elements are complex in the vapor state, then there is all the more reason for believing that they are likewise complex in the liquid

[16] Brockmöller, *Z. physik. Chem.*, **81**, 129 (1912).
[17] Dobbie and Fox, *Proc. Roy. Soc.*, **98A**, 149 (1921).
[18] Biltz and V. Meyer, *Z. physik. Chem.*, **4**, 263 (1889).

state and, indeed, in this state their complexity may be expected to be relatively high. It is well known that sulfur in the liquid state is complex and certain facts indicate that selenium and tellurium in the solid state are complex. Naturally, the complexity of vapors varies as a function of the temperature, the more complex molecules breaking down at higher temperatures. It appears probable that in these complexes the various atoms are not identically involved.

Z. Phys. Chem., vol. 154A, 1931, p. 1

8

Translated Summary

Salzartige Verbindungen und Intermetallische Phasen des Natriums in Flüssigem Ammoniak

E. ZINTL,
J. GOUBEAU, and
W. DULLENKOPF

Following up studies of C. A. Kraus, F. W. Berstrom, *et al.*, metallic sodium in liquid ammonia was treated with elements which, in the long form of the Periodic Table, occur one to seven positions before the noble gases. A series of products, ranging from the sodium halides to intermetallic phases of sodium (such as those with mercury or gold), was obtained.

To establish the composition of compounds in the sodium–lead system (which we shall use as an illustrative example), the sodium solution was treated, not with metallic lead, but with an anhydrous ammonia solution of lead iodide. The reaction then proceeded quite rapidly and could be carried out as a kind of titration. Compound formation was indicated by conductivity or potential measurements during the titration.

Compounds prepared in this way from sodium and elements one to four places before the noble gases are generally readily soluble in ammonia and have extraordinarily intense colors. They frequently have compositions entirely different from those of the corresponding compounds obtained from melts and they show all the structure types of the polyhalides and polysulfides. By analogy to the latter compounds, the compound shown to be Na_4Pb_9 in ammonia should have the constitution of a saltlike sodium enneaplumbide, $Na_4[Pb(Pb)_8]$. Indeed, according to Kraus, a lead-saturated sodium solution conducts a current. The conditions under which Faraday's Law could be tested on the solutions were discussed. The studies of F. H. Smyth show that, in the electrolysis of sodium–lead solutions, the amount of lead deposited at the anode corresponds quantitatively to our established composition Na_4Pb_9. Faraday's Law is therefore obeyed, and consequently the saline character of the compounds in ammoniacal solution was proven. The designation "polyanionic salts" was proposed for such compounds.

By the electrolysis of a solution of sodium iodide in liquid ammonia between lead electrodes, the dual character of lead was demonstrated, i.e., its ability to go into solution both cathodically and anodically.

The stepwise formation of polyanionic salts in the titration studies, for instance the reaction of lead iodide and sodium solutions, was discussed.

Our preceding remarks regarding the polyanionic salts refer only to their ammoniacal solutions; there they are assumed to be in the form of ammines, corresponding to formulations such as $[Na(NH_3)_n]_4^+[Pb_9]^{4-}$. The "parent compounds" of the polyanionic salts (Na_2S, Na_2Te, Na_2As, and Na_3Bi) are difficultly soluble in ammonia, in accord with the usually observed relationship between ionic size and solubility.

In sharp contrast to the elements one to four places before the noble gases, metals five to seven places before the noble gases react with sodium in ammonia to form, not soluble saltlike compounds, but rather insoluble ammonia-free alloy phases with atomic lattices. The question of the significance to be ascribed to the turning points in the titration curves in such cases was discussed.

The ability to form polyanionic salts coincides with the ability to form volatile hydrides. For example, on going from lead to thallium, the solubility of the sodium compounds decreases markedly.

An apparatus for quantitatively carrying out the reactions in liquid ammonia was described; the results obtained with this apparatus were described for the systems Na–S, Na–Se, Na–Te, Na–As, Na–Sb, Na–Bi, Na–Sn, Na–Pb, Na–Tl, Na–Zn, Na–Cd, Na–Hg, Na–Cu, Na–Ag, and Na–Au.

Thermodynamic Properties of the Ammoniated Electron

Laitinen and Nyman were the first to measure the standard potential of the "electron electrode" in liquid ammonia—that is, the potential (relative to the hydrogen electrode) for the following reversible half-reaction,

$$e^-_{am} \rightleftharpoons e^- \qquad E^\circ_{-36^\circ} = 1.9\ V$$

From this potential one can calculate the free energy change for the reaction

$$e^-_{am} + NH_4^+{}_{am} \rightleftharpoons \tfrac{1}{2} H_{2\,g} + NH_{3\,am}$$

Coulter and Maybury, using an apparatus and technique originally described by Kraus and Schmidt,[1] determined the heat of the latter reaction. These free energy and heat data permitted the calculation of the entropy of the ammoniated electron and were useful for calculating thermodynamic data for reactions involving the ammoniated electron. The papers by Laitinen and Nyman and by Coulter and Maybury are reprinted in the following pages.

[1]C. A. Kraus and F. C. Schmidt, *J. Amer. Chem. Soc.*, **56**, 2297 (1934).

[Reprinted from the Journal of the American Chemical Society, **70**, 3002 (1948).]

9

[CONTRIBUTION FROM THE NOYES CHEMICAL LABORATORY OF THE UNIVERSITY OF ILLINOIS]

Polarography in Liquid Ammonia. II. The Electron Electrode[1]

BY H. A. LAITINEN AND C. J. NYMAN[1a]

The electrode reactions which might occur at a cathode on electrolysis of anhydrous liquid ammonia solutions have been summarized by Makishima.[2] For a platinum electrode, metal deposition, reduction in valence state of the metal ion, hydrogen discharge, and electron dissolution are possible processes. The dissolution of electrons was observed by several investigators when a salt solution, whose ions were non-reducible, was electrolyzed with a platinum cathode. On electrolysis of solutions of tetramethylammonium chloride and hydroxide and tetraethylammonium chloride between platinum electrodes at $-34°$, Palmaer[3] obtained in each case a blue solution, which he attributed to the electron or to the "free alkylammonium radical." Similar observations were made by Kraus[4] and by Schlubach.[5] Forbes and Norton[6] have prepared solutions of tetramethyl-, tetraethyl-, tetrapropyl-, tetrabutyl-, trimethylbutyl-, triethylbutyl-, tripropylbutyl-, tributylmethyl-, tributylethyl- and diethyldibutylammonium ions and electrons by electrolysis of the iodides of these ions between platinum electrodes. Cady[7] and Kraus[8] found on electrolysis of sodium solutions between platinum electrodes that the transfer of electrons from the electrode to the solution was the only cathodic process which occurred and that the opposite transfer was the only anodic process which occurred.

All the processes listed for a platinum cathode can also occur at a dropping mercury cathode. In addition, there is the possibility of amalgamation of the cation of the salt solution, as is known to be the case with the alkali metal ions.[1] For electrons to enter solution from the surface of a dropping mercury cathode, the potential of electron dissolution must be more positive than the amalgamation potential of the cation of the salt solution.

(1) Presented before the Division of Physical and Inorganic Chemistry at the Chicago Meeting of the American Chemical Society, April, 1948. For paper I, see THIS JOURNAL, **70**, 2241 (1948).

(1a) Present address: Department of Chemistry, The State College of Washington, Pullman, Wash.

(2) S. Makishima, *J. Faculty Eng. Tokyo Imp. Univ.*, **21**, 115 (1938).

(3) W. Palmaer, *Z. Elektrochem.*, **8**, 729 (1902).

(4) C. A. Kraus, THIS JOURNAL, **35**, 1732 (1913).

(5) H. H. Schlubach, *Ber.*, **53**, 1689 (1920).

(6) G. S. Forbes and C. E. Norton, THIS JOURNAL, **48**, 2278 (1926).

(7) H. P. Cady, *J. Phys. Chem.*, **1**, 707 (1897).

(8) C. A. Kraus, THIS JOURNAL, **30**, 1323 (1908); **36**, 864 (1914).

When a transfer of electrons from the electrode to the solution, or the reverse process, takes place, the electrode can be considered to be an electron electrode. It was found possible to study these processes polarographically with both platinum and dropping mercury electrodes.

Experimental

The electrolysis cells and apparatus for preparing the anhydrous liquid ammonia solutions, as well as the thermostat, have been described previously.[1] Unless otherwise specified, the cell with the internal reference electrode was used.

The dropping mercury electrode had the following characteristics in a saturated solution (0.0057 M) of tetrabutylammonium iodide in liquid ammonia at −36°. At a pressure of 20 cm. of mercury, the drop time t was 5.1 sec. (open circuit), and the mass of mercury flowing through the capillary was 1.184 mg./sec.

The platinum electrode was of a type described by Laitinen and Kolthoff,[9] and consisted of a platinum disc sealed to the end of a glass tube in such a manner that the disc was vertical when the electrode was in place. The area of the electrode was about 3.1 sq. mm.

The reference electrode used in this investigation was a mercury pool in 0.005 M iodide solution.

A Sargent Model XX polarograph was used. The calibration of this instrument was checked by inserting a known resistance in place of the electrolysis cell and measuring accurately the applied potential. With these data the current flowing through the circuit was calculated and compared with that indicated by the recorder. All applied potential values were checked by means of a student type potentiometer.

The ammonia, obtained from the Matheson Company, was dried by condensation on sodium before being distilled into the electrolysis cell.

The alkali metal salts were C. P. materials of commerce and were dried at 110° before use.

The tetraalkylammonium iodides were prepared in an organic preparations course at the University of Illinois. The tetramethyl- and tetraethylammonium iodides were recrystallized from 95% ethanol, the tetrapropylammonium iodide from 50% ethanol-50% ethyl acetate solution, and the tetrabutylammonium iodide from absolute ethyl acetate. Residual current curves of 0.005 M solutions of these salts in liquid ammonia showed no waves at potentials which corresponded to alkali metal reduction, and the salts were therefore considered free of such impurities.

Sodium amide was prepared by the reaction of sodium with liquid ammonia in the presence of a trace of ferric nitrate hexahydrate as described by Greenlee and Henne.[10]

Theoretical

A polarogram made with a dropping mercury electrode on a 0.001 molar solution of sodium iodide in a saturated solution of tetrabutylammonium iodide is shown in Fig. 1. It is proposed that the increase in current at potentials more negative than 2.3 volts was due to the dissolution of electrons. Assuming, for the present, that this was the process which occurred, an equation for the current–voltage curve obtained can be derived.

The potential of an electrode behaving as an electron electrode should be given by a Nernst expression of the type

$$E_e = E_e^0 - \frac{RT}{F} \ln C_e \qquad (1)$$

where E_e is the potential of the electrode; E_e^0 is the standard potential of the electron electrode; and C_e is the concentration of the electron at the surface of the electrode in moles per liter. Assuming that the Ilkovic equation describes the behavior of an electron with regard to its diffusion from a dropping mercury electrode, then

$$C_e = i/6.05 \times 10^5 D^{1/2} m^{2/3} t^{1/6} \qquad (2)$$

where i is the current which flows (microamperes); D is the diffusion coefficient of the electron; m is the mass of mercury flowing through the capillary (mg./sec.); and t is the drop time of the capillary in seconds. On substitution of C_e from equation 2 into equation 1, the expression

$$E_e = E^0 - \frac{RT}{F} \ln i/6.05 \times 10^5 D^{1/2} m^{2/3} t^{1/6} \qquad (3)$$

is obtained. Equation 3 can be simplified by writing

$$E_e = A - \frac{RT}{F} \ln i \qquad (4)$$

where

$$A = E_e^0 + \frac{RT}{F} \ln 6.05 \times 10^5 D^{1/2} m^{2/3} t^{1/6} \qquad (5)$$

Equations 3 and 4 give the potential of a dropping mercury electrode when a current of i microamperes flows due to the dissolution of electrons. The constant A represents the potential of the electrode when a current of one microampere is flowing. A depends on E_e^0 and on the characteristics of the electrode in a manner given by equation 5. If the process is reversible, a plot of E *versus* log i should have a slope of −0.047 at −36°. The value of A can be determined most accurately from a plot of this type.

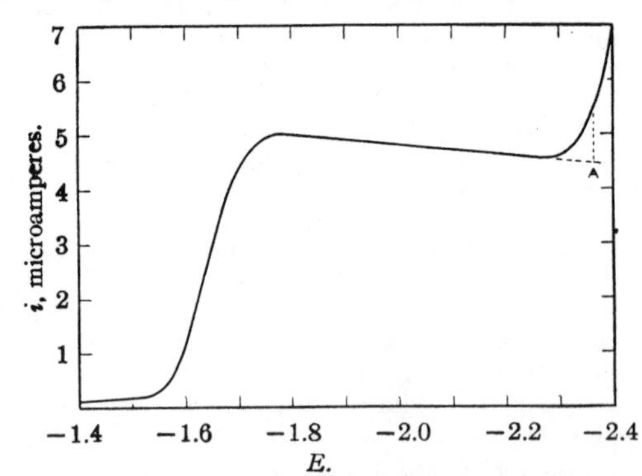

Fig. 1.—Polarogram of 0.001 M NaI in saturated $(C_4H_9)_4$-NI, dropping mercury electrode.

An expression similar to equation 4 would also describe the current–voltage curve obtained when using a platinum electrode. The constant A would change for different electrodes, because the diffusion of electrons would vary with the area and

(9) H. A. Laitinen and I. M. Kolthoff, THIS JOURNAL, 61, 3344 (1939).

(10) K. W. Greenlee and A. L. Henne, "Inorganic Syntheses," 2, 128 (1946).

shape of the electrode and with the orientation of the electrode surface.

Data and Discussion

On electrolysis of solutions of uni-univalent salts with a dropping mercury electrode, the three possible cathodic reactions which must be considered are hydrogen discharge, amalgamation of the cation, and electron dissolution. Since it has been previously reported that the alkali metal ions can be reduced polarographically in liquid ammonia, salts of these ions cannot be used as electrolytes in a study of mercury as an electron electrode. Tetrabutylammonium iodide was chosen as the one to be studied in this connection because it gives a very negative break in water solution, and it was hoped that it would prove to be nonreducible in liquid ammonia even at the most negative potentials. From the polarogram in Fig. 1 it is seen that the value of A is 0.72 volt more negative than the sodium half-wave potential. In the discussion which follows, it will be shown that this break is caused by the transition of electrons from the electrode to the solution and not by one of the other two processes.

If the increase in current at potentials more negative than 2.3 volts was caused by hydrogen discharge, then the value of A should be a function of ammonium ion concentration. Pleskov and Monosson[11] found the dissociation constant of liquid ammonia to be 1.9×10^{-33} at $-50°$. Assuming it to be the same at $-36°$, the ammonium ion and amide ion concentrations are equal to 4.4×10^{-17} mole per liter in a neutral solution. If the amide ion concentration were increased to 10^{-4} mole per liter, the ammonium ion concentration would be decreased to 1.9×10^{-29} mole per liter. From the Nernst equation, this would cause hydrogen discharge to be more difficult and the constant A to be 0.58 volt more negative than in a neutral solution. A polarogram was run on a solution which contained 10^{-4} mole of sodium amide per liter and which was saturated with tetrabutylammonium iodide. In this solution the value of A was found to be 0.70 volt more negative than the sodium half-wave potential. Since there was no shift in the value of A with a decrease in ammonium ion concentration, the process observed could not have been hydrogen discharge.

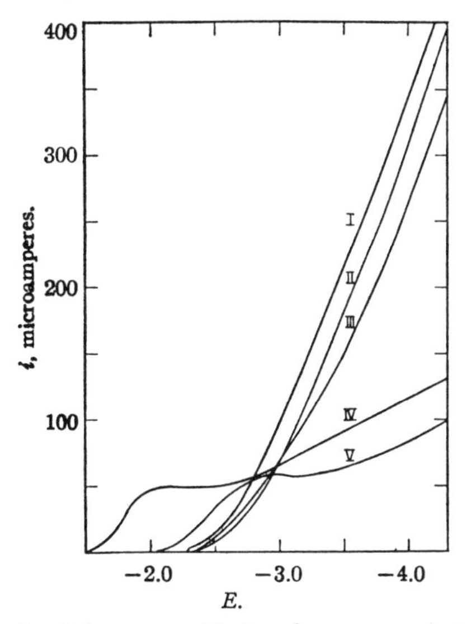

Fig. 2.—Polarograms with dropping mercury electrode: curve I, 5×10^{-3} M $(C_2H_5)_4NI$; curve II, satd. $(C_4H_9)_4NI$; curve III, 5×10^{-3} M $(C_3H_7)_4NI$; curve IV, 5×10^{-3} M NaI; curve V, satd. $(CH_3)_4NI$.

If the increase in current was due to reduction of the cation to form an amalgam, a limiting current should have been observed, such as the one obtained on the polarographic reduction of 0.005 molar sodium iodide solution (Curve IV, Fig. 2). On the other hand, if no amalgamation occurred, the only other cathodic process which could have occurred was the dissolution of electrons. A polarogram for a saturated solution (0.0057 M) of tetrabutylammonium iodide (Curve II, Fig. 2) showed no limiting current even at potentials as negative as 4 volts. This was a good indication that the process was not amalgamation of the tetrabutylammonium ion.

If the process was that of electron dissolution, then the cation of the salt solution should have no effect on the potential of electron dissolution, and, therefore, the tetrabutylammonium ion may be replaced by any ion which is not reduced at these potentials. In this connection, a series of experiments was carried out using tetramethyl-, tetraethyl-, tetrapropyl- and tetrabutylammonium iodides. A polarogram (Curve V, Fig. 2) of a saturated solution (0.0042 M) of tetramethylammonium iodide showed a limiting current similar to that shown by sodium ion. This was not unexpected since McCoy[12,13] and co-workers had previously prepared tetramethylammonium amalgam at $-34°$ by electrolysis of the chloride salt in absolute methanol. They were unable to prepare tetraethylammonium amalgam under similar conditions. A polarogram (Curve I, Fig. 2) of a solution of tetraethylammonium ion showed no limiting current, but the constant A was more positive than for the tetrabutylammonium iodide solution. This can be seen more conveniently from Fig. 3, which is plotted on a larger scale. The tetraethylammonium iodide curve was very erratic and not nearly so well defined as the tetrabutylammonium iodide curve. It appears possible that both electron dissolution and amalgamation of the cation might have occurred with the tetraethylammonium iodide solution. The current–voltage curve (Curve III, Fig. 2) for the tetrapropyl-

(11) V. A. Pleskov and A. M. Monosson, *Acta Physicochim. U. R. S. S.*, **1**, 713 (1935).

(12) H. N. McCoy and W. C. Moore, THIS JOURNAL, **33**, 273 (1911).

(13) H. N. McCoy and F. L. West, *J. Phys. Chem.*, **16**, 261 (1912).

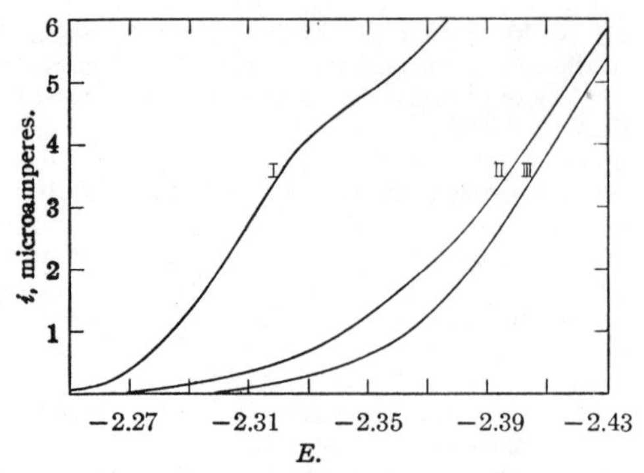

Fig. 3.—Polarograms with dropping mercury electrode: curve I, 5×10^{-3} M $(C_2H_5)_4NI$; curve II, 5×10^{-3} M $(C_3H_7)_4NI$; curve III, satd. $(C_4H_9)_4NI$.

ammonium iodide was almost identical in every respect with the current-voltage curve of tetrabutylammonium iodide. This curve is also shown in Fig. 3. The values of A obtained from these curves by plots of E *versus* log i are recorded in Table I along with the slopes of such plots. The fact that these quantities are almost identical for tetrapropyl- and tetrabutylammonium iodide solutions and the fact that no limiting current was obtained indicated that the same process, electron dissolution, took place in both solutions.

It might be expected that the current-voltage curves for sodium ion and tetramethylammonium ion would show a sharp increase in current at −2.36 volts if that is the potential at which electron dissolution occurs. One reason that a sharp rise is not encountered is the fact that the concentration of positive ions at the surface of the electrode is vanishingly small, because they are being amalgamated as fast as they reach the electrode surface. If the cations are being amalgamated, then the solution surrounding the electrode would have a very high resistance as compared to the bulk of the solution, and the flow of current would be impeded. From the slopes of the current-voltage curves at the more negative potentials, the maximum resistances of the solutions at the electrode surface can be calculated. The maximum resistances of the solutions of 0.005 M tetraethyl-, 0.005 M tetrapropyl- and saturated tetrabutylammonium iodide solutions were about 5,000 ohms, and the resistances of saturated tetramethyl ammonium iodide and 0.005 M sodium iodide solutions were about 20,000 ohms. A gradual rise in current is observed at very negative potentials for the latter two salts on electrolysis, and this must correspond to some cathodic process which in all probability is electron dissolution.

It has already been mentioned that liquid ammonia solutions of cations and electrons can be prepared electrolytically using a platinum cathode, if the cations are stable in the presence of the solvated electron. As in the case of the dropping mercury electrode, changing the type of cation of

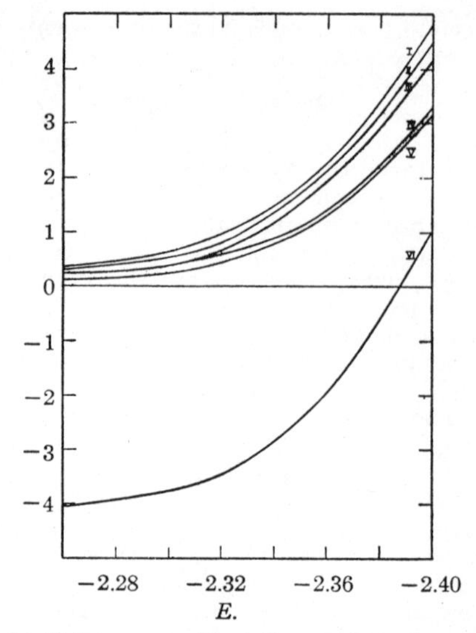

Fig. 4.—Polarograms with stationary electrodes: curve I, satd. $(C_4H_9)_4NI$ with mercury surface; curve II, 5×10^{-3} M NaI, Pt surface; curve III, satd. $(C_4H_9)_4NI$, Pt, surface; curve IV, 5×10^{-3} M LiI, Pt surface; curve V, satd. $(CH_3)_4NI$, Pt surface; curve VI, 6×10^{-3} M Na, Pt surface.

the salt solution should have no effect on the potential of electron dissolution. Polarograms of solutions of lithium, sodium, tetramethylammonium and tetrabutylammonium iodides using a platinum electrode are shown in Fig. 4, and the similarity of the curves bears out the fact that the process is the same in each case. The values of A were nearly constant, but the slopes of the E *versus* log i plots differed from the theoretical value by a considerable margin. These quantities are also recorded in Table I. The difference between the observed slopes and the theoretical ones may be due to poor diffusion conditions in the solutions near the electrode surface or possibly to a potential drop caused by the resistance of the solution. In any event, diffusion to a stationary electrode is not so well defined as to a dropping mercury electrode.

TABLE I

Electrode	Salt	Concn., m./l.	A, v.	Slope E vs. log i
Drop Hg	$(C_2H_5)_4NI$	0.005	−2.26	.....
Drop Hg	$(C_3H_7)_4NI$	.005	−2.34	−0.064
Drop Hg	$(C_4H_9)_4NI$	.0057	−2.36	− .063
Pt	LiI	.005	−2.35	− .104
Pt	NaI	.005	−2.34	− .120
Pt	$(CH_3)_4NI$	.0042	−2.35	− .105
Pt	$(C_4H_9)_4NI$	.0057	−2.34	− .098
Stationary Hg	$(C_4H_9)_4NI$	.0057	−2.34	− .093
Pt	Na metal	.006	−2.35	− .073

To determine whether the electrode material affected the potential of electron dissolution, a polarogram was run on a tetrabutylammonium iodide solution using a stationary mercury electrode. The platinum electrode used in the experiments described previously was plated with copper and then with mercury until a smooth film of mercury covered the electrode surface. If the same process occurred at the mercury surface as at the platinum surface on electrolysis of a tetrabutylammonium iodide solution, then the values of A and the slopes of the E *versus* log i plots should be comparable, because the shape and area of the electrodes were approximately the same. This was observed experimentally and, therefore, the process was the same. These experiments also indicate that there is no detectable overvoltage to the dissolution of electrons either from a platinum or mercury surface. Curve I in Fig. 4 is the polarogram of a tetrabutylammonium iodide solution using the stationary mercury electrode, and the data obtained from it are recorded in Table I.

The platinum electrode used above was plated with copper and, as in the case of the stationary mercury electrode, it should have the same area and shape as the platinum electrode. It was found on electrolysis of a saturated solution of tetrabutylammonium iodide using the copper electrode that A was more positive than for a platinum electrode, and that it varied between -1.5 and -1.8 volts. Reproducible values could not be obtained. Apparently, hydrogen discharge occurred at the copper surface more readily than the dissolution of electrons. This would indicate that freshly plated copper has a low hydrogen overvoltage in liquid ammonia, while bright platinum and mercury have high hydrogen overvoltages. Pleskov[14] has reported that mercury has a higher hydrogen overvoltage in liquid ammonia than in water.

Platinum black electrodes have been reported by Pleskov and Monosson[15] to function as reversible hydrogen electrodes in liquid ammonia. This indicates that such electrodes have little or no hydrogen overvoltage. When a platinum black electrode was used as the cathode for obtaining a current–voltage curve of a solution of tetrabutylammonium iodide, a large current was observed as soon as a negative potential was applied. At a constant potential, 0.1 volt more negative than the reference electrode, the current flowing slowly decreased. This would be the state of affairs if the process were hydrogen discharge from an unbuffered solution. That hydrogen discharge is not the process which takes place at the surface of a bright platinum electrode is indicated by the fact that solutions of electrons can be prepared by electrolysis using platinum cathodes and that metal solutions are stable in the presence of bright platinum electrodes.

In addition to having a cathodic process with the formation of an electron solution, there should be a corresponding anodic process with the transfer of electrons from the solution to the electrode. The anodic transfer can take place at any potential more positive than that of electron dissolution. On electrolysis of a solution of sodium metal, an anodic current was observed at the platinum electrode, and the polarogram obtained is shown by Curve VI in Fig. 4. An H-type cell, with a sintered glass disc to separate the anode and cathode compartments, was used for this experiment. One compartment contained the sodium solution and the platinum electrode, and the other contained the mercury reference electrode in a saturated solution of tetrabutylammonium iodide. A trace of mercuric iodide was added to the tetrabutylammonium iodide solution to depolarize the reference electrode. The potential at which the anodic current has decreased by one microampere in this experiment should be the same as the potential at which the cathodic current has increased by one microampere on electrolysis of a salt solution. The value of A observed for the sodium solution was very close to that observed for salt solutions as can be seen from Table I. The fact that the curve goes smoothly through the zero current line indicates that there is no overvoltage for either the dissolution process or the reverse process.

An unsuccessful attempt was made to obtain an anodic curve using the dropping mercury electrode in a sodium solution. The failure to obtain a polarogram was attributed to the fact that sodium solutions are not stable in the presence of mercury because of sodium amalgam formation.

A solution of tetrabutylammonium ions and electrons was prepared by electrolyzing solutions of tetrabutylammonium iodide in an H-type cell between mercury pool electrodes. Anodic current–voltage curves were obtained for solutions prepared in this manner, but the results were not reproducible. An anodic current was observed at potentials as negative as -2.3 volts.

Makishima[3] calculated the standard potential of the electron electrode to be 2.45 volts more negative than the standard potential of a lead electrode or 2.04 volts more negative than the standard potential of the hydrogen electrode at $-34°$. Forbes and Norton[6] have measured the oxidation potentials of the solutions of tetraalkylammonium ions and electrons mentioned earlier as well as the oxidation potentials of solutions of lithium, sodium and potassium. The concentrations of cations and electrons in these solutions were of the order of 0.005 M, and the oxidation potentials of all were found to be 2.590 ± 0.015 volts more negative than the potential of a silver electrode in a saturated solution of silver nitrate at $-78°$. There was no general trend in the variation of potential with concentration of electron, and it may be possible that their analytical method for determining the concentration of electron was not accurate enough to establish such a trend.

(14) V. A. Pleskov, *Acta Physicochim.*, *U. R. S. S.*, **11**, 305 (1939).

(15) V. A. Pleskov and A. M. Monosson, *ibid*, *U. R. S. S.*, **1**, 871 (1935).

By a consideration of the current–voltage curve for a solution of sodium metal, the standard potential of the electron electrode can be evaluated. The potential at which the current–voltage curve crossed the zero current axis is a function of the surface concentration of electron, and since at this potential no current is flowing, the concentration at the surface of the electrode should be equal to the concentration in the bulk of the solution. The current–voltage curve crossed the zero current axis at −2.388 volts, and the concentration of electron at the electrode surface was 0.006 mole per liter. On substitution of these data into equation 1, the standard potential of the electron electrode was estimated to be 2.49 volts more negative than the mercury reference electrode. It is possible to relate this potential to the standard potential of the hydrogen electrode in liquid ammonia. The difference between the potential of the mercury reference electrode and the potential of a lead electrode in an 0.1 N lead nitrate solution was previously reported[1] as 0.318 volt at −36°. Pleskov and Monosson[15] have reported the difference of potential between a lead electrode in 0.1 N lead nitrate solution and a hydrogen electrode in 0.1 N ammonium nitrate to be 0.352 volt at −50°. The partial pressure of hydrogen in the system was 450 mm. Neglecting the temperature coefficient of the above cell and the activity coefficient of the ammonium ion, the difference of potential between a lead electrode in 0.1 N lead nitrate solution and the standard potential of the hydrogen electrode was calculated to 0.310 volt. Combining these cells, the standard potential of the electron electrode is 1.86 volts more negative than the standard potential of the hydrogen electrode.

Equation 5 can be used to calculate the standard potential of the electron electrode, since the remaining terms can be evaluated. The diffusion coefficient can be calculated from the equivalent conductance of the electron at infinite dilution, and the other terms are experimental quantities. Kraus found that the limiting conductance of the electron at infinite dilution was 910 ohm^{-1}-cm.^{2}-equiv.$^{-1}$. Using the equation

$$D = RT\lambda^0/zF^2 \qquad (6)$$

derived by Nernst,[16,17] the diffusion coefficient D of the electron can be calculated. R is 8.317 volt-coulombs, T is the absolute temperature, λ^0 is the equivalent conductance of the electron at infinite dilution, z is the charge of the electron, and F is 96,500 coulombs. At −36°, D is 1.93 × 10^{-4} sq. cm./sec. The drop time of the electrode was 1 sec., and m was 1.184 mg./sec. For polarograms of saturated solutions of tetrabutylammonium iodide, the average value of the constant A was −2.36 volts with reference to the potential of a mercury pool in a saturated solution of tetrabutylammonium iodide. Substituting these data into equation 5, the standard potential of the electron electrode was estimated to be 2.55 volts more negative than the mercury reference electrode and 1.92 volts more negative than the standard potential of the hydrogen electrode at −36°.

These two values for the standard potential of the electron electrode are in fair agreement with the theoretical values calculated by Makishima.

The potential at which the dissolution of electrons occurs has an important theoretical and practical significance. It represents the limit to the negative potentials which can be applied to an electrode for polarographic measurements in liquid ammonia.

It is possible to relate the standard potential of the electron electrode to the solubility of metals in the following way. When a metal dissolves in liquid ammonia to form solutions of cations and electrons, according to the reaction

$$M \longrightarrow M^{n+} + ne^-$$

an anodic and cathodic process are taking place simultaneously. In the dissolution of metal ions, the potential of the metal is given by the expression

$$E_1 = E^0_M + \frac{RT}{nF}\ln C_{M^{n+}} \qquad (7)$$

For the dissolution of electrons, the potential of the metal is given by the expression

$$E_2 = E^0_e - \frac{RT}{nF}\ln (C_{e^-})^n \qquad (8)$$

Since both processes are taking place at the same electrode, the potential E_1 must be equal to E_2 at all times, and therefore

$$E^0_M - E^0_e = -\frac{RT}{nF}\ln C_{M^{n+}}(C_{e^-})^n \qquad (9)$$

The logarithmic term $C_{M^{n+}}(C_{e^-})^n$ is equivalent to the solubility product of the metal, and the solubility of the metal is given by $C_{M^{n+}}$. These relationships have been pointed out previously by Makishima[2] and were used by him in evaluating the standard potential of the electron electrode. Equation 9 implies that univalent metals whose standard electrode potentials are more negative than that of the electron electrode will be more soluble than one mole per liter. Conversely, univalent metals whose standard potentials are more positive than the standard potential of the electron electrode have a maximum solubility less than one mole per liter. Equation 9 has been found to be valid, at least qualitatively, for those alkali metals and alkaline earth metals whose solubilities and standard electrode potentials are known. Theoretically all metals would be soluble to an extent governed by equation 9. The most positive metals would have an infinitesimally low solubility.

Summary

1. Platinum and mercury electrodes were found to function as electron electrodes in liquid ammonia solutions when the cation of the in-

(16) W. Nernst, *Z. physik. Chem.*, **2**, 613 (1888).

(17) J. J. Lingane and I. M. Kolthoff, THIS JOURNAL, **61**, 825 (1939).

different electrolyte was non-reducible. An equation was derived for the polarograms obtained and for estimation of the standard potential of the electron electrode from polarograms of salt solutions.

2. Two different types of measurements gave values of −1.86 and −1.92 volts for the standard potential of the electron electrode *versus* the standard potential of the hydrogen electrode at −36°. These values are roughly in agreement with a theoretical value of −2.04 volts calculated by Makishima.

3. A relationship between the solubility of metals in liquid ammonia and the standard potential of the electron electrode was presented.

Urbana, Ill. Received April 15, 1948

[Reprinted from the Journal of the American Chemical Society, **71**, 3394 (1949).]

10

[Contribution from the Chemical Laboratory of Boston University]

The Heats of Reaction of Lithium, Sodium, Potassium and Cesium with Ammonium Ion in Liquid Ammonia at −33°

By Lowell V. Coulter and Robert H. Maybury

The thermochemistry of oxidation–reduction reactions involving the alkali metals in liquid ammonia is of twofold interest. From a long range view the heats of reaction of the metals with ammonium ion in liquid ammonia, for example

$$M + NH_4^+ = M^+ + NH_3 + {}^1/_2H_2 \quad (1)$$

may provide basic thermal values which when combined with corresponding free energy changes permit evaluation of relative partial molal ionic entropies in this solvent. In addition to the use of ionic entropies in the calculation of oxidation–reduction potentials not measurable directly, these ionic properties also permit evaluation of relative entropies of solvation of ions in liquid ammonia in a manner developed by Latimer[1] for water solutions. The existing scarcity of accurate free energy data for reactions occurring in liquid ammonia imposes temporary restrictions on the development of a set of reliable ionic entropies for this medium. However, the presumably simpler nature of liquid ammonia as compared with water arising from weaker hydrogen bonding would appear to simplify somewhat the theoretical treatment o the solvation process in this medium and therefore justify the exploration of this solvent.

Of immediate interest is the utilization of thermal data for the above reaction for the comparison of the nature of liquid ammonia solutions of the ammonia soluble metals. Prevailing concepts of these systems, while differing with regard to the equilibria involved in the more concentrated solutions and though incomplete as to the exact nature of the ammoniated electron, agree that the dilute solutions consist of solvated metal ions and single electrons resulting from essentially complete ionization in the dilute range of concentration. An exact similarity for these solutions has been observed by Gibson and Argo,[2] who have reported identical absorption spectra for dilute solutions of lithium, sodium, potassium and cesium. It is to be expected that these systems would likewise possess identical thermochemical properties for reactions involving the solvated electron of these solutions with a common oxidizing agent, as for example

$$e^-_{(am)} + NH_4^+{}_{(am)} = {}^1/_2H_{2(g)} + NH_{3(l)}$$

Direct measurement of the thermal effect associated with this reaction has not seemed experimentally feasible. It may be obtained indirectly, however, as the difference between the heat of solution of the metal in pure liquid ammonia and the heat of reaction of the metal with ammonium ion in liquid ammonia. The latter heat of reaction for solid lithium, sodium, potassium and cesium determined in this research combined with the literature values for the heats of solution of the metals has made possible the evaluation of the heat of this reaction. For dilute solutions of each of these metals we have obtained heats of reaction ranging from 39.7 kcal. for potassium to 41.6 kcal. for cesium with a mean of 40.4 kcal. This we regard as indicative of a common reaction for these solutions involving the solvated electron which energetically appears identical within experimental error in all cases.

(1) Latimer, *Chem. Rev.*, **18**, 349 (1936).

(2) Gibson and Argo, This Journal, **40**, 1327 (1918).

Experimental

The calorimeter employed was essentially of the type used by Kraus and Schmidt[3] but differed in size and method of collecting the gaseous products of the reaction, hydrogen and vaporized ammonia. The calorimeter I of Fig. 1 consisted of a Pyrex vacuum-jacketed test-tube capped with a removable top G by means of a ground glass joint. The top was connected through F to a manifold and served as a point of suspension for the thermocouple well E, sample crushing rod K and the reciprocating stirrer B-J suspended by the spring A. Intermittent energizing of the solenoid C provided a stirring rate of 60 strokes per minute.

Sample bulbs of the metals were held near the bottom of the calorimeter in a platinum stirrup L located directly below the sample crushing rod K to which the holder was loosely attached. The crushing rod extended beyond the calorimeter top through a gas tight seal at D. A fine mesh platinum gauze enclosed the sample bulbs and served as a bubble trap whereby continuous contact of metal and solution was prevented and a controlled reaction rate obtained. The calorimeter was thermostated by boiling liquid ammonia contained in the closed Dewar vessel M of Fig. 1.

The gas collecting system consisted of a six-liter flask P thermostated at 25.0 ± 0.1°, and a manometer Q which were connected to the manifold through a small metal needle valve O. Withdrawal of gas from the calorimeter into the evacuated collecting system was made by manual adjustment of the valve during the reaction period. The entire gas collecting system had a total volume of 6715 ml. The connecting lines which were not thermostated amounted to about 1% of the total volume.

For each thermochemical measurement a weighed amount of ammonia, usually between 110 and 120 g., was introduced at N and condensed in the calorimeter to a liquid depth of about 14 cm. as indicated by the dotted line in Fig. 1. Temperature measurements were made with a copper–constantan thermocouple, calibrated at the sublimation point of carbon dioxide and the freezing point of mercury. Potentials were measured with a Type K-2 Leeds and Northrup potentiometer with a matched galvanometer having a sensitivity of 20 mm. per microvolt for a scale at a distance of 15 ft. from the galvanometer.

Commercial anhydrous liquid ammonia of 99.2% purity, according to the supplier, was used for the thermostating liquid. Stock ammonia for the calorimeter reaction medium was prepared by distillation of the ammonia from the large commercial cylinders into a small cylinder containing sodium metal as a drying agent.

The metals employed in the research were of the following grades as indicated by the supplier and were used without further purification except for cesium which underwent a distillation in the process of sample preparation: lithium, Lithaloy, low sodium quality (typical analysis as furnished by the supplier, N, 0.15; Si, 0.05; Fe, 0.02; Al, 0.01; Ca, 0.10; Na, 0.15; K, 0.05%); sodium, Baker, Analytical Reagent; potassium, Baker, C. P.; cesium, Fairmount Chemical Co.; C. P. Baker Analytical Reagent ammonium chloride and bromide dried at 100° were the oxidizing agents in the reactions. Samples of oxide-free sodium and potassium for the reaction were prepared in small thin-walled glass bulbs in the manner described by Kraus.[4] The cesium samples were prepared by a high vacuum distillation of the metal from the shipping ampoules into the weighed sample bulbs. Lithium samples with clean surfaces were prepared by cutting cylinders of the metal from the ingots under oil followed by removal of the oil with dried benzene rinses.

Procedure.—The net measured heat effect associated with reaction (1) was a composite of heat absorbed by the vaporization of ammonia during the reaction and the thermal change resulting from the temperature change of the calorimeter and contents. The experimental determination of the first of these quantities involved the withdrawal

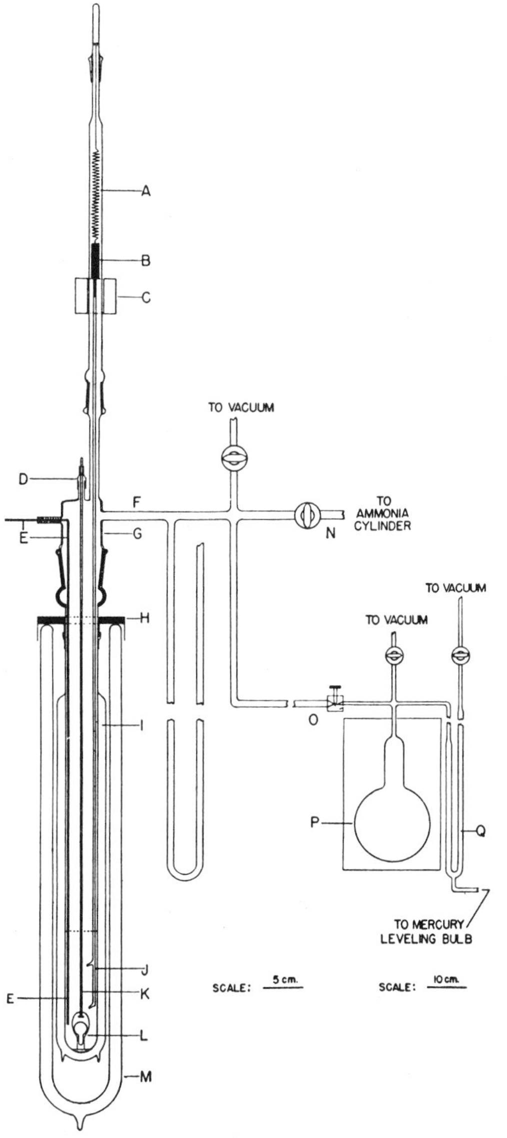

Fig. 1.

of gas during the reaction period from the calorimeter manifold into the calibrated gas collecting system through the manually controlled needle valve at such a rate as to maintain a constant pressure in the calorimeter. The second of these quantities was obtained from the measured temperature change of the calorimeter, as determined by the thermocouple measurements, and the total heat capacity of the calorimeter and contents.

Following the assembly of the calorimeter containing the reactants and liquid ammonia the temperature of the calorimeter was allowed to approach a steady state value.

When a constant temperature or temperature drift was attained, usually less than a thousandth of a degree per minute the sample bulb was fractured by a downward thrust of the crushing rod which initiated the reaction. During the reaction period withdrawal of vaporized ammonia and product hydrogen from the manifold into the

(3) Kraus and Schmidt, THIS JOURNAL, 56, 2297 (1934).

(4) Kraus, *ibid.*, 30, 1197 (1908).

TABLE I

SUMMARY OF HEAT EFFECTS

Expt.	G. atoms of metal or moles of salt	Moles of NH_4 salt in soln.	$-\Delta T$, cor.	C, (total ht. cap.)	$q_1 = C \times \Delta T$	G. of NH_3 vaporized	q_2 = Sp. ht. vap. × G. vap.	Net heat $q_1 + q_2$	Reacn. time, min.	$-\Delta H$ kcal. per g. atom
	Li and NH_4Br									
F	0.00240	0.00257	0.561	140.1	− 78.6	0.6070	199	120	5	50.3
E	.00248	.00248	.361	140.0	− 50.5	0.5406	177	126	2	51.1
G	.00463	.00528	.995	139.7	−139	1.133	371	232	5	50.3
C	.00615	.00640	1.01	136.8	−138	1.381	452	314	1.5	51.2
15	.01326	.0227	1.083	131.8	−142.7	2.450	799.3	656.6	8	49.7
16	.01447	.0227	1.123	140.5	−157.8	2.695	882.7	724.9	7	50.3
	Na and NH_4Cl									
3	.004961	.0283	0.525	149.6	− 78.5	0.8238	269.2	190.7	1.5	38.4
4	.00958	.0115	1.316	148.6	−195.6	1.748	572.4	376.8	24	39.3
5	.01050	.0115	1.418	143.7	−203.8	1.865	612.3	408.5	19	38.8
2	.01317	.0283	1.433	150.3	−215.4	2.227	728.1	512.7	30	38.9
6	.02220	.0254	1.474	142.9	−210.6	3.242	1061.8	851.2	13	38.3
	Na and NH_4Br									
9	.01718	.0227	1.450	145.2	−210.5	2.665	871.8	661.3	10	38.5
8	.02072	.0227	1.391	146.4	−203.6	3.059	1001.7	798.1	6.5	38.5
	K and NH_4Br									
10	.01053	.0227	1.218	146.6	−178.6	1.829	598.6	420.0	9	39.9
11	.01209	.0227	1.278	146.6	−187.4	2.038	666.9	477.7	8.5	39.6
12	.01542	.0227	1.331	146.4	−194.9	2.461	805.4	610.5	9.5	39.6
	Cs and NH_4Br									
H	.005843	.00641	1.13	137.6	−156	1.221	400	244	18	41.8
I	.008911	.00986	1.17	139.3	−163	1.626	531	368	28	41.3
	Li									
J	.0718	.0000	−0.089	138.8	12.4	2.064	673.7	686.1	6	9.55
	NH_4Br									
K	.04380	.0000	−0.25	137.0	34.0	1.172	382	416	10	9.50

gas collecting system was accompanied by approximate temperature readings of the calorimeter in order to establish the temperature–time pattern from which the graphically determined mean temperature was calculated. During the post reaction period regular temperature measurements were again recorded for a period of fifteen to thirty minutes during which time a steady temperature drift was apparent. The corrected temperature change taking into account radiation heat gain was calculated from the temperature patterns so obtained in the usual manner for this type of calorimeter.[5] Although the reactions studied were exothermic, a temperature lowering of the calorimeter always occurred because of the release of gaseous hydrogen within the liquid. The mean calorimeter temperature was approximately −33°.

From the known volume, temperature and pressure of the gas collected in the gas collecting system the weight of ammonia vaporized was calculated by means of Berthelot's equation of state and the critical constants for ammonia.[6] The vaporization heat effect was determined from the net amount of ammonia vaporized and the specific heat of vaporization of ammonia[7] at the mean calorimeter temperature. Pressures were determined by a mercury manometer read by a cathetometer to 0.05 mm. Correction was made for the hydrogen gas liberated by the reaction and collected along with the ammonia in the manifold or gas collecting system on the assumption that the hydrogen gas behaved ideally. A correction was also applied to the weight of ammonia vaporized in those experiments having a net pressure change in the calorimeter system.

The heat capacity of the calorimeter was determined in a fashion similar to a heat measurement except that an electric heater was substituted for a sample of metal. To reproduce the temperature pattern as nearly as possible, hydrogen gas was measured into I through a side arm not shown in Fig. 1 and allowed to bubble through the solvent ammonia in the calorimeter simultaneously with the introduction of measured quantities of electrical energy. The dependence of the calorimeter heat capacity on depth was determined electrically at 25° with water in the calorimeter. For a depth of 15 cm. the heat capacity of the calorimeter was 21 cal. per degree at −33°.

Specific heats for liquid ammonia and the alkali metals were taken from the work of Overstreet and Giauque[8] and the compilation by Kelley.[9] A specific heat of 0.2 cal./g. was assumed for Pyrex glass in evaluating the small contributions of sample containers and glass rod to the total heat capacity of the calorimeter.

The heat effect associated with the crushing of an evacuated sample bulb containing no sample was determined in a manner identical with the reaction heat measurements. No pressure or temperature change (±0.01°) was found associated with the process.

Discussion

The observed heat effects for each of the thermochemical measurements are summarized in Table I. The total heat capacity and corresponding corrected temperature change of the calorime-

(5) W. P. White, "The Modern Calorimeter," Chemical Catalog Co. (Reinhold Publishing Corp.), New York, N. Y., 1929, pp. 40–42.

(6) "International Critical Tables," McGraw–Hill Book Co., Inc., New York, N. Y., Vol. III, 1928, p. 234.

(7) Osborne and Van Dusen, *Bur. of Standards, Bull.* **14**, 439 (1917).

(8) Overstreet and Giauque, THIS JOURNAL, **59**, 254 (1937).

(9) Kelley, *Bur. of Mines, Bull.*, 434 (1940).

ter resulting from the reaction are tabulated in columns 4 and 5. The thermal effect, q_1, associated with the temperature change follows in column 6. The vaporization effect, q_2, listed in column 8 has been calculated from the grams of ammonia vaporized, tabulated in column 7 and the specific heat of vaporization of ammonia at the mean reaction temperature. The total net heat effect per sample and the change in heat content per gram atom for eq. (1) appear in columns 9 and 11, respectively. The values for the latter have been corrected only in the case of lithium for an effective 0.5% impurity.

Comparison of the heats of reaction fail to reveal any significant dependence of the reaction heat on concentration in the range investigated. In the case of lithium the mean of the values for experiments F and E is almost identical with the mean of G and C in which the concentrations were more than doubled. A further increase in concentration does appear, however, to give a slightly decreased reaction heat as evidenced by the mean of experiments 15 and 16 which is about 1.5% less. An analogous trend is to be observed for sodium. Experiment 6, for example, has a heat of reaction about 2% smaller than experiments 4 and 5 in which the dilution was greater. However, in view of an experimental error of at least 1% which must be assigned these values, we are inclined at the present to regard the dilution heat effects to be within the final experimental error to be assigned these measurements. By analogy with similar aqueous systems it is to be noted that dilution heat effects arising from differences in relative heat contents at these concentrations, approximately 400 moles of solvent per mole of solute, are considerably less than 1% of the heats of reaction measured in this research. We shall, consequently, employ these values at this time as the heats of reaction for infinite dilution. On this assumption the heats of reaction for each metal have been averaged and summarized in column 2 of Table II.

TABLE II

Reactants	ΔH_1, kcal. per g. atom	ΔH_2, kcal. per g. atom	ΔH_3, kcal. per equiv.
Li and NH_4Br-1	−50.5	(−8.0)[a]	(−42.5)
Li and NH_4Br-2	−50.5	−9.6[b]	−40.9
Na and NH_4Br	−38.5	+1.40[a]	−39.9
Na and NH_4Cl	−38.8	+1.40[a]	−40.2
K and NH_4Br	−39.7	0[c]	−39.7
Cs and NH_4Br	−41.6	0[c]	−41.6

[a] Kraus and Schmidt, THIS JOURNAL, 56, 2298 (1934). [b] This research. [c] Schmidt, Studer and Sottysiak, *ibid.*, 60, 2780 (1938).

The heats of solution of the alkali metals in pure liquid ammonia at low concentrations have been determined by Kraus and Schmidt,[10] and Schmidt, Studer and Sottysiak[11] at −33° and are tabulated for convenience in column 3 of Table II as ΔH_2 values. Subtraction of the solution reaction

$$M_{(s)} = M^+_{(am)} + e^-_{(am)} \quad (2)$$

and the associated heat of solution, ΔH_2, from our measured heats of reaction, ΔH_1, has given for each metal solution the molar heat effect, ΔH_3, for the common reaction

$$e^-_{(am)} + NH_4^+{}_{(am)} = NH_{3(l)} + {}^1/_2H_{2(g)} \quad (3)$$

These values are tabulated in the last column of Table II.

It has been assumed in writing the foregoing equations that essentially complete dissociation of the metals and electrolytes involved occurs in liquid ammonia. That such may not be the case with electrolytes is indicated by their conductance in liquid ammonia. Gur'yanova and Pleskov[12] have found the Debye–Onsager theory inadequate for these systems and have accounted for conductance properties with some success on the basis of ion pairs. However, as is to be seen from a comparison of the reaction heats of sodium with the two ammono acids, ammonium chloride and ammonium bromide, appreciable differences in ionic association leading to energy effects are not revealed in the concentration range investigated. This is also evident from the apparent independence of the heats of reaction on concentration in general.

Calculation of ΔH_3 for the lithium solution based on the heat of solution reported by Kraus and Schmidt,[10] and Schmidt and co-workers,[11] led to the value −42.5 kcal. designated as lithium–ammonium bromide-1 in Table II, differing by about 6% from the mean obtained for sodium and potassium. Since this appeared to exceed a reasonable experimental error of 1 to 2% on the basis of a similarity for all alkali metal solutions we redetermined the heat of solution of lithium in pure liquid ammonia and obtained a heat of solution of 9.55 kcal. (see expt. J of Table I) which when combined with ΔH_2 gave for ΔH_3 −40.9 kcal. in better agreement with the values for sodium and potassium solutions. Further measurements in connection with other work now in progress substantiate this observation. We have not been able to account for this discrepancy in the heat of solution of lithium in a satisfactory manner.

The possibility that a heat of amide formation was contributing to our heat of solution does not appear likely in view of the observed stability of the solution. The nature of the temperature–time pattern for the heat of solution of the metal in comparison with other reactions likewise gave no indication of a significant side reaction.

Since our disagreement with previous workers on the heat of solution of lithium could be accounted for on the basis of instrumental and procedural differences, we remeasured the heat of solution of ammonium bromide which has also

(10) Kraus and Schmidt, THIS JOURNAL, 56, 2298 (1934).

(11) Schmidt, Studer and Sottysiak, *ibid.*, 60, 2780 (1938).

(12) Gur'yanova and Pleskov, *J. Phys. Chem.* (*U. S. S. R.*), 8, 345 (1936).

been studied by Schmidt and co-workers[13] in the same calorimeter employed for the lithium measurements. We have obtained 9.50 kcal. for the exothermic heat of solution of ammonium bromide at a concentration of 147 moles of ammonia per mole of ammonium bromide. This value is 2.5% lower than the interpolated value of 9.76 obtained from a large scale plot of values obtained by Schmidt. Since these two values are essentially in agreement on the basis of a 1% error for each value, it does not appear that the difference for lithium amounting to about 20% can be ascribed to instrumental differences. It is our intention, however, to redetermine the heat of solution of this metal along with others in a larger calorimeter now under construction which will afford an opportunity to examine these heat effects at much lower concentrations and at the same time eliminate any radiation effect on the solution reaction which might conceivable be involved with the partially jacketed calorimeters employed so far in liquid ammonia studies.

The slightly higher values obtained for ΔH_3 of the cesium solution, -41.5 kcal., which exceeds the mean for lithium, sodium and potassium, -40.2 kcal., by 3% may be indicative of some difference between this solution and the other alkali metals. However, in view of the fact that a few tenths of a per cent. impurity of one of the lighter alkali metals or calcium can account for this observed deviation, it does not appear appropriate to assign any uniqueness to the dilute ammonia solutions to cesium at this time.

In general, then, it appears that all of the dilute alkali metal solutions investigated in this research are energetically identical within about 1 kcal. Weighting all values equally we obtain for the reaction represented by eq. (3) at $-33°$: $\Delta H = -40.4 \pm 1$ kcal.

This heat of reaction will now furnish a basis for the investigation of the nature of solutions of the alkaline earth metals, calcium in particular. On the basis of the absorption spectra[2] and the magnetic susceptibility[14] of dilute calcium ammonia solutions, a difference appears to exist between calcium and the alkali metals in the first instance and between calcium and barium in the second. Measurements are now in progress to determine whether or not the heat of reaction of the calcium solution with ammonium ion indicates normal ionization of calcium or the formation of Ca_2^{++} and one electron ionization per gram atom of calcium as proposed by Yost and Russell.[15]

The relative partial molal ionic entropies of the alkali metal ions in liquid ammonia may now be calculated from the measured heats of reaction for eq. 1 and the corresponding free energy changes derived from cell measurements by Pleskov and Monoszon.[16] These values appear in column 3 of Table III along with our measured heats of reaction from which the partial molal ionic entropies relative to $\bar{S}^0_{NH_4^+} = 0$ at $-33°$ have been calculated in the usual manner.[1] It does not appear that the free energy changes employed in the calculation are above criticism because of the presence of liquid junction potentials in the cells employed by Pleskov and Monoszon and in view of the approximation of activity coefficients made for the solutes in the cell solutions in obtaining the standard electrode potentials.

TABLE III

IONIC AND SOLVATION ENTROPIES IN LIQUID AMMONIA AT $-33°$

Ion	ΔH^0 kcal.		ΔF^0 kcal.	$\bar{S}^0_{240}$°K. cal./deg.	ΔS of solvation cal./deg.
Li^+	$(-66.380)^a$	-50.6	-51.4	-26.8	-52
Na^+	$(-57.520)^a$	-38.7	-42.4	-9.5	-38
K^+	$(-60.340)^a$	-39.7	-45.6	$+2.6$	-27
Rb^+	$(-61.210)^a$	-40.0^b	-44.4	-2.4	-34
Cs^+	$(-62.040)^a$	-41.6	-44.9	-3.0	-42

[a] Corresponding values for the reaction $M + H_3O^+_{(aq)} = M^+_{(aq)} + {}^1/_2H_2 + H_2O$ at 25° [Latimer, *Chem. Rev.*, 18, 349 (1936)]. [b] Estimated from value for potassium.

Comparison of the ionic entropies obtained for the metal ions indicates a departure of $\bar{S}^0_{Rb^+}$ and $\bar{S}^0_{Cs^+}$ from the consistent and expected trend for lithium, sodium and potassium. This is emphasized in a consideration of the relative solvation entropies of each of the ions in the last column of Table III, where ΔS of solvation $= \bar{S}_{M^+} - S_{(gas\ ion)}$. Although smaller negative values for the solvation entropy of Rb^+ and Cs^+ are expected in comparison with K^+ because of the larger ionic radius and consequently less ordering effect on the solvent, the reverse occurs. The admitted uncertainties in the standard potentials undoubtedly will account for these inconsistencies. It is of interest that solvation entropies of the lithium, sodium and potassium ions appear to be linearly dependent on the reciprocal of the ionic radius as already observed for aqueous ions by Latimer[1] and Buffington[17] and that the slope of the curve is approximately the same as obtained for these ions in water solution.[1]

For comparison purposes we have included in Table III the reaction heats of the alkali metals with hydronium ion in water solution. These values appear in parentheses with the corresponding values in column 2 obtained in this research for the reaction represented by eq. (1). It is to be noted that in both systems a minimum value is obtained for $-\Delta H$ for the sodium reaction. Although the difference between the reaction heats in the two systems is about constant at 20 to 21 kcal. for potassium, rubidium and cesium, the dif-

(13) Schmidt, Sottysiak and Kluge, THIS JOURNAL, 58, 2509 (1936).

(14) Freed and Sugarman, *J. Chem. Phys.*, 11, 354 (1943).

(15) Yost and Russell, "Systematic Inorganic Chemistry," Prentice-Hall, Inc., New York, N. Y., 1944, p. 148.

(16) Pleskov and Monoszon, *J. Phys. Chem.* (*U. S. S. R.*), 4, 696 (1933); *Acta Physicochim.* (*U. R. S. S.*), 2, 615 (1935).

(17) Latimer and Buffington, THIS JOURNAL, 48, 2297 (1926).

ference in the case of lithium amounts to only 16 kcal., thereby indicating a relatively greater interaction energy of the lithium ion with ammonia than with water.

Acknowledgment.—We wish to express our appreciation to the Research Corporation for a Frederick Cottrell grant in aid of research which made this investigation possible. We are also indebted to the Sigma Xi for an earlier grant supporting exploratory work. We acknowledge the assistance of Mr. Sumner P. Wolsky who performed some of the calibration experiments.

Summary

The heats of reaction of lithium, sodium, potassium and cesium metals with dilute solutions of ammonium bromide and ammonium chloride have been determined in a liquid ammonia calorimeter at $-33°$. Combination of these heats of reaction with the known heats of solution of the metals in pure liquid ammonia has given the heat of reaction of each dilute meta solution with ammonium ion. Exothermic heats of reaction varying from 39.7 to 41.6 kcal. (mean 40.4 ± 1 kcal.) indicate within experimental error an identical reaction in each case and, therefore, a close similarity in the nature of the dilute solutions of the alkali metals in liquid ammonia. A redetermination of the heat of solution of lithium in liquid ammonia gave -9.6 kcal. for ΔH instead of the reported value -8.0 kcal. From the measured heats of reaction and corresponding free energy changes relative partial molal ionic entropies and entropies of solvation of the alkali metal ions in liquid ammonia have been calculated.

BOSTON, MASS. RECEIVED JANUARY 5, 1949

Potentiometric Titrations in Liquid Ammonia

The following paper by Watt and Otto describes an extension of the work of Zintl, Goubeau, and Dullenkopf (p. 48). The apparatus described here was used by Watt and his co-workers to investigate the reaction of a wide variety of materials with alkali metal–ammonia solutions. A simpler, coulometric method for carrying out such reactions has recently been described.[1]

---

[1]W. L. Jolly and E. Boyle, *Anal. Chem.*, **43**, 514 (1971).

Reprinted from JOURNAL OF THE ELECTROCHEMICAL SOCIETY
Vol. 98, No. 1, January, 1951

Potentiometric Titrations Involving Solutions of Metals in Liquid Ammonia[1]

11

GEORGE W. WATT AND JOHN B. OTTO, JR.[2, 3]

Department of Chemistry, The University of Texas, Austin, Texas

ABSTRACT

A relatively simple apparatus for use in potentiometric titrations in which the titrating reagents consist of solutions of alkali metals (or alkaline earth metals) in liquid ammonia at −38°C is described, together with complete procedural and operational details. The utility of these methods has been demonstrated in the study of the formation of polyanionic salts by reactions between polysulfides of sodium, potassium, rubidium, and cesium in liquid ammonia and liquid ammonia solutions of the corresponding metals.

INTRODUCTION

Nearly twenty years ago, Zintl and co-workers (1) described a method for carrying out potentiometric titrations involving solutions of metals in liquid ammonia. These workers also reported results obtained by titrating ammonia solutions of compounds of elements in Groups I to VI in the periodic table with ammonia solutions of sodium. Of particular interest in connection with the present investigation is the fact that they reported evidence for the formation of salts of homopolyatomic anions (e.g., Na_2S_x, where x = 1-6 inclusive) by the nontransitional elements of Groups IV to VI inclusive. Probably owing to the complexity of the equipment described by Zintl, *et al.*, together with experimental difficulties inherent in the method, there appears to have been no subsequent application of this potentially very useful technique.

The present investigation was undertaken with a view to developing simplified equipment and procedures for such titrations. It was hoped thereby to make available a method which should be useful in the detection of intermediate oxidation states of the elements, in the study of the mechanism of the reduction of both inorganic and organic substances in liquid ammonia (2), in the study of acid-base reactions in ammonia, and numerous other applications. A secondary objective was to clear up uncertainties in the work of Zintl, *et al.*, particularly with reference to the existence of certain of the sodium polysulfides.

EXPERIMENTAL

All chemicals employed in this work were strictly anhydrous reagent grade materials which were analyzed by standard methods. All transfers were made in an anhydrous oxygen-free atmosphere.

Titration Equipment

The apparatus is shown diagrammatically and approximately to scale in Fig. 1. Several scales were used in this drawing because of space limitations. The cross-sectional view of the titrator[4] is approximately $\frac{1}{4}$ actual size; all tubing in the remainder of the drawing is about $\frac{1}{8}$ size. The vertical dimensions of the unfolded view of the titrator are approximately $\frac{1}{8}$ scale, and for all other parts, about $\frac{1}{14}$ scale. Capillary tubes K1, K2, and K3 are 2-mm bore.

The titrator is shown in Fig. 1 in both a semi-schematic unfolded view, and in a separate cross-sectional view in which the components are shown in the manner in which they are arranged with respect to the Dewar flask D, which is the housing for the essential parts of the titrator[4]. This flask (indicated by a dashed line in the unfolded view of the titrator) has the inside dimensions of about 7 x 30 cm and is used to contain refrigerant liquid ammonia and serves as the cooling bath for the titrator. The flask is provided with a liquid ammonia inlet, a fritted glass gas dispersion tube for stirring the refrigerant with a stream of predried air, a low temperature thermometer, and a gas outlet.

V is a calibrated container used to make up solutions of metals in liquid ammonia to known concentrations. This vessel is provided with a ground glass joint J1, through which the alkali metal sample is introduced after removal of cap C2, and a fritted glass gas dispersion cylinder F1, which serves to disperse the gas used to stir the solution and to filter the solution when it is transferred into buret B via T1 and K1. B is the titration buret from which metal solutions are added to the titration vessel R via T2 and K2. Vessel R, which contains a liquid ammonia solution of the substance being titrated, may be opened at J2 for the introduction of samples, and is provided with a fritted glass filter disc F2 which serves to disperse the gas used to

[1] Manuscript received July 11, 1950.

[2] Humble Oil and Refining Company Fellow, 1948-1949.

[3] Present address: Mound Laboratory, Monsanto Chemical Company, Miamisburg, Ohio.

[4] I.e., the volumetric vessel V, buret B, and reaction vessel R.

stir the solution. Electrodes E_i and E_r are connected to the minus and plus terminals, respectively, of a Leeds & Northrup No. 7651 potentiometer; the voltage measurement circuit is wired in the conventional manner and includes a calibrated (U.S.N.B.S.) standard cell.

Attached to the titrator are burets P1 and P2 which are essentially pressure regulating devices by means of which metal solutions are transferred from V to B, and thence into R.

Ammonia and nitrogen gas introduced into the system (at pressures indicated by manometer M) is controlled in relation to the rate of bubbling through L, which contains mineral oil, and which serves not only to seal the exit gas system against back diffusion of air but also to equalize outlet pressures. A by-pass to S6 provides for evacuation without drawing the mineral oil back into the other parts of the system; trap N is simply an added safeguard.

Joint J3 provides for attachment of a small glass still used for purification of alkali metals by distillation *in vacuo* and the filling of small glass ampoules (1) with the metal samples that are sub-

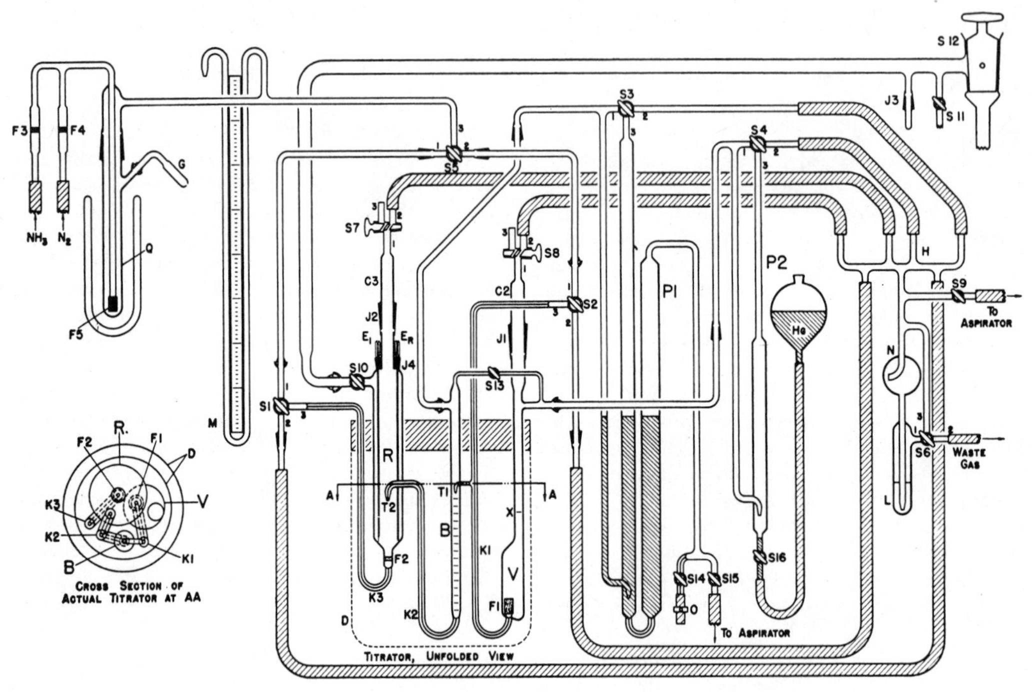

FIG. 1. Titrator and accessories

pass through fritted glass discs F3 and F4, respectively, and are scrubbed through Q (which contains a liquid ammonia solution of potassium) in order to remove the last traces of oxygen. Scrubber Q consists of a straight tube bearing a side arm addition bulb G for the introduction of solid potassium, and a gas dispersion cylinder F5; Q is cooled by means of liquid ammonia contained in a small Dewar flask which is provided with liquid ammonia inlet and gas outlet tubes.

All gases leave the system via manifold H from which they are discharged, after passing through bubbler L, either directly into the atmosphere or after scrubbing through absorbers containing acid to remove ammonia. The rate of discharge of gas

sequently introduced into V. J3 serves also as a point of attachment for a device (Fig. 2) used for filling the reference electrode compartment with anhydrous ammonia gas.

Certain other auxiliary equipment is not shown in Fig. 1. This includes a vacuum system which provides a means of evacuating the titrator and gas burets P1 and P2, the gas inlet and/or outlet system, and any other equipment attached at J3. Attachment of the main body of the apparatus to the vacuum system is via S11 (to a McLeod gauge) and via S12 to an oil pump supplemented by a mercury vapor diffusion pump. Liquid ammonia used as a refrigerant in the Dewar flasks surrounding Q and R, B, and V is drawn from a commercial cylinder, collected for temporary storage in an intermediate Dewar flask, and dispensed in the manner described by Johnson and Fernelius (3).

The electrode system finally selected for use in the present work[5] consists of a clean platinum wire (E_i) which is immersed in the solution being titrated and a differential electrode (E_r) consisting of a modification of the type of electrode originally described by Muller (4) and shown in detail in Fig. 2*a*. For each titration the reference electrode compartment C is filled with a portion of the solution to be titrated; this is done in the manner described below.

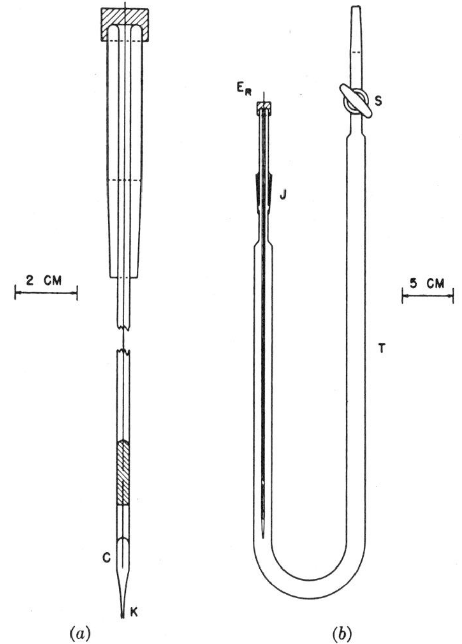

Fig. 2. (*a*) Reference electrode, E_r. (*b*) Device used for filling reference electrode with gaseous ammonia.

Preparation of Apparatus for Titration

Prior to the introduction of the reference electrode E_r into reaction vessel R of Fig. 1, the electrode compartment is filled with anhydrous ammonia gas as follows: E_r is inserted into tube T at ground joint J as shown in Fig. 2*b*, and the assembled apparatus shown in Fig. 2*b* is attached at J3 of Fig. 1, evacuated to about 10^{-6} mm, baked with a heat lamp to assist in degassing, and finally filled with anhydrous ammonia gas at a pressure of 1 atm via stopcock S. The electrode is maintained in this condition until needed for use in a titration, whereupon it is inserted into R (Fig. 1) in the manner described under *Procedure*. Except for the ammonia gas retained in the apparatus beyond stopcock S in Fig. 2*b*, that introduced into the system for the purpose of filling the electrode compartment is removed by means of a water aspirator via S9 of Fig. 1.

Approximately 2 g of potassium cut into small pieces is introduced into G, after which G is attached to Q.

With stopcocks S1, S2, S3, S4, and S5 in position 1,2,3; S6 in position 1,3; S7 and S8 in position 1,2; S10, S11, S12, S13, and S14 in the open position; and S9, S15, and S16 in the closed position, the entire system is evacuated (with the oil pump only) to a pressure of about 0.05 mm. Next, with stopcocks S1, S2, S3, and S4 changed to position 1,3, and S7 and S8 to the off position, the titrator, gas burets P1 and P2, and the gas inlet side of the system are evacuated to about 0.01 mm. Stopcock S5 is then changed to the 1,2 position and nitrogen is admitted to the gas inlet system to a pressure of 1 atm. Meanwhile, evacuation of the titrator and gas burets is continued by means of both the oil pump and the mercury vapor pump. At the same time, the titrator is baked with a heat lamp to facilitate drying. Stopcocks S11 and S12 are closed and nitrogen is admitted through S5 to all parts of the system except the gas exit system. Thereupon, S10 is closed and nitrogen is admitted to the gas exit system by putting S8 in position 1,2. This reduces the pressure in the titrator to less than atmospheric, but the pressure therein is brought back to atmospheric by adding more nitrogen through S5; S8 is then turned to the off position.

In this manner, the apparatus is prepared for introduction of samples into V and R. Stopcock S5 is placed in position 3,2; stopcock S8 is turned to position 1,2; and S6 is changed to position 1,2. While nitrogen is bubbled fairly rapidly through L, cap C2 is removed, S8 is turned to the off position, and in the shortest possible time the glass ampoule containing the alkali metal is lowered to the bottom of V by means of a wire slipped through a glass hook on the ampoule. The wire is quickly withdrawn and a crushing rod long enough to extend above J1 is inserted into V. S8 is changed to position 1,3 and C2 is reattached at J1. C2 is then swept out with nitrogen for at least 5 min, after which S8 is turned to the off position and the flow of nitrogen is cut off. S5 is turned to position 1,2 and the titrator and burets P1 and P2 are evacuated with the oil pump to about 0.005 mm, after which nitrogen is

[5] The use of bimetallic electrode systems (e.g., Pt vs. 90% Pt-10% Rh, and Pt vs. W) in preliminary experiments involving the titration of solutions of ammonium salts or alkali polysulfides with solutions of alkali metals gave results that were unreliable, as judged in terms of known end-points that could be detected visually with an error of the order of ±0.5 per cent.

again let into the system through S5 to a pressure of 1 atm.

By an identical sequence of operations, cap C3 is removed and the substances needed to make up the solution to be titrated in R are introduced. In the present studies these consist of a known weight of sulfur and a known weight of the appropriate alkali metal iodide contained in a fragile glass ampoule. The alkali iodides, which are readily soluble in ammonia, are added to increase the conductivity of the solution in R.

By the same procedure used in introducing the alkali metal into V, C2 is removed at J1, the bulb containing the alkali metal is quickly crushed, C2 is replaced, and the system is again evacuated and subsequently filled with nitrogen as before. By a similar procedure the bulb in R is crushed, but the system is not thereafter evacuated because this operation tends to carry finely divided solids up into S10, C3, and S7. With S5 in position 1,3 and S1 in position 1,2, refrigerant ammonia is transferred to the Dewar flask surrounding Q. While this is being done, nitrogen is put through the remainder of the system at a rate such as to prevent drawing a vacuum when Q is cooled by the refrigerant. Some or all of the potassium is transferred from G to Q and ammonia is condensed in Q until it is approximately ⅔ full of potassium solution. The flow of ammonia is turned off, S1 is changed to position 1,3; S7 to position 1,2, and nitrogen is put through while refrigerant ammonia is being brought into D. The refrigerant in D is stirred with dry air at a rate such that the temperature is lowered to about −36° C, and the nitrogen is replaced by ammonia which is condensed in R against the pressure head in L. After about 45 ml of ammonia are condensed, the ammonia is shut off, and nitrogen is bubbled through the solution in R until the space above the solution is filled with nitrogen saturated with ammonia; the flow of nitrogen is then stopped.

To condense ammonia in V, S5 is changed to position 3,2; S8 to position 1,2; and S7 and S13 are closed so that the path of gas flow is through V via F1. Ammonia is condensed in V to a point about 2 cm below the calibration mark X. Ammonia in the gas inlet stream is replaced by nitrogen which carries with it (from Q) enough residual ammonia to bring the volume up to X. This is an extremely delicate operation which requires adjustment of the temperature of D to the working temperature of −38° C, while at the same time manipulating the temperature of Q and the rate of gas flow so that the final solution in V fills it to, and only to, the calibration mark X when the gas stream.is shut off and the metal solution rises in K1. In approaching completion of the condensation, trial readings are taken by opening S13; this allows solution to flow back into K1, with a meniscus at X. The total volume in V can be determined if the level of solution in K1 is within the range of the calibration marks on B, in terms of which K1 is calibrated[6].

The only additional operations necessary in preparation for the titration consist of filling gas burets P1 and P2 with nitrogen saturated with ammonia. This is done by changing S4 to position 3,2, closing S8, and opening S13 which allows the ammonia-free nitrogen in P2 to be replaced by nitrogen which becomes saturated with ammonia in its passage through Q. S3 is changed to position 1,2 and S4 is returned to position 1,3. A vacuum is pulled on P1 via S15 (with S14 closed) by means of a water aspirator in order to lower the level of the mercury in P1 below the level of the tip which enters P1 near the bottom by way of a ring seal. Then S15 is closed and the aspirator turned off. S3 is then turned to position 3,2, whereupon the ammonia-free nitrogen in P1 is replaced by nitrogen saturated with ammonia. S3 is returned to position 1,2; air is admitted through S14 until the level of the mercury in the right hand member of P1 is approximately 15 cm above the level in the left hand member[7]. Upon completion of these operations, S8 is turned to position 1,2; S3 is returned to position 1,3; and the flow of nitrogen is stopped; S5 is then moved to position 3,1 and S7 is placed in position 1,2. This permits the solution in R to be stirred by the nitrogen admitted via S1. At this stage the stopcock arrangement is as follows: S1, S2, S3, S4, and S5 in position 1,3; S6, S7, and S8 in position 1,2; S13 in the open position; and S9, S10, S11, S12, S14, S15, and S16 in the closed position, and the system is in condition for the actual titration.

Titration Procedure

The procedure given below is for the titration of a liquid ammonia solution of an alkali metal polysulfide, M_2S_x, with an ammonia solution of the corresponding alkali metal, M. This necessarily requires the *in situ* preparation of the polysulfide.

About one third of the metal solution in V is transferred to B for accurate measurement by placing S3 in position 1,2 (which allows gas in B to escape when it is displaced by solution), closing S8 and S13 and opening S16 slowly so as to admit mercury from the leveling bulb into P2 which compresses the gas therein and hence above the metal

[6] F1 does not fill with metal solution; hence, the true volume of solution in V is obtained by subtracting the volume of F1 from the volume of V read from a calibration curve in which the volume in V is plotted as ordinate against the buret reading express in milliliters.

[7] This condition is necessary to obviate too rapid transfer of metal solution from B to R.

solution in V. Metal solution flows from V through T1 via K1 into B. Flow is stopped by closing S16 and the pressure between B and V is equalized by opening S8 first, and then S13 which allows the solution in K1 to drop to a level equal to that in V; S3 is then returned to position 1,3.

The alkali polysulfide is next prepared by transferring from B to R a quantity of metal solution sufficient to give a composition M_2S_x where $x > 6$. This is done by closing S8 and S13 and cautiously opening S14 which allows air to enter the right hand member of P1 through O[8] and compresses the gas in the left member of P1 and hence above the solution in B and causes it to flow up K2 to T2, where the flow is stopped abruptly by closing S14. (If any drops fall from T2, which is drop-calibrated, they are counted, recorded, and subsequently added to the volume of metal solution added to R.) The reading of buret B is then taken with the solution in K2 just at the opening of tip T2. S14 is again cautiously opened and the desired quantity of metal solution is delivered into R. S14 is closed and the buret is again read with the solution in K2 at tip T2 as before. The pressure between R and B is equalized by opening S8 first, then S13, which allows the solution in K2 to drop from T2 to a level equal to that in B. The flow of nitrogen used to stir the solution in R is not interrupted during either of these transfer operations.

The polysulfides form very slowly and, on the scale of operation used in the present work, complete dissolution of the sulfur required as much as 12 hr. All of the alkali polysulfide solutions are red in color and appear to be perfectly homogeneous.

Upon completion of the interaction of alkali metal and sulfur, reference electrode E_r is introduced by first turning off the nitrogen stream used to stir the contents of R, placing S8 in position 1,3, and then turning S7 to position 1,3 and removing the plug from J4 of Fig. 1. E_r is quickly transferred from J (Fig. 2*b*) to J4 (Fig. 1), whereupon the condensation of the gaseous ammonia in compartment C (Fig. 2*a*) of electrode E_r causes this compartment to become filled with polysulfide solution from R via capillary K. S7 and S8 are returned to position 1,2 and the stirring of the contents of R by nitrogen gas is resumed.

The titration is conducted in conventional manner by transferring small portions of alkali metal solution from B to R by the procedure already described, and measuring the potential after each addition. In this work, very small portions (usually around 0.3 ml) were added and every precaution was taken to insure the re-establishment of equilibrium before potentiometer readings were recorded. In some cases this required only a few minutes; in others, as much as one hour was required for constant potentiometer readings after each addition of alkali metal solution.

[8] O is a short section of rubber tubing inside of which is a wire and which is nearly closed by a pinch clamp whereby the rate of air flowing through the rubber tubing and hence through S14 is controlled.

Experimental Results

All potentiometric titrations reported in this paper were carried out at $-38° \pm 0.5°C$. Alkali polysulfides having compositions within the range M_2S_{12} to M_2S_6 (where M = Na, K, Rb, or Cs) were prepared *in situ* in the presence of the corresponding alkali iodides and titrated with standard solutions of the corresponding alkali metals. A feature common to all of these titrations was an initial rise in potential amounting to approximately 100 mv; this was observed regardless of the initial composition of the solution titrated. In all cases the formation of a precipitate was observed just past the M_2S_4 inflection point. The time required for the actual potentiometric titrations ranged from 25 to 40 hours.

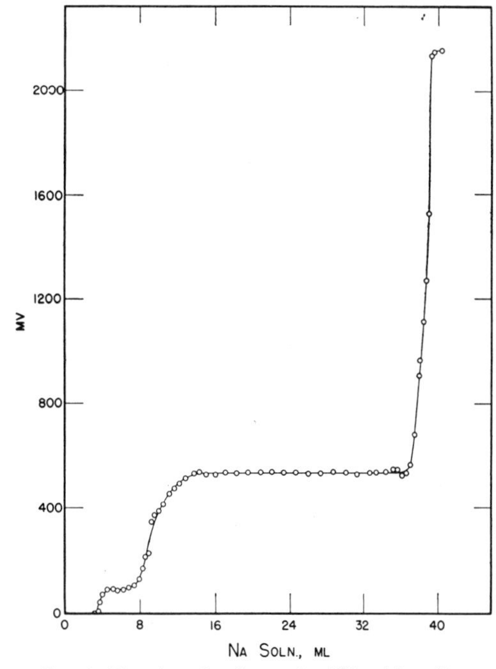

Fig. 3. Titration of sodium polysulfide with sodium

Essential data relative to typical experiments are given in Table I; the actual titration data are shown graphically in Fig. 3 to 7, inclusive. Fig. 5 represents a case that differs from the others in that it involves the titration of *potassium* polysulfide with potassium

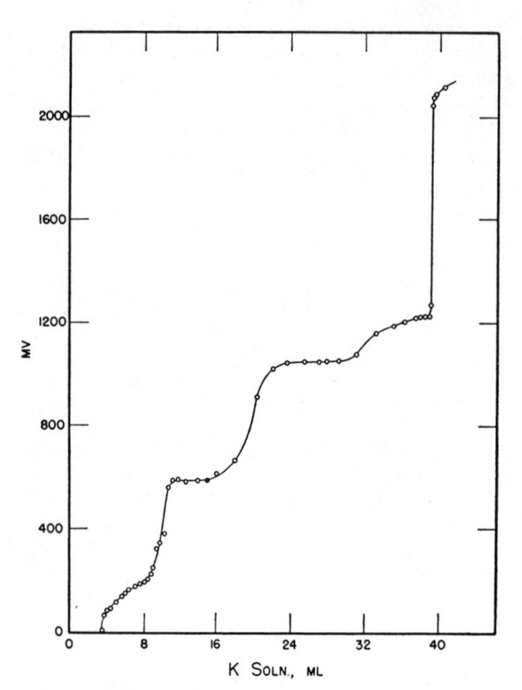

FIG. 4. Titration of potassium polysulfide with potassium (in the presence of potassium iodide).

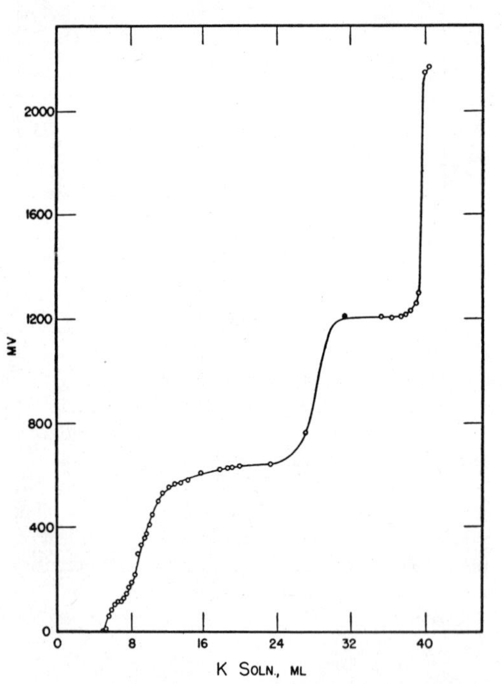

FIG. 5. Titration of potassium polysulfide with potassium (in the presence of sodium iodide).

in the presence of *sodium* iodide. In Table I, the inflection points and the change in potential, ΔV, were obtained graphically from the titration curves shown in Fig. 3 to 7.

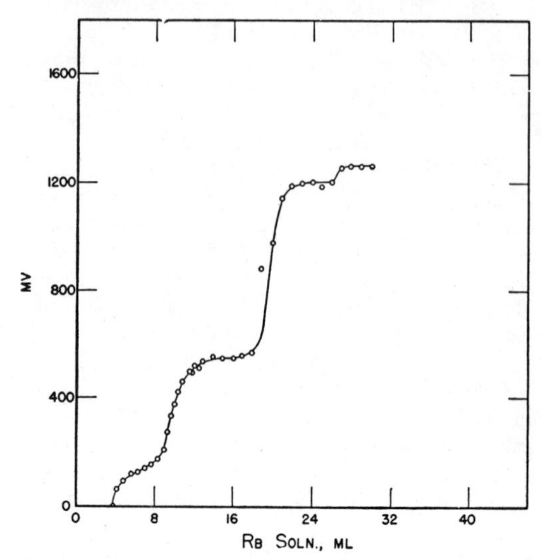

FIG. 6. Titration of rubidium polysulfide with rubidium

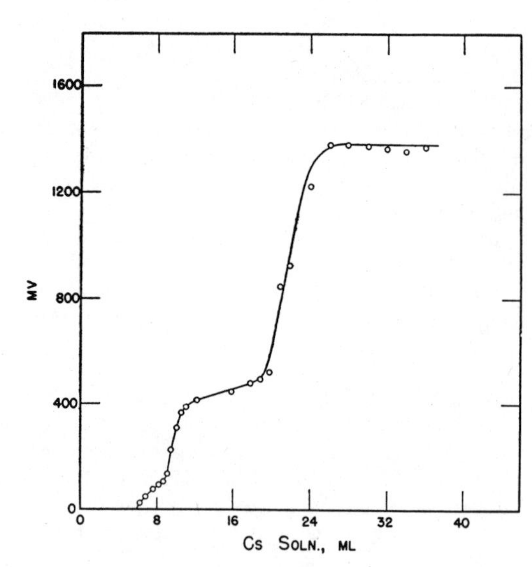

FIG. 7. Titration of cesium polysulfide with cesium

DISCUSSION

The data represented by Fig. 3 show clearly the formation of the compounds Na_2S_4 and Na_2S. On the basis of his studies of reactions carried out at $-33°$ C in a manner just the reverse of the procedure employed in the present work (i.e., addition of sulfur to ammonia solutions of sodium to form Na_2S, followed by further additions of sulfur), Bergstrom (5) reported the formation of Na_2S, Na_2S_2, and Na_2S_4, but indicated that considerable difficulty was encountered in detecting end-points. The fact that the present work provided no evidence whatever for the formation of Na_2S_2, together with a careful inspection of Bergstrom's published data, suggests that the composition attributed to Na_2S_2 may have been a mixture of Na_2S and Na_2S_4. Through the use of methods quite similar to those employed in the present investigation, but involving titrations at $-60°C$, Zintl and co-workers (1) found evidence for the formation of Na_2S_6, Na_2S_5, Na_2S_4, Na_2S_3, Na_2S_2, and Na_2S. However, only the inflection points corresponding to Na_2S_6, Na_2S_4, and Na_2S were reproducible, and this fact was clearly brought out by Zintl, *et al.*, but has apparently been overlooked by others (6). Since the present work confirms the formation of Na_2S_4 and Na_2S, there remains the question relative to Na_2S_6, and the broader question of the existence—in ammonia—of polysulfides of the type $Na_2S_{>4}$.

It has already been pointed out that, in the present work, each titration showed an initial rise in potential amounting to approximately 100 mv *regardless of the initial composition.* In most of their titrations, Zintl and co-workers used initial polysulfide compositions approximating Na_2S_6; hence it is not surprising that their first observed change in potential was erroneously attributed to this particular species. It is our opinion that the initial polysulfide solutions of high sulfur content consist of a species that is the precursor of Na_2S_4 together and in equilibrium with sulfur in solution in liquid ammonia. It is well known that sulfur exhibits appreciable solubility in ammonia and that it reacts at least to a limited extent to form "derivatives of ammonium sulfide and sulfur nitride" (7). At least on a qualitative basis, the results of the present work are consistent with the view that the initial 100-mv change may be attributable to the conversion of dissolved excess sulfur or the products of its interaction with the solvent to that species which is the precursor of Na_2S_4. Finally, it should be recognized that Hugot (8) has reported the formation of Na_2S_5 by the direct interaction of sodium and sulfur in liquid ammonia; however, Bergstrom (5) attributes this result to failure to attain equilibrium.

The data shown in Fig. 4 are entirely in agreement with results obtained earlier by Bergstrom (5) and show unequivocally the formation of K_2S_4, K_2S_2, and K_2S. The slight increase in potential which occurs after addition of approximately 30 ml of potassium solution does not correspond to any reasonable composition. It is of particular interest to observe, by comparison of Fig. 4 and 5, that different results are obtained when sodium iodide is substituted for potassium iodide in the titration of potassium polysulfide with potassium. In the presence of sodium iodide, there is no end-point corresponding to K_2S_2 but there is formed a compound corresponding to the formula M_6S_4. This result, as well as the absence of K_2S_2, could well arise because of altered solubility relationships owing to the presence of sodium ion.

Table I. *Data relative to the potentiometric titration of alkali polysulfides with alkali metals in liquid ammonia at $-38°C$*

Fig. No.	Soln. titrated[a]		Normality of metal soln.	Inflection point			Sulfur/metal	
	Sulfide[b]	MI, mg		Metal soln., ml	Approx. ΔV, mv	Compound indicated	Calcd.	Found
3	Na_2S_{12}	100.6	0.0534	9.2 38.9	440 1620	Na_2S_4 Na_2S	2.0 0.50	2.1 0.50
4	K_2S_{11}	111.1	0.0516	10.0 20.0 39.2	355 455 .860	K_2S_4 K_2S_2 K_2S	2.0 1.0 0.50	1.9 0.95 0.49
5	K_2S_8	100.5[c]	0.0443	9.2 28.5 39.6	440 550 950	K_2S_4 M_6S_4[d]	2.0 0.67 0.50	2.1 0.67 0.49
6	Rb_2S_{10}	142.6	0.0663	10.0 19.8	340 640	Rb_2S_4 Rb_2S_2	2.0 1.0	1.9 0.95
7	Cs_2S_6	173.2	0.0410	9.7 21.8	255 870	Cs_2S_4 Cs_2S_2	2.0 1.0	2.0 0.89

[a] The approximate volume of each initial sulfide solution was 45 ml.
[b] Approximate composition assuming complete interaction of sulfur and metal (see DISCUSSION).
[c] Sodium iodide.
[d] Where M = K and/or Na.

The formation of mixed alkali metal salts in liquid ammonia has been reported previously (9), and it seems likely that the compound M_6S_4 is of this type.

As shown in Fig. 6 and 7, both rubidium and cesium polysulfides are converted successively to the tetrasulfides and disulfides when titrated with solutions of the corresponding metals. It proved impractical to continue these titrations to the monosulfide end-points because the dense yellow crystals of the apparently quite insoluble disulfides reacted so slowly with metal solutions that several hours were required for completion of the reaction with 1-ml portions. Klemm and co-workers (10) have prepared the monosulfides of rubidium and cesium by the direct interaction of the elements in liquid ammonia.

Summary

Relatively simple apparatus and procedures for carrying out potentiometric titrations in liquid ammonia have been developed and described in detail. The utility of these methods has been demonstrated in studies involving the titration of liquid ammonia solutions of alkali polysulfides with liquid ammonia solutions of alkali metals.

It has been shown that polysulfides of sodium, potassium, rubidium, and cesium react with solutions of the corresponding metals to form salts of the type M_2S_4 in all cases, M_2S_2 in the cases of potassium, rubidium, and cesium, and M_2S in the cases of sodium and potassium. No evidence for the existence of several sulfides of sodium previously reported as products of similar reactions could be obtained.

Any discussion of this paper will appear in a Discussion Section, to be published in the December 1951 issue of the JOURNAL.

References

1. E. Zintl, J. Goubeau, and W. Dullenkopf, *Z. physik. Chem.*, **A154**, 1 (1931).
2. G. W. Watt, *Chem. Revs.*, **46**, 289, 317 (1950).
3. W. C. Johnson and W. C. Fernelius, *J. Chem. Education*, **6**, 441 (1929).
4. E. Muller, *Z. physik. Chem.*, **135**, 102 (1928).
5. F. W. Bergstrom, *J. Am. Chem. Soc.*, **48**, 146 (1926).
6. H. J. Emeleus and J. S. Anderson, "Modern Aspects of Inorganic Chemistry," p. 448, D. Van Nostrand Company, New York (1939).
7. O. Ruff and E. Geisel, *Ber.*, **38**, 2659 (1905).
8. C. Hugot, *Compt. rend.*, **129**, 299 (1899).
9. E. B. Maxted, *J. Chem. Soc.*, **111**, 1016 (1917).
10. W. Klemm, H. Sodomann, and P. Langmesser, *Z. anorg. u. allgem. Chem.*, **241**, 281 (1939).

The Reaction of Oxygen with Metal–Ammonia Solutions

The following paper by Thompson and Kleinberg was significant because it describes both the first synthesis of the very unstable compound lithium superoxide and a relatively early measurement of optical absorption spectra of liquid ammonia solutions.

[Reprinted from the Journal of the American Chemical Society, **73**, 1243 (1951).]

12

[CONTRIBUTION FROM THE CHEMICAL LABORATORY OF THE UNIVERSITY OF KANSAS]

The Oxidation of Lithium and the Alkaline Earth Metals in Liquid Ammonia

BY JOSEPH K. THOMPSON AND JACOB KLEINBERG

Solutions of the metals (lithium, calcium, strontium, and barium) were rapidly oxidized by two methods, which are designated as the "one-cell method" and the "two-cell method." The former consisted of oxidation by dropping the metal sample directly into oxygen-saturated liquid ammonia; whereas the latter method consisted of dissolving the metal in ammonia and then slowly passing this solution into a cell containing oxygen-saturated liquid ammonia so that oxidation was instantaneous. All the metals react to yield chiefly monoxide and small percentages of peroxide; no higher oxides are found in the products. All the metals yield small amounts of amide as indicated by the presence of nitrite in the oxidation products. In the case of the alkaline earth metals the products contain traces of unreacted metal as shown by the evolution of hydrogen upon addition of water. Lithium, when rapidly oxidized in liquid ammonia at $-78°$, forms a bright lemon-yellow solution. Measurements of the absorption spectra of oxidized solutions of lithium, sodium and potassium show that all have absorption bands reasonably close to the same wave length, 380 mμ. Since sodium and potassium are known to form superoxides upon rapid oxidation in liquid ammonia, the similarity of their absorption spectra with that of lithium supports the postulate that the latter forms a superoxide which is stable in liquid ammonia solution at $-78°$. Electrolysis of magnesium bromide in liquid ammonia between a magnesium anode and a platinum cathode yields blue solutions of the ionized metal. Oxidation of such solutions results in the formation of small amounts of peroxide; faint, but positive tests for nitrite are also obtained.

The literature on the oxidation of liquid ammonia solutions of lithium and the alkaline earth metals is scanty. Pierron[1] reported a 23% yield of peroxide when oxygen is bubbled through solutions of lithium in liquid ammonia. Guntz and Mentrel[2] reported the formation of gelatinous mixed oxide precipitates containing 7–9% peroxide when barium is treated similarly.

The oxidation in liquid ammonia of the alkali metals other than lithium has been thoroughly studied.[3] According to Kraus and Whyte[4] slow oxidation of potassium, and presumably of sodium also, results in the initial formation of peroxide which is reduced by excess metal to the monoxide. Ammonolysis of the latter yields amide which is subsequently oxidized to nitrite. Rapid oxidation of potassium produces chiefly the peroxide, provided the process is interrupted as soon as free metal disappears; the peroxide may then be converted to superoxide by continued oxidation. There is practically no amide formation when potassium is rapidly oxidized at $-50°$.[5] Schechter, Thompson and Kleinberg[6] have demonstrated that amide formation can be avoided during the rapid oxidation of sodium only under carefully controlled conditions. The purpose of the work described in this report was to study the behavior of lithium and the alkaline earth metals under the conditions which inhibited amide formation in the case of sodium.

Experimental

Method of Oxidation.—Solutions of lithium and the alkaline earth metals were oxidized by two methods, which will be designated as the "one-cell method" and the "two-cell method." The apparatus used was essentially the same as that designed and used by Schechter, Thompson and Kleinberg[6] in their studies on the oxidation of sodium in liquid ammonia. In the "one-cell method" the sample of metal was dropped into a cell containing about 75 cc. of liquid ammonia (refrigeration grade) through which dry oxygen was bubbling at a rate of about 40 cc. per minute. In the "two-cell method" the metal sample was dissolved in about

(1) P. Pierron, *Bull. soc. chim.*, [5] **6**, 235 (1939).
(2) A. Guntz and Mentrel, *ibid.*, [3] **29**, 585 (1903).
(3) G. W. Watt, *Chem. Revs.*, **46**, 293 (1950).
(4) C. A. Kraus and E. F. Whyte, THIS JOURNAL, **48**, 1781 (1926).
(5) C. A. Kraus and E. F. Parmenter, *ibid.*, **56**, 2384 (1934).
(6) W. H. Schechter, J. K. Thompson and J. Kleinberg, *ibid.*, **71**, 1816 (1949).

40 cc. of ammonia in one cell, and then the solution was passed slowly into the second cell which contained about 75 cc. of liquid ammonia through which dry oxygen was bubbling at a rate of about 40 cc. per minute. The solution was passed slowly enough that no blue color persisted in the second cell; the metal was oxidized immediately as the solution entered the oxidation cell. The samples of metal, a few milliequivalents in weight, were cut and scraped in a dry-box and then transferred to the cell in glass capsules which fitted into the cell by means of ground glass joints.[7]

Method of Analysis.—After oxidation was complete by either method, the ammonia was allowed to evaporate, and the residue was analyzed. Analysis for peroxide consisted of catalytic decomposition of the product with manganese dioxide suspension and measurement of the volume of oxygen evolved, followed by titration of the dissolved residue with standard acid to determine the total base content. The evolved gas was ignited to determine the amount of hydrogen from the reaction of unoxidized metal with the catalyst suspension. The dissolved residue was tested for nitrite by means of a spot test involving the dye formation between α-naphthylamine and sulfanilic acid in the presence of nitrite.[8] Since the products contained a yellow material, it was concluded that no superoxide was present.

Absorption Spectra Measurements.—Absorption spectra measurements were made at −78° on liquid ammonia solutions of the oxidation products of lithium, sodium and potassium with a Coleman Model 11 Universal Spectrophotometer which had been modified to accommodate liquid ammonia solutions. The conversion was accomplished by substituting a quartz Dewar flask and absorption cell similar to those described by Gibson and Argo[9] for the conventional absorption cell and carrier.[10] No change was made in the optical or electrical system of the instrument. In using this instrument it was necessary to measure the apparent transmission (indicated by the galvanometer reading) over the entire wave length range separately for solvent and solution. At any particular wave length the difference between the transmission of the solvent and that of the solution divided by the transmission of the solvent gives the absorption of the solution. Multiplication of this quantity by 100% gives the % absorption of the solution.

Results and Discussion

Oxidation Studies.—Lithium dissolves readily at temperatures near −33° to form a deep blue solution. There is no evidence that an insoluble phase is formed at such temperatures. Oxidation of the blue solution at this temperature by either the "one-cell method" or the "two-cell method" gives a milky-white suspension, which subsequently turns light yellow in color. As the ammonia evaporates, a white residue is left.

Lithium does not dissolve readily in liquid ammonia at −78°, but the metal can be oxidized at this temperature by dropping the sample into liquid ammonia through which oxygen is bubbling. The metal gradually dissolves, becoming oxidized immediately upon solution. No intermediate blue solution of metal is formed. The color of the solution after oxidation at this temperature is a bright lemon-yellow. When the solution is permitted to warm up toward −33°, the yellow color fades, giving way to a white suspension. The color cannot be brought back by cooling the cell again. Whether the metal is oxidized at −33 or −78°, the product after evaporation of the ammonia is the same white flaky substance. Studies of the absorption spectra of these solutions and their implications will be discussed later.

Calcium does not dissolve appreciably when placed in liquid ammonia at −78°. At −33° the metal is readily soluble, forming blue solutions which separate into two liquid phases of approximately the same density. Oxidation of these solutions gives a white or gray residue.

The behavior of strontium with regard to its solubility in liquid ammonia is identical with that of calcium. Oxidation gives a white suspension; the product after evaporation of the ammonia is a gray powder.

Barium dissolves the most readily of the three alkaline earth metals, forming the usual deep blue solution. When the metal is added to ammonia at −78°, there appear to be two liquid phases; but as the temperature is raised, the solution becomes homogeneous. Oxidation of this solution gives the usual white suspension. The product after evaporation of the ammonia is always a gray powder.

The results of the oxidation studies are summarized in Table I. Lithium and the alkaline earth metals behave nearly alike when oxidized in liquid ammonia. The product in every case contains chiefly monoxide and a relatively low percentage of peroxide. This behavior is to be contrasted with that of the other alkali metals, all of which give high percentages of superoxide when oxidized under these conditions. It must be pointed out that monoxide is probably not formed by any process other than that of direct oxidation. The mechanism for monoxide formation found for potassium, namely, reduction of peroxide by metal,[4] would not appear to apply here since essentially the same quantity of monoxide is obtained in both the "one-cell" and "two-cell" methods of oxidation. In the latter method there is never any excess of metal in the solution being oxidized and consequently no possibility for reduction of peroxide.

TABLE I

PRODUCTS FOUND AFTER OXIDATION OF LITHIUM AND THE ALKALINE EARTH METALS IN LIQUID AMMONIA AT −33°

	One-cell oxidation			Two-cell oxidation		
Metal	Peroxide, %	Nitrite, %	Unreacted metal, %	Peroxide, %	Nitrite, %	Unreacted metal, %
Lithium	26 ± 6 (17 runs)	5	0.5	26 ± 6 (23 runs)	5	0.5
Calcium	12 ± 2 (8 runs)	3–10	1	12 ± 2 (6 runs)	3–10	1
Strontium	20 ± 2 (5 runs)	4	2	25 ± 3 (6 runs)	4	2
Barium	37 ± 6 (9 runs)	4	8	25 ± 5 (7 runs)	2	1

The tendency to form peroxide in the alkaline earth group increases in the order of increasing

(7) A few exploratory tests were made with magnesium. It was not possible to get magnesium into solution directly, but it was found that, when solutions of magnesium bromide in liquid ammonia were electrolyzed between a platinum cathode and a magnesium anode, blue solutions of ionized metal were formed. This behavior was quite analogous to that described for aluminum in a previous report from this Laboratory (A. W. Davidson, J. Kleinberg, W. E. Bennett and A. D. McElroy, THIS JOURNAL, **71**, 377 (1949)). The products of oxidation of such solutions of magnesium gave positive qualitative tests for peroxide and for nitrite. Quantities were too small for any quantitative determinations to be made.

(8) F. Feigl, "Qualitative Analysis by Spot Tests," Nordemann Publishing Company (Elsevier-Amsterdam), New York, N. Y., 1937, p. 203.

(9) G. E. Gibson and W. L. Argo, THIS JOURNAL, **40**, 1327 (1918).

(10) The authors are indebted to Mr. R. R. Miller, of the Naval Research Laboratory, Washington, D. C., for the loan of these quartz items.

ionic size. This order is only qualitative, and it should not be inferred that a direct proportionality exists. Lithium, which is smaller than any of the alkaline earth metals, does not fall into such an order on the basis of ionic size alone.

The tendency for unreacted metal to appear in the product of oxidation increases with increasing ionic size. This is true especially in runs made by the "one-cell" method. There is little evidence of unreacted metal in the products of "two-cell" runs.

Oxidation by either the "one-cell" or the "two-cell" process gives about the same yield of peroxide for lithium and calcium. For strontium the "two-cell" process gives slightly higher amounts than the "one-cell" process. For barium, on the other hand, the latter method gives greater yields than the former. No explanation can be offered for these variations.

In every case the oxidation product contains nitrite equivalent to a few per cent. of the total metal used. This fact indicates that there is a strong tendency for amide formation to occur. One source of amide formation may be the ammonolysis of monoxides, as described by Kraus and Whyte[4]; another source[11] may be the decomposition of ammoniates according to the equation

$$M(NH_3)_6 \longrightarrow M(NH_2)_2 + H_2 + 4NH_3$$

The exact mechanism of nitrite formation is unknown and is worthy of further investigation.

Absorption Spectra Studies.—It has been previously mentioned that lithium is oxidized at −78° to form a lemon-yellow solution. Sodium and potassium give yellow solutions in which some solid yellow matter is suspended. The color of the solutions is less intense than that formed from lithium due to limited solubility of the superoxides. Absorption studies were made on oxidized solutions of lithium, sodium and potassium after filtration of solid matter. The % absorption of the solutions relative to the pure solvent is shown in the graph in Fig. 1.

Comparison of the absorption curves in the graph shows that in every case the solution has an absorption maximum in the violet region of the spectrum. When allowance is made for the 35

(11) E. Botolfsen, *Bull. soc. chim.*, [4] **31**, 561 (1922).

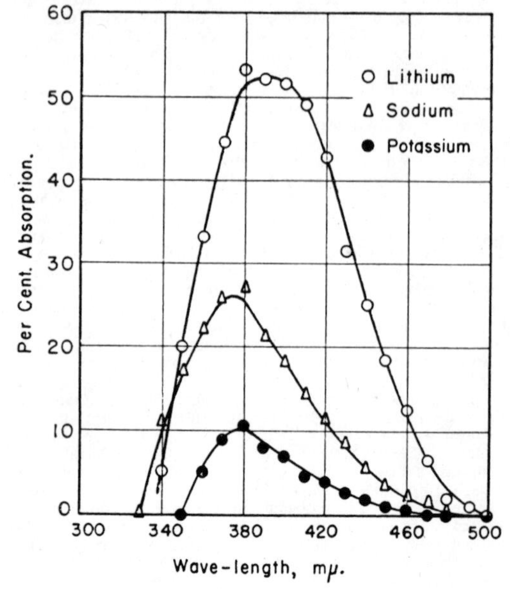

Fig. 1.—Absorption spectra of products of oxidation of alkali metals in liquid ammonia.

mμ band width of the instrument, the maxima in the absorption curves appear to occur reasonably close to the same wave length, 380 mμ. The relative heights of the absorption maxima decrease in the order lithium, sodium and potassium. Though no quantitative measurement of solubility was made, this order appears to be that of decreasing solubility of the substance responsible for the absorption. Since sodium and potassium are known to form superoxides upon rapid oxidation in liquid ammonia, the similarity of these absorption spectra supports the postulate that lithium forms a superoxide which is stable in liquid ammonia solution at −78°.

Acknowledgment.—The authors are indebted to the Office of Naval Research for financial aid which made this investigation possible.

LAWRENCE, KANSAS RECEIVED SEPTEMBER 1, 1950

The Reduction of Transition Metal Compounds

The following note by Behrens describes some examples of an important type of reaction: the reduction of transition-metal compounds to unusual low oxidation states by metal–ammonia solutions.

13

Neuartige Reaktionen der Metallcarbonyle mit Alkalimetallen

Von Helmut Behrens

Anorganisch-chemisches Laboratorium
der Technischen Hochschule München

(Z. Naturforschg. **7 b**, 321—322 [1952]; eingegangen am 9. Mai 1952)

Durch Einwirkung von Alkalimetallen, besonders in ihrer außerordentlich reaktionsfähigen, tiefblauen Lösung in flüssigem Ammoniak, auf Metallcarbonyle entstehen Alkalimetall-Metallcarbonyle, die bei Eisen- und Kobaltcarbonylen den Alkalisalzen der betr. Carbonylwasserstoffe entsprechen. Auch Metallcarbonylhalogenide sind derartigen Reaktionen zugänglich, desgl. lassen sich Umsetzungen von Metallcarbonylen mit Amalgamen, speziell der Alkalimetalle, durchführen.

Bereits früher [1] konnte über die Reaktionen von Eisen- und Kobaltcarbonylwasserstoff mit Alkali- und Erdalkalimetallen *in flüssigem Ammoniak* berichtet werden, bei denen sich unter Abgabe von Wasserstoff die Alkali- und Erdalkalisalze der betreffenden Carbonylhydride bilden.

Es war nun von besonderem Interesse, das *Verhalten der reinen Metallcarbonyle und Metallcarbonylhalogenide gegenüber den außerordentlich reaktionsfähigen Metalllösungen in flüssigem Ammoniak* zu untersuchen. Während unter normalen Bedingungen weder die *monomeren* noch die *mehrkernigen reinen Metallcarbonyle* mit Natrium reagieren [2], sind sie — wie neue Versuche jetzt ergeben haben — mit der bekannten blauen Natriumlösung in flüssigem Ammoniak sehr leicht zur Umsetzung zu bringen. Dabei zeigt sich, daß bei denjenigen Metallen, von denen Carbonylwasserstoffe bekannt sind, sowohl bei den monomeren wie den mehrkernigen Typen die Natriumsalze der betreffenden Carbonylwasserstoffe gebildet werden, in denen die Edelgasschale im Metallcarbonylanion erreicht ist.

Ist die Zahl der CO-Gruppe im Carbonyl *gleich* derjenigen im Carbonylwasserstoff, so erfolgen die Umsetzungen nach dem Schema:

$$[Me(CO)_n]_x + y\,Na = x\,Na_y\,[Me(CO)_n].$$

So erhält man aus Kobalttetracarbonyl $[Co(CO)_4]_2$ bzw. Eisentetracarbonyl $[Fe(CO)_4]_3$ die Natriumsalze $[Co(CO)_4]Na$ bzw. $[Fe(CO)_4]Na_2$.

Ist die Zahl der CO-Gruppen jedoch *größer* als die im Carbonylhydrid, so erfolgt teilweise CO-Substitution durch Na. In quantitativer Umsetzung entsteht so aus Eisenpentacarbonyl und Natrium Eisencarbonylnatrium:

$$Fe(CO)_5 + 2\,Na = [Fe(CO)_4]Na_2 + CO.$$

Auch das Enneacarbonyl $Fe_2(CO)_9$ setzt sich im gleichen Sinn quantitativ mit Natrium um.

Schließlich wurde der Fall, daß das Carbonyl *weniger* CO als der betreffende Carbonylwasserstoff besitzt, am Beispiel des Kobalttricarbonyls $[Co(CO)_3]_4$ untersucht. Nach der Gleichung:

$$[Co(CO)_3]_4 + 3\,Na = 3\,Na[Co(CO)_4] + Co$$

wird hier ebenfalls vorwiegend Kobaltcarbonylnatrium neben metallischem Kobalt gebildet.

Besonders interessant sind die Reaktionen der Alkalimetalle mit den *Hexacarbonylen der Chromgruppe* sowie mit *Nickelcarbonyl*, da ja von diesen Metallen Carbonylwasserstoffe oder deren Alkaliderivate bisher nicht isoliert werden konnten. In allen Fällen erfolgt auch hier Reaktion unter teilweiser Abgabe von CO. Wie weit es sich hier um Substitutionen von CO durch Na handelt, müssen weitere Versuche noch klären. Im Falle des $Cr(CO)_6$ scheint das gelbe, in flüssigem Ammoniak lösliche Reaktionsprodukt bei der Hydrolyse zu ähnlichen Substanzen zu führen, wie sie bei der Basenreaktion des $Cr(CO)_6$ mit alkoholischer Lauge bei höheren Temperaturen anfallen [3]. Bei der Umsetzung von $Ni(CO)_4$ mit Natrium erhält man eine tiefrote Lösung; indessen handelt es sich hier nicht um ein Alkalimetall-Derivat, sondern wohl um ein Ammoniakat des Carbonyls.

Auch *Metallcarbonylhalogenide* setzen sich in ähnlicher Weise mit Natrium um. Läßt man die blaue Natriumlösung auf festes $Fe(CO)_4J_2$ einwirken, so erhält man in fast quantitativer Ausbeute $[Fe(CO)_4]Na_2$:

$$Fe(CO)_4J_2 + 4\,Na = [Fe(CO)_4]Na_2 + 2\,NaJ.$$

Die Ausdehnung dieser Versuche auf die Carbonyle und Carbonylhalogenide des *Rheniums* und der *Edelmetalle* ist vorgesehen. Sie ermöglicht gegebenenfalls die Darstellung von neuen Derivaten dieser Carbonyle, die bisher noch nicht zugänglich waren. Es sind weiter Untersuchungen über die *Einwirkung von Amalgamen*, speziell der Alkalimetalle, auf Metallcarbonyle im Gang, nachdem sich bereits gezeigt hat, daß auch *Natriumamalgam mit Lösungen von Kobaltcarbonyl* in indifferenten Mitteln im oben erwähnten Sinn reagiert.

Die Arbeiten, an denen die Herren Dipl.-Chem. F. Lohöfer und R. Weber beteiligt sind, werden in jeder Richtung fortgesetzt.

[1] H. Behrens, Angew. Chem. **61**, 444 [1949].

[2] Vgl. Abeggs Handbuch d. anorg. Chemie, IV. Bd., 3. Abt., 4. Tl. (Nickelband), S. 809.

[3] Vgl. vorstehende Mitteilung von W. Hieber.

The Synthesis of Methyl Phosphines

In the following paper, Wagner and Burg describe the synthesis of methylphosphine and dimethylphosphine by reactions which are closely related to those studied by Joannis in 1894. (See p. 17.) It is significant that a definite hydridic reaction was shown by $(CH_3)_2PH$ in the presence of amide ion. A similar ammonolytic side reaction was later observed in the reaction of germane with metal–ammonia solutions.[1]

---

[1]D. Rustad and W. L. Jolly, *Inorg. Chem.,* **6**, 1986 (1967).

[Reprinted from the Journal of the American Chemical Society, **75**, 3869 (1953).]

14

[CONTRIBUTION FROM THE DEPARTMENT OF CHEMISTRY, UNIVERSITY OF SOUTHERN CALIFORNIA]

Liquid Ammonia Chemistry of the Methyl Phosphines

BY ROSS I. WAGNER[1] AND ANTON B. BURG

RECEIVED FEBRUARY 3, 1953

Good yields of CH_3PH_2 and $(CH_3)_2PH$ are obtained by liquid ammonia reactions of the type $2e^- + 2PH_3 \rightarrow 2PH_2^- + H_2$; $CH_3Cl + PH_2^- \rightarrow Cl^- + CH_3PH_2$. The conversion of CH_3PH_2 to $(CH_3)_2PH$ by this method is not always efficient, for CH_3PH_2 is a weaker protic acid than PH_3, and may have some hydridic tendency. A definite hydridic reaction is shown by $(CH_3)_2PH$ in the presence of amide ion, with which it forms an orange-colored complex, and then ammonolyzes to a salt from which the bis-phosphinoamine $(CH_3)_2P(NH)P(CH_3)_2$ (m.p. 39.5°, v.t. 3.2 mm. at 27°) is obtained by action of NH_4Br. This amine has a small tendency to disproportionate, yielding ammonia and other products, and may have been formed through disproportionation of $(CH_3)_2PNH_2$. The orange complex is easily decomposed into $(CH_3)_2PH$ and amide, or on treatment with CH_3Cl gives a 92% yield of $(CH_3)_3P$.

In relation to studies of the phosphinoborine polymers[2] it was necessary to develop new methods of synthesis of methyl and dimethyl phosphines, since the earlier bomb tube methods[3,4] gave only very small yields relative to the required effort. Of a number of approaches which were tried, the best was based upon the use of metals in liquid ammonia, according to the equations (1) $2PH_3 + 2e^- \rightarrow 2PH_2^- + H_2$; (2) $CH_3Cl + PH_2^- \rightarrow CH_3PH_2 + Cl^-$; (3) $2CH_3PH_2 + 2e^- \rightarrow 2CH_3PH^- + H_2$; (4) $CH_3Cl + CH_3PH^- \rightarrow (CH_3)_2PH + Cl^-$. Step (1) has long been known[5] and the others were planned by extrapolation from similar studies of arsine.[6] Either sodium or calcium was employed as the source of electrons for steps (1) and (3), and the corresponding chlorides were precipitated in steps (2) and (4). The yields of CH_3PH_2 by this method were not far from quantitative, with 87% recovery; for $(CH_3)_2PH$ the best was 67%, based upon CH_3PH_2.

The observations of H. C. Brown, *et al.*, on the sharp increase of electron-donor bonding power of methylphosphines with increasing methylation,[7] led to the expectation that the removal of a proton from a phosphine would become more difficult in the same order. In fact, reaction (1) is fast, but (3) is definitely slower and harder to complete without side-reactions leading to an impairment of the yield of dimethylphosphine. Furthermore, all attempts to form an authentic salt of the $(CH_3)_2P^-$ ion from sodium and $(CH_3)_2PH$ failed on account of an ammonolysis reaction in which the P–H bond acted not as a proton source, but as a hydride, much as the Si–H bond does in the reaction of $(C_2H_5)_3SiH$ with amide in liquid ammonia.[8]

The hydridic reaction of dimethylphosphine with sodium in liquid ammonia occurred in such a manner as to suggest that the first step was an accelerated amide formation, and the next an amide-promoted ammonolysis of the P–H bond. Some light was thrown on the latter aspect by the observation that sodium amide and dimethylphosphine (each virtually insoluble in liquid ammonia) together dissolve in liquid ammonia at −78° to form a deep-orange-colored solution, which then slowly gives off one H_2 at higher temperatures, without reprecipitation of sodium amide. The resulting sodium salt evidently is some kind of phosphinamide—either $Na(CH_3)_2PNH \cdot xNH_3$ or $Na(CH_3)_2PNP(CH_3)_2 \cdot xNH_3$—for its treatment with ammonium bromide leads through a product of uncertain

(1) This paper is based primarily upon parts of the M.S. thesis and Ph.D. dissertation submitted by Ross Irving Wagner to the Graduate Faculty of the University of Southern California. Generous support of this work by the Office of Naval Research is gratefully acknowledged. An item of post-dissertation research, supported by a grant kindly provided by the American Potash and Chemical Corporation, also is included.

(2) A. B. Burg and R. I. Wagner, THIS JOURNAL, **75**, 3872 (1953).

(3) A. W. Hofmann, *Ber.*, **4**, 605 (1871).

(4) N. R. Davidson and H. C. Brown, THIS JOURNAL, **64**, 718 (1942).

(5) A. Joannis, *Compt. rend.*, **119**, 557 (1894).

(6) W. C. Johnson and A. Pechukas, THIS JOURNAL, **59**, 2068 (1937); also, private communications from W. C. Johnson are gratefully acknowledged.

(7) H. C. Brown, E. A. Fletcher, E. Lawton and S. Sujishi, Abstracts of Papers Presented at the 121st National Meeting of the American Chemical Society, p. 9N (1951).

(8) C. A. Kraus and W. K. Nelson, THIS JOURNAL, **56**, 195 (1934).

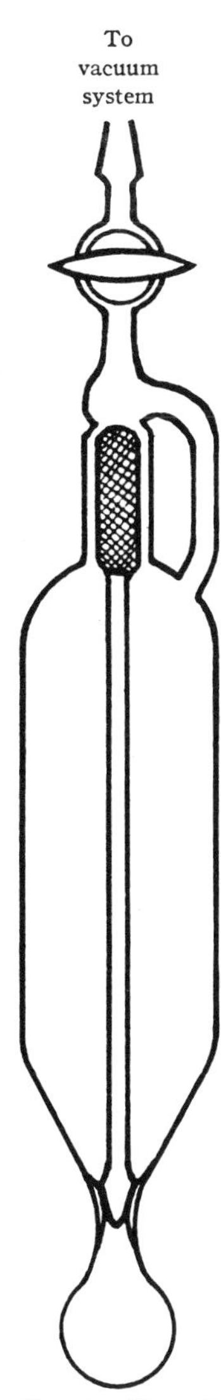

Fig. 1.—Magnetic separatory funnel.

constitution to a volatile bisphosphinoamine, $(CH_3)_2P(NH)$-$P(CH_3)_2$, which undergoes a partial disproportionation at moderately elevated temperatures.

The constitution of the original deep-orange solute is difficult to judge. Since both components, $NaNH_2$ and $(CH_3)_2PH$, are only very slightly soluble in ammonia, it is suggested that amide attaches to phosphorus, forming a complex ion in which the electron-octet of P is exceeded; then the high electron-density on P would promote the hydridic behavior of the P–H bond. Such a complex ion would not be expected to be very stable, and in fact the formation of the orange solution is very definitely reversible even at $-78°$.

The orange solution reacts with CH_3Cl to give a very high yield of $(CH_3)_3P$, and one might explain this result either as an addition of CH_3^+ groups to the hypothetical $(CH_3)_2P^-$ ion or as a similar reaction with the amide-complex ion of $(CH_3)_2PH$, with a loss of H^+ and NH_2^- from the resulting aggregate, to form ammonia. The $(CH_3)_2P^-$ hypothesis lacks verisimilitude because all attempts to isolate a pure salt of that anion, by several different approaches, have failed; but an ammoniate of it still could be present as a minor component of an elaborate equilibrium system. On the other hand, the rapid addition of CH_3Cl to $(CH_3)_2PH$ (but not to NH_2^-) at $-78°$ would require a fairly special mechanism. Possibly the attachment of NH_2^- to $(CH_3)_2PH$ strongly activates the unshared electrons of P for displacement of Cl^- from CH_3Cl. The state of this solution evidently deserves further study.

It is possible that CH_3PH_2 is involved with amide in a similar situation, less favorable to the hydridic reaction but tending to impair step (3) in the preparation of $(CH_3)_2PH$. If so, it might be possible to obtain nearly quantitative yields of $(CH_3)_2PH$ by adding CH_3PH_2 to $NaNH_2$ or KNH_2 in liquid ammonia at $-78°$, and treating the product with methyl chloride quickly, allowing little time for any hydridic reaction to occur.

Experimental Part

Preparation of Phosphine.—Pure phosphine was prepared by dropwise addition of aqueous potassium hydroxide to a resublimed sample of phosphonium iodide. For storage, it was condensed from the high vacuum apparatus into a steel cylinder.

Methylation of Phosphine.—In the first experiment, a solution of 3.65 g. of sodium in 250 ml. of liquid ammonia was treated with enough phosphine to discharge the blue color (in an apparatus adapted from the literature)[9] and the light-yellow solution was treated with the calculated proportion of CH_3Cl, discharging the yellow color and precipitating NaCl. The resulting two-phase liquid system was distilled off and treated with sodium, introduced in 50-mg. capsules. At $-78°$ the first 100 mg. of sodium was used up in ten minutes, but three hours were required to fade the blue color of 100 mg. of sodium after 2.9 g. had been added, corresponding to conversion of 79% of the CH_3PH_2. The chrome-yellow solution now was treated with CH_3Cl in amount equivalent to the sodium used up at this point, and the denser of the two liquid phases, chiefly $(CH_3)_2PH$, was separated by means of the magnetically operated separatory funnel indicated in Fig. 1; this worked very well except that some of the oily $(CH_3)_2PH$ adhered to the walls above the cut-off. Since $(CH_3)_2PH$ has some small solubility in liquid ammonia, some but not all of the dissolved product was recovered by column-fractionation of the ammonia phase. The yield of $(CH_3)_2PH$ actually formed was estimated as 25–30%, based upon the original phosphine. Its purity was indicated by its vapor tension of 342 mm. at 0° (lit. 338).[4]

The next method began with the formation of $Ca(PH_2)_2 \cdot 6NH_3$ from Ca and PH_3 in liquid ammonia.[10] After removal of the solvent, the salt was decomposed *in vacuo* at 50° to form CaPH, PH_3 and NH_3, with 82% recovery of the indicated phosphine. Then the CaPH, which in the dry state would not react with CH_3Cl, was suspended in stirred liquid ammonia and converted by CH_3Cl to $CaCl_2 \cdot 8NH_3$ and $(CH_3)_2PH$; yield 33%. This method had two major disadvantages: the uncertainty of getting CaPH cleanly, without forming Ca_3P_2 or retaining $Ca(PH_2)_2$, and the tedious character of the heterogeneous final step.

Finally, the original process was tried again, using 40 g. of calcium (instead of sodium) in 2 liters of liquid ammonia. This time the CH_3PH_2 was isolated after step (2), in 87% yield, by means of a fractionating column having a special liquid-separatory reflux head (shown in Fig. 2), operating at $-78°$. Ordinary fractionation would not work well because the immiscible methylphosphine underwent ammonia vapor distillation and so preceded the bulk of the ammonia. Hence the head was arranged to collect a pool of CH_3PH_2 while returning the less dense liquid ammonia. At the end of the process the pool was drained by cooling the side-arm receiver; the relatively small amount of ammonia which followed it was removed by means of the magnetic separatory funnel. Another difference from the original process was in step (3): the calcium (37 g.) was first dissolved in the liquid ammonia (1.7 liters) and the CH_3PH_2 was distilled in as rapidly as feasible, at $-78°$. Then the methyl chloride was introduced rapidly (on a stream of dry nitrogen to avoid stoppage of the delivery tube) so that the dimethylphosphine was formed before any serious disturbance by side reactions could occur. The product was isolated by means of the separatory-head column and the magnetic funnel; the yield of $(CH_3)_2PH$ after purification was 72.5 g. (1.17 moles, or 67.2%, based upon CH_3PH_2).

The Reaction of $(CH_3)_2PH$ with Sodium in Liquid Ammonia.—In an attempt to form the salt $NaP(CH_3)_2$, 18.1 cc.[11] of $(CH_3)_2PH$ and 17.4 mg. (16.95 cc.) of Na, in 0.5 ml. of liquid ammonia, changed from blue to deep orange during 3.5 hours at -40 to $-35°$. The evolved hydrogen now was measured as 11.06 cc., or 130% of that expected from the equation $2Na + 2(CH_3)_2PH \rightarrow H_2 + 2NaP(CH_3)_2$. The volatile components were distilled out *in vacuo* and separated, with recovery of 6 cc. of $(CH_3)_2PH$. Hence the reaction had produced nearly one H_2 per $(CH_3)_2PH$ used up, or more if the recovery of $(CH_3)_2PH$ was incomplete. The purity of the hydrogen was demonstrated by combustion over CuO; hence the production of so much must be ascribed to a hydridic reaction. A similarly high yield of hydrogen was found in two other experiments at different concentrations of reactants.

The Reaction of $(CH_3)_2PH$ with $NaNH_2$.—Sodium amide was made from 25.2 mg. (24.5 cc.) of sodium by means of a trace of ferric nitrate in 0.5 ml. of liquid ammonia: H_2,

(9) C. A. Kraus and C. L. Brown, This Journal, **52**, 4034 (1930).

(10) C. Legoux, *Compt. rend.*, **209**, 47 (1939).

(11) Throughout this paper, the term "cc." refers to the volume which the designated substance would occupy as a gas at standard conditions.

12.20 cc. (99.6%). Then 24.5 cc. of $(CH_3)_2PH$ was condensed upon the frozen $NaNH_2$–NH_3 system, and allowed to mix in at the melting point, forming a solution having the deep orange color noted before. At −30° this solution produced 1.0 cc. of H_2 in the first 15 minutes. Toward the end of the reaction, as the orange color was fading toward yellow, the temperature was raised to −10°, and after four days the total hydrogen amounted to 24.47 cc.—just one H_2 per $(CH_3)_2PH$ or $NaNH_2$. Two similar experiments on a far larger scale, and at temperatures up to 25°, gave the same conclusion: that $(CH_3)_2PH$ with an equivalent amount of $NaNH_2$ yields one H_2 per mole.

Formation of $(CH_3)_2P(NH)P(CH_3)_2$.—Sodium amide made from 472.1 mg. of Na (460 cc.) was treated with 405 cc. of $(CH_3)_2PH$ in 9 ml. of liquid ammonia, in a heavy-walled bomb tube. Effervescence was seen at room temperature, and during three days the deep orange color faded to yellow. The hydrogen now was pumped off and measured as 407.3 cc., or 1.005 H_2 per $(CH_3)_2PH$. The solution was treated with 1.8 g. of NH_4Br, which discharged the yellow color, with precipitation of NaBr. Then all material volatile at 65° was distilled off *in vacuo* and the ammonia fraction was removed through a reflux-head at −78°, leaving a volatile residue which melted on warming to room temperature. Repeated fractional condensation, using traps at 0, −78, and −196°, brought out more ammonia and finally yielded a volatile white solid which melted sharply at 39.5°. The vapor density of this product indicated the molecular weight to be 137.6; calcd. for $(CH_3)_2P(NH)P(CH_3)_2$, 137.1. Microanalysis: 35.15% C, 9.57% H and 11.06% N; calcd. 35.04, 9.56 and 10.22.

The vapor tensions of this solid bis-phosphinoamine: (3.21 mm. at 27.1°, 4.13 at 30.0°, 5.40 at 33.5°, and 7.22 at 37.0°) agreed well enough with the equation $\log_{10} p_{mm.} = 11.240 - 3221/T$ (calcd. values 3.27, 4.14, 5.47 and 7.18) to indicate a single substance, as suggested also by the sharpness of the melting point; however the vapor tension curve ($\log p$ *vs.* $1/T$) for the liquid showed such an extreme downward-concavity that a Nernst-approximation equation based upon points in the range 8–32 mm. went through a maximum of 54 mm. at 120°. Such behavior could be explained by a limited formation of a more volatile substance at medium temperatures—in amounts which became a less important fraction of the vapor at higher temperatures.

For a direct test of this effect, a 15.4-cc. sample of the compound was heated in a sealed tube for 15 hours at 145°, yielding 1.4 cc. of ammonia and a trace of liquid impurity which could not readily be separated from the main solid sample. A disproportionation clearly had occurred to a limited extent—presumably forming ammonia and a tris-phosphinoamine $[(CH_3)_2P]_3N$, but not necessarily limited to these substances. The question whether the original product of the NH_4Br treatment was a mono-phosphinoamine $(CH_3)_2PNH_2$, disproportionating easily to form the bis-phosphinoamine and ammonia, or only an ammoniated bis-phosphinoamine, was not decided.

Reversal of the $(CH_3)_2PH$-NH_2^- Addition Reaction.—Sodium amide was made from 176.5 mg. of Na (172.0 cc.) in 2 ml. of liquid ammonia; then the equivalent amount of $(CH_3)_2PH$ was added at −78° and the mixture was stirred magnetically for one hour to ensure completion of the addition reaction. Now the solvent ammonia was distilled off at −78°, leaving a more and more viscous residue, which finally was pumped at −78°, toward a trap at −196°, for some 30 hours. The H_2 now amounted to 0.75 cc. and the $(CH_3)_2PH$, 100.8 cc.—isolated from ammonia by addition of $B(CH_3)_3$ and repeated fractional condensation through a trap at −23°, which passed the $(CH_3)_2PH{\cdot}B(CH_3)_3$ but held back the $H_3N{\cdot}B(CH_3)_3$. A further 25.6 cc. of $(CH_3)_2PH$ was recovered by pumping the now solid residue for 30 minutes at −23°, and then 30 minutes at 25°; total $(CH_3)_2PH$ now 73.4% of the original; total H_2, 1.31 cc.

Conversion of $(CH_3)_2PH$ to $(CH_3)_3P$.—The solid which remained, after removal of 73.4% of the $(CH_3)_2PH$ from the amide-addition product in the preceding experiment, was redissolved in liquid ammonia and treated with 183.0 cc. of CH_3Cl at −78°. After six hours 0.56 cc. of H_2 and 25.0 cc. of $(CH_3)_3P$ were isolated; yield of the latter, 55%, based upon the unrecovered $(CH_3)_2PH$.

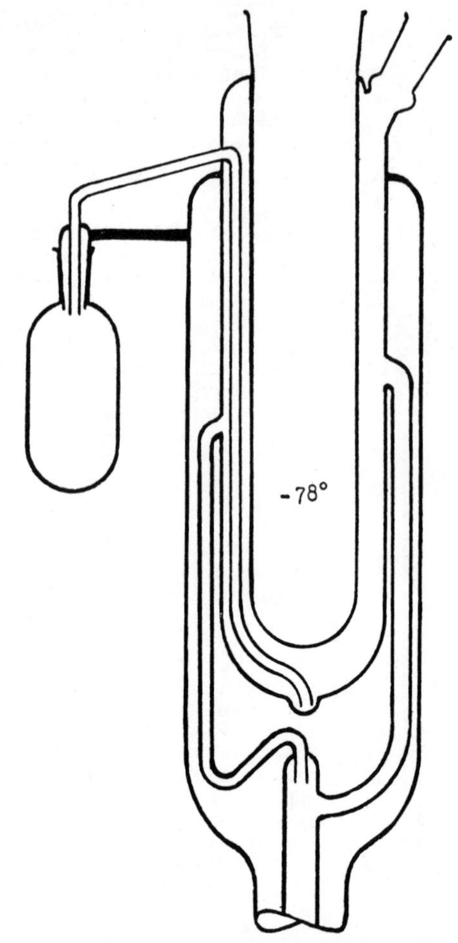

Fig. 2.—Liquid separatory reflux head.

For a nearly quantitative result, a solution of 75.5 cc. of $(CH_3)_2PH$ and 74.3 cc. of sodium amide in liquid ammonia was treated with 75.5 cc. of CH_3Cl, with 30 minutes of stirring at −78°. All material volatile at room temperature now was distilled off, with isolation of 0.30 cc. of H_2 and 68.1 cc. of $(CH_3)_3P$. The latter was characterized by its melting point (−88°, lit. −85°)[12] and vapor tension at 0° (162.0 mm., lit. 161.3).[13] The yield was nearly 92%, based upon the least component, the sodium amide.

LOS ANGELES 7, CALIFORNIA

(12) N. R. Davidson and H. C. Brown, THIS JOURNAL, **64**, 319 (1942).

(13) E. J. Rosenbaum and C. R. Sandberg, *ibid.*, **62**, 1622 (1940).

The Reaction of Tetraalkylammonium Ions with Ammoniated Electrons

For many years solutions containing tetraalkylammonium ions and ammoniated electrons (the latter being introduced as alkali metals or by electrolysis) were the objects of fascinated study. The question was: do these solutions contain NR_4 radicals, or can the properties of the solutions be completely accounted for by assuming the coexistence of NR_4^+ and e_{am} ions? In the following paper, Hazlehurst, Holliday, and Pass describe their study of the chemical reactions of such solutions. A more recent study of such reactions has been reported by Grovenstein and Stevenson.[1] For a study of the absorption spectra of tetraalkylammonium solutions, see the reprinted paper on p. 351.

[1]E. Grovenstein, Jr., and R. W. Stevenson, *J. Amer. Chem. Soc.*, **81**, 4850 (1959).

15

891. *The Reaction of Tetra-alkylammonium Halides with Potassium in Liquid Ammonia.*

By D. A. HAZLEHURST, A. K. HOLLIDAY, and G. PASS.

The cation of a tetra-alkylammonium salt NR_4X undergoes N–C bond fission with potassium in liquid ammonia, giving a hydrocarbon RH and an amine NR_3. With a tetramethylammonium salt the yield of methane decreases in the order X = Cl, Br, I, and C–H fission may occur to yield hydrogen and ethylene. The rate of reaction at $-78°$ is R = Pr^n > Me > Et; unsymmetrical ions R_3NMe^+ react very rapidly, yielding methane. Amide ion resulting from N–C fission or solvolysis gives olefin if R contains a β-carbon atom.

THE reaction of tetramethylammonium iodide with a solution of potassium in liquid ammonia was first investigated by Thompson and Cundall;[1] the products at room temperatures were trimethylamine, ethane, and potassium iodide. Schlubach and Ballauf[2] found that tetraethylammonium chloride and potassium yielded more gas than that expected from the equation $2K + 2NEt_4Cl = 2KCl + 2NEt_3 + C_4H_{10}$. Recently, Jolly[3] has found that tetraethylammonium bromide and potassium in liquid ammonia at $-36°$ yield triethylamine and variable amounts of ethylene, ethane, and hydrogen. In some of Jolly's experiments, ferric nitrate was added to catalyse the reaction between the ammoniated electron (written as e^- here) and solvent, *i.e.*, $2e^- + 2NH_3 = 2NH_2^- + H_2$. It is known that the reaction of the ions NEt_4^+ and NH_2^- yields ethylene, and it is not clear how much of Jolly's ethylene was formed in this way.

One object of all the above investigations was to find evidence for the existence of the tetra-alkylammonium radical NR_4. In the reaction $NR_4X + K$ (X = halogen), in liquid ammonia, KX is precipitated; hence there are in solution NR_4^+ ions and solvated electrons, e^-. The latter may undergo solvolysis to give amide ion and hydrogen; or they may unite with the NR_4^+ ions to give radicals NR_4; or they may attack NR_4^+ ions and break bonds.

[1] Thompson and Cundall, *J.*, 1888, **53**, 761.
[2] Schlubach and Ballauf, *Ber.*, 1921, **54**, 2811.
[3] Jolly, *J. Amer. Chem. Soc.*, 1955, **77**, 4958.

Valency considerations make the existence of the radical NR_4 seem improbable, and hence breakdown of the ion NR_4^+ is likely; Birch [4] has shown that electrons of metal–ammonia solutions can produce fission of the N–C bonds in quaternary ammonium ions (*e.g.*, $Ph{\cdot}NMe_3^+ \longrightarrow PhH + NMe_3$). Kantor and Hauser [5] have shown that fission of a C–H bond can occur by reaction of a base B^- (*e.g.*, NH_2^-) in liquid ammonia, *e.g.*,

$$R_3\overset{+}{N}{\cdot}\underset{|}{\overset{|}{C}}H + NH_2^- \longrightarrow NH_3 + R_3\overset{+}{N}{\cdot}\underset{|}{\overset{-}{C}}{-} \quad \text{(I)}$$

The carbanion (I) may then re-arrange to give a tertiary amine, or undergo alkylation. If a β-carbon atom with attached hydrogen is present, then β-elimination is preferred, giving an olefin, *e.g.*,

$$R_3\overset{+}{N}{\cdot}CH_2{\cdot}CH_3 + NH_2^- \longrightarrow R_3N + C_2H_4 + NH_3$$

In the present work, the reaction of NR_4^+ ions and ammoniated electrons has been investigated *in vacuo* at $-78°$, and as many reaction products as possible have been analysed quantitatively so that the reaction mechanism might be understood.

Experimental

All manipulations involving liquid ammonia were made in a vacuum apparatus. The ammonia was dried over sodium and stored as gas. Quaternary ammonium halides were either obtained from B.D.H. or made by usual methods; they were dried *in vacuo* and analysed for purity before use. The dry salt was introduced into the reaction tube, followed by the potassium (weighed under light petroleum), and the tube was then attached to the vacuum system and pumped for some hours. About 10 c.c. of liquid ammonia were condensed into the tube, and the latter was sealed off and left at $-78°$, with occasional shaking, until the blue colour disappeared. The tube was then placed in liquid nitrogen and opened to the vacuum system; non-condensable gases were removed with the Töpler pump and analysed by combustion. The ammonia was carefully removed from the reaction tube, and condensable hydrocarbon products were then separated by fractionation through U-traps; absorption of the large amount of ammonia on calcium chloride assisted separation of hydrocarbons of similar vapour pressure to ammonia. Condensable and non-condensable hydrocarbon products were analysed by combustion, by absorption in bromine water, and by infrared absorption spectra; vapour-density measurements were also made in some cases. Identification of the small amount of amine formed ($<0{\cdot}1$ g.) in presence of much ammonia was carried out by one or more of the following methods: (*a*) allowing amine to accumulate in the solvent ammonia from successive experiments and then extracting the hydrochloride with chloroform (Watt and Otto's method [6]); (*b*) using larger quantities of reactants, and then separating the amines by fractional distillation, using the micro-still described by Craig; [7] (*c*) separating the hydrochlorides by paper chromatography.[8]

After removal of all volatile material from the reaction tube, the solid residue was dissolved in water, warmed to remove any traces of ammonia and amine, and analysed for (i) alkalinity (corresponding to amide ion content of the residue), (ii) potassium and quaternary ammonium ion concentration, by use of sodium tetraphenylboron, and (iii) halide content, as a check on the amount of salt added initially. From (ii) and (iii) the amount of quaternary salt used in the reaction could be estimated. Determinations in the residues were occasionally precluded by loss of solid on removal of solvent ammonia.

Results.—These are given in Tables 1, 2, and 3; units are mmoles unless otherwise stated. Expts. 1—7, Table 1, were with potassium and tetraethylammonium chloride. The results of expts. 1 and 2 show that very variable ethane : ethylene ratios were obtained under apparently similar conditions, and that the time of reaction (*i.e.*, the time to become colourless) also varied. It may be noted that solutions of potassium with no quaternary salt present became colourless after very variable periods of time, and that these times were of the same order as those obtained

[4] Birch, *Quart. Reviews*, 1950, **4**, 69.
[5] Kantor and Hauser, *J. Amer. Chem. Soc.*, 1951, **73**, 4122.
[6] Watt and Otto, *ibid.*, 1947, **69**, 836.
[7] Craig, *Ind. Eng. Chem. Anal.*, 1936, **8**, 219.
[8] Schwyzer, *Acta Chem. Scand.*, 1952, **6**, 219.

when tetraethylammonium chloride was present. In expts. 3 and 4 excess of potassium was used but similar variations occurred; the use of excess of quaternary salt (expt. 5) or of a higher reaction temperature (expt. 6) did not produce any significant changes except that the reaction time was much shorter at the higher temperature. In expt. 7, the reaction tube was opened after a short period of reaction to find the relative rates of production of ethane and

TABLE 1.

Experiment	1	2	3	4	5	6	7	8 †	9 †
Temp.	−78°	−78°	−78°	−78°	−78°	−33°	−78°	−78°	−78°
NEt_4Cl added	1·00	1·06	1·06	0·55	2·15	1·03	1·05	1·10	1·28
NEt_4Cl used	—	—	1·06	0·55	1·00	—	—	0·75	1·19
K added	0·92	0·99	3·96	1·45	1·08	0·92	0·94	0·91	1·16
Products: H_2	0·03	0·27	1·38	0·60	0·05	—	0·01	0·00	0·00
C_2H_6	0·45	0·22	0·48	0·12	0·54	0·42	0·08	0·00	0·00
C_2H_4	0·45	0·66	0·55	0·39	0·55	0·46	0·06	0·29	1·04
NH_2^-	—	0·08	—	0·87	0·05	—	—	0·61	0·08
Time (hr.)	120	186	230	160	160 ‡	20	6 *	24	168

* Reaction incomplete.
† Potassium–ammonia solution colourless before addition of NEt_4Cl.
‡ Left at −78° for a further 384 hours.

TABLE 2.

Experiment	10	11	12	13	14 †	15 †
Halide (X)	Cl	Cl	Br	I	Cl	Cl
NMe_4X added	1·32	6·26	1·48	1·17	2·62	1·92
,, used	0·56	0·70	—	0·27	0·03	0·51
K added	1·14	1·35	1·31	1·15	1·94	1·10
Products: H_2	0·06	0·23	0·34	0·35	0·00	0·00
CH_4	0·49	0·44	0·03	0·20	0·00	0·00
C_2H_6/C_2H_4	trace	trace	trace	trace	0·06 *	0·00
NH_2^-	0·54	0·78	—	0·79	1·81	0·54
Time (hr.)	45	41	168	140	44	312

* Ethylene only.
† Potassium–ammonia solution colourless before addition of NMe_4Cl.

TABLE 3.

Experiment	16	17	18	19	20	21
Cation	$\overset{+}{N}Pr^n_4$	$Et_3\overset{+}{N}H$	$Et_3\overset{+}{N}Me$	$Et_3\overset{+}{N}Me$	$Pr^n_3\overset{+}{N}Me$	$Pr^n_3\overset{+}{N}Me$
added	1·56	1·37	1·36	1·32	1·43	1·32
used	0·98	—	1·07	0·94	1·43	1·00
K added	1·07	1·17	1·07	0·96	1·16	1·12
Products: H_2	0·02	0·54	0·00	0·02	0·00	0·00
CH_4	0·00	0·00	0·45	0·38	0·34	0·52
C_2H_6	0·00	0·00	trace	trace	trace	trace
C_2H_4	0·00	0·00	0·42	0·40	0·00	0·00
C_3H_8	0·48	0·00	0·00	0·00	0·00	0·00
C_3H_6	0·52	0·00	0·00	0·00	0·45	0·47
NH_2^-	0·15	0·07	0·11	0·10	0·28	0·10
Amines Found	Pr^n_3N —	Et_3N —	Et_3N Et_2NMe	Et_3N Et_2NMe	Pr^n_3N Pr^n_2NMe	Pr^n_3N Pr^n_2NMe
Time (hr.)	<20	<0·01	4	4	0·6	1·6

ethylene. In expts. 8 and 9, the reaction was between potassium amide and tetraethylammonium chloride, and the reaction times were arbitrary. Ethylene was the only hydrocarbon product of these amide reactions.

In the above experiments, triethylamine was the only amine product identified, and no higher hydrocarbons or methane was found.

The results of potassium–tetramethylammonium halide reactions are given in Table 2, expts. 10—13. Methane was the major hydrocarbon product, and trimethylamine was the only amine identified. The trace of noncondensable hydrocarbon was found (in expt. 10) to be mainly ethane, containing a little olefin, probably ethylene. In expts. 10 and 11, with the chloride, the effect of increasing the concentration of the latter was to increase the hydrogen yield, but there was little change in reaction time. Expts. 12 and 13 show that the hydrocarbon yield was reduced when the bromide and iodide were used. In expts. 14 and 15, potassium amide was used and the reaction times were arbitrary. Expt. 14 (where a little ethylene was found) was a single experiment, whereas 15 is typical of several similar experiments in which no ethylene was found.

The reaction of tetra-*n*-propylammonium chloride and potassium is given as expt. 16, Table 3. The other results in this table are for unsymmetrical quaternary salts. The reaction with trimethylammonium chloride was very rapid and hydrogen was the only gaseous product. Triethylmethylammonium chloride also reacted fairly rapidly, and reproducible results were obtained; little or no hydrogen was formed and there were two amine products as well as methane, ethylene, and a trace of ethane. Expts. 20 and 21, by use of methyltri-*n*-propylammonium chloride gave similar results, and propylene instead of ethylene.

Discussion

From a quaternary ion NR_4^+, the hydrocarbon RH and the amine NR_3 are always obtained. This clearly indicates N–C bond fission as postulated by Birch,[4] and suggests the scheme

$$NR_4^+ + e^- \longrightarrow NR_3 + R \quad (1)$$

$$R + e^- \longrightarrow R^- \quad (2)$$

$$R^- + NH_3 \longrightarrow RH + NH_2^- \quad (3)$$

Amide ion is also produced by electron solvolysis, *i.e.*,

$$NH_3 + e^- \longrightarrow NH_2^- + \tfrac{1}{2}H_2 \quad (4)$$

The tetraethylammonium ion reacts slowly, and the variation in ethane : ethylene ratio does not appear to be determined by reactant ratio or temperature. The absence of butane suggests that ethyl radicals formed by reaction (1) immediately undergo reaction (2), so that disproportionation of ethyl radicals is unlikely as a source of ethylene. The latter must be formed by the reaction

$$NEt_4^+ + NH_2^- \longrightarrow NEt_3 + C_2H_4 + NH_3 \quad (5)$$

If amide ion were formed only by (3), then equal amounts of ethane and ethylene would be expected—provided that reaction (5) is complete when the solution becomes colourless. The results of expts. 8 and 9 verify reaction (5), but show that the reaction is slow at $-78°$. Expt. 7 shows that formation of ethylene is somewhat less than that of ethane at an early stage in the reaction, and expt. 5 shows that while a $C_2H_6 : C_2H_4$ ratio may exceed 1 after 160 hr. or so, the ultimate ratio is 1 after long storage, *i.e.*, when reaction (5) is complete. But there are experiments in which the $C_2H_6 : C_2H_4$ ratio is much less than 1, and here more ethylene must be formed by amide from the solvolysis reaction (4). The solvolysis reaction can proceed at a rate comparable with that of the ethane-forming reaction; but this rate is subject to considerable variation because of the possible presence of a trace of catalyst. Hence, if the solvolysis reaction happens to be catalysed, then extra ethylene and hydrogen are produced at the expense of ethane. The stoicheiometric basis for this reasoning can be understood by reference to expt. 2. Here, 0·22 mmole of C_2H_4 was produced by amide from reaction (3); hence 0·44 mmole must have been produced by amide from the solvolysis reaction. This corresponds to 0·22 mmole of hydrogen, but the amount actually found was 0·27 mmole. Hence $0·05 \times 2 = 0·10$ mmole of amide was produced, by solvolysis, in excess of that used to produce ethylene, and should have remained as such. In fact 0·08 mmole was found in the residue. A similar check can be applied in the other cases where the data are available. In expt. 2, the potassium required to produce ethane and ethylene was 0·44 mmole, for the " excess " of ethylene 0·44 mmole, and as residual amide 0·08 mmole, a total of 0·96 mmole compared with 0·99 mmole actually added.

Tetramethylammonium chloride (Table 2) reacts more rapidly than the tetraethylammonium salt, and any effect of solvolysis is therefore reduced. However, the results show that the yield of methane is reduced and that of hydrogen increased by using the bromide or iodide, or by using a large excess of the chloride (expt. 11). Now the amount of quaternary salt added in these experiments exceeds that required to saturate the liquid ammonia at $-78°$. The effective concentration of quaternary ion is therefore determined primarily by the extent to which the anion X^- is removed as insoluble KX; and as the solubilities of KX are in the order KI > KBr > KCl, the quaternary-ion concentration (and yield of methane) will be low in the case of the bromide and iodide. The unused electrons in these cases take part in slow solvolysis and the hydrogen yield therefore increases. The amount of methane and hydrogen is satisfactorily accounted for by the equations (1)—(4). Thus, in expt. 10, 0·98 mmole of potassium is required for the methane and 0·12 mmole for the hydrogen, *i.e.*, 1·10 mmoles. The remainder of the potassium added (0·04 mmole) can be accounted for in terms of the trace of condensable hydrocarbon, mainly ethane. The latter may originate by combination of two methyl radicals, from reaction (1), or by a reaction

$$NMe_4^+ + Me^- \longrightarrow NMe_3 + C_2H_6 \quad . \quad . \quad . \quad . \quad . \quad . \quad . \quad . \quad . \quad (6)$$

Porter [9] has shown that methyl radicals are produced from tetramethylammonium amalgam on heating, and the recovery of ethane from Thompson and Cundall's tetramethylammonium iodide–potassium reaction at ordinary temperature [1] suggests that radical combination may have occurred in their experiments. But reaction (6) is also a possible source of ethane, since the CH_3^- ion acts as a base in liquid ammonia and so can attack the quaternary ion. In the case of tetraethylammonium salts no corresponding reaction to give butane occurred, probably because of preferential attack on the NEt_4^+ ion by NH_2^- rather than by $C_2H_5^-$ ions.

The results of Table 2 show that some amide and quaternary ions have reacted, but without giving hydrocarbon products. Franklin [10] refers to a reaction

$$NMe_4^+ + NH_2^- \longrightarrow [NMe_4{\cdot}NH_2] \longrightarrow NMe_3 + Me{\cdot}NH_2$$

and the results of expt. 15 show that this reaction occurs slowly at $-78°$. There remains the single experiment 14, where a small amount of ethylene was produced from the NMe_4^+–NH_2^- reaction, and the probability that a trace of ethylene was formed in the NMe_4^+–K reactions also. Formation of ethylene would require, first, fission of a C–H bond by a base B^-, thus : [5]

$$Me_3\overset{+}{N}{\cdot}Me + B^- \longrightarrow Me_3\overset{+}{N}{\cdot}\overset{-}{C}H_2 + BH \quad . \quad . \quad . \quad . \quad . \quad . \quad . \quad . \quad . \quad (7)$$

The carbanion formed by this reaction has been prepared in ether solution by Wittig and Wetterling.[11] They found it to be stable; it did not rearrange to give ethyldimethylamine, nor did it decompose directly to give ethylene. Reaction with water gave tetramethylammonium hydroxide. Hence, if ethylene is to be formed from this carbanion, it must be by alkylation to $Me_3N^+{\cdot}CH_2Me$ followed by β-elimination. Now if the base B^- in reaction (7) is NH_2^-, then the reaction produces ammonia. In liquid ammonia as solvent, the reverse of reaction (7) seems probable, by analogy with the reaction with water. Hence little or no ethylene would be expected from a NMe_4^+–NH_2^- reaction. But if B^- is identified as e^-, then reaction (7) becomes

$$\overset{+}{N}Me_4 + e^- \longrightarrow Me_3\overset{+}{N}{\cdot}\overset{-}{C}H_2 + \tfrac{1}{2}H_2 \quad . \quad . \quad . \quad . \quad . \quad . \quad . \quad . \quad . \quad (8)$$

and hydrogen is produced : and if the carbanion is then destroyed by ammonia, the other reaction product is amide, so that the overall result is

$$e^- + NH_3 \longrightarrow NH_2^- + \tfrac{1}{2}H_2$$

i.e., the same as the solvolysis reaction. It is therefore possible that carbanion formation by reaction (8) can yield a trace of ethylene by alkylation, but the greater amount of carbanion must undergo solvolysis and produce amide. In expt. 11, a large excess of

[9] Porter, *J.*, 1954, 760.
[10] Franklin, " The Nitrogen System of Compounds," Reinhold, New York, 1935, p. 63.
[11] Wittig and Wetterling, *Annalen*, 1947, **557**, 193.

tetramethylammonium chloride produced more hydrogen and amide, and less hydrocarbon, than the normal amount. It may be that the production of amide and hydrogen was here catalysed by the carbanion mechanism; equally, the effect may have been a heterogeneous catalysis of the solvolysis reaction by the large amount of undissolved quaternary salt.

The reaction of tetra-*n*-propylammonium chloride and potassium (expt. 16, Table 3) proceeds rapidly and little solvolysis occurs. Propane and propene are found in approximately equal amounts as would be expected if amide ion formed by reaction (3) reacts completely with the $Pr^n{}_4N^+$ ion to yield propene. However, the latter reaction is not complete, since some amide and unchanged quaternary ion are found in the residues. It is therefore possible that all the propane formed was not recovered from the ammonia, *i.e.*, the ratio propane : propene is too low.

The rate of reaction of tetra-alkylammonium ions and potassium is in the order $R = Pr^n > Me > Et$. The rapid reaction of the tetrapropylammonium ion may be attributed to two factors—the greater solubility of the quaternary salt, and the greater ease of fission of the N–Pr bond compared with that of N–Et or N–Me. The more rapid reaction of tetramethyl- than of tetraethyl-ammonium ion cannot be attributed to these factors; but it may well be that reactions (2) and (3) occur more readily with methyl than with ethyl radicals.

With the unsymmetrical quaternary ions of Table 3, N–H or N–C fission occurred to give the tertiary amine, *i.e.*, hydrogen was obtained from trimethylammonium ions and methane from triethyl- or tri-*n*-propyl-methylammonium ions. The amide produced attacks the group with a β-carbon atom if available, *i.e.*,

$$Et_3NMe^+ + NH_2^- \longrightarrow Et_2NMe + C_2H_4 + NH_3$$
$$Pr^n{}_3NMe^+ + NH_2^- \longrightarrow Pr^n{}_2NMe + C_3H_6 + NH_3$$

so that two amines are formed in the products. The order of ease of fission here is $H > Me >$ larger alkyl, and the rapidity of reaction of the unsymmetrical ions may be due to a combination of steric effects, the presence of an easily removed hydrogen atom or methyl group, and the ready occurrence of reactions (2) and (3). The appearance of a trace of ethane from the ions Et_3NMe^+ and $Pr^n{}_3NMe^+$ is of interest. Formation by a reaction

$$R_3NMe^+ + Me^- \longrightarrow R_3N + C_2H_6$$

is unlikely, since methyl ions undergo rapid solvolysis; and in any case, the Me^- ion is a base and the reaction would therefore be (for Et_3NMe^+)

$$Et_3NMe^+ + Me^- \longrightarrow Et_2NMe + MeH + C_2H_4$$

It is more likely that ethane is formed here by union of methyl radicals since these will be released more readily from triethyl- or tripropyl-methylammonium than from tetramethylammonium ions.

Thanks are due to Professor G. E. Coates for some helpful suggestions, and to Mr. P. Kennedy for assistance with some of the experimental work. One of us (G. P.) is indebted to the Department of Scientific and Industrial Research for a maintenance grant.

Department of Inorganic and Physical Chemistry,
The University of Liverpool. [*Received, June 28th*, 1956.]

Birch Reductions

Metal–ammonia solutions are very useful reagents for the organic chemist. Birch and his co-workers have extensively studied organic reductions effected by these reagents, and such reactions are often called "Birch reductions." In the following paper, we have reprinted the first half of a general review article on this subject by Birch and Smith. (The second half of the article describes applications of the reagents to particular syntheses and structure determinations, and was not considered appropriate for this volume.)

16

REDUCTION BY METAL–AMINE SOLUTIONS: APPLICATIONS IN SYNTHESIS AND DETERMINATION OF STRUCTURE

By A. J. Birch and Herchel Smith
(Chemistry Department, Manchester University)

Work since the last reviews of the subject [1, 2] has been concerned chiefly with the exploitation of metal–amine solutions in synthesis and in the investigation of natural products. Theoretical developments have been mainly incidental and have added little to what was already known.[1] We are here concerned with the practical aspects but an appreciation of the theoretical background is essential for use of the reagents to the best purpose, since variations in technique are possible. So references to theoretical aspects are made below where necessary, but overlapping with earlier reviews has been avoided as far as possible.

1. Reduction by metal–ammonia and metal–amine solutions

Two main types of reduction are observed: (*a*) partial or complete saturation of a wide variety of unsaturated substances including polycyclic aromatic compounds, dienes, and trienes which are conjugated or are rendered so under the alkaline conditions of the reaction, and in some cases of simple olefins; and (*b*) reductive fission of alkyl aryl or diaryl ethers and sulphides and the hydrogenolysis of various groups attached to nitrogen, oxygen, and sulphur. Which particular reagent is used depends on the nature of the substrate and on how far it is desired that reduction shall proceed.

Metal and an "Acid" in Liquid Ammonia.—The reagents so far examined consist of an alkali metal and an "acid" such as methanol or ammonium chloride in liquid ammonia, sometimes with co-solvents such as ether or tetrahydrofuran. The reagents are powerful if the "acid" is an alcohol, and are then capable of reducing a terminal double bond or an isolated benzene ring. In this they differ from ammonia reagents lacking the alcohol, although solutions of lithium in ethylamine are also capable of reducing benzene rings and terminal double bonds (see below). The alcohol also has the effect of buffering the reaction mixture, preventing accumulation of strongly basic NH_2. This explains the comparative simplicity of the results since base-catalysed rearrangements of double bonds are usually avoided. Reduction of a benzene ring leads to the $\alpha\delta$-dihydro-derivative, unlike the lithium–ethylamine reagent where the $\alpha\delta$-dihydro-derivative which is probably formed initially is rearranged to the conjugated $\alpha\beta$-dihydro-derivative which is then rapidly reduced further. As will be seen, similar results to the latter are obtained with calcium hexammine and by reduction with sodium and ethanol in liquid ammonia followed by

[1] Birch, *Quart. Rev.*, 1950, **4**, 69. [2] Watt, *Chem. Rev.*, 1950, **46**, 317.

an excess of sodium in ammonia.[3] The $\alpha\delta$-hydrogen atoms added by the metal-alcohol-ammonia reagent avoid carbon atoms carrying dimethyl-amino-, alkoxy-, or alkyl groups in that order, and are attracted to positions carrying carboxyl groups. The latter effect outweighs the others ; carboxyl groups labilise *o*- and *p*-methoxyl groups [4] to hydrogenolysis.

The requirement for an added source of protons is now interpreted [5] in terms of the following equilibria :

$$\mathrm{Ar} + \varepsilon \rightleftharpoons \mathrm{Ar}^{-*} + \mathrm{HOR} \rightleftharpoons \mathrm{Ar^{*}H} + \mathrm{OR}^- \xrightarrow{\varepsilon} \mathrm{ArH}^- \xrightarrow{\mathrm{ROH}} \mathrm{ArH_2} + \mathrm{OR}^-$$

The anion-radical Ar^{*-} formed initially must add a proton in order that reduction may be completed ; it appears not to be sufficiently basic to abstract this proton from ammonia and requires a more acidic proton source. If the acidity of this source is high, as with ammonium salts, the predominant reaction is evolution of gaseous hydrogen unless the substance is very rapidly and readily reduced ; alcohols seem to have about the optimum $\mathrm{p}K_a$ for the reduction of benzene rings.

The reduction with sodium and methanol or ethanol in liquid ammonia of *o*-xylene,[6] naphthalene,[7] tetralin,[8] and 1 : 4-dihydronaphthalene [7] has been subjected to a rigorous re-examination. The results confirm the originally defined reducing properties of the reagent. Diphenyl has been reduced in both rings, giving a product showing no selective ultraviolet absorption which is either compound (I) or, more probably, compound (II).[9]

(I) 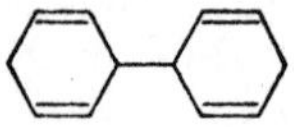 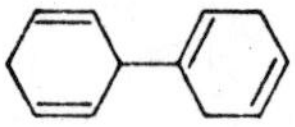 (II)

An important modification in technique has been the use of lithium instead of sodium or potassium for the reduction of aromatic rings. The greater solubility of lithium in liquid ammonia enables larger proportions of co-solvents to be used without the formation of two-phase systems ; consequently, difficulties arising from the low solubility of substrates in ammonia systems can be overcome. Higher yields of reduction products may be associated with the high concentration of metal and also with the higher normal reduction potential of lithium in ammonia (—2·99 v) compared with that of potassium or sodium (—2·59 and —2·73 v respectively).[10] An earlier device for increasing the solubility of phenyl ethers in ammonia systems, *viz.*, the formation of glyceryl or 2-hydroxyethyl ethers,[11, 12] enables the cheaper sodium or potassium to be used. This technique has not yet been fully investigated, but in reduction of 3-2′-hydroxyethoxyœstra-

[3] Birch, *J.*, 1946, 593. [4] Ref. 1, p. 88.
[5] Birch, *J. Roy. Inst. Chem.*, 1957, **80**, 100.
[6] Hückel and Wörffel, *Chem. Ber.*, 1955, **88**, 338.
[7] Hückel and Schlee, *ibid.*, p. 346.
[8] *Idem*, *ibid.*, p. 2098.
[9] Hückel and Schwen, *ibid.*, 1956, **89**, 150.
[10] Wilds and Nelson, *J. Amer. Chem. Soc.*, 1953, **75**, 5360.
[11] Birch and Mukherji, *J.*, 1949, 2531.
[12] Birch, *J.*, 1950, 367.

1 : 3 : 5-trien-17β-ol [13] the yield of 19-nortestosterone is of the same order as by the lithium method.[14] In cases where there are serious losses of alkoxyl groups by reductive fission the method may be superior in that it inhibits hydrogenolysis through alkoxide formation by the side-chain hydroxyl group. It is possible also that reduction is facilitated by cyclic donation of a proton to an anionic intermediate.

Terminal double bonds may be reduced by the alcohol-containing reagent. This has been observed with various dialkylallylamines [15] and with hex-1-ene, which has been converted into hexane in 41% yield by two atomic proportions of sodium and methanol in liquid ammonia.[16] In the last case no reduction occurs in the absence of an alcohol or when ammonium bromide is used as proton donor. Of particular interest is the observation that 2-*cyclo*propylpent-1-ene gives 2-*cyclo*propylpentane [16] and no ring-open products with the sodium–ammonia–methanol reagent, whereas methyl *cyclo*propyl ketone affords a mixture of methyl propyl ketone and pentan-2-ol with sodium and ammonium sulphate in liquid ammonia.[17]

A further illustration of the control exerted by the acid strength of the proton donor over the reduction products is the ultimate formation of aldehydes rather than alcohols when ammonium acetate is substituted for ethanol as the proton source in the sodium–ammonia reduction of amides.[18] Alcohol production is ascribed to the ethoxide-catalysed decomposition and further reduction of the intermediate 1-amino-alcohol (aldehyde–ammonia). The buffering of the medium by use of the more acidic ammonium acetate as proton source avoids this decomposition and the amino-alcohol is converted into the aldehyde during the working up:

$$\mathrm{R{\cdot}CO{\cdot}NH_2} \longrightarrow \mathrm{R{\cdot}\overset{O^-}{\overset{|}{C}}H{\cdot}NH_2} \xrightarrow{H^+} \mathrm{R{\cdot}CH(OH){\cdot}NH_2} \longrightarrow \mathrm{R{\cdot}CHO}$$

$$\mathrm{R{\cdot}CH(OH){\cdot}NH_2} \xrightarrow{OEt^-} \mathrm{NH_3 + R{\cdot}CHO} \longrightarrow \mathrm{R{\cdot}CH_2{\cdot}OH}$$

Aldehydes may also be ultimately obtained by the sodium–ammonia–*alcohol* reduction of amidines (even arylamidines), presumably *via* the 1 : 1-diamines.[18] This result is due to the lower acidity of NH than of OH and also to absence of hydrogenolysis of the C–N bonds. In contrast to the reductive fission of C–O bonds which in general occurs readily in benzyl ethers and in aryl acetals [19] and ketals,[20] C–N bonds conjugated with aromatic nuclei are not split, because nitrogen is less electrophilic than oxygen.

Pyridine compounds are reduced more readily than the hydrocarbons

[13] Birch and Bauer, unpublished work.
[14] Wilds and Nelson, *J. Amer. Chem. Soc.*, 1953, **75**, 5366.
[15] King, *J.*, 1951, 898.
[16] Greenfield, Friedel, and Orchin, *J. Amer. Chem. Soc.*, 1954, **76**, 1258.
[17] Volkenburgh, Greenlee, Derfer, and Boord, *ibid.*, 1949, **71**, 3595.
[18] Birch, Cymerman-Craig, and Slaytor, *Austral. J. Chem.*, 1955, **8**, 512.
[19] Birch, Hextall, and Sternhell, *ibid.*, 1954, **7**, 256.
[20] Pinder and Smith, *J.*, 1954, 113.

of the same ring size because the heteroatom is better able than carbon to stabilise a negative charge. Pyridine and quinoline compounds, for example, readily give 1 : 4-dihydro-derivatives ; [21] di-, tri-, and tetra-meric dihydro-compounds are also produced. Reduction can be effected even in the absence of added proton sources, but the products may then contain larger amounts of polymers.

Thiophen with sodium and ethanol in liquid ammonia gives a complex mixture of 2 : 3- and 2 : 5-dihydrothiophen, but-2-ene-1-thiol, but-1- and -2-ene, and hydrogen sulphide.[22] Presumably reduction occurs to 2 : 3- and 2 : 5-dihydrothiophen, which then undergo further reactions because of the known ready reductive fission of C–S bonds. Pyrrole and furan rings appear to be unaffected.

Sodium, Potassium, or Lithium in Liquid Ammonia.—In a number of instances little difference would be expected from the " buffered " and protonated reagents mentioned above. For example, (—)-α-phellandrene (mentha-1 : 5-diene) is reduced by sodium or by sodium and ethanol in liquid ammonia to the same mixture of (—)- (60%) and (+)-menth-1-ene (40%),[23] showing a large proportion of αβ-reduction. The reduction of other 1 : 3-dienes to the 1 : 4-dihydro-derivatives, *e.g.*, 2-methyl-, 2 : 3-dimethyl-, and 1 : 1 : 3-trimethyl-butadiene to the but-2-enes in yields of 98—99, 93—94, and 72% respectively,[24] should not be altered by the presence of alcohols. In other cases, if addition of protons to the intermediate anions is avoided, by ensuring the absence of proton sources other than ammonia, the occurrence of further reduction is inhibited by the negative charges present. Proton-addition occurs during the working-up. Sodium–ammonia reduction of diphenyl has been shown to be analogous to that of naphthalene,[25] two atoms of sodium being added, to give a deep red sodium salt decomposed by ammonium chloride to 1 : 4-dihydrodiphenyl.[9] The phenyl group here exerts an effect similar to that of carboxyl,[27] as is to be expected from its ability to stabilise an adjacent anionic charge. Similarly, fluorene yields an unstable dihydrofluorene of undetermined constitution which readily disproportionates to fluorene and 1 : 4 : 11 : 12-tetrahydrofluorene.[28] In related work it was shown that *cyclo*pentadiene and indene give *cyclo*pentene and indane respectively.[9]

The reduction of αβ-unsaturated ketones to the saturated ketones, of which examples are given below, also illustrates the protective effect of a negative charge in an intermediate in permitting eventual isolation of the saturated ketone. Similar reactions have been carried out with unsaturated esters and acids. The protection of allyl alcohols against hydrogenolysis

[21] Ref. 2, p. 362 ; Birch, unpublished work.
[22] S. F. Birch and McAllan, *Nature*, 1950, **165**, 899.
[23] Birch, unpublished work.
[24] Levina, Svarchenko, Kostin, Treschova, and Okimevich, *Sborkin obshchei Khim.*, *Akad. Nauk S.S.S.R.*, 1953, **1**, 355 : *Chem. Abs.*, 1955, **49**, 829.
[25] Ref. 1, p. 81.
[26] Benkeser, Arnold, Lambert, and Thomas, *J. Amer. Chem. Soc.*, 1955, **77**, 6042.
[27] Ref. 1, p. 86 ; Birch, Hextall, and Sternhell, *Austral. J. Chem.*, 1954, **7**, 256.
[28] Hückel and Schwen, *Ber.*, 1956, **89**, 481.

and of acetylenes against reduction can be achieved by initial formation of the salts.

Lithium in Alkylamines.—Solutions of lithium in amines of low molecular weight, such as methylamine, ethylamine, and the propylamines, constitute reducing agents of very great power if little selectivity. The amines are, in general, more powerful solvents for organic substances than ammonia and have higher boiling points ($C_2H_5{\cdot}NH_2$, b.p. 16·5° ; NH_3, b.p. −33°). Accordingly their use may avoid a common and serious difficulty often encountered in liquid ammonia reductions, namely, the low solubility of the substrate in the solvent system. The higher working temperature also undoubtedly facilitates the initial steps in the reduction and favours the conjugation and therefore further reduction of the primary products. The annexed examples illustrate typical reductions by these reagents. Simple

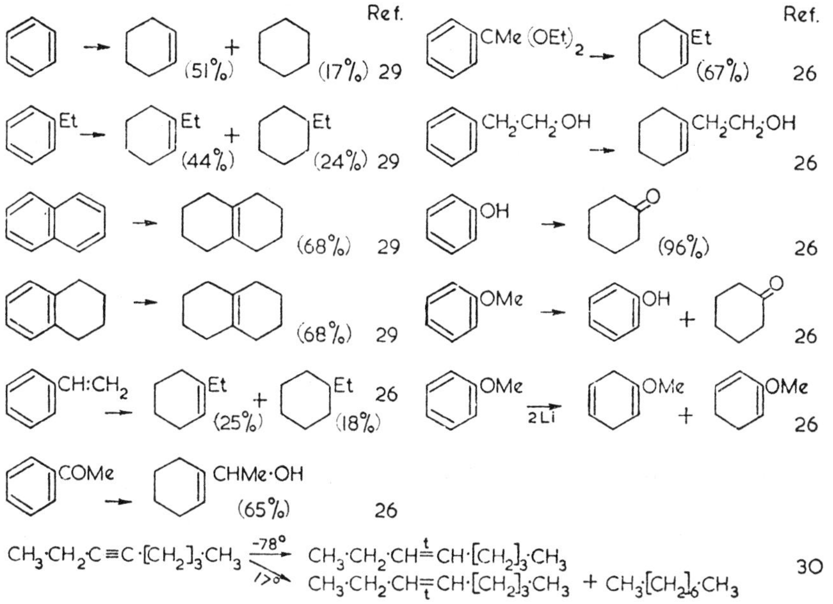

$$CH_3{\cdot}CH_2{\cdot}C{\equiv}C{\cdot}[CH_2]_3{\cdot}CH_3 \xrightarrow{-78^\circ} CH_3{\cdot}CH_2{\cdot}CH\overset{t}{=}CH{\cdot}[CH_2]_3{\cdot}CH_3$$
$$\xrightarrow{>0^\circ} CH_3{\cdot}CH_2{\cdot}CH\underset{t}{=}CH{\cdot}[CH_2]_3{\cdot}CH_3 + CH_3{\cdot}[CH_2]_6{\cdot}CH_3 \quad 30$$

olefins may be saturated, tetrasubstituted double bonds being least readily reduced in accord with the view that the process involves initial electron addition.[1] A similar reduction of double bonds, which must however be terminally situated, occurs with sodium and methanol in liquid ammonia (see above). Inhibition of the reduction of di- and tri-substituted double bonds by working at low temperatures (*e.g.*, −78°) has been observed.[26, 30]

Isolated benzene rings are reduced to the tetrahydro-state or partly to the hexahydro-state depending on the conditions. Acetophenone gives the allylic alcohol 1-1′-hydroxyethyl*cyclo*hexene, whereas its diethyl ketal gives 1-ethyl*cyclo*hexene in agreement with the view that hydrogenolysis of the

[29] Benkeser, Robinson, Sauve, and Thomas, *J. Amer. Chem. Soc.*, 1955, **77**, 323.
[30] Benkeser, Schroll, and Sauve, *ibid.*, p. 3378.

hydroxyl group in the intermediate benzyl alcohol is inhibited by salt formation. Similarly, reduction of phenol (more correctly lithium phenoxide) is very largely stopped at the *cyclo*hexanone stage, probably by protection of the carbonyl group as the enol anion. In contrast, formation of the phenoxide anion or the acetophenone enol anion [20] is sufficient to inhibit reduction by metal–ammonia–alcohol reagents. Use of excess of lithium favours dealkylation of anisole since the product then consists of phenol and *cyclo*hexanone ; the theoretical amount of lithium leads to a mixture of 2 : 5-dihydroanisole (the initial product) and the conjugated 2 : 3-dihydro-isomer.

Although allyl (and benzyl) alcohols may resist hydrogenolysis owing to salt-formation, allyl ethers, for which salt-formation is impossible, are readily cleaved. The course of the reaction is similar to that with the alkali metal–ammonia reagent, but alkylamine systems offer the practical advantage of greater solvent power and reactivity. The cleavage of *cis*-(+)-carvotanacetyl methyl ether with lithium and ethylamine yields (±)-*p*-menth-1-ene,[31] in agreement with the view that the reaction proceeds through a symmetrical intermediate, probably the mesomeric anion produced together with alkoxide ion by the addition of two electrons.[32] It is known that compounds capable of producing very stable anions in a fission reaction are reduced readily,[33] so it would be expected that allyl acetates and benzoates should be cleaved more easily than the free alcohols provided the first stage is not reduction of the ester-carbonyl group. This has been demonstrated in the steroid series, where, for example, lithium in ethylamine converts 3β-acetoxycholest-4-ene (III) to cholest-4-ene, and 4-β-acetoxycholest-5-ene

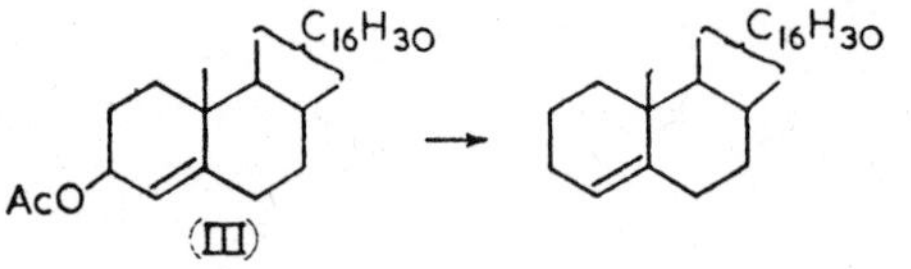

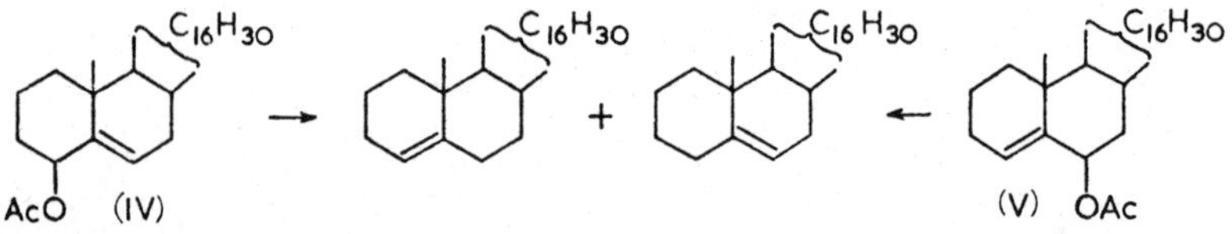

(IV) and 6-β-acetoxycholest-4-ene (V) both give the same mixture of cholest-4- and -5-ene.[31] There is no evidence of formation of a thermodynamically unstable isomer since 3-β-acetoxycholest-1-ene gives cholest-2-ene and none of the less stable cholest-1-ene.[31]

Work on the fission of steroid epoxides has confirmed that the direction of reductive ring opening, as with propylene oxide,[34] is consistent with a potential-determining stage involving the addition of 2 electrons. The

[31] Hallsworth, Henbest, and Wrigley, *J.*, 1957, 1969.
[32] Ref. 1, p. 71.
[33] Dean and Berchet, *J. Amer. Chem. Soc.*, 1930, **52**, 2823.
[34] Birch, *J. Proc. Roy. Soc. New South Wales*, 1949, **83**, 245.

planar geometrical requirements for the transition state[35] in such cases ensure that axial alcohols are produced stereospecifically. Thus 5α : 6α-epoxides (VI) are converted into 5α-alcohols, and 2α : 3α-, 7α : 8α-, and

(VI) 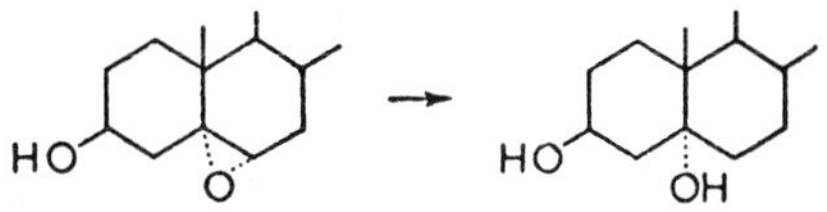 HO … O → HO … OH

9α : 11α-epoxides similarly give 3α-, 8α-, and 9α-alcohols, respectively.[36] Lithium aluminium hydride may reduce steroid *vic*-epoxides similarly,[37] but it has no effect on the sterically hindered 7α : 8α- and 9α : 11α-epoxides, thereby illustrating the power and low steric hindrance associated with the metal–amine reagents.

Calcium Hexammine.—The reducing capabilities of calcium hexammine have been but little investigated during the period since the last review. Reductions are usually carried out with a suspension of the reagent in an inert solvent (*e.g.*, ether, dioxan, 1 : 2-dimethoxyethane, tetrahydrofuran) so that solubility may usually be achieved by choosing a suitable solvent. Under sufficiently vigorous conditions the reagent saturates the double bond in simple olefins (2 : 5-dimethylhex-2-ene, for example, gives 2 : 5-dimethylhexane[38]) and reduces isolated benzene rings to the dihydro- or tetrahydro-state.[39, 40] Methoxyl groups may be cleaved from the ring by fission of intermediates, *e.g.*, methyl *m*-tolyl ether gives 1-methyl*cyclo*hexene,[3] and 2-methoxynaphthalene gives a mixture of hexahydronaphthalenes having homoannular conjugated diene systems.[40] Conjugation is probably due to the presence of calcium amide.

The Protection of Functional Groups.—The protection of reducible groups in some cases can be accomplished by salt formation to give anions. This has already been illustrated for allyl alcohols and carbonyl compounds. Ethynyl groups may be similarly protected ; thus, the sodium salt of undeca-1 : 7-diyne is converted into undec-7-en-1-yne by sodium in liquid ammonia.[41] This method which is simple in operation is only useful when the alcohol, enol, or acetylene is considerably the strongest acid present. The presence of an acid of comparable strength (*e.g.*, ethanol) permits reduction to occur.[20] Conversion of a carbonyl into a non-reducible group can be achieved through formation of an acetal, ketal, or enol ether provided the alkoxyl groups are not in an allylic or a benzyl position and the enol-ether double bond is unconjugated. Accordingly there are difficulties when the carbonyl group is in the α- or β-position to an aromatic ring.[19, 20] The problem has been solved for benzaldehyde derivatives by converting them

[35] See, *e.g.*, Barton and Cookson, *Quart. Rev.*, 1956, **10**, 67.
[36] Hallsworth and Jenbest, *J.*, 1957, 4604.
[37] *E.g.*, Plattner, Heusser, and Kulkarni, *Helv. Chim. Acta*, 1949, **32**, 265 ; Plattner, Fürst, Koller, and Kuhn, *ibid.*, 1954, **37**, 258.
[38] Kazanskii and Gostunskaya, *J. Gen. Chem.* (*U.S.S.R.*), 1955, **25**, 1659.
[39] Kazanskii and Glushev, *ibid.*, 1938, **8**, 642 ; *Bull. Acad. Sci. U.R.S.S.*, 1938, 1061, 1065, and earlier papers.
[40] Birch and Dunstan, unpublished work.
[41] Dobson and Raphael, *J.*, 1955, 3558.

into tetrahydroglyoxalines, *e.g.*, (VII), which resist hydrogenolysis whilst the aromatic nucleus is reduced.[18] The aldehyde group can then be regenerated by acid treatment. The presence of other reducible groups is

$$\text{C}_6\text{H}_5\text{CHO} \rightarrow \text{C}_6\text{H}_5\text{-HC}\begin{matrix}\text{NR—CH}_2\\ |\\ \text{NR—CH}_2\end{matrix} \rightarrow \text{(C}_6\text{H}_9\text{)-HC}\begin{matrix}\text{NR—CH}_2\\ |\\ \text{NR—CH}_2\end{matrix} \rightarrow \text{(C}_6\text{H}_9\text{)CHO}$$

(VII)

undesirable, *e.g.*, tetrahydro-1 : 2 : 3-triphenylglyoxaline (VII ; R = Ph) undergoes fission [18] because the *N*-phenyl groups stabilise the negative charge on the intermediate nitrogen anion. Hence dialkyltetrahydro-glyoxalines are generally used. Unfortunately the method is inapplicable to ketones, which fail to react with *NN'*-dialkyethylenediamines.

Stereochemical Aspects of Reduction.—A review [42] of the products formed by the reduction of various multiple bond systems by dissolving metal reagents which act through the formation of intermediate carbanions has indicated that where stereoisomeric products are possible the thermo-dynamically stable ones are usually formed. The rule holds in many cases, *inter alia*, the reduction of acetylenes to *trans*-ethylenes,[41, 43] ketones to secondary alcohols, oximes to amines, conjugated dienes and trienes to olefins, and $\alpha\beta$-unsaturated ketones, esters, and acids to the corresponding saturated derivatives, although recent work has shown that it is not univer-sally applicable. The case of $\alpha\beta$-unsaturated ketones is of particular importance. The reduction has been interpreted [42] as following the reaction

$$\underset{\text{(VIII)}}{-\overset{\beta}{\text{C}}=\overset{\alpha}{\text{C}}-\text{C}=\text{O}} \xrightarrow{2\varepsilon} \underset{\text{(IX)}}{-\bar{\text{C}}-\text{C}=\text{C}-\bar{\text{O}}} \xrightarrow{\text{H}^+} -\text{CH}-\text{C}=\text{C}-\bar{\text{O}}$$

path delineated. The stability of the enol anion (IX) in the absence of an excess of " acid " of comparable strength permits the eventual isolation of the saturated ketone. In a further discussion [44] of this type of reaction it has been pointed out that (VIII) should add a proton at the very basic β-position by a process involving little activation energy, in which case the nature of the product is determined by the most stable conformation of the anion rather than the least hindered approach of the proton donor. This accounts for the fact that when the β-position is capable of yielding stereo-isomers, the most stable one is invariably formed. For the α-position other considerations apply and the nature of the final product may depend on whether ketonisation of the enol anion is thermodynamically or kinetically controlled. Kinetic control at this stage has been observed to give the thermodynamically unstable isomer in reduction of the ketone (X), which gives the *cis*-isomer [44] (XI), readily convertible into the *trans*-isomer. A closely related example is formation of the *cis*-product (XIII) by lithium-ammonia reduction of the styrene (XII).[45] The mechanism of reduction of

[42] Barton and Robinson, *J.*, 1954, 3045. [43] Campbell, *Chem. Rev.*, 1942, **31**, 77.
[44] Birch, Smith, and Thornton, *J.*, 1957, 1339.
[45] Johnson, Ackerman, Eastham, and Dewalt, *J. Amer. Chem. Soc.*, 1956, **78**, 6303.

such styrene compounds is not clear, but if it involves an 8 : 9-dianion (steroid numbering) the first proton should be added at the more reactive 9-position. The formation of the less stable isomer in this case has been ascribed to the influence of the 5α-hydroxyl group which by alkoxide formation induces the negative charge at $C_{(9)}$ to adopt the β-position. When the less stable isomer is formed there is apparently a requirement that an aromatic ring shall be attached to the ring system, since reduction of Δ^9-octal-1-one gives only the *trans*-decalone.[46] A possible explanation for this is that the aromatic ring reduces the energy difference between the *cis*- and the *trans*-form of an anion such as (XIV) and thereby ensures an appreciable concentration of the former at equilibrium. In a more complex case in the steroid series the aromatic ring may not be necessary, provided that approach of the proton-donor required for the formation of the more stable product is sufficiently hindered : this may be so in the reduction of 3β-acetoxyergosta-8 : 22-dien-7-one [46] (XV).

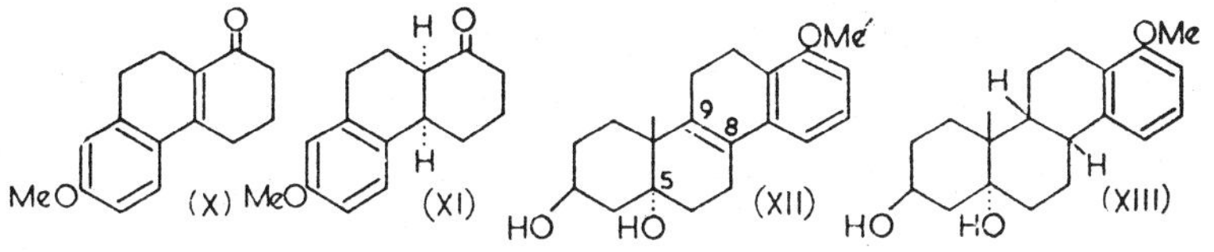

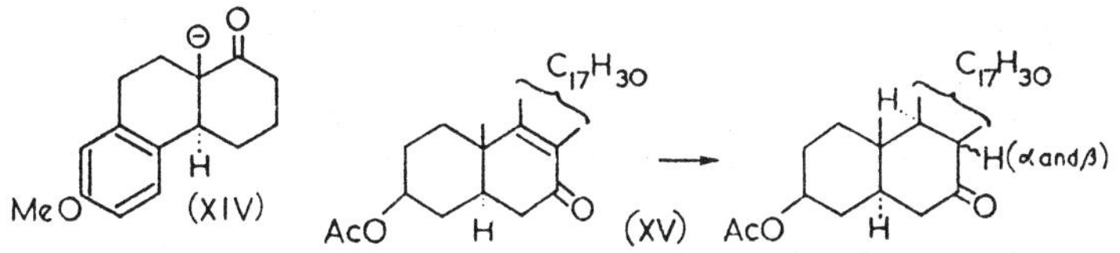

It has recently been shown that the sodium-ammonia reduction of dideuteroacetylene gives *trans*-dideuteroethylene,[47] in line with the fact that this method of producing *trans*-olefins from acetylenes is completely stereospecific.[43, 41] It is noteworthy that equilibrium mixtures contain appreciable amounts of *cis*-ethylenes.[48]

[46] Birch, Smith, and Wilson, unpublished work.
[47] Rabinowitz and Looney, *J. Amer. Chem. Soc.*, 1953, **75**, 2652.
[48] Kilpatrick, Prosen, Pitzer, and Rossini, *J. Res. Nat. Bur. Stand.*, 1946, **36**, 559.

The Reversible Decomposition of Metal–Ammonia Solutions

Metal–ammonia solutions are unstable toward decomposition into metal amide and hydrogen:

$$e^-_{am} + NH_3 \rightarrow \frac{1}{2} H_2 + NH_2^-$$

Although this reaction ordinarily is slow, it can be accelerated by suitable catalysts. Indeed, the iron oxide-catalyzed reaction is often used for the synthesis of alkali metal amides.[1] Obviously this decomposition reaction can frustrate studies of metal–ammonia solutions. It is important for those who work with metal–ammonia solutions to know what materials are catalytic and how to avoid them. In this respect, a paper by Jackman and Keenan[2] (which we were unable to reproduce for this volume) should be consulted. These authors made a very definitive study of the glass-catalyzed reaction.

In the following paper, Kirschke and Jolly show that the decomposition reaction is reversible and that the ammoniated electron can exist in equilibrium with elemental hydrogen and the amide ion. The equilibrium constant for this reaction permits an accurate evaluation of the free energey of formation of e^-_{am}.

---

[1]R. Levine and W. C. Fernelius, *Chem. Revs.*, **54**, 449 (1954).
[2]D. C. Jackman and C. W. Keenan, *J. Inorg. Nucl. Chem.*, **30**, 2047 (1968).

[Reprinted from Inorganic Chemistry, 6, 855 (1967).]

17

CONTRIBUTION FROM THE DEPARTMENT OF CHEMISTRY OF THE UNIVERSITY OF CALIFORNIA AND THE INORGANIC MATERIALS RESEARCH DIVISION OF THE LAWRENCE RADIATION LABORATORY, BERKELEY, CALIFORNIA 94720

The Reversibility of the Reaction of Alkali Metals with Liquid Ammonia

BY ERNEST J. KIRSCHKE AND WILLIAM L. JOLLY

Received December 27, 1966

Ammoniated electrons exist in solutions of the alkali metal amides in liquid ammonia which have reached thermodynamic equilibrium with hydrogen gas. By using both electron spin resonance and optical spectroscopy to measure the electron concentration, an equilibrium constant of 5×10^4 was measured for the reaction

$$e_{am}^- + NH_3 = NH_2^- + {}^1/_2H_2$$

and 3×10^9 for the reaction

$$Na^+ + e_{am}^- + NH_3 = NaNH_2(s) + {}^1/_2H_2$$

at 25°. From the temperature coefficients of these reactions, the approximate $\Delta H°$ values of −16 and −12 kcal/mole, respectively, were obtained.

Introduction

Solutions of alkali metals in liquid ammonia decompose slowly (rapidly in the presence of certain catalysts) to form the metal amides and hydrogen. For potassium, rubidium, and cesium, the reaction may be written as

$$e_{am}^- + NH_3 = NH_2^- + {}^1/_2H_2 \quad (1)$$

For lithium and sodium (whose amides are sparingly soluble in ammonia) the reactions may be written as

$$Li^+ + e_{am}^- + NH_3 = LiNH_2(s) + {}^1/_2H_2 \quad (2)$$

$$Na^+ + e_{am}^- + NH_3 = NaNH_2(s) + {}^1/_2H_2 \quad (3)$$

These reactions can be useful for the preparation of the amides, but more often they are undesirable side reactions in the study and use of metal–ammonia solutions. Kraus[1] suggested that a systematic study of these reactions was a necessary prelude to the study of more stable solutions. No one has directly measured the equilibrium constants of these reactions, but it is possible to correlate thermodynamic data for electrolytes in liquid ammonia and thereby to calculate indirectly the equilibrium constants. In this way one can calculate, for 25°, the approximate values $K = 1.2 \times 10^6$ for reaction 1,[2] $K = 7 \times 10^{12}$ for reaction 2,[2,3] and $K = 1.3 \times 10^{10}$ for reaction 3.[2,4] The purpose of this work was to demonstrate the reversibility of these reactions and to measure directly the equilibrium constants and their temperature coefficients.

The experiments were of two types, corresponding to two methods for determining the electron concentration. In one type of experiment, the electron concentration was determined spectrophotometrically. In the other type of experiment, the electron concentration was determined by electron spin resonance.

Experimental Section

Description of the Optical Cell.—The high pressure optical cell is shown in Figures 1 and 2. The heavy stainless steel body of the cell resisted attack by the solutions to be studied, contained the high pressures anticipated, and acted as an efficient heat reservoir to dissipate the heat absorbed from the infrared source by the sample. The windows were made of Pyrex glass and were approximately 1 cm thick. The O rings were of a special polyethylene–propylene composition (Porter Seal Co.) selected for resistance to attack by metal ammonia solutions. A stainless steel bellows seal valve was used as a master valve for the same reason. Stainless steel, V seated, packed valves were used in the manifold. Stainless steel tubing and Swagelock fittings were used in the construction of the manifold and gauge connections. When assembled, the unit could hold 136 atm of helium gas with no detectable leakage for a period in excess of 24 hr.

Temperature control was accomplished by circulating thermostated water through copper heat-exchange coils soldered to the body of the cell. The temperature was measured with a copper–constantan thermocouple located in a well a few millimeters from the sample cavity, in conjunction with a Rubicon potentiometer. The temperature was maintained at 25 ± 0.1°.

(1) C. A. Kraus, *J. Chem. Educ.*, **30**, 83 (1953).

(2) W. L. Jolly in "Solvated Electron," Advances in Chemistry Series, No. 50, American Chemical Society, Washington, D. C., 1965, p 30.

(3) "Selected Values of Chemical Thermodynamic Constants," National Bureau of Standards Circular 500, U. S. Government Printing Office, Washington, D. C. The free energy of formation of $LiNH_2$ was calculated from the heat and an estimated entropy.

(4) L. V. Coulter, J. R. Sinclair, A. G. Cole, and G. C. Roper, *J. Am. Chem. Soc.*, **81**, 2986 (1959).

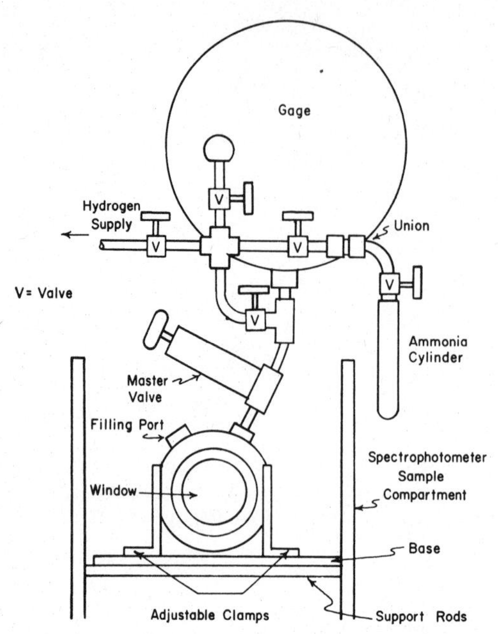

Figure 1.—Optical cell viewed along axis of the light beam.

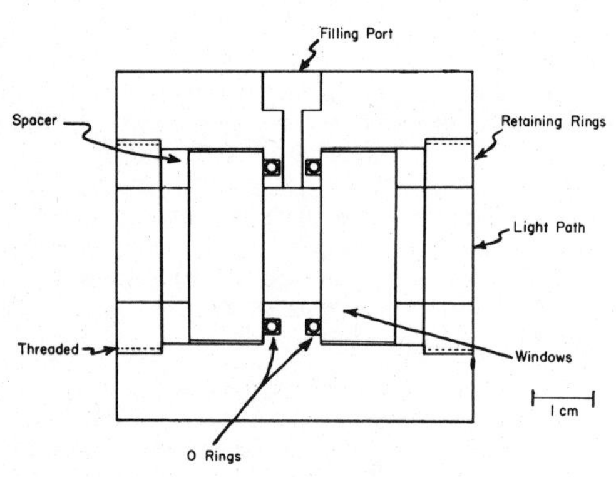

Figure 2.—Cross section of optical cell.

Pressure was determined with a stainless steel Bourdon gauge with a range of 0–2000 psig (136 atm) divided in increments of 20 psig (1.36 atm). By fitting the tip of the gauge pointer with a thin piece of flat metal mounted perpendicular to the gauge face, readings could be estimated to ± 0.1 division (± 0.14 atm).

The cell was mounted on a base plate with adjustable clamps designed to permit accurate, reproducible positioning of the unit in the sample compartment of a Cary Model 14 recording spectrophotometer. Once the proper position had been established, the cell could be repositioned by simply ensuring that two edges of the base plate were in firm contact with two walls of the compartment.

Calibration of the Optical Cell.—The light path length of the optical cell was mechanically measured with a micrometer caliper at various pressures. The path length at zero pressure was 0.992 cm and increased about 0.012%/atm increase in pressure. These measured values were used in the calculations. The light path length was checked against a cell of known path length using the 12,000 A absorption band of a solution of methanol in ethyl ether. A value of 0.996 cm was obtained which, considering the possible errors in this technique, is in good agreement with the mechanical measurement.

The cell was filled with liquid NH_3 and the effect of pressure on the spectrum was checked at several pressures up to about 130 atm. Hydrogen gas was used to pressurize the cell. The results indicated an increase in absorbance with pressure. At 100 atm this increase was 1.7%, which was larger than predicted from the increase in path length. Maybury and Coulter[5] measured the adiabatic compressibilities of liquid ammonia solutions, and from their data we estimate an increase in the density of liquid ammonia of 0.5% per 100 atm increase in pressure. We conclude that the measured increase in absorbance was the result of changes in both path length and density.

The volume of the cell was checked on a vacuum line by filling the cell to a known pressure of nitrogen and then pumping the nitrogen into a gas buret of known volume using a Toepler pump. The volumes of the gauge and the manifold were determined in the same manner.

The Bourdon gauge was calibrated on a dead weight tester and a calibration chart prepared. Above 10 atm the gauge read 0.1 atm high. All readings were corrected by the use of the calibration chart. Periodically the gauge was checked against a second calibrated gauge; the calibration did not change over the period the gauge was in use.

Procedure for Obtaining Optical Data.—Prior to use, the optical cell was carefully cleaned, the metal surfaces of the sample compartment coated with 0.05 *M* chloroplatinic acid, and the whole unit dried in an oven at 120°. This procedure left a thin coating of platinum in the cell to act as a catalyst. The cell was assembled and leak tested with helium at 136 atm.

The cell was attached to a vacuum line and evacuated. The entire system was wrapped with heating tape and the temperature raised to about 100°. Care was taken to loosen the retaining rings during the bake out to prevent cracking of the windows due to the different coefficient of expansion of steel and Pyrex. The cell was pumped out for a minimum of 8 hr and then filled with dry, oxygen-free nitrogen or argon. Ammonia, dried with sodium metal, was then distilled into a small steel cylinder (10 ml) equipped with a valve. This valve was closed; the cylinder was removed from the vacuum line, weighed, and then attached to the manifold of the optical cell. A small hole was cut in the end of a plastic glove bag (I²R Co., Cheltenham, Pa.) and this was sealed to the filling port of the cell. The bag was continuously flushed with nitrogen or argon which had passed, successively, through a column packed with BTS catalyst (BASF Colors and Chemicals Inc.) to remove traces of oxygen and a column of magnesium perchlorate to remove traces of water. In a run involving sodium or potassium, a piece of metal was cleaned of all oxide and placed in an extrusion press in the dry bag. A length of wire was extruded and placed in a small tared weighing bottle which had been baked in an oven at 200° and then cooled in the dry gas atmosphere. The sample was weighed and returned to the dry bag, and the bag was thoroughly flushed with dry gas. The filling port of the cell was opened and a known amount of metal introduced.

Cesium was much more difficult to handle because of its semiliquid state and extreme reactivity. We found that we could deliver accurate volumes of clean cesium into the cell with a micropipet heated to 30°. The density of cesium at this temperature was estimated at 1.84 g/cc. The metal at 30° was drawn up into the pipet which was kept at the same temperature. The tip of the pipet was then placed into the filling port and dry nitrogen was used to force the metal from the pipet into the cell. When the pipet had been properly cleaned and dried, no reaction occurred, and the pipet emptied cleanly. The entire operation took place in a well-flushed glove bag.

After the introduction of metal, the port was closed, and the cell was evacuated. The valve to the vacuum line was closed; cold water was circulated through the cell heat exchanger, and ammonia was distilled into the cell from the small steel cylinder.

(5) R. H. Maybury and L. V. Coulter, *J. Chem. Phys.*, **19**, 1326 (1951).

When the cell appeared full, the master valve was closed, and ammonia in the manifold was condensed back into the steel cylinder using liquid nitrogen. The manifold was again opened to the vacuum line, and the pressure was measured with a thermocouple gauge to check that no ammonia was left in the manifold. The cylinder was removed and weighed.

In the spectral range in which the electron absorption band occurs, ammonia has intense absorption bands. It was necessary to take readings at wavelengths between these bands, where the ammonia absorption is low. At zero hydrogen pressure, the concentration of the electron was zero. Therefore, the absorbance due to the electron was taken to be the difference in the total absorbance at pressure $P > 0$ and that at $P = 0$.

$$A_{e^-} = A_T(P > 0) - A_T(P = 0)$$

In early experiments the absorbance at zero hydrogen pressure was determined after removing the hydrogen produced by the reaction of the metal with ammonia. After the hydrogen pressure dependence of the reaction had been thoroughly established, this step was eliminated. Instead, the total absorbance was plotted against the square root of the hydrogen pressure and the line extrapolated to zero hydrogen pressure.

Water at 25° was circulated through the heat exchanger, and the system was allowed to come to chemical equilibrium. This usually took 1–2 hr, depending upon the concentration of the alkali metal. The cell was positioned in the sample compartment of a Cary Model 14 spectrophotometer and the absorbance measured. Because of the sensitivity of the electron absorption band and the equilibrium constant to temperature, the sample was exposed to the light source for periods of less than 5 sec, and time was allowed to dissipate any heat absorbed before taking the next reading. To ensure that equilibrium had been reached, the procedure was repeated at 20-min intervals until three successive readings were identical.

At this point hydrogen gas at approximately 150 atm was added to the manifold. This in turn was carefully bled into the gauge section until a pressure near that desired was registered on the gauge. The valve to the manifold was closed; the master valve was opened, and the pressure was allowed to equilibrate. This normally took 5–10 min. The master valve was then closed and the system allowed to reach chemical equilibrium as established by three successive absorbance readings at 20-min intervals. This equilibration took from 1 to 3 hr, depending on the concentration of the amide and the pressure used. This over-all procedure was repeated until the desired data had been obtained. To test for reversibility in several runs, the process was reversed, that is hydrogen was removed from the system. It was impossible in this step to prevent some ammonia from coming off with the hydrogen and (with the exception of saturated solutions) increasing the amide concentration. This ammonia was collected by passing the gas being removed very slowly through two traps at liquid nitrogen temperature. The hydrogen was pumped away and the amount of ammonia determined by either condensing it into a gas bulb and weighing or by absorbing it in a known quantity of acid and titrating. The amide concentration was corrected accordingly.

Throughout the procedure, extreme care had to be taken to prevent liquid ammonia from condensing above the master valve in the gauge section, resulting in a change in the amide concentration. This was accomplished both by heating the gauge to a temperature about 5° greater than the cell and by keeping the master valve closed except when taking a pressure reading.

Determination of the Amide Concentration in the Optical Studies.—The concentrations of amide in the solutions in the optical cell were determined from the weights of the metals used and the volumes of the ammonia. The volume of the ammonia solution was determined from the weight of ammonia used, corrected for the amount of ammonia in the gas phase. Three approximations were made in this calculation: first, that the vapor pressure of the solution could be estimated using Raoult's law; second, that knowing the pressure and volume of the ammonia gas, the number of moles of gas could be calculated using van der Waal's equation for ammonia;[6] third, that the effect of the amide concentration on the volume of the solution could be estimated by adding the volume of the amide formed (calculated from the density of the solid amide[7]) to the volume of the liquid ammonia calculated from its weight and density.[8]

Preparation of the Esr Samples.—The esr samples were prepared by sealing measured amounts of alkali metal and ammonia in Pyrex glass tubes and allowing chemical equilibrium to be reached. In all of the samples studied, the intense blue color of the alkali metal completely faded and was replaced either by the yellow color of the amide ion (in the cases of potassium and cesium) or a precipitate of metal amide (in the cases of sodium and lithium).

The epr sample tubes consisted of 3-mm Pyrex tubing which had been selected for uniformity of wall thickness and bore. Each tube was fitted with a pyrex O ring joint and annealed. The cross section of each tube was determined gravimetrically using mercury. The tubes were thoroughly cleaned, rinsed with a 0.001 *M* solution of chloroplatinic acid, and placed in an oven at 200° for 2 hr. This procedure left a very fine coating of platinum on the walls to act as a catalyst for the preparation of the amide from the metal and liquid ammonia. The presence of the catalyst did not interfere with the spectrum, and without it the reaction to form the amide was extremely slow.

The tubes were placed on the vacuum line, evacuated, and flamed. They were then filled with dry nitrogen, removed, capped, and weighed. The tubes were transferred to a dry bag and charged with samples of the metal which had been extruded from a hand press. The tubes were carefully reweighed, placed back on the vacuum line, and evacuated. Dry liquid ammonia was then distilled onto the samples. From the weight of metal used and the cross section of the tube, the approximate level of ammonia to give a desired concentration could be determined. The tubes were sealed off, and both sections were weighed to determine the weight of ammonia used. The tubes were set aside and allowed to reach chemical equilibrium.

Certain experiments called for the addition of sodium bromide, potassium bromide, or both salts to the solution. The salts were dried for several hours at 180° and allowed to cool in a desiccator. Short lengths, 2–3 cm, of melting point capillary were weighed on a microbalance, filled with an estimated amount of the desired salt, and reweighed. The capillary with the salt was placed in the esr sample tube and the whole unit evacuated and gently flamed. The sample tubes were filled with dry gas, and the samples were prepared as described above.

Method for Determining the Esr Spectrum.—After the reaction had reached equilibrium (usually a few days after the blue color of the metal solution had disappeared) the spectrum was recorded on a Varian V-4502 epr spectrometer. A small crystal of diphenylpicrylhydrazyl (DPPH) was placed in the cavity near the sample to act as a reference and intermediate standard. The temperature was controlled to ±0.1° by passing dry air through a heat exchanger and then through a small dewar in which the sample was positioned. Figure 3 shows the arrangement. The temperature could be varied by changing the temperature of the heat exchanger bath and by controlling the flow of air. The temperature was measured with a copper–constantan thermocouple.

The electron concentration was determined by indirectly comparing the sample signal with that of a solution of recrystallized vanadium(IV) oxyacetylacetonate of known concentration.

The *Q* value of the microwave cavity is seriously affected by the introduction of samples of varying dielectric strength. A lowering of the *Q* value decreases the sensitivity of the spectrometer and results in a reduced signal intensity. Such sensitivity

(6) "Handbook of Chemistry and Physics," 46th ed, Chemical Rubber Co., Cleveland, Ohio, 1966.

(7) R. Levine and W. C. Fernelius, *Chem. Rev.*, **54**, 449 (1954).

(8) C. Cragoe and D. Harper, National Bureau of Standards Scientific Papers No. 420, U. S. Government Printing Office, Washington, D. C., 1921, p 313.

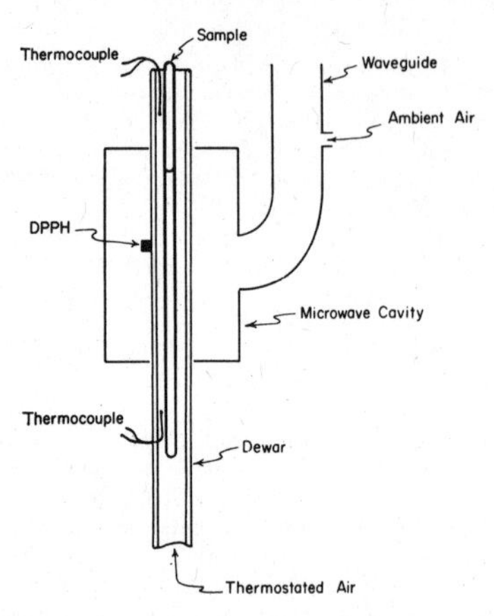

Figure 3.—Schematic drawing of esr microwave cavity, showing arrangement of the sample and the DPPH intermediate reference standard.

changes were compensated for by means of the DPPH intermediate standard. When the spectra of each sample and the standard sample were recorded, the spectrum of the DPPH was also recorded. One of the samples was selected as a reference, and the ratio of the signal intensity of each of the other samples ($I_{(obsd)}$) to that of the reference ($I_{R(obsd)}$) was corrected by the ratio of the DPPH signal intensities (D_R/D) according to the equation $I/I_R = (I_{(obsd)}/I_{R(obsd)})D_R/D$. Then by determining the concentration of spins in the reference sample by means of the vanadium(IV) solution, the electron concentration of all of the samples could be calculated. This technique also accounts for any change in cavity match or radiofrequency power level.[9] Readings on a given sample could be duplicated to within $\pm 10\%$.

The temperature-dependence studies were carried out with the sample left in place in the spectrometer. After establishment of thermal equilibrium, the spectrum was taken at 10-min intervals. Equilibrium was considered to have been reached when three successive readings were of the same intensity. The temperature was then changed slightly in a direction opposite to the main direction in which the temperature had been shifted, and the change in intensity noted. A signal change in the direction predicted by this temperature change indicated that we had been at equilibrium.

Determination of the Hydrogen Pressure and Amide Concentration of the Esr Sample.—The total volume of each prepared sample tube was determined from its internal length and cross section. An adjustment ($\sim$0.2%) was made for the rounded ends. The volume of liquid was determined either by directly measuring the height in the tube or by determining the weight and estimating the density of the solution. When possible, both methods were used, and they agreed within 2%.

After the spectrum had been recorded, the sample tube was placed in a specially built glass vessel which could be evacuated and which permitted breaking the tube and determining the hydrogen volumetrically. Using the solubility data of Wiebe and Tremearne[10] and the volume data, the pressure of hydrogen in the tube was calculated. For every 0.5 mole of hydrogen formed, 1 mole of amide forms, and therefore for the potassium and cesium runs it was possible to calculate the amide concentrations.

Results of the Optical Study

The optical spectra of solutions of $NaNH_2$, KNH_2, and $CsNH_2$ were recorded at 13,650 A and at hydrogen pressures up to approximately 100 atm. Some of the results are shown in Table I. As previously stated, readings had to be taken at a wavelength where the

(9) Varian Technical Information Publication 87-114-000, pp 5-7–5-11.

(10) R. Wiebe and T. H. Tremearne, *J. Am. Chem. Soc.*, **56**, 2357 (1934).

Table I

Equilibrium Data from Optical Studies at 13,650 A

Molarity of amide	Hydrogen press., atm	Total absorbance	Net absorbance	Concn of absorbing species, 10^5M
2.01	0	0.325	...	...
	12.5	0.585	0.260	2.37
	25.9	0.730	0.405	3.69
	40.8	0.835	0.509	4.62
	50.3	0.900	0.573	5.20
	72.1	1.000	0.672	6.10
	87.2	1.061	0.732	6.62
	100	1.125	0.796	7.20
1.80	100.0	1.045[a]	0.770	6.98
1.70	0	0.340	...	...
	15.0	0.640	0.300	2.75
	38.1	0.800	0.459	4.18
	64.0	0.928	0.585	5.30
	88.5	1.020	0.676	6.12
	99.7	1.043	0.698	6.32
1.85[b]	64.5	0.940	0.599	5.42
1.99[b]	42.2	0.817	0.475	4.32
2.13[b]	17.0	0.680	0.340	3.12
1.41	100.0	0.994[a]	0.652	5.90
1.21	100	0.898	0.572	5.18
1.00	0	0.216	...	...
	4.0	0.329	0.113	1.04
	11.6	0.415	0.199	1.83
	25.2	0.495	0.278	2.53
	49.0	0.618	0.401	3.65
	82.7	0.720	0.502	4.54
	99.6	0.775	0.556	5.02
	100	0.776[a]	0.557	5.05
0.93	100	0.778[a]	0.542	4.91
0.80	100	0.740	0.537	4.87
0.72 ($CsNH_2$)	100	0.700[a]	0.447	4.05
0.68	0	0.270	...	...
	8.85	0.402	0.132	1.21
	28.6	0.488	0.218	1.99
	49.6	0.525	0.254	2.31
	68.4	0.640	0.368	3.34
	88.0	0.660	0.388	3.51
	100	0.688	0.416	3.77
0.73[b]	72.0	0.644	0.372	3.37
0.79[b]	47.0	0.591	0.320	2.91
0.90[b]	18.3	0.503	0.233	2.14
0.58	100	0.641[a]	0.394	3.57
0.55	0	0.300[a]	...	...
	0.7	0.330	0.030	0.28
	18.4	0.465	0.165	1.51
	45.2	0.557	0.276	2.51
	73.5	0.640	0.339	3.07
	95.0	0.691	0.388	3.52
	100	0.705	0.403	3.65
0.50	100	0.665	0.363	3.29
0.40	100	0.613	0.333	3.01
0.28	100	0.552	0.281	2.54
0.26	100	0.558[a]	0.289	2.62
0.20 ($CsNH_2$)	0	0.260[a]	...	...
	0.8	0.270	0.010	0.09
	11.6	0.345	0.085	0.78
	25.9	0.400	0.140	1.28
	60.5	0.462	0.201	1.83
	100	0.520[a]	0.259	2.34
	104	0.522	0.260	2.35
0.19	100	0.540[a]	0.247	2.23
0.16	100	0.611	0.298	2.70
0.15	100	0.496[a]	0.185	1.67
0.067	100	0.438[a]	0.167	1.51
0.027 ($NaNH_2$)	0	0.256[a]	...	...
	25.8	0.263	0.007	0.06
	49	0.270	0.014	0.13
	67	0.276	0.020	0.18
	100	0.277	0.021	0.19
	58[b]	0.274	0.018	0.17
	24.5[b]	0.260	0.004	0.04

[a] Determined by extrapolation of the measured absorbance. [b] The hydrogen pressure was reduced to test the reversibility of the equilibrium. When necessary, the amide concentration was adjusted for loss of ammonia.

ammonia absorbance exhibits a minimum. It was desirable that this minimum be fairly broad (about 500 A) and at a wavelength for which the extinction coefficient of the electron is high. We found that the minimum at 13,650 A satisfied both of these conditions and gave the most reproducible results. Some measurements were made at 11,200, 12,500, 16,000, and 17,500 A; the results were similar, but the data were more scattered.

The concentration of the electron was calculated by substituting the values of the net absorbance A_{e^-}, the extinction coefficient ϵ, and the path length d (Table I) into the Beer-Lambert equation, $(e_{am}^-) = A_{e^-}/\epsilon d$. We used an extinction coefficient of $1.1 \times 10^4\ M^{-1}\ cm^{-1}$ at 13,650 A, taken from the data of Corset and Lepoutre.[11,12] Figure 4 shows the relationship between the electron concentration and the hydrogen pressure for several amide concentrations. The data for all concentrations studied exhibited the same linear dependence on the square root of the hydrogen pressure. To determine that it was specifically the hydrogen gas pressure which caused the increase in absorbance, the following experiment was performed. The spectrum at 25 atm hydrogen pressure was recorded, and then, using argon, the pressure was increased by 25 atm and the spectrum again recorded. The experiment was repeated using helium and nitrogen. No increase in absorbance other than that due to changing path length and density was detected. To ensure that nothing had affected the samples during the experiments, hydrogen was again added to the system and the spectrum recorded. An increase in absorbance was detected in every case.

From the data in Table I at 100 atm of hydrogen pressure, assuming the amide to be completely dissociated, we calculated equilibrium quotients for reaction 1 from the equation

$$K_c = \frac{[NH_2^-]P_{H_2}^{1/2}}{[e_{am}^-]}$$

The values of K_c are plotted against amide concentration in Figure 5. Obviously K_c is not constant with changing amide concentration. To try to resolve this difficulty we treated the data in two ways. First, a modified Debye–Hückel treatment was employed combining the properties of KNH_2 in liquid ammonia calculated by Fuoss and Kraus[13] with the calculations for an alkali metal made by Arnold and Patterson.[14] This treatment required a series of approximations, not the least of which was that the Debye–Hückel theory was valid at these concentrations. Unfortunately this treatment failed to produce a significant improvement in the data.

(11) J. Corset and G. Lepoutre in "Solutions Metal-Ammoniac: Propriétés Physico-Chimiques," G. Lepoutre and M. J. Sienko, Ed., W. A. Benjamin, New York, N. Y., 1964, p 186.

(12) G. Lepoutre, private communication. Errata to ref 11: Figure 3, p 188; the abscissa reads 13,000 under the 1 of 13,000 and 10,000 under the 1 of 10,000. Page 188, line 3, second paragraph; change 13,000 cm^{-1} to 12,000 cm^{-1}.

(13) R. M. Fuoss and C. A. Kraus, *J. Am. Chem. Soc.*, **55**, 476 (1933).

(14) E. Arnold and A. Patterson, *J. Chem. Phys.*, **41**, 3089 (1964).

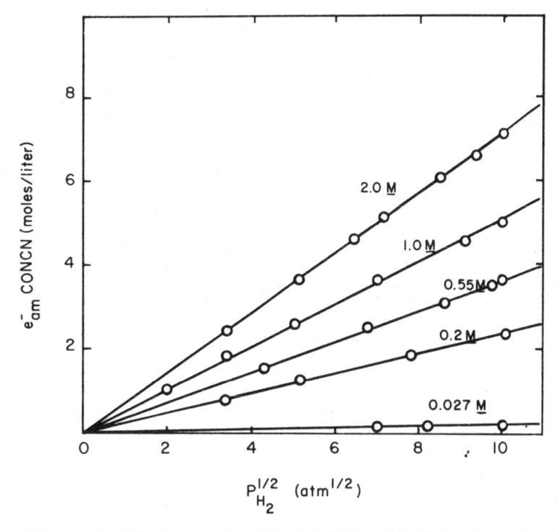

Figure 4.—Electron concentration (as determined by optical measurements) *vs.* the square root of the hydrogen pressure at several amide concentrations.

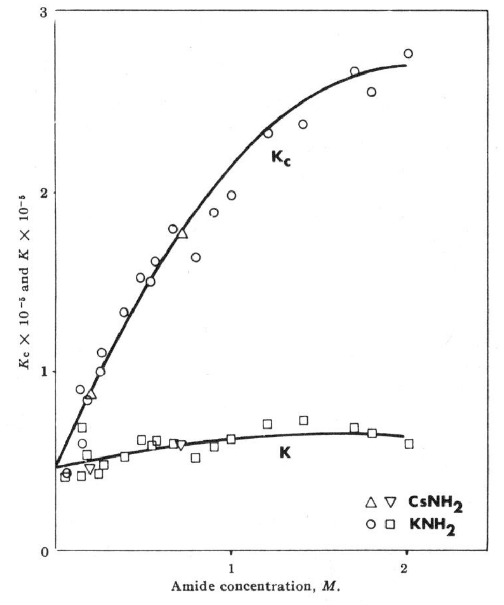

Figure 5.—K_c and K *vs.* amide concentration for reaction 1. Data from optical measurements.

Second, an equilibrium constant was calculated from the equation

$$K = \frac{P^{1/2}[NH_2^-]\gamma_\pm^2(KNH_2)}{[e_{am}^-]\gamma_\pm^2(K)} = K_c \frac{\gamma_\pm^2(KNH_2)}{\gamma_\pm^2(K)}$$

in which $\gamma_\pm(KNH_2)$ and $\gamma_\pm(K)$ are the empirically determined mean ionic activity coefficients for potassium amide and potassium, respectively. Since no activity coefficient data have been reported for any of the alkali amides, values were estimated from the measurements on NH_4Cl at 25° by Ritchie and Hunt.[15]

(15) H. W. Ritchie and H. Hunt, *J. Phys. Chem.*, **43**, 407 (1939).

The activity coefficients for potassium were taken from the work of Marshall.[16] The latter values were measured at −33° and were left uncorrected. The principle of ionic strength was assumed valid. The results of this calculation appear in Table II, and the calculated values of K are plotted in Figure 5. Clearly the K values are much more constant than the K_c values.

The value of K for reaction 1 obtained by extrapolation to zero concentration is approximately 5 × 10⁴. A value of K = 3 × 10⁹ was calculated for reaction 3 from the data for a saturated $NaNH_2$ solution, using the expression

$$K = \frac{P_{H_2}^{1/2}}{[Na^+][e^-]\gamma_\pm^2(Na)}$$

The $\gamma_\pm(Na)$ value was estimated from the data of Marshall.[16]

Table II

Calculated Values of the Equilibrium Constant Using Estimated Activity Coefficients; P_{H_2} = 100 atm

Amide concn, M	$10^2\gamma_\pm$ of amide[15]	$10^2\gamma_\pm$ of metal[16]	Electron concn, 10^5M	$K \times 10^{-4}$
2.0	1.35	2.90	7.20	6.0
1.8	1.50	2.95	6.98	6.7
1.7	1.57	3.00	6.32	7.3
1.4	1.82	3.25	5.90	7.5
1.2	1.97	3.55	5.18	7.2
1.0	2.25	4.00	5.05	6.3
0.93	2.37	4.25	4.91	5.9
0.80	2.65	4.70	4.87	5.2
0.72[a]	2.91	5.05	4.05	5.9
0.68	3.03	5.25	3.77	6.0
0.58	3.55	5.75	3.57	6.2
0.55	3.75	6.05	3.68	5.8
0.50	4.05	6.35	3.29	6.2
0.40	4.45	7.05	3.01	5.3
0.28	6.10	9.25	2.54	4.8
0.26	6.35	9.50	2.62	4.4
0.20[a]	7.72	10.5	2.34	4.6
0.19	8.70	11.0	2.23	5.3
0.16	9.80	11.7	2.70	4.1
0.15	10.5	12.0	1.67	6.9
0.067	18.0	18.5	1.51	4.2
0.027[b]	...	27.0	0.18	300,000

[a] $CsNH_2$. [b] $NaNH_2$.

Results of the Esr Measurements

Determination of the Equilibrium Constant.—In an earlier study,[17] using electron spin resonance to determine the total concentration of unpaired electrons, we observed that the equilibrium quotient (K_c) for reaction 1 varied markedly with the concentration of potassium amide. At the time it was believed that a change in the sensitivity of the esr spectrometer was responsible for the inconstancy of K_c. More recently we have used smaller diameter sample tubes in an attempt to correct the problem. Table III shows the results. Obviously the K_c values are still not constant with changing amide concentration. However, a marked improvement in the data is achieved by calculating equilibrium constants (K) using activity coefficients of potassium and potassium amide (as estimated for the constants obtained from optical data). The values for K_c and K are plotted against amide concentration in Figure 6. Extrapolation to zero concentration yields a value of approximately 5 × 10⁴, which is fortuitously in agreement with the value obtained from optical data.

Table III

Equilibrium Data from Epr Experiments

Concn of amide, M	Hydrogen press, atm	Concn of electrons, 10^6M	$K_c = \frac{P_{H_2}^{1/2}[NH_2^-]}{[e^-]}$	$K^a = K_c \times \frac{\gamma_\pm^2(KNH_2)}{\gamma_\pm^2(K)}$
0.269	7.52	4.50	1.65 × 10⁵	0.7 × 10⁵
0.361	18.9	8.05	1.96 × 10⁵	0.8 × 10⁵
0.815	19.2	9.24	3.87 × 10⁵	1.2 × 10⁵
1.00	19.4	9.92	4.44 × 10⁵	1.4 × 10⁵
1.08	16.8	8.45	5.26 × 10⁵	1.6 × 10⁵
2.40	21.5	13.0	8.55 × 10⁵	1.4 × 10⁵

[a] $\gamma_\pm(KNH_2)$ estimated from ref 15; $\gamma_\pm(K)$ from ref 16.

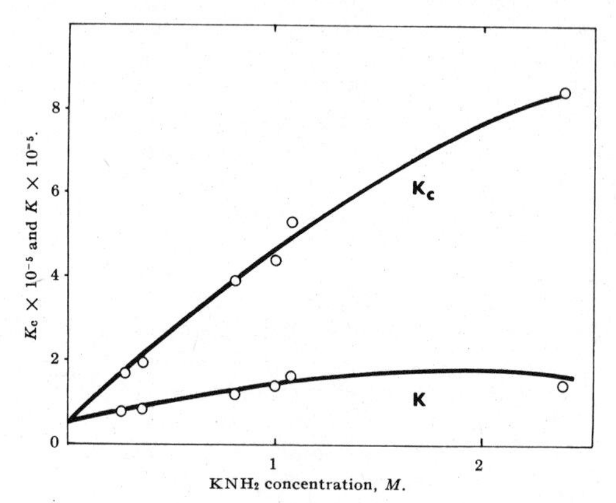

Figure 6.—K_c and K *vs.* amide concentration for reaction 1. Data from esr measurements.

Two sets of experiments were made using esr to verify the hydrogen pressure dependence results found by optical methods. In the first, 1 M KNH_2 solutions were prepared in such a way as to vary the hydrogen pressure. This was accomplished by simply keeping the weight of potassium and the volume of ammonia in each sample constant and varying the total volume of the tube. The second set of experiments was run using saturated $NaNH_2$ solutions. Here the volume of the tubes and the quantity of ammonia were kept constant and the quantity of metal varied. The results of the two sets of experiments are shown in Table IV. In both cases the intensity of the spectrum was linearly dependent on the square root of the hydrogen pressure.

Temperature Dependence of the Equilibrium.—The temperature dependences of the equilibrium constants of reactions 1 and 3 were obtained by measuring signal intensities for systems equilibrated at various temperatures. We did not calculate absolute equilibrium constants for each temperature, but rather calculated the ratios of the equilibrium constant at room

(16) P. R. Marshall, *J. Chem. Eng. Data*, **1**, 399 (1962); ref 11, p 97.
(17) E. J. Krischke and W. L. Jolly, *Science*, **147**, 45 (1965).

TABLE IV
HYDROGEN PRESSURE DEPENDENCE

Hydrogen press, atm	Signal intensity	Electron concn, $10^6 M$	10^{-5} × equilibrium constant
	1.0 M KNH_2 Solution		
35.5	5.70	13.1	1.4[a]
29.0	4.92	11.3	1.5[a]
19.4	4.32	9.92	1.4[a]
16.8	3.68	8.43	1.5[a]
10.0	2.87	6.60	1.5[a]
	Saturated $NaNH_2$ Solution (0.027 M)		
38	0.418	0.94	33,000[b]
34	0.382	0.87	34,000[b]
25	0.310	0.71	36,000[b]
22	0.316	0.71	34,000[b]
19	0.281	0.64	34,000[b]
14	0.243	0.56	34,000[b]

[a] From the expression $K = [NH_2^-]\gamma_\pm^2 P_{H_2}^{1/2}/[e^-]\gamma_\pm^2$. [b] From the expression $K = P_{H_2}^{1/2}/[Na^+][e^-]\gamma_\pm^2$. The activity coefficients were estimated from ref 15 and 16.

temperature to that at each temperature, K_{rt}/K. The signal intensity value had to be corrected for the change in the Boltzmann distribution of the energy levels with changing temperature. At a given temperature the distribution ratio is given by the expression $N_2/N_1 = \exp(-\Delta E/kT)$. By expanding and neglecting higher order terms we obtain the relation $N_2 - N_1 = -N_1\Delta E/kT$. The quantity $N_2 - N_1$ is proportional to the signal intensity so we corrected the relative intensity values S/S_{rt} by multiplying by the ratio T/T_{rt}.

The ratio of the equilibrium quotients for reaction 1

$$\frac{K_{rt}}{K} = \frac{[e_{am}^-]}{[e_{am}^-]_{rt}} \frac{[NH_2^-]_{rt}}{[NH_2^-]} \frac{P_{rt}^{1/2}}{P^{1/2}}$$

had to be corrected for the change in the hydrogen pressure with temperature and the change in the NH_2^- concentration due to the change in the density of the solution. The pressure correction was made by assuming that hydrogen gas behaves ideally and by taking into consideration the change in solubility of hydrogen in liquid ammonia with changing temperature and pressure.[10] The NH_2^- concentration is directly proportional to the density of the solution; therefore, density was substituted for the NH_2^- concentration in the above expression for K_{rt}/K. As a fair approximation for the density of the solution we used the density of liquid ammonia.[8] The complete expression for reaction 1 was

$$\frac{K_{rt}}{K} = \frac{S}{S_{rt}} \frac{T}{T_{rt}} \frac{D_{rt}}{D} \left(\frac{P_{rt}}{P}\right)^{1/2}$$

The ratio K_{rt}/K is plotted against $1/T$ in Figure 7. It will be noted that all the points for several different amide concentrations fall fairly well on one straight line. From the slope we calculate $\Delta H° = -15.7$ kcal/mole for reaction 1.

The ratio of the equilibrium constants for reaction 2 is

$$\frac{K_{rt}}{K} = \frac{[e_{am}^-][Na^+]P_{rt}^{1/2}}{[e_{am}^-]_{rt}[Na^+]_{rt}P^{1/2}}$$

We approximated the terms involving the electron and sodium ion as shown above and obtained

$$\frac{K_{rt}}{K} = \frac{S}{S_{rt}} \frac{T}{T_{rt}} \frac{D}{D_{rt}} \left(\frac{P_{rt}}{P}\right)^{1/2}$$

This ratio is plotted against $1/T$ in Figure 8; the slope yields $\Delta H° = -12.3$ kcal/mole for reaction 3.

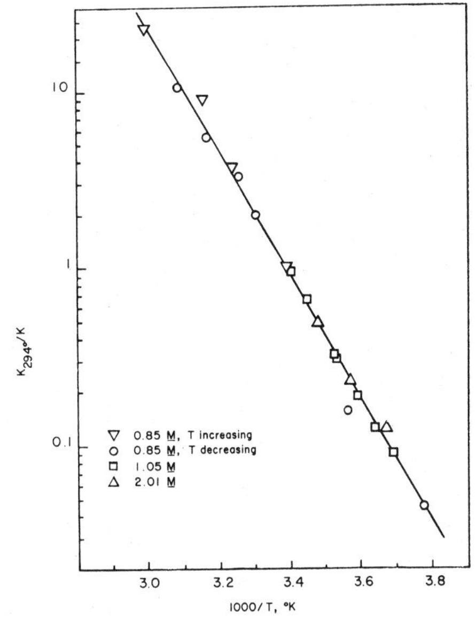

Figure 7.—Logarithm of the ratio $K_{294°}/K$ vs. 1000/T for solutions of KNH_2 in equilibrium with hydrogen.

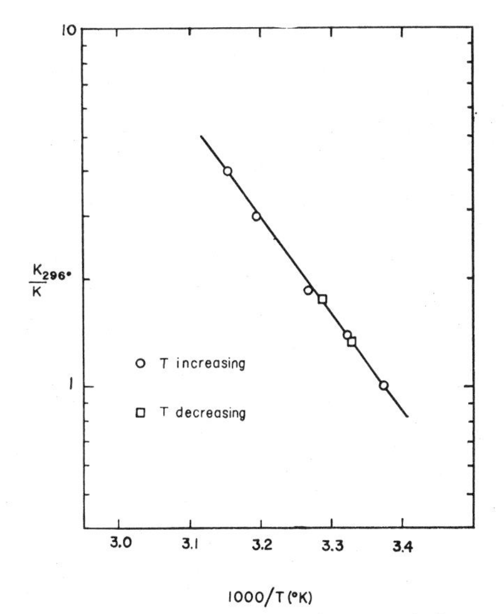

Figure 8.—Logarithm of the ratio $K_{296°}/K$ vs. 1000/T for a saturated solution of $NaNH_2$ in equilibrium with hydrogen.

Discussion

It seems certain that chemical equilibrium was achieved in our studies of reactions 1 and 3. For systems in which the amide concentration was held constant, the concentration of electrons (whether determined optically or by esr) was proportional to the square root of the hydrogen pressure.

Equilibrium quotients calculated from data for reaction 1 were found to increase markedly with increasing concentration of amide ion. However when these quotients were corrected to "constants" by use of estimated activity coefficients, the values were reasonably constant with changing amide concentration. The equilibrium constant at 25° corresponding to the most dilute solutions, $K = 5 \times 10^4$, is in fair agreement with that calculated indirectly from the known free energies of formation[2] of the species in reaction 1, $K = 1.2 \times 10^6$. Our experimental value of the heat for reaction 1, $\Delta H° = -15.7$ kcal/mole, is probably in reasonable agreement with that calculated indirectly from calorimetric data,[2] $\Delta H° = -11.5$ kcal/mole, particularly in view of the inaccuracies associated with heats determined from temperature coefficients. An equilibrium constant of 3×10^9 was calculated from data for reaction 3, using estimated activity coefficients. This value may be compared with the value 1.3×10^{10} calculated indirectly from the known free energies of formation of the species in reaction 3.[2,4] In view of the fact that our experimental value of the heat for reaction 3, $\Delta H° = -12.3$ kcal/mole, was obtained from data in a very narrow temperature interval, it is in remarkably good agreement with the heat calculated indirectly from calorimetric data, $\Delta H° = -11.7$ kcal/mole.[2,4]

Of the various species in reactions 1 and 3, that for which the free energy of formation is least accurately known is the electron, e_{am}^-. Using the tabulated[2,4] free energies for the other species and our directly determined equilibrium constants for reactions 1 and 3, we calculate the values 44.1 and 45.1 kcal/mole, respectively, for $\Delta F_f°(e_{am}^-)$ at 25°. We recommend use of the average value, 44.6 kcal/mole.

Acknowledgments.—This research was supported in part by the United States Atomic Energy Commission. The assistance and advice of Professor Rollie J. Myers and his associates is gratefully acknowledged. E. J. K. was supported by the U. S. Navy Postgraduate Educational Program.

“Chemical” Generation of the Ammoniated Electron

The reaction of hydrogen with the amide ion to yield the ammoniated electron, discussed in the preceding paper, is just one example of several methods known for preparing the ammoniated electron from chemicals other than electropositive metals. In the following paper, Salot and Warf report the generation of the ammoniated electron by the reaction of ytterbium(II) solutions with potassium amide:

$$Yb^{2N} + 3NH_2^{Y} \qquad Yb(NH_2/_3 + e^{-}_{am}$$

Analogous examples of the latter type of reaction are described in papers by Bergstrom[1] and Allbut and Fowles.[2]

---

[1]F. W. Bergstrom, *J. Amer. Chem. Soc.,* **47**, 2317 (1925).
[2]M. Allbut and G. W. A. Fowles, *J. Inorg. Nucl. Chem.,* **25**, 67 (1963).

[Reprinted from the Journal of the American Chemical Society, **90**, 1932 (1968).]

18

Chemical Generation of the Ammoniated Electron *via* Ytterbium(II)

Sir:

We have found that substantial concentrations of the solvated electron in liquid ammonia can be prepared by treating ytterbium(II) solutions with potassium amide.

When ytterbium metal was dissolved in liquid ammonia[1] and allowed to stand at room temperature until the blue color was discharged, the solid formed was found to be virtually insoluble in liquid ammonia and to contain paramagnetic Yb(III). The solid appeared to consist of white and rust-colored constituents, and to have an over-all N/Yb ratio of 2.8; we regard the solid as an ytterbium(III) imide–amide mixture.

In attempts to prepare pure $Yb(NH_2)_2$, a light yellow-orange solution of ytterbium(II) iodide (from Yb metal and NH_4I) was mixed at 25° in a pressurized glass system with pale yellow potassium amide solution. An intense blue solution resulted, regardless of whether the YbI_2 or KNH_2 was in excess. A white precipitate was also formed, which on analysis gave an N/Yb ratio of 3.5; we regard this as ammoniated $Yb(NH_2)_3$. Ytterbium(II) thiocyanate, substituted for the iodide, also gave the blue solution. The imide–amide mixture, from ytterbium metal and ammonia, was found to yield no blue coloration when treated with KNH_2 in liquid ammonia; this is in harmony with the presence of Yb(III) and the absence of Yb(II).

The evidence that the blue solutions genuinely were of the ammoniated electron is as follows: (1) when decanted into a separate bulb and evaporated, first the characteristic golden color of concentrated metal solutions appeared, followed by the silvery metal itself; (2) extraction of the blue solution with mercury yielded a potassium amalgam; (3) ytterbium in the tripositive state was a product; (4) we observed the absorption spectrum of the blue solution in the near infrared and found the asymmetric band peaking at 5400 cm^{-1} at 24°, which is characteristic of ammoniated electrons, as reported by Corset and Lepoutre;[2] and (5) the absorbance of the solutions decayed by zero-order kinetics as the e_{am}^- was converted to H_2 and NH_2^-, as demonstrated by Warshawsky.[3]

Europium behaved in a manner analogous to that of ytterbium. Samarium, when treated similarly, exhibited only a faint blue coloration, and we believe that this can be attributed to the presence of a europium impurity (1.5% spectrographically). The evidence is that when Sm is dissolved in NH_3 solutions of NH_4I, it is oxidized to the tripositive state.

Kirschke and Jolly[4] showed that the equilibrium constant for the reaction $0.5H_2 + NH_2^- = NH_3 + e_{am}^-$ is 2×10^{-5}. It is understandable, therefore, that Yb(II) and Eu(II), being stronger reducing agents than hydrogen, should produce the ammoniated electron by interaction with amide ion. The oxidation potentials are further enhanced by the exceedingly low solubility of the resulting rare earth triamides. The over-all reaction might be represented: $Yb_{am}^{2+} + 3NH_2^- = Yb(NH_2)_3 + e_{am}^-$. We do not yet have the equilibrium constants or potentials which apply.

(1) J. C. Warf and W. L. Korst, *J. Phys. Chem.*, **60**, 1590 (1956).
(2) J. Corset and G. Lepoutre, "Solutions Métal-Ammoniac," G. Lepoutre and M. S. Sienko, Ed., W. A. Benjamin, Inc., New York, N. Y., 1963, p 186 ff.
(3) I. Warshawsky, ref 2, p 167 ff.
(4) E. S. Kirschke and W. S. Jolly, *Inorg. Chem.*, **6**, 855 (1967).

Stuart Salot, James C. Warf
Department of Chemistry, University of Southern California
Los Angeles, California 90007
Received February 1, 1968

The Reaction of Water and Alcohols with the Ammoniated Electron

In view of the synthetic importance of metal–ammonia solutions as deprotonating agents, kinetic studies of the reaction of the ammoniated electron with protonic acids are of practical, as well as theoretical, importance. The following paper by Dewald and Tsina describes the kinetics and mechanism of the reaction of the ammoniated electron with water and *t*-butyl alcohol.

A Kinetic Study of the Reaction of Water and *t*-Butyl Alcohol with Sodium in Liquid Ammonia[1]

by Robert R. Dewald and Richard V. Tsina[2]

Department of Chemistry, Tufts University, Medford, Massachusetts 02155 *(Received June 3, 1968)*

19

The reaction of two weak acids, water and *t*-butyl alcohol, with sodium in liquid ammonia has been studied conductometrically at $-33.9°$. The kinetic data are consistent with the general mechanism: $ROH + NH_3 \underset{k_2}{\overset{k_1}{\rightleftharpoons}} NH_4^+ + RO^-$ and $NH_4^+ + e_{am}^- \xrightarrow{k_3} NH_3 + 0.5H_2$, where R is *t*-butyl and H. For *t*-butyl alcohol, k_1 was found to be $(2.5 \pm 0.7) \times 10^{-6}\ M^{-1}\ sec^{-1}$ and $k_2/k_3 = 1.3 \pm 0.3$. The values found for water were $k_1 = (7 \pm 2) \times 10^{-6}\ M^{-1}\ sec^{-1}$ and $k_2/k_3 = 0.55 \pm 0.16$. The bimolecular rate constant, k_3, is estimated to be $(4 \pm 2) \times 10^6\ M^{-1}\ sec^{-1}$ at $-33.9°$.

Introduction

Alcohols react readily with solutions of alkali metals in liquid ammonia with the evolution of hydrogen and the formation of alkoxides.[3,4] Kraus and White,[4] in work of a qualitative nature, reported that the reaction between sodium and ethanol in liquid ammonia was vigorous at first but rapidly slowed down, with the result that after 2.5 hr the reaction had gone to only 70% completion. They noted that extensive precipitation occurred. Later, Kelly, *et al.*,[5] reported kinetic data for the ethyl alcohol–sodium ammonia system. Using ammonia labeled with tritium, these workers followed the progress of the reaction by monitoring the volume and activity of the evolved gas (H_2 and HT). They found that the reaction was initially first order in EtOH and zero order in Na. Between 25 and 50% completion, the reaction order was found to be indefinite, but during the last 50%, the reaction was slower and followed second-order kinetics, first order in both metal and alcohol. Only one set of characteristic data, however, was reported. Recently, Jolly[6,7] evaluated these data using a Powell plot based on the presumed mechanism

$$ROH + NH_3 \underset{k_2}{\overset{k_1}{\rightleftharpoons}} NH_4^+ + RO^- \quad (1)$$

$$NH_4^+ + e_{am}^- \xrightarrow{k_3} NH_3 + 0.5H_2 \quad (2)$$

where R is ethyl. If the ammonium ion concentration is at a low steady-state value, the rate law

$$\frac{-d(e_{am}^-)}{dt} = \frac{k_1k_3(e_{am}^-)(NH_3)(ROH)}{k_2(RO^-) + k_3(e_{am}^-)} \quad (3)$$

is obtained. Jolly found that the mechanism, (1) and (2), was at least qualitatively compatible with the data reported. The best Powell plot fit to the data was obtained by setting $k_1 = 8 \times 10^{-3}\ sec^{-1}$ and $k_2/k_3 = 6 \times 10^3$. It was found initially,[6] however, that in the region $f = 0.10$–0.40, where f is the fraction of the reaction, the experimental points deviated unaccountably from the theoretical curve, but the fit was later[7] considerably improved when the formation of a complex between ethanol and the ethoxide ion was introduced. Jolly[7] pointed out that this assumption is quite arbitrary. One might further note that Kelly, *et al.*, employed relatively large concentrations of ethyl alcohol (0.2–0.4 *M*) and sodium (0.2–1.0 *M*), since fairly large amounts of gas were needed in order to perform an accurate analysis. The observations of Chablay[3] and Kraus[4] lead one to suspect that some precipitation of sodium ethoxide probably occurred,[7] the effect of which on the observed reaction kinetics is not *a priori* predictable.

We have studied the reaction kinetics of sodium with the two weak acids, water and *t*-butyl alcohol, with the aim of obtaining kinetic data in a system in which precipitation was not occurring. Water has been shown to react readily with sodium–ammonia solutions.[8] Pleskov[9] reported that solutions of sodium in liquid ammonia may be employed as a rapid method for the determination of small quantities of water. The ob-

(1) Presented in part before the Physical Chemistry Division at the 154th National Meeting of the American Chemical Society, Chicago, Ill., Sept 1967.

(2) This paper is part of a thesis submitted to Tufts University in partial fulfillment of the requirements for the degree of Doctor of Philosophy.

(3) E. Chablay, *Ann. Chim.* (Rome), **8**, 145 (1917).

(4) C. A. Kraus and G. F. White, *J. Amer. Chem. Soc.*, **45**, 768 (1923).

(5) E. J. Kelly, H. V. Secor, C. W. Keenan, and J. F. Eastham, *ibid.*, **84**, 3611 (1962).

(6) W. L. Jolly, "Non-Aqueous Solvent Systems," T. C. Waddington, Ed., Academic Press, New York, N. Y., 1965, p 39.

(7) W. L. Jolly, Advances in Chemistry Series, No. 50, American Chemical Society, Washington, D. C., 1965, p 27.

(8) W. C. Fernelius and G. W. Watt, *Chem. Rev.*, **29**, 195 (1937).

(9) V. A. Pleskov, *Zavodskaya Lab.*, **6**, 177 (1937); *Chem. Abstr.*, **31**, 61351 (1937).

served reaction, which is almost certainly due to solvated electrons, e_{am}^-, can be written

$$e_{am}^- + H_2O \longrightarrow 0.5H_2 + OH^- \qquad (4)$$

in which NaOH precipitates readily.[10] Chablay[3] noted that tertiary alkoxides are more soluble than secondary or primary alkoxides. Consequently it was felt that precipitation might best be avoided by using *t*-butyl alcohol as one of the weak acids in this study. By employing low reactant concentrations, we have successfully avoided precipitation during the course of the reaction.

Experimental Section

Ammonia (Matheson) was condensed *in vacuo* in a trap containing sodium metal. The blue solution was stored in this trap in contact with an iron magnet until the extent of the autodecomposition reaction

$$Na + NH_3 \longrightarrow NaNH_2 + 0.5H_2 \qquad (5)$$

was considerable. About half of the ammonia was distilled into a second trap which had been evacuated and flamed until the pressure stabilized at less than 2×10^{-6} torr. The ammonia was then distilled back into the trap containing the sodium. The blue solution was next frozen with liquid nitrogen and degassed. The second trap was again evacuated and flamed. The cycle of distilling back and forth, freezing, degassing, and flaming was repeated at least five times. The ammonia was then stored in contact with sodium, distilling enough for one or two experiments into the second trap just prior to use. Sodium (United Mineral and Chemical Co.) was distilled twice *in vacuo* and was stored in Pyrex capillaries. Ammonium bromide (Fisher reagent) samples were prepared by placing the salt in break-seal tubes which were sealed off under vacuum after evacuation to about 10^{-6} torr.

The reaction of the two weak acids with sodium was followed conductometrically in an apparatus similar to that described elsewhere.[11] The reaction vessel was constructed of Pyrex and had a calibrated bulb for volume determination. The conductance cell[11] and break-seal tubes containing samples of the reactant and solid NH_4Br were sealed onto two side arms. The electrodes were either gold-plated platinum or in a few cases gold-plated tungsten. The cells were calibrated with standard KCl solutions using the data of Jones and Bradshaw.[12] Either a high-precision ac bridge similar to that described elsewhere[13] or a Wayne Kerr Universal bridge B 221A was used. The procedure followed in measuring the resistance of the solution is also described elsewhere.[11]

The *t*-butyl alcohol (Fisher Certified) was first distilled through a 3-ft packed column, and the middle fraction was introduced into a trap. The alcohol was next degassed by repeated freezing and evacuation. It was then distilled *in vacuo* into a second trap into which sodium metal had been distilled. After the sodium had completely reacted, the alcohol was degassed by repeated freezing and pumping. The alcohol was distilled into tared break-seal tubes which were then sealed off under high vacuum and weighed. Water samples (starting with doubly distilled conductance water) were also prepared as described above, except the sodium metal was omitted. During the course of our investigation it became apparent that very small quantities of water would be necessary for some experiments (around 5×10^{-6} mol or less). These samples were prepared by equilibrating the water reservoir at an appropriate, thermostated temperature with evacuated bulbs of known volume on the vacuum line. The contents of the bulbs were then transferred to break-seal tubes, held at liquid nitrogen temperature, and then sealed under high vacuum.

The procedure followed for the sodium–ammonia solution preparation in the Pyrex apparatus is described elsewhere.[11] We feel that this technique is instrumental in removing traces of water from the walls of the apparatus as well as dissolved H_2 gas from the solution. After the blue solution had been thoroughly degassed, the apparatus was disconnected from the vacuum line and immersed in a refrigerated bath at $-33.9 \pm 0.2°$ (Dow Corning No. 200 silicone fluid was used as the bath liquid.) The resistance of the sodium–ammonia solution was subsequently monitored as a function of time for approximately 1 hr to ensure stability. Next, the break-seal containing the *t*-butyl alcohol or water sample was broken, the solution was mixed, and the resistance of the mixture was followed as a function of time.

Bright platinum electrodes were found to be unsuitable for making measurements after either the alcohol or water was added to the sodium–ammonia solution, probably because of contamination of the electrodes by hydrogen evolved during the reaction. The gold-plated electrodes, however, were found to be quite suitable for taking conductance measurements in the reacting systems. This observation is consistent with the findings of Windwer and Sundheim[14] in a study of the solutions of the alkali metals in ethylenediamine.

In the experiments with *t*-butyl alcohol, after the reaction rate became negligible (unbleached solution), the evolved H_2 gas was pumped through three liquid nitrogen traps using mercury leveling bulbs and was collected in a calibrated gas buret in which the volume and pressure of the gas were measured. During the

(10) M. Skossarewsky and N. Tchitchinadze, *J. Chem. Phys.*, **14**, 11 (1916).

(11) R. R. Dewald and J. H. Roberts, *J. Phys. Chem.*, **72**, 4224 (1968).

(12) G. Jones and B. C. Bradshaw, *J. Amer. Chem. Soc.*, **55**, 1780 (1933).

(13) G. E. Smith, Ph.D. Thesis, Michigan State University, 1963.

(14) S. Windwer and B. R. Sundheim, *J. Phys. Chem.*, **66**, 1254 (1962).

pumping, the blue solution was frozen and kept at liquid nitrogen temperature. Next, solid ammonium bromide was added, *via* a break-seal, to the blue solution, and the remainder of the H_2 gas was collected. Finally, the total amount of metal could be determined from the stoichiometry of the reactions

$$NH_4Br + Na \longrightarrow 0.5H_2 + NH_3 + NaBr \quad (6)$$

$$ROH + Na \longrightarrow 0.5H_2 + RONa \quad (7)$$

in which R is *t*-butyl. The initial sodium concentration, as determined from the evolved H_2, was in good agreement with the value determined from the initial resistance of the sodium–ammonia solution.

In this work, the conductivity measured for the reacting mixture was taken to be due to sodium alone, and the contribution to the conductance by either of the weak acids was considered to be negligible. Support for this procedure is apparent from the following data: for a 10^{-2} *M* solution of water in liquid ammonia at −33.9°, the specific conductance[15] is 2.45×10^{-7} ohm^{-1} cm^{-1} compared with 1.06×10^{-4} ohm^{-1} cm^{-1} for a 10^{-4} *M* sodium–ammonia solution[11] at the same temperature. Similarly, the contribution to the total conductivity due to alkoxides and hydroxides[10] produced was considered negligible in the initial phase of the reaction from which data were used in calculating rate constants. Conductance–concentration data obtained in this laboratory[11] for sodium–ammonia solutions were used in the calculations.

Results

a. The Reaction of t-Butyl Alcohol with Sodium in Liquid Ammonia. Table I contains a summary of the results obtained from the reaction of sodium with *t*-butyl alcohol in liquid ammonia at −33.9°. Figure 1 shows three sodium concentration *vs.* time plots. It should be noted that the alcohol is present in excess. As shown in Figure 1, the sodium reacts rapidly at first and then the reaction rate levels off to an almost negligible change with time, even though the solution remained blue. Other experiments using different initial concentrations gave similar results.

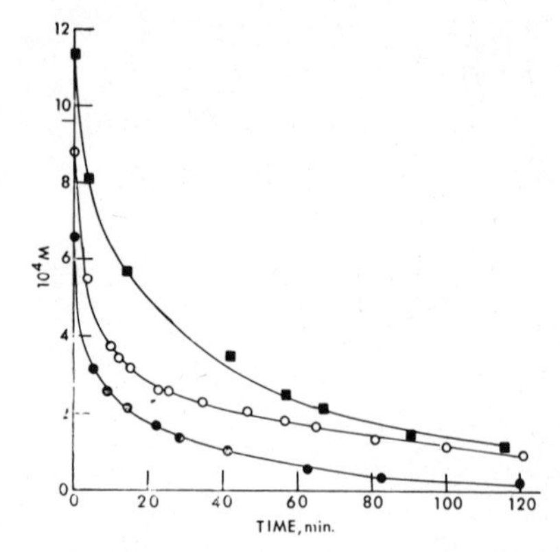

Figure 1. Sodium concentration, *M*, *vs.* time plots for the reaction of sodium with *t*-butyl alcohol at −33.9° in liquid ammonia: ■, run 114A; ○, run 28C; ●, run 113A.

When sodium amide was present, as determined by following the conductance of a decomposing sodium solution (for 24 hr), no large drop in the initial sodium concentration was observed upon addition of the alcohol. In this experiment the initial concentrations were about 1.2×10^{-3}, 4.2×10^{-4}, and 2.66×10^{-3} *M* sodium, sodium amide, and *t*-butyl alcohol, respectively. No precipitation was observed during the course of the reaction of sodium with *t*-butyl alcohol in any of the experiments. This observation is consistent with the reported greater solubility of tertiary alkoxides.[3]

b. The Reaction of Water with Sodium in Liquid Ammonia. Table II lists the kinetic data obtained for the reaction of sodium with water in liquid ammonia. It can be seen that the range of concentrations covered is indeed very large. In experiments 31C, 45B, and 46B (high concentrations) precipitation was

Table I: Kinetic Data for the Reaction of Sodium with *t*-Butyl Alcohol in Liquid Ammonia at −33.9°

Run no.	10^6k_1, M^{-1} sec^{-1}	k_2/k_3	10^3[Na], *M*	10^2[$(CH_3)_3COH$], *M*
11B	3.3	1.1	6.07	5.14
28C	1.7	1.8	0.882	1.54
30C	3.8	1.5	1.34	4.57
102A	1.7	1.3	5.69	4.43
104A	1.7	0.78	5.46	2.14
113A	3.1	1.6	0.657	1.85
114A	2.1	1.2	1.13	1.74
142A	2.3	0.97	0.405	0.805
Av	2.5 ± 0.7[a]	1.3 ± 0.3[a]	...	...

[a] Average error.

Table II: Kinetic Data for the Reaction of Sodium with Water in Liquid Ammonia at −33.9°

Run no.	10^6k_1, M^{-1} sec^{-1}	k_2/k_3	10^4[Na], *M*	10^3[H_2O], *M*
29B	4.0	0.63	0.400	0.268
38B	5.6	0.87	0.175	0.116
34C	6.5	0.49	0.714	0.118
35C	8.0	0.32	0.535	0.0757
27B	12	0.21	1.53	0.123
46B	*a*	0.76	82.3	89.3
45B	*a*	0.58	28.7	275
31C	*a*	0.55	84.7	18.8
Av	7 ± 2[b]	0.55 ± 0.16[b]	...	...

[a] Data insufficient to calculate k_1. [b] Average error.

(15) R. R. Dewald and R. V. Tsina, unpublished work

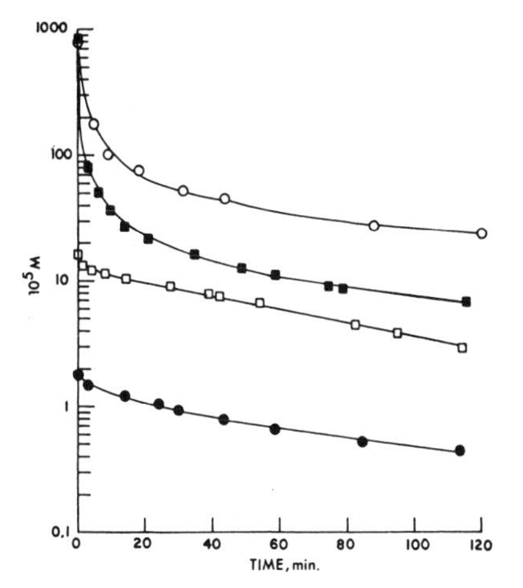

Figure 2. Semilogarithmic plots of sodium concentration, *M*, *vs.* time for the reaction of sodium with water at −33.9° in liquid ammonia: ○, run 46B; ■, run 31C; □, run 27B; ●, run 38B.

observed during the course of the reaction. Owing to the rapid initial slope change in the above three experiments, the data were insufficient to calculate k_1, although reasonable values of k_2/k_3 could be obtained. The five runs, 25C, 27B, 29B, 34C, and 38B, proceeded without observable precipitation and were used in calculating both k_1 and k_2/k_3. Figure 2 shows examples of semilogarithmic plots of the sodium concentration *vs.* time for the reaction of sodium with water in liquid ammonia at −33.9°.

Discussion

For both weak acids, water and *t*-butyl alcohol, the kinetic data can be best explained in terms of reactions 1 and 2. This mechanism is similar to that suggested for other systems as outlined by Russell.[16] For the scheme

$$\mathrm{ROH} \underset{k_2}{\overset{k_1}{\rightleftharpoons}} \mathrm{RO^-} + \mathrm{NH_4^+} \tag{8}$$

$$\mathrm{NH_4} + \mathrm{e_{am}^-} \xrightarrow{k_3} \mathrm{NH_3} + 0.5\mathrm{H_2} \tag{9}$$

in which k_1 pertains to a pseudo-first-order step, if the concentration of NH_4^+ is small so that at any time

$$\frac{-\mathrm{d(ROH)}}{\mathrm{d}t} = \frac{-\mathrm{d(e_{am})}}{\mathrm{d}t} = \frac{2\mathrm{d(H_2)}}{\mathrm{d}t} \tag{10}$$

the rate expression becomes

$$\frac{-\mathrm{d(e_{am})}}{\mathrm{d}t} = \frac{k_1k_3(\mathrm{ROH})(\mathrm{e_{am}^-})}{k_2(\mathrm{RO^-}) + k_3(\mathrm{e_{am}^-})} \tag{11}$$

Equation 11 may be rewritten in the form[17]

$$\frac{\mathrm{d}x}{\mathrm{d}t} = \frac{40k_1(a - x)(b - x)}{b - x + (k_2/k_3)x} \tag{12}$$

in which $a = (\mathrm{ROH})_i$, $b = (\mathrm{e_{am}^-})_i$, $x = (\mathrm{ROH})_i - (\mathrm{ROH}) = (\mathrm{RONa})$, and 40 is the molar concentration of the solvent, ammonia, in the pseudo-first-order step, eq 8. Ingold, *et al.*,[17] have shown that when V_2 and V_3 (V_n is the velocity of step n) are similar and $(\mathrm{RO^-})_i = 0$, eq 12 may be integrated to yield

$$40k_1t = \left[1 - \frac{a(k_2/k_3)}{a - b}\right] \ln \frac{a}{a - b} + \left[\frac{b(k_2/k_3)}{a - b}\right] \ln \frac{b}{b - x} \tag{13}$$

The above expression was used to evaluate k_1 and k_2/k_3 for the *t*-butyl alcohol–sodium and water–sodium runs in Tables I and II. It might be noted, however, that Ingold, *et al.*,[17] have pointed out that eq 13 is not very sensitive to the choice of k_2/k_3, a choice of 1.0 in some of their data being about as good as a choice of 1.2. The values of k_2/k_3 obtainable by this method therefore cannot be expected to be reliable to better than about 20%. Our precision ranges from 23% in the case of *t*-butyl alcohol experiments to 29% in the case of water experiments. In the latter case, the low concentrations needed to avoid precipitation are probably reponsible for the somewhat higher imprecision.

For the case of the partly decomposed sodium solution in which a substantial quantity of amide ion was initially present, it would be expected that the ammonium ion concentration would be small since

$$\mathrm{NH_4^+} + \mathrm{NH_2^-} \longrightarrow 2\mathrm{NH_3} \tag{14}$$

is known to proceed far to the right.[18] In this experiment a slow initial rate for the disappearance of sodium was observed upon addition of the alcohol. The slow reaction rate in this case would seem to substantiate the proposed two-step mechanism.

The equilibrium constant for the ammonolysis

$$\mathrm{H_2O} + \mathrm{NH_3} \rightleftharpoons \mathrm{NH_4^+} + \mathrm{OH^-} \tag{15}$$

in wet ammonia at −33.9° is unreported. We performed a conductance experiment[15] in which known quantities of water vapor were added to carefully purified ammonia (as determined from its conductivity), raising the water concentration from 3.56×10^{-5} *M* to a final value of 8.69×10^{-3} *M*. We were able to estimate an equilibrium constant at −33.9° for reaction 15 of about $(4 \pm 2) \times 10^{-12}$, which is comparable with the value estimated at 25° by Clutter and Swift.[19]

(16) G. R. Russell in "Techniques of Organic Chemistry," S. L. Friess, E. S. Lewis, and A. Weissberger, Ed., Vol. 8, Interscience Publishers, New York, N. Y., 1961, Part 1, p 383.

(17) E. D. Hughes, C. K. Ingold, S. Patai, and Y. Pocker, *J. Chem. Soc.*, 1230 (1957).

(18) H. Smith, "Organic Reactions in Liquid Ammonia," Interscience Publishers, New York, N. Y., 1960, p 39.

This result allows us to calculate a tentative value for the rate constant, k_3, of reaction 2. We find, using the k_1 and k_2/k_3 averages tabulated for the water–sodium reaction in Table II, a value of $k_3 = (4 \pm 2) \times 10^6$ M^{-1} sec^{-1} at $-33.9°$ in liquid ammonia. We may compare this result with the results found for the comparable reaction of the hydrated electron, e_{aq}^-, with the ammonium ion in water at zero ionic strength, as summarized by Rabani.[20] The rate constants reported vary from 1.1×10^6 to 1.8×10^6 M^{-1} sec^{-1} at ambient (presumably) temperature. Our estimate of k_3 appears to be in agreement with the work in the aqueous system[20] and is also consistent with the rate of the reaction between cesium and ethylenediammonium ions in ethylenediamine.[21]

Our kinetic data for the reaction of the two weak acids with sodium in liquid ammonia support the conclusion that the mechanism for reaction 4 is reactions 1 and 2.[22] The average values of k_1 and k_2/k_3 determined in the present work are in generally poor agreement with the comparable constants deduced from the data of Kelly, *et al.*,[5] by Jolly.[6,7] We feel that this discrepancy may be due to unavoidable precipitation occurring when ethyl alcohol and sodium react at the concentrations employed by the former workers. We are now reinvestigating the kinetics of the reaction of ethyl alcohol with sodium in liquid ammonia. Also, the value reported in this work for the rate constant of the ammonium ion-solvated electron reaction is within the range of the stopped-flow method, and an effort is now being made in this laboratory to measure this constant directly.

Acknowledgment. This research was supported by the National Science Foundation under Grant No. GP 6239.

(19) D. R. Clutter and T. J. Swift, *J. Amer. Chem. Soc.*, **90**, 601 (1968).

(20) J. Rabani, Advances in Chemistry Series, No. 50, American Chemical Society, Washington, D. C., 1965, p 242.

(21) L. H. Feldman, R. R. Dewald, and J. L. Dye, Advances in Chemistry Series, No. 50, American Chemical Society, Washington, D. C., 1965, p 163.

(22) R. R. Dewald and R. V. Tsina, *Chem. Commun.*, 647 (1967).

The Kinetics of Fast Reactions of the Ammoniated Electron

Many reactions of the ammoniated electron are too rapid to measure by conventional kinetic techniques. When it is desired to measure the rates of several closely related rapid reactions, it is often useful to carry out these reactions simultaneously in the same solution and to measure the *relative* rates of the competing reactions by determining the ratio of reaction products. Applications of this technique are described in the first two of the following three notes, i.e., those by Jacobus and Eastham and Dewald, Brooks, and Trickey. In the third note, Brooks and Dewald describe a direct stopped-flow determination of the rate constant for the fast reaction of NH_4^+ and e^-_{am}. It will be noted that their reported rate constant, $k = 1.2 + 10^6\ M^{-1}\ sec^{-1}$ at $-35°$, is quite different from the inaccurate value ($\sim 10\ M^{-1}\ sec^{-1}$) that Jacobus and Eastman took from an earlier study. Thus the quantitative conclusions of Jacobus and Eastman should be appropriately revised.

20

The Mechanism of Metal–Ammonia Reduction of Alkyl Halides

By JOHN JACOBUS* and JEROME F. EASTHAM

(*Department of Chemistry, Princeton University, Princeton, N.J., 08540; and Department of Chemistry, University of Tennessee, Knoxville, Tennessee)*

WHILE reduction of organic halides (RX) by alkali metals (M) in ammonia is an old, well known reaction,[1] and one whose mechanism has received some study,[2,3] the kinetics of these reductions have never been examined. In order to study this reaction we devised an apparatus which

$$RX + 2M + NH_3 \rightarrow RH + MX + MNH_2 \qquad (1)$$

would allow the controlled addition of blue metal-in-ammonia solutions from a burette to clear ammonia solutions of a variety of organic halides in a stirred reactor. The apparatus (burette and reactor) was vacuum-jacketed and the solutions were maintained at $-33°$.

With all halides studied, with one exception, the reductions were stoicheiometric according to equation (1) and were extremely fast, occurring as rapidly as the metal solution was added. The blue metal solution acted as its own coloured indicator. In other words, this colour disappeared in the reactor immediately, until an equivalent of metal solution had been added; with the next drop of metal solution, the solution in the reactor became a persistent blue. Saturated chlorides, bromides, and iodides and unsaturated fluorides, chlorides, bromides, and iodides all were reduced immediately. The one exception to this reduction being extremely fast was with saturated fluorides. Neither alkyl fluorides nor polyfluoroalkanes were reduced by the metal-in-ammonia solutions.

Because the reduction (1) of alkyl halides proved to be too fast to be amenable to kinetic analysis by common techniques, competitive reactions were employed to ascertain approximate rate constants. Liberation of hydrogen from ammonium salts [equation (2)] was found

$$NH_4X + M \rightarrow NH_3 + \tfrac{1}{2}H_2 \qquad (2)$$

to be of the same order of magnitude in reaction rate as the reduction. Collected in the Table are data obtained for the reduction by lithium of cyclohexyl chloride and bromide in liquid ammonia in the presence of ammonium salts as competitive reactants.

The reaction of $e^-(NH_3)$ with ammonium ion is diffusion-controlled.[4] As can be seen from the data in the Table (experiments 1—4), in competition with ammonium ion the relative rate of the reaction of cyclohexyl chloride is in the range 0·5—1·0, *i.e.* the rate constant is *ca.* $10^9 < k < 10^{10}$, while for cyclohexyl bromide the relative rate is in the range 0·1—0·2, *i.e.* the rate constant is *ca.* 10^9.

Competitive reductions[a]

Experiment number	1	2	3	4	5	6
X in $C_6H_{11}X$	Cl	Cl	Cl	Cl	Br	Br
X in NH_4X	Cl	Br	Cl	Br	Cl	Br
$C_6H_{11}X$	25·3	65·8	22·5	22·8	23·4	29·0
NH_4X	24·9	65·8	44·9	45·7	46·8	58·0
Li	48·7	131·1	45·4	47·3	46·7	60·6
H_2 (produced)	11·7	21·0	18·1	18·0	21·4	25·6
C_6H_{12} (produced)	12·2[b]	44·2[b]	4·6[b]	5·5[b]	2·9[c]	4·5[d]

[a] All values in millimoles. [b] Sole organic product except for recovered halide. [c] Product contained 0·2 mmole of cyclohexene plus recovered halide. [d] Product contained 0·25 mmole of cyclohexene plus recovered halide.

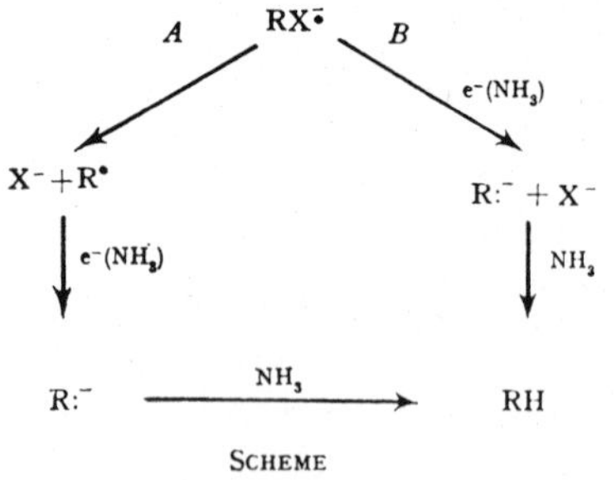

SCHEME

A priori, two mechanisms by which the reaction might proceed are a free-radical pathway (*A*) or a carbanionic pathway (*B*) (Scheme). In a series of papers, Wepster *et al.*[3] have considered both pathways and have concluded, mainly on stereochemical grounds, that the reduction proceeds *via* pathway *B*., since "A study of data in the literature which in reason must be assumed to relate to reactions proceeding *via* free radicals leads to the conclusion that free radicals always undergo racemization before any other reaction. . .can take place."[3a]

Bartlett *et al.* have demonstrated that free radicals can be trapped prior to loss of stereochemical integrity;[5] the 9-decalyl radical has been successfully trapped with oxygen ($k > 3 \times 10^7 M^{-1}$ sec.$^{-1}$). Garst[6] has shown that the reaction of hex-5-enyl radical with sodium naphthalenide (e^-) has $k > 10^8 M^{-1}$ sec.$^{-1}$. The reaction of alkyl carbanions with proton donating solvents such as NH_3 should be diffusion controlled.[7] In other words, it is reasonable that the reaction of an intermediate free radical with $e^-(NH_3)$ has a rate constant of the order 10^8—$10^{10} M^{-1}$ sec.$^{-1}$ to form a relatively configurationally stable carbanion and that protonation of the carbanion thus formed is also rapid; hence, the stereochemical results are not contrary to pathway *A*, the free radical pathway. We believe that the data of Wepster *et al.* can be interpreted as the first evidence of an optically active free radical. Experiments currently under investigation will distinguish between pathways *A* and *B*.

(*Received, September 10th*, 1968; *Com.* 1234.)

[1] See, for example, the review by H. Smith, "Organic Reactions in Liquid Ammonia" Part 2, Interscience, New York, **1963, pp. 196 ff.**
[2] A. Beverloo, M. C. Dielemann, P. E. Verkade, K. S. DeVries, and B. M. Wepster, *Rec. Trav. chim.*, **1962, 81**, 1033.
[3] P. E. Verkade, K. S. DeVries, and B. M. Wepster, *Rec. Trav. chim.*, **1964, 83**, (a) p. 367; (b) p. 1149.
[4] W. L. Jolly and L. Prizant, *Chem. Comm.*, **1968, 1345.**
[5] P. D. Bartlett, R. E. Pincock, J. H. Rolston, W. G. Schindel, and L. A. Singer, *J. Amer. Chem. Soc.*, **1965, 87, 2590.**
[6] J. F. Garst, P. W. Ayers, and R. C. Lamb, *J. Amer. Chem. Soc.*, **1966, 88, 4260.**
[7] D. J. Cram, "Fundamentals of Carbanion Chemistry," Academic Press, New York, 1965, p. 14.

21

Some Competitive Rate Studies of the Ammoniated Electron with Metal Ions: Evidence for Transient Intermediates

By R. R. DEWALD,* J. M. BROOKS, and M. A. TRICKEY

(*Department of Chemistry, Tufts University, Medford, Massachusetts* 02155)

Summary Competitive rate studies of the reaction of the solvated electron in liquid ammonia with metal ions indicate the formation of transient intermediates.

SINCE its identification, the rate of the reaction of the hydrated electron with inorganic cations, anions, and complex ions has been extensively studied.[1] Many of the reactions of inorganic substrates with hydrated electrons are extremely rapid, their reactions being essentially diffusion controlled.[1] On the other hand, little is known about the kinetic behaviour of ammoniated electrons, produced by the dissolution of alkali metals in liquid ammonia.

We have measured the rate of the reaction of the solvated electron with the ammonium ion in liquid ammonia by using the stopped-flow technique and taking advantage of the kinetic salt effect.[2] The rate constant was found to be $6 \pm 3 \times 10^6$ M^{-1} s^{-1} at $-35°$ and zero ionic strength. This value is in good agreement with the rate reported for the reaction of hydrated electrons with ammonium ions.[3] On the other hand, we found that the rate of the reaction of ammoniated electrons with silver ions (AgCN being rather soluble in liquid ammonia[4]) is too rapid to be studied by the flow technique at $-35°$. We then attempted to obtain information about the rate of the $Ag^+ + e^-_{am}$ reaction by competitive methods. In these experiments solutions containing about equal concentrations of NH_4I and AgCN (*ca.* 5×10^{-3} M) were mixed with a sodium–ammonia solution (e^-_m *ca.* 3×10^{-3} M) and the amount of H_2 produced was measured. Assuming the H_2 was only produced by reaction (1) then the ratio of the rates, k_2/k_1, based on the total H_2 yield was found to be about 1·2.

$$e^-_{am} + NH_4^+ \xrightarrow{k_1} NH_4 \rightarrow \tfrac{1}{2} H_2 + NH_3 \quad (1)$$

$$e^-_{am} + Ag^+ \xrightarrow{k_2} Ag^0 \quad (2)$$

However, this observation is not consistent with the results of the stopped-flow experiments which indicated that k_2 is considerably greater than k_1. Similar competitive experiments using solutions containing NH_4I and other added cations (HgI_2, CdI_2, PtI_4, ZnI_2, and ZrI_2) and assuming complete ammonolysis of the metal halides,[5] yielded ratios of the rates, $k(M^{n+} + e^-_m)/k(NH_4^+ + e^-_m)$, between 1 and 10. Moreover, the reactions of the above ions and their complexes with hydrated electrons have all been reported to be essentially diffusion controlled[1] with the ratio $k(M^{n+} + e^-_{aq})/k(NH_4^+ + e^-_{aq}) \simeq 10^4$. Our competitive kinetic results indicate that the amount of hydrogen gas from the competing systems is much greater than would be expected if the rates of reactions (1) and (3) differ by a factor of 10^4. However, if the species formed by reaction (3) is itself a

$$M^{n+} + e^-_{am} \xrightarrow{k_3} M^{(n-1)+} \quad (3)$$

reactive transient such that it reacts with ammonium ions, perhaps as in reaction (4), followed by reaction (5), or as in

$$M^{(n-1)+} + NH_4^+ \rightarrow M^{n+} + NH_4 \quad (4)$$

$$2\,NH_4 \rightarrow H_2 + 2NH_3 \quad (5)$$

the sequence suggested elsewhere,[6] then we can account for the amount of H_2 produced in these systems even though $k_3/k_1 \approx 10^5$, *i.e.*, reaction (3) being essentially diffusion controlled in liquid ammonia. The formation of Mn^+, Co^+, Ni^+, Zn^+, and Cd^+ in aqueous solution has been demonstrated[7] through their u.v. absorption spectra as products of the e^-_{aq} reactions with the bivalent metal ions. Chemically reactive transients of the lower oxidation states of gold[8] and silver[9] have been obtained by reducing solutions of Au^+ and Ag^+ with hydrated electrons. Therefore, we suggest that the results of our competitive kinetic studies might be viewed as chemical evidence for reactive transient species being formed by reduction of the metal cations by solvated electrons in liquid ammonia.

This work was supported by the National Science Foundation.

(*Received, June 8th,* 1970; *Com.* 882.)

[1] M. Anbar and J. Neta, *Internat. J. Appl. Radiation Isotopes*, 1965, **16**, 235; M. Anbar, *Quart. Rev.*, 1968, **22**, 578; M. Anbar and E. J. Hart in "Radiation Chemistry—I," Advances in Chemistry Series, No. 81, American Chemical Society, Washington, 1968, p. 79.
[2] R. R. Dewald and K. W. Browall, *Chem. Comm.*, 1968, 1511.
[3] J. Jortner, M. Ottolenghi, J. Rabini, and G. Stein, *J. Chem. Phys.*, 1962, **37**, 2488.
[4] G. Jander, "Chemie in wasserfreiem flüssigem Ammoniak," Interscience, New York, 1966, p. 162.
[5] G. W. A. Fowles, *Progr. Inorg. Chem.*, 1964, **6**, 1.
[6] W. L. Jolly and L. Prizant, *Chem. Comm.*, 1968, 1345.
[7] J. H. Baxendale, E. M. Fielden, and J. P. Keene, *Proc. Roy. Soc.*, 1965, *A*, 286, 320; J. P. Keene, *Radiation Res.*, 1964, **22**, 1; J. H. Baxendale in "Pulse Radiolysis," eds. M. Ebert, J. P. Keene, A. J. Swallow, and J. H. Baxendale, Academic Press, New York, 1965, p. 15; G. E. Adams, J. H. Baxendale, and J. W. Boag, *Proc. Chem. Soc.*, 1963, 241.
[8] J. H. Baxendale, E. M. Fielden, and J. P. Keene, ref. 7, p. 207.
[9] A. S. Ghosh-Mazumdar and E. J. Hart, ref. 3, p. 193.

22

Rate of the Reaction of the Ammoniated Electron with the Ammonium Ion at −35°[1]

by J. M. Brooks and R. R. Dewald*

Department of Chemistry, Tufts University, Medford, Massachusetts 02155 *(Received December 7, 1970)*

Publication costs assisted by the National Science Foundation

Kinetic studies of the reactions of sodium with weak acids in liquid ammonia have been interpreted in terms of the following scheme

$$HA + NH_3 \underset{k_2}{\overset{k_1}{\rightleftharpoons}} NH_4^+ + A^- \quad (1)$$

$$e_{am}^- + NH_4^+ \xrightarrow{k_3} NH_3 + {}^1/_2H_2 \quad (2)$$

where HA represents water,[2] urea,[3] hydrazine,[4] ethanol,[5] and *tert*-butyl alcohol.[2] Analyses of the kinetic data for these reactions yield values for k_1 and k_2/k_3. If the equilibrium constants were known with certainty,[2,5-8] then one could calculate k_2 and k_3. However, estimates of k_3 have varied from 10^6 to 10^{10} M^{-1} sec^{-1}.[2,3] We have measured k_3 directly at −35° by reaction of solutions of sodium with ammonium bromide in liquid ammonia using the stopped-flow technique.

Experimental Section

The thermostated stopped-flow apparatus used in this study has been described elsewhere.[9] All operations including solution make-up, solution transfer, rapid mixing, and observation were performed *in vacuo*. The progress of the reaction was followed by monitoring the absorbance decay of the solvated electron at 1000 nm where the extinction coefficient is about 10^4.[10] The path length of the observation tube was calibrated by comparing the transmittance of acetone solutions of Kodak Q-switching dye no. 9740 as determined on a Perkin-Elmer 450 spectrophotometer and on the flow apparatus. The effective path was about 0.85 mm for a nominal diameter of 1 mm.

Ammonia was purified using procedures described elsewhere[11] and the concentration of the metal–ammonia solutions was determined conductometrically using methods and data published elsewhere.[12,13] Solutions of ammonium bromide (Fisher, reagent) were prepared by adding an appropriate amount of the aqueous salt solution to the make-up vessel followed by evaporation of the water on the vacuum line. In some experiments sodium bromide (J. T. Baker, USP, experiments 5–7; Alfa Inorganics, ultrapure, experiments 17, 18) or sodium iodide (Matheson Coleman and Bell, reagent, experiments 2–4; Alfa Inorganics, ultrapure, experiment 18) was added to the ammonium bromide solution in order to take advantage of the kinetic salt effect to slow the reaction rate.

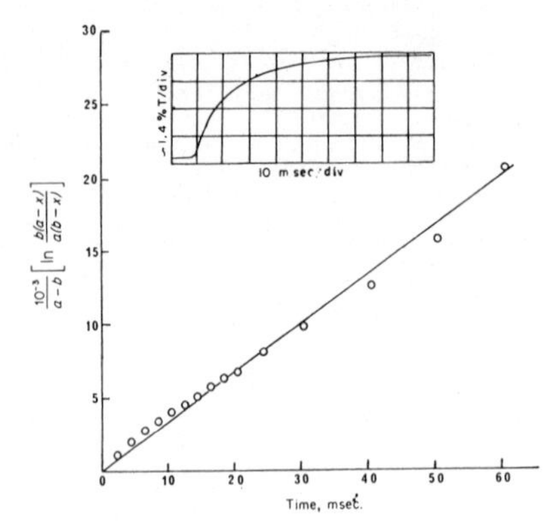

Figure 1. A typical oscilloscope trace and a plot of $[1/(a - b)]$ ln $[b(a - x)/a(b - x)]$ *vs.* time for experiment no. 17.

Results and Discussion

If the reaction is first order with respect to each of the reactants then a plot of $1/(a - b)$ ln $[b(a - x)/a(b - x)]$ *vs.* t should give a straight line with zero intercept, where a and b are the initial concentrations of ammonium ions and ammoniated electrons and x is the electron concentration at time t. The linearity of a typical second-order plot is shown in Figure 1. In experiments in which the ammonium ion concentration was sufficiently greater than the sodium concentration, the appropriate pseudo-first-order plots were also linear. In all cases the plots were linear for at least 3 half-lives or about 88% of the reaction, indicating that the reaction is second order overall. The value of the observed k_3, eq 2, was computed for each experiment. Tables I and II summarize the rate experiments.

Comparison of ammoniated electron concentration obtained from conductivity data with the concentration estimated spectrophotometrically provides a good estimate of the amount of decomposition of the sodium–ammonia solution that occurred during solution trans-

(1) Presented in part before the Physical Chemistry Division at the 160th National Meeting of the American Chemical Society, Chicago, Ill., Sept 1970.

(2) R. R. Dewald and R. V. Tsina, *J. Phys. Chem.*, **72**, 4520 (1968).

(3) W. L. Jolly and L. Prizant, *Chem. Commun.*, 1345 (1968).

(4) J. Belloni, *Int. J. Radiat. Phys. Chem.*, **1**, 411 (1969).

(5) R. R. Dewald, "Metal-Ammonia Solutions, International Conference," Butterworths, London, 1970, p 193.

(6) M. Herlem, *Bull. Soc. Chim. Fr.*, 1687 (1967).

(7) D. R. Clutter and T. J. Swift, *J. Amer. Chem. Soc.*, **90**, 601 (1968).

(8) M. Alei and A. E. Florin, *J. Phys. Chem.*, **73**, 857 (1969).

(9) R. R. Dewald and J. M. Brooks, *Rev. Sci. Instrum.*, **41**, 1612 (1970).

(10) M. Gold and W. L. Jolly, *Inorg. Chem.*, **1**, 818 (1962).

(11) R. R. Dewald and J. H. Roberts, *J. Phys. Chem.*, **72**, 4224 (1968).

(12) R. R. Dewald, *ibid.*, **73**, 2615 (1969).

(13) R. R. Dewald and K. W. Browall, *Chem. Commun.*, 1511 (1968).

Table I: Summary of the Kinetic Data for the Reaction of the Ammoniated Electron with the Ammonium Ion in Liquid Ammonia at −35° (Experiments with No or Low Concentration of Added Salt)

Expt no.	$10^4[NH_4Br]$, M	$10^6[e_{am}^-]$, M	10^3[salt], M	$10^{-6}k_{obsd}$ M^{-1} sec^{-1}	$10^{-6}k_{corr}$ M^{-1} sec^{-1}
6	2.7	12.0[a]	67.0[b]	1.6	1.4
12	0.38	3.4	None	11.0	1.4
15	1.6	5.0	None	9.3	1.3
16	2.0	1.8	None	6.2	0.91
17	2.0	2.0	5.3[b]	2.8	0.95
18	2.0	3.4	5.8[b]	1.5	1.2
19	2.0	3.2	6.3[c]	3.6	1.4
Av					1.2 ± 0.2[d]

[a] Cs metal; in all other experiments, Na was used. [b] NaBr. [c] NaI. [d] Standard deviation.

Table II: Summary of the Kinetic Data for the Reaction of the Ammoniated Electron with the Ammonium Ion in Liquid Ammonia at −35° (Experiments with High Concentrations of Added Salt)

Expt no.	$10^4[NH_4Br]$, M	$10^4[e_{am}^-]$, M	10[salt], M	$10^{-6}k_{obsd}$, M^{-1} sec^{-1}	$10^{-6}k_{corr}$, M^{-1} sec^{-1}
2	5.8	0.68	2.0[b]	3.3	5.9
3	5.6	0.77	2.1[b]	2.2	4.0
4	5.1	0.51	2.0[b]	3.1	5.6
5	5.0	2.6	3.6[c]	1.5	2.7
7	5.8	1.5[a]	4.2[c]	1.1	2.1
Av					4.1 ± 1.7[d]

[a] Cs metal; in all other experiments, Na was used. [b] NaI. [c] NaBr. [d] Standard deviation.

fer. This decomposition was found to be sufficiently small such that it did not affect the values of the observed rate constant, *i.e.*, within experimental error (±20%).

The observed rate constants (k_{obsd}) were corrected for the kinetic salt effect[13,14] according to

$$\log (k_{cor}/k_{obsd}) = 9.77\mu^{1/2}/(1 + 3.14\mu^{1/2}) \qquad (3)$$

Equation 3 was derived from Brönsted–Bjerrum theory of ionic reaction combined with the extended Debye–Hückel theory assuming a mean value for the distance of closest approach of 4.5 Å, a density of 0.6825 g/ml, and a dielectric constant of 21.8 for liquid ammonia at −35°. The ionic strength was estimated by using an ion-pairing dissociation constant of 2.9×10^{-3} for NaBr[15] and 5.6×10^{-3} for NaI. The value used for NaI was estimated by using the Fuoss theory.[16]

The corrected values of the rate constant given in Table II (high added salt concentration) vary from about 2 to 6×10^6 M^{-1} sec^{-1}, while the values given in Table I (little or no added salt) are fairly consistent with an average value of about 1.2×10^6 M^{-1} sec^{-1}. We feel that the discrepancy between the values given in the two tables probably results from the fact that the equations and constants derived or measured for dilute solutions cannot be expected to yield very reliable results in the high concentration region. It should also be noted that the added salts in the experiments listed in Table II might contain sufficient impurities that could result in an increase in the observed rate. Therefore, we feel that the best value for the corrected second-order rate constant is the value given in Table I, *i.e.*, $1.2 \pm 0.2 \times 10^6$ M^{-1} sec^{-1} at −35°.

The simplest interpretation of our results assumes that the rate-limiting step is a → b in the scheme

$$\overset{a}{(e_{am}^-) + NH_4^+} \longrightarrow \overset{b}{NH_4} \dashrightarrow \overset{c}{{}^1/_2H_2 + NH_3} \qquad (4)$$

The ammonium radical might be expected to react further as

$$2NH_4 \longrightarrow 2NH_3 + H_2 \qquad (5)$$

The second-order rate constant obtained in this study is in good agreement with the value reported for the reaction of the hydrated electron with the ammonium ion in aqueous solutions,[17] the value estimated by Dewald and Tsina,[2] and the value determined from pulse radiolysis studies in liquid ammonia.[18]

Acknowledgement. This research was supported by the National Science Foundation.

(14) G. V. Buxton, F. S. Dainton, and M. Hammerli, *J. Chem. Soc.*, 1191 (1967).

(15) V. F. Hnizda and C. A. Kraus, *J. Amer. Chem. Soc.*, **71**, 1565 (1949).

(16) R. M. Fuoss, *ibid.*, **80**, 5059 (1958).

(17) M. Anbar and P. Neta, *Int. J. Radiat. Isotopes*, **18**, 493 (1967)

(18) J. L. Dye, M. G. de Backer, and L. M. Dorfman, *J. Chem. Phys.*, **52**, 6251 (1970).

Physical Properties and Models of Metal–Ammonia Solutions

II

Franklin's Esteem of Cady

During the first half of this century, Franklin and Kraus probably did more to elucidate the chemistry of liquid ammonia solutions than everybody else combined. It is not well known that their work was prompted by the work of Cady, a remarkably modest man who chose not to continue working in this field. To help clarify the early history of liquid ammonia research in America, we present an excerpt from the introduction of Franklin's book *The Nitrogen System of Compounds.*[1]

[1]This and the following excerpt are the only exceptions to the chronological order of the publications.

23

The Nitrogen System of Compounds

E. C. FRANKLIN

In order to give credit to a man for what may properly be regarded as a brilliant idea the following bit of history is interjected here. In the autumn of 1896 Hamilton P. Cady, then an undergraduate, and now Professor of Chemistry at the University of Kansas, was working at the regulation course in quantitative analysis. Observing after a time that the young man was becoming bored with his task, the writer, at the time giving instruction in analytical chemistry, proposed to him that he prepare several of the cobaltamine salts and confirm the composition of one or two of them by analysis. Some days later with a beautifully crystallized specimen of one of these interesting salts in his hand Cady stated that the ammonia in these and other salts containing ammonia must function in a manner very similar to that of water in salts with water of crystallization.[2] He suggested furthermore that liquid ammonia would probably be found to resemble water in its physical and chemical properties. As a direct consequence of Cady's suggestion has followed all the work done in this country on liquid ammonia.

[2] The parallelism between water of crystallization and ammonia of crystallization was noted many years ago by Robert Kane and by Rose [*Ann. Chim. Phys.*, 72, 342, 346 (1839); Remsen, "Inorganic Chemistry," 2d. Ed., 593 (1890)].

Cady's Account of His Own Work

Cady and Franklin were guests of honor at a dinner of the Division of Physical and Inorganic Chemistry on April 15, 1936, during a meeting of the American Chemical Society in Kansas City, Mo. We have reprinted an excerpt from the address given by Cady on that occasion. This excerpt gives further insight into the early work in ammonia chemistry.

24

H. P. CADY

My father died when I was eleven. My mother was wonderful. She had five children, and though my father's estate netted nothing to anyone but the lawyers, she gave us each a good education by taking boarders and teaching music. As you can see there was very little money in our house. But the neighbors had lawns — large ones — which I could mow for a dime. These went for chemicals and books.

I found an old chemistry book of my grandfather's, a Comstock, which in some ways was more detailed than the Youman, but both were sketchy. I added a very much used copy of Wurtz's Chemistry, a good book, covering inorganic, organic, and the beginnings of physical chemistry. Some wise friend, I think Professor Jewett of Oberlin, advised me to get Remsen's *Inorganic Chemistry* and Prescott and Johnson's *Qualitative Analysis*. These were wonderful books. And I really learned them almost literally word for word. I subscribed for the *American Chemical Journal*. It cost me a lot, $4.00 a year, so it had to be read. I did read and reread each article until I felt I understood it. Among them was a paper by Remsen on double halides of heavy metals and the alkalies, with special reference to their water of crystallization. Never since that time have I lost interest in this type of compounds. In his advanced chemistry, Remsen pointed out that ammonia forms many compounds with salts corresponding closely to those of water with the same salts.

When I came to the University of Kansas in '94, I read everything available on water of crystallization and ammonia and its compounds, with special attention to liquid ammonia and the ammonia of crystallization. I made many of the compounds just to see what they looked like. The great similarity in every way of water and ammonia impressed itself more and more strongly upon me. I finally laid my information before Franklin with a suggestion that liquid ammonia might reasonably be expected to be an ionizing solvent.

Franklin was instantly all enthusiasm. He persuaded Professor Bailey to invest a large part of the meager fund available to the chemistry department for research, in a cylinder of liquid ammonia. He made me some vacuum jacketed test tubes. There were no vacuum jacketed vessels on the market anywhere in the world in those days.

Most unfortunately for me, Franklin was under

contract to go to Central America for a gold mining company and left before the ammonia arrived.

It was a heart-trying moment when I drew out some liquid ammonia into one of the vacuum tubes and tested it to see if it was a conductor of electricity. Finding the resistance to be very high, as I expected, I then dropped in a piece of potassium iodide, which Gore had shown to be soluble. I shall never again experience a thrill equal to that which I felt when the solution became a good conductor and electrolysis took place!

Just imagine! Here was a chain of reasoning extending back through the years, brought to a successful conclusion. I had at last found something the world did not know, and I had added a second to its list of ionizing solvents!

I was young; sleep was unimportant. Before the next day's duties began I had found out a lot and you know the joy I had in doing it. If you ever read my first paper on ammonia, I hope you will remember it was done by an undergraduate without direction, for Franklin was away and Professor Bailey's training was on entirely different lines. The equipment was crude, but the best Professor Bailey could give me. Library facilities were also non-existent. I was in deep water, but swam out the best I could.

However, I did block out the field, and I do not know of any major errors of conclusion that I made.

The two following years I spent at Cornell. I would have been glad to continue to work with ammonia, but Bancroft was interested in the phase rule, so that became my weakness, too.

Franklin returned to the University of Kansas, asked, and of course received, permission to work on ammonia. He and Kraus did a lot of beautiful work. After my return to the University, Franklin and I worked together for a time. I've never been happier. Franklin is the best chemist, and the dearest man! It was a sad day for me when he told me he was going to Stanford. When he actually had gone and I was moving my things into his office desk, as I laid his old pipes to one side in a drawer where they would be ready for him when he came back to visit, I felt as I did when my mother handed me my father's watch; and I said to myself:

"Franklin, you enjoy the ammonia work so much I am going to leave the field to you and find something else." I don't think I ever told Franklin this. He jumped me many times for not working on it more. And I do occasionally get back to certain phases of it, but Franklin and Kraus have done such splendid work that there is little excuse for mine.

7

The Electrolytic Behavior of Liquid Ammonia Solutions

The following paper by Cady describes what was perhaps the first physical chemical study of liquid ammonia solutions. Cady's conclusions regarding the nature of metal–ammonia solutions were remarkably astute.

25

THE ELECTROLYSIS AND ELECTROLYTIC CONDUCTIVITY OF CERTAIN SUBSTANCES DISSOLVED IN LIQUID AMMONIA

BY HAMILTON P. CADY

The remarkable similarity existing between certain crystalline salts containing water of crystallization and similar salts containing ammonia, has suggested the idea that perhaps water and ammonia might be analogous in some of their other properties. Acting on this suggestion it was thought best to test the dissociative power of liquid ammonia, on dissolved substances.

The only record that could be found of any work bearing directly on this subject, was a statement by Dr. L. Bleekrode,[1] that liquid ammonia was a good conductor of electricity, and that while the current was passing the liquid turned blue, and became colorless again when the current ceased, Weyl,[2] and G. Gore,[3] investigated the action of various substances, dissolved in liquid ammonia, upon each other, but they seem not to have tried the conductivity of such solutions.

The liquid anhydrous ammonia used in the following experiments, was the ordinary commercial ammonia used in the manufacture of ice, and proved to be of sufficient purity for the purpose. The experiments were made in vacuum-jacketed test tubes, which were made in our laboratory by Prof. E. C. Franklin, and answered admirably for the purpose. This may be illustrated by the fact that 15 cc. would not all evaporate inside of three hours, at the ordinary temperature. The test tubes were fitted with a doubly-bored cork stopper, through one of the holes of which the ammonia was led in,

[1]Phil. Mag. [5] **5**, 384 (1878).
[2]Pogg. Ann. **121**, 601 ; **123**, 350 (1864).
[3]Proc. Roy. Soc. **20**, 441 (1872).

while the uncondensed gas was allowed to escape through a long glass tube, inserted into the other opening and bent at right angles. This arrangement protected the ammonia from moisture while it was being drawn from the cylinder. During the experiment the ammonia was protected by a calcium chloride tube, filled with soda lime.

The electrodes used in the experiments on electrolysis, were simply platinum plates held in position by glass rods, fused to the leading-in wires. When pure ammonia was submitted to the action of a battery of six storage cells having a potential of 12 Volts, no signs of a current could be detected. Even with a potential of 110 Volts and electrodes 1 cm. apart, and having an area of 25 square cm., only a few hundredths of an ampere passed through. The liquid did not turn blue, but simply boiled vigorously. If however a small quantity of a soluble salt be added to the ammonia the solution becomes an excellent conductor, and in the case of the sodium or potassium salts, the solution turns blue, if the 110 volt current be used, but it becomes colorless again when the current is shut off.

This furnishes an explanation for the difference between these results and those obtained by Dr. Bleekrode, l. c. His ammonia was prepared as follows: «The chlorides of silver and calcium were saturated with the gas; and with them quicklime and sodium were enclosed in the condensation tube, in order, on expelling the gas by heating, to remove the last traces of water. The separated liquid ammonia was several times poured back over the sodium, by inverting the tube, and redistilled». Evidently the ammonia had become contaminated with some of the sodium hydroxide, formed by the action of water on sodium. This makes a solution that conducts readily and the sodium set free at the cathode dissolves, forming a bright blue solution, the color disappearing when the current is shut off. Dr. Bleekrode ascribed the blue coloration to the formation of free ammonium, NH_4, according to the reaction $2NH_3 = NH_4 + NH_2$.

Weyl l. c. describes the formation of ammonium by the action of a solution of sodium upon ammonium chloride or sulphate. He obtained a bright blue liquid that rapidly decomposed into ammonia and hydrogen. Since the solution of sodium in ammonia is also bright blue, it would appear to be difficult to say just when the color was due to sodium and when it was caused by free ammonium. If an excess of ammonium bromide or iodide, both of which are very soluble in ammonia, be added to a solution of sodium in that menstrum, the blue color is instantly destroyed, and at the same time hydrogen is evolved. With ammonium chloride or sulphate the action is the same only slower, as these salts are not very soluble in ammonia.

If an electric current of 110 volts be passed through a solution of an ammonium salt in ammonia, there is a violent evolution of gas, but no signs of a blue coloration. In neither of these experiments could any evidence of a blue coloration that might be ascribed to the presence of free ammonium, be detected.

If a current of varying strength, up to 110 volts be passed through solutions of salts of silver, copper, or barium, the metals are deposited upon the cathode, but there is no sign of a blue color. If sodium be added to a solution of one of these salts the metal is precipitated but no blue color appears until an excess of sodium is present. Weyl, thought that he had obtained a blue solution of ammonium compounds of these metals by the action of sodium solution on their chlorides according to the reaction—

$$N_2H_6Na_2 + 2NH_3 + BaCl_2 = N_2H_6Ba + 2NH_3NaCl.$$

According to Weyl these ammonium compounds easily decompose into ammonia and the metal. Their existence is not confirmed however by the above experiments. The blue color observed by Weyl was due perhaps to the presence of free sodium. In every case he used chlorides of the metals, and these being only slightly soluble the action of the sodium would be relatively slow.

If potassium iodide be dissolved in ammonia and a current of from six to twelve volts be passed through the solution, electrolysis takes place, hydrogen is evolved and a deposit collects on each electrode. On the cathode the deposit is dark gray and reacts violently with water, giving a hissing sound and ammonia and potassium hydroxide are formed. The deposit probably consisted of potassium amide KNH_2, but a sufficient quantity for analysis could not be obtained. If the current be passed in series through the potassium iodide solution and through an apparatus for the generation of electrolytic gas, the ratio of the hydrogen from potassium iodide solution, to the mixed electrolytic gas was not that required by theory, but the volume of hydrogen was always too small. That the gas was hydrogen was proven by exploding it with oxygen.

The deposit on the anode varied from bluish black to olive green. It was insoluble in water but soluble in ether, alcohol and chlroform, and in potassium iodide solution, in each case with the evolution of gas. It explodes violently on heating, by friction and on contact with acids. The same substance was formed in the electrolysis of several other iodides. It was impossible to obtain enough of this compound to use for analysis, but it agrees closely in physical and chemical properties with the compound formed by the action of iodine on liquid ammonia, and which has the composition HN_3I, as noted in a previous article.[1]

Mercuric iodide dissolves easily in ammonia. The solution conducts readily, and the products of electrolysis are mercury and the explosive compound of iodine noted above.

FIG. 1 FIG. 2

Silver nitrate is also readily soluble and the solution is a good conductor. Metallic silver is deposited on the cathode.

Lead nitrate is easily soluble and the solution is a moderately good conductor, while metallic lead is deposited.

The behavior of a solution of sodium in ammonia is quite remarkable; the solution is an excellent conductor. There is no deposit on the electrodes; no gas is evolved, and the blue color of the sodium is not altered by the passage of an enormous quantity of electricity. If only a little sodium is present the color becomes more intense around the cathode. There is no polarization current.

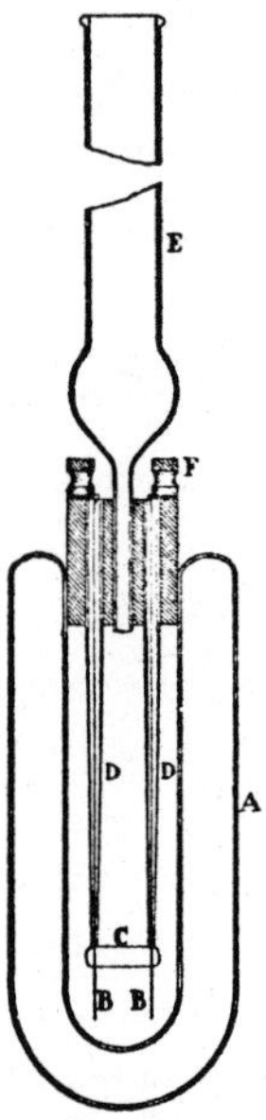

In the experiments on conductivity the following apparatus was used :—

"A" is a vacuum jacketed test-tube. It is graduated so that

[1] Kan. Univ. Quar. Vol. VI, No. 2, p. 71 (1897).

the volume of the ammonia can be read directly. B B are the electrodes, consisting of a stout platinum foil welded to strong platinum wires and held in position by the glass rod C which is fused on to the wires. The wires that conduct the current are insulated by the slender glass tubes D D, the lower ends of which are fused on to the wires, and the latter are connected with the binding posts F F. E is a calcium chloride tube filled with soda lime. Fig. 2 is a side view of the electrodes and connections.

The results given below are preliminary and are not regarded as possessing any very great accuracy; they are however close enough to settle the question as to whether or not ammonia possesses the power of dissociating dissolved substances to a degree comparable with water. The chief sources of error that have not been fully overcome, are the impossibility of preventing boiling at the electrodes which increases the apparent resistance and consequently lowers the molecular conductivity; the difficulty of accurately controlling the dilution on account of the evaporation of the ammonia; and finally the slight impurity of the ammonia. This latter of course increases the error in the determination of substances that are only slightly dissociated.

The presence of a small amount of water does not seem to have a measurable effect on either the conductivity of ammonia alone or of solutions of substances dissolved therein. The ammonia used had a conductivity of 71×10^{-7}.

Mercuric nitrate is not completely soluble, as a yellow compound resembling basic nitrate separates out, so no determination could be made.

In the case of the sodium solution, no signs of a polarization current could be detected with a sensitive galvanometer, and it will be noticed that the molecular conductivity rises with the concentration, contrary to that of electrolytes. As has been mentioned above no gas is given off when the solution is submitted to electrolysis. The sodium is not affected by the current, nor is there any deposit on the electrodes. In spite of the fact that the sodium is in solution, the solution seems to conduct like a metal and not like an electrolyte.

In this connection it is interesting to note that J. J. Thomson has shown that the vapor of sodium is a conductor, a result that we had also obtained before learning of his researches.

Salt used	Dissolved in ammonia at -34° V	μ	Dissolved in water V	μ	
Potassium chloride	Not sufficiently soluble				
Potassium iodide,	80.	169.	100.	116.	at 18°
	100.	178.			
	110.	179.			
Potassium bromide,	100.	169.	128.	117.	at 18°
	120.	179.			
	135.	181.			
Potassium nitrate,	80.	123.	100.	114.	at 18°
	100.	124.			
	120.	131.			
Ammonium chloride,	40.	96.5	55.7	105.	at 18°
	50.	98.5			
	55.	99.			
	61.5	103.			
Ammonium bromide,	40.	124.			
	50.	132.			
	57.5	143.			
	67.5	144.			
Ammonium iodide,	40.	146.			
	50.	155.			
	60.	173.			
Sodium bromide,	140.	154.	128.	115.8	at 25°
	150.	158.			
Sodium iodide,	150.	166.	128.	112.3	at 25°
Silver nitrate,	140.	147.	166.	103.3	at 18°
Mercuric iodide,	150.	102.			
Mercuric cyanide,	130.	39.			
Lead nitrate,	105.	77.			
	130.	88.			
Metallic sodium,[1] 23 being taken as the molecular weight,	4.28	393.			
	3.97	413.			
	3.8	448.			

[1]These results are somewhat low, because the ammonia contained some water. This of course would react with the sodium and form sodium hydroxide, which has a much lower conductivity than sodium.

In conclusion it may be said that ammonia seems to possess the power of dissociation of dissolved substances to as great an extent as water, and in most cases the ions seem to travel even faster in it than in water. It would seem, furthermore, that water and ammonia do resemble each other in their power to dissociate dissolved substances, as well as in their ability to unite directly with certain metallic salts.

It is proposed to continue this work with improved apparatus, and it is hoped that greater accuracy may be attained.

The Transference Numbers of the Ions in Sodium–Ammonia Solutions

In the following paper, Kraus describes his determinations of the potentials of concentration cells of solutions of sodium in liquid ammonia over an extended concentration range. The data were of interest because they yielded approximate values of the transference numbers for the positive and negative current carriers; the ratio of the negative to the positive transference number was found to range from about 7 in dilute solutions to 280 in concentrations slightly below 1 *M*. It was clear that pure electronic conduction, characteristic of metallic behavior, is approached on going to concentrated sodium–ammonia solutions. Thus metal–ammonia solutions were recognized as a medium for studying the transition between electrolytic and metallic behavior.

26

[CONTRIBUTIONS FROM THE RESEARCH LABORATORY OF PHYSICAL CHEMISTRY OF THE MASSACHUSETTS INSTITUTE OF TECHNOLOGY, NO. 98.]

SOLUTIONS OF METALS IN NON-METALLIC SOLVENTS. V. THE ELECTROMOTIVE FORCE OF CONCENTRATION CELLS OF SOLUTIONS OF SODIUM IN LIQUID AMMONIA AND THE RELATIVE SPEED OF THE IONS IN THESE SOLUTIONS.[1]

BY CHARLES A. KRAUS.

Received March 11, 1914.

Introduction.

It was shown in the fourth paper of this series that the conduction process in solutions of the metals in ammonia is an ionic one. The positive carrier is identical with the positive ion of the salts of the same metal, while the negative carrier, which appears to be the same for different metals, can consist only of the negative electron e^-, either free or in association with ammonia. The characteristic properties of the metal solutions are due to the presence of the negative carrier.

With nonsoluble electrodes, such as platinum, the negative carrier passes into and out of the solution without observable material effects. That portion of the current which is carried through the solution by the negative carrier is, therefore, similar to the current in a metal, and in passing a current from a solution of one concentration to a solution of another concentration the only work involved is that of transferring the positive carrier.[2]

[1] Previous papers of this series have appeared as follows: "I," THIS JOURNAL, 29, 1557 (1907); "II," *Ibid.*, 30, 653 (1908); "III," *Ibid.*, 30, 1197 (1908); "IV," *Ibid.*, 30, 1323 (1908). The sentence beginning on line 25, p. 1332 of the fourth paper should read: "According to this hypothesis, the negative carrier should move more *slowly* in dilute than in concentrated solutions, ..."

[2] This is true for very dilute and very concentrated solutions. At intermediate concentrations the influence of the solvent envelope must be taken into account, as will be described below.

In the metal solutions, Faraday's laws do not hold true in their ordinary sense; for the passage of one equivalent of electricity is not accompanied by a transformation or transfer of one equivalent of matter. Under given conditions, the material change accompanying the current is a definite fraction of that predicted by Faraday's law. To be applicable to the metal solutions, Faraday's laws must be extended to include not only ordinary matter, but also such forms of matter as we find in the negative electron. Under this generalization, Faraday's law would merely state that, in a medium carrying a current, the amount of electricity passing a given cross-section is equal to the number of carriers crossing the section multiplied by the number of charges on one carrier and the value of the unit charge. This view is that commonly accepted in the present-day theories of metallic conduction.

In dilute solutions of the metals in ammonia we thus have what is equivalent to a mixture of metallic and electrolytic conduction. Of course, in these solutions, the negative electron is in all probability surrounded with an envelope of solvent molecules; however, the work involved in carrying solvent from a solution of one concentration to that of another, approaches zero as a limit as the dilution increases indefinitely.

It is the purpose of this paper to adduce further evidence as to the nature of the negative carriers in the metal solutions. If the hypothesis that the negative carrier is associated with ammonia in dilute solutions is correct, we should expect the negative carrier to possess a speed comparable with that of ordinary ions in ammonia. Moreover, we should expect that at higher concentrations the size of the envelope would decrease or even disappear entirely. Consequently its speed should increase at higher concentrations.

As was shown in the fourth paper of this series, the positive ions in the solution of a metal are identical with the positive ions of the salts of the same metal. As has been shown by the work of Franklin and Cady,[1] the speed of ordinary ions in ammonia does not change greatly with concentration. We may, therefore, determine the variation of the speed of the negative carrier as a function of concentration by comparison with the speed of the positive carrier.

The ratios of the mean speeds of two carriers present in a solution are usually determined by means of the moving boundary or of the Hittorf method. Unfortunately, such measurements in the metal solutions are extremely difficult. It was therefore determined to employ the method of measuring the electromotive force of concentration cells, which likewise involves the transference numbers of the ions. In the case of the solutions of the metals in ammonia, this method has much to recommend it; for not only does the electromotive force method make it possible to

[1] THIS JOURNAL, 26, 499 (1904).

work with very small quantities of material over large ranges of concentration, but both anode and cathode constitute reversible electrodes which are peculiarly free from disturbances.

Experimental.

Description of the Apparatus.—The apparatus employed in these experiments is outlined in Fig. 1. The solutions whose electromotive force was to be determined were prepared in the tubes *B* and *I*. These two tubes were connected by a tube, *M*, and the valve *L*. This valve was made by grinding a glass rod, *H*, into a conical seat in the glass tube at the bottom of *K*. The valve was polished with rouge and was quite impervious to liquids. A lubricant could not be employed. The plunger of this valve was sealed to a tube, *F*, of larger diameter, which was joined to the cell-arm *I* by means of a rubber tube, *G*. The greater portion of the rubber tube was covered with an impervious cement. No serious leakage was observed due to this connection. The valve was opened by raising the plunger *H* slightly. A clamp was provided to hold the plunger when desired. Normally, the valve was closed. The tube *B* was provided with a branch capillary, *N*, through which the metal was introduced in the manner described in previous papers.[1] The upper portion of the cell-arm *KI* consisted of a graduated tube which was calibrated with the plunger in position. The other cell-arm was provided with a narrower graduated tube, *A*, likewise calibrated. The volumes of the two cell-arms, up to fixed points on the graduated tubes, were determined. The lower portions of

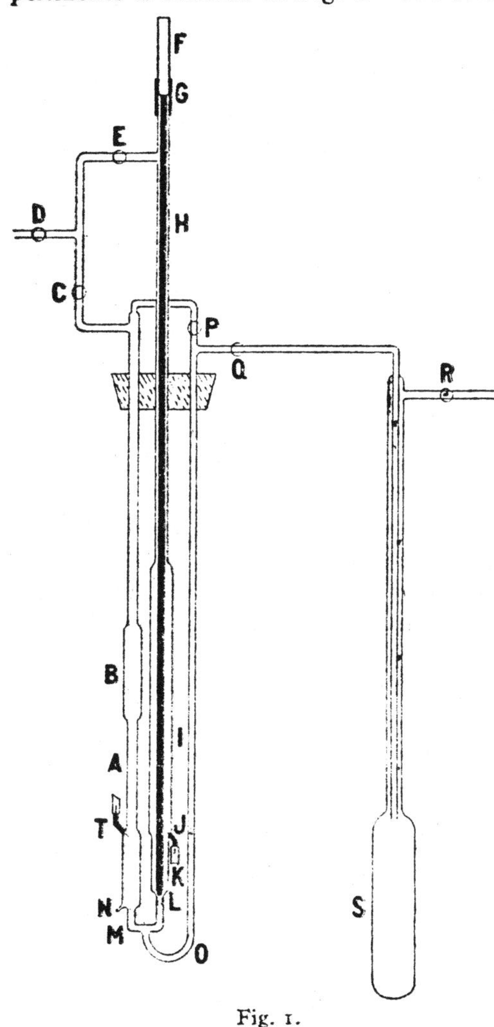

Fig. 1.

[1] For a description of this operation, see THIS JOURNAL, 30, 1206 (1908).

the two cells each had a volume of about 4 cc. The tube *O*, which was joined to the bottom of the connecting tube *M*, was employed in changing the volume of the solution in the tubes *BI*. The electrode connections are shown at *JT*.

The procedure in manipulating the cell was somewhat as follows: After introducing the sodium, the cell was surrounded by boiling ammonia, connection was made through *DCE* with the source of ammonia, and liquid was condensed until its level reached the bottom of the smaller graduated tube *A*. The valve *L* was open in order that the liquid should be at the same level in both arms. The contents of the cell were thoroughly mixed. This was done by removing the bath, closing the cock *E*, and warming either of the tubes *BI* with the hand. The solution was thus driven into the other arm, and so back and forth until homogeneous. After replacing the bath, the volumes were read off on the graduated tubes when equilibrium had been established. Ammonia was now condensed into both tubes until the volume in the tube *KI* had been approximately doubled, the liquid levels being kept the same in the two arms. After mixing the solutions separately, the volumes were again read. Knowing the original volumes and the added volumes, it was possible to determine the relative concentrations of the solutions in the two tubes. The experiment was so carried out that the volumes were approximately in the ratio 1 : 2.

The cocks *CEP* being open, the valve *L* was opened cautiously, and the electromotive force was measured. In general, the level in *A* was originally kept a little below that in *I*, so that on opening *L*, solution from *I* ran into the connecting tube, and the contact between the two solutions was made at a point where mixing was a minimum.

After determining the electromotive force, the valve *L* was closed, as were also the cocks *EP*. The tube *S*, which had previously been exhausted, was surrounded with ammonia boiling under reduced pressure. When the temperature was well below the normal boiling point of ammonia, the cock *Q* was opened and the solution in *B* was forced out through the tube *O* by the pressure of its own vapor. When the solution had been transferred, the cock *Q* was closed and the cocks *PE* and valve *L* were opened. Half the solution in *I* now ran into *B*, after which *L* was closed and the volumes were read. The cocks *CDE* were opened and fresh ammonia was distilled into the cell as before. The ammonia in *S* was meanwhile evaporated through *R*, the sodium being left behind. After determining the liquid volumes, the valve was again opened and the electromotive force was determined as before. In this way the more concentrated solution always occupied the arm *B*, and the dilute solution of one experiment formed the concentrated solution of the next succeeding one.

A correction, of course, had to be applied for the volume of the small graduated tube A.

When the dilution had been carried sufficiently far, the entire contents of the cell were run into S and evaporated. The apparatus was exhausted, the tube S disconnected, and the metal was dissolved in alcohol. This solution was then washed out and titrated to determine the sodium. Thus all the data were obtained for determining the concentrations of the different solutions. After a series of experiments, the entire apparatus was taken down, washed, and dried. It was then set up again and operations carried out as before.

The electromotive force was determined by comparing with a standard cell, a sensitive galvanometer of 3000 ohms resistance serving as null instrument. The sensitiveness, of course, decreased in very dilute solutions where the resistance of the cell became comparable with that of the galvanometer. In the more concentrated solutions a change of 0.1 millivolt caused a deflection of about 6.0 scale divisions, while in the most dilute solutions it produced a change of only about 1.0 division. The percentage accuracy in the dilute solutions did not decrease in this proportion for the reason that the total electromotive force measured was greater in these solutions.

Although the electrodes employed were very small, the constancy of the electromotive force was remarkable. Within the limits of sensitiveness of the galvanometer, no fluctuations could be observed except in the most dilute solutions, where electrode effects appeared.

Calculations.—We have seen that, when a current passes through a metal solution, the metal is carried from anode to cathode. On the other hand, the anion apparently consists merely of a negative electron or of an electron associated with ammonia.

In the solution we have the equilibrium

$$M^+ + e^- \rightleftarrows Me$$

where M^+ is the positive ion, e^- the negative ion (consisting of the negative electron) and Me represents the neutral metal atom. According to the results of a previous paper, work is involved only in the transfer of the positive metal ions from one solution to another. Equating the osmotic and electrical work and solving for the electromotive force, we have

$$E_1 = \frac{2nRT}{F} \log_e \frac{(M^+)_1}{(M^+)_2}, \qquad \text{I}$$

where E_1 is the electromotive force of the cell due to the transfer of solute, F is the electrochemical equivalent, n is the fraction of the current carried by the positive carriers, R is the gas-constant, T the absolute temperature, and $(M^+)_1$ and $(M^+)_2$ are the activities of the positive ions in the two solutions, respectively.

As stated in the preceding section, the negative carrier consists of the negative electron, either free or associated with the solvent. The electrons themselves contribute nothing to the work of the cell. If, however, they are associated with the solvent, the solvent will be carried from a solution of higher to a solution of lower vapor-pressure. In concentrated solutions, we must, therefore, take this factor into account.[1] If m molecules of ammonia are associated with one electron, then for every equivalent of electricity, $m(1 - n)$ mols of ammonia will be carried from the dilute solution of concentration c_2 and vapor-pressure p_2 to the concentrated solution of concentration c_1 and vapor-pressure p_1, the factor $1 - n$ being the fraction of the current carried by the negative carrier. Since the negative carrier moves from the dilute to the concentrated solution, the solvent moves from higher to lower vapor-pressures, osmotic work is done, and the electromotive force due to the transfer of solvent will be in the same direction as that due to the transfer of metal. The solvent carried by the negative carrier will, therefore, contribute the electromotive force

$$E_2 = \frac{m(1-n)RT}{F} \log_e \frac{p_2}{p_1}. \qquad \text{II}$$

We are assuming here that the amount of ammonia associated with the carriers in the two solutions is constant, an assumption which must be very nearly true, since the concentrations differ only in the ratio of 1 : 2. The ammoniation of the metallic ions does not enter into the osmotic work, since the metal does not lose its solvent at the electrode; for the metal in any case remains in solution. The only factor introduced here is due to change in the ammoniation of the positive ion, which may be neglected. Moreover, it will be shown that, even if this were very large, it would not produce an appreciable result in the electromotive force of the cell.

We have, therefore, for the electromotive force E of the cell in the direction from solution 1 to solution 2,

$$E = E_1 + E_2 = \frac{2nRT}{F} \log_e \frac{(M^+)_1}{(M^+)_2} + m(1-n)\frac{RT}{F} \log_e \frac{p_2}{p_1}. \qquad \text{III}$$

If the concentrations are made sufficiently small the activities $(M^+)_1$ and $(M^+)_2$ become equal to the concentrations M_1^+ and M_2^+, respectively, and the second term vanishes, since p_2 and p_1 approach equality.

We have then

$$E = \frac{2nRT}{F} \log_e \frac{M_1^+}{M_2^+}, \qquad \text{IV}$$

from which n, the transference number of the metal, may be calculated.

It has been shown in a preceding paper[2] that Raoult's law does not

[1] Lewis, THIS JOURNAL, 30, 1355 (1908).

[2] *Ibid.*, 30, 1197 (1908).

hold for solutions of sodium in ammonia at moderate concentrations. At these concentrations, therefore, Equation IV does not apply strictly. Nevertheless, it may reasonably be assumed that the equation will yield results of a correct order of magnitude. In dilute solutions, however, as will be shown in the next paper, the mass-action law applied to solutions of metals in ammonia. For these solutions, therefore, Equation IV is exact and the resulting value of n yields a precise measure of the transference number of the metal ion.

As we shall see presently, the transference numbers of the carriers in the more concentrated solutions change rapidly with concentration. This factor also has been neglected in the preceding equation. It does not seem worth while to introduce a correction term for this factor, since so many other uncertainties underlie the application of the equation to concentrated solutions. The values obtained for n may be looked upon as mean values between the concentrations in question. Throughout the experiments the concentration ratio of approximately 1 : 2 was maintained between the solutions measured. Over this concentration-interval the ionization does not change greatly, so that the ratio of total salt concentrations may be employed in place of the ratio of ion concentrations. When the conditions of experiment warrant a greater precision, a correction will later be made.

It has also been assumed that the equilibrium existing in the solution is that of a binary electrolyte. The correctness of this assumption will be shown in another paper.

Experimental Results.—The results obtained in three independent series of experiments are embodied in the following table. The first column gives the number of the observation; the second and third columns give the concentration of the solutions in the cell, while the fourth column gives the logarithm (to the base 10) of the ratio of these concentrations. Under c_a is given the mean concentration of the two solutions and under $E \times 10^3$ the observed electromotive force in millivolts. Finally, in the seventh column, are given the values of the ratio $(1 - n)/n$, where n is the transference-number of the sodium ion Na^+, calculated from Equation IV, the absolute temperature being 240°.

Discussion.

The fraction of the current carried by the positive carrier Na^+ is n. That carried by the negative carrier is $1 - n$ and the ratio $(1 - n)/n$ is the ratio of the amount of the current carried by the negative carrier to that carried by the positive carrier. It is to be noted that in the case of the negative carrier this ratio does not necessarily represent the speed of any one negative carrier relative to that of the positive carrier, for the speed of all the negative carriers is not necessarily of the same order of magnitude. The ratio, therefore, involves the average value for all

TABLE I.—RELATION BETWEEN ELECTROMOTIVE FORCE AND CONCENTRATION.

No.	C_1.	C_2.	log c_1/c_2.	c_a.	$E \times 10^3$.	$(1 - n)/n$.
			Experiment I.			
1	0.4560	0.1928	0.3741	0.325	0.85	41.0
2	0.1770	0.0863	0.3123	0.131	0.96	39.0
3	0.0805	0.0460	0.2434	0.063	1.00	21.4
4	0.0420	0.0148	0.3274	0.031	1.40	20.6
5	0.0177	0.0065	0.4373	0.012	2.55	15.0
			Experiment II.			
2	0.7628	0.3390	0.3522	0.551	0.609	54.0
3	0.3020	0.1892	0.2029	0.295	0.491	38.2
4	0.1625	0.0897	0.2580	0.126	0.790	32.0
5	0.0837	0.0490	0.2322	0.066	0.867	24.4
6	0.0447	0.0222	0.3046	0.033	1.355	20.4
7	0.0205	0.0121	0.2307	0.016	1.300	15.4
			Experiment III.			
1	1.014	0.6266	0.2344	0.870	0.080	277.6
2	0.9738	0.4693	0.3160	0.732	0.328	90.6
3	0.4571	0.3181	0.1525	0.387	0.336	42.2
4	0.4339	0.2310	0.2739	0.335	0.620	41.2
5	0.2224	0.1600	0.1437	0.191	0.384	34.6
6	0.2135	0.1153	0.2674	0.164	0.72	33.4
7	0.1109	0.0770	0.1580	0.094	0.50	29.2
8	0.1064	0.0570	0.2690	0.081	0.86	28.8
9	0.0548	0.0376	0.1639	0.046	0.65	23.2
10	0.0526	0.0274	0.2934	0.040	1.07	25.0
11	0.0263	0.0186	0.1510	0.023	0.78	17.4
12	0.0253	0.0140	0.2576	0.020	1.38	16.4
13	0.0135	0.0091	0.1582	0.011	0.92	11.4
14	0.0130	0.0070	0.2658	0.010	1.80	13.2
15	0.0065	0.0034	0.2779	0.0050	2.60	9.2
16	0.0031	0.0016	0.2883	0.0024	3.40	7.0

the negative carriers present. In the case of the sodium ion we may safely assume that all the ions move with the same mean speed, and that the speed is independent of concentration. Any variation of the ratio, therefore, represents the change in mean speed of the negative carriers.

In Fig. 2, the values of the ratio $(1 - n)/n$ are plotted as ordinates against those of the logarithms of the mean dilution ($\log 1/c_a$) as abscissae. The different points are in excellent agreement, both for points of the same and of different series. This shows that the results obtained are reproducible and represent equilibrium conditions. Such variations as appear are due partly to experimental error and partly to the approximations underlying the calculations.

It will be seen on inspection that in dilute solutions the negative carrier carries about seven times as much current as the positive carrier. The fraction of the current carried by the negative carrier increases with concentration, particularly as normal concentration is approached. In the most concentrated solutions (mean concentration 0.87 normal) the negative carrier carries approximately 280 times as much current as the

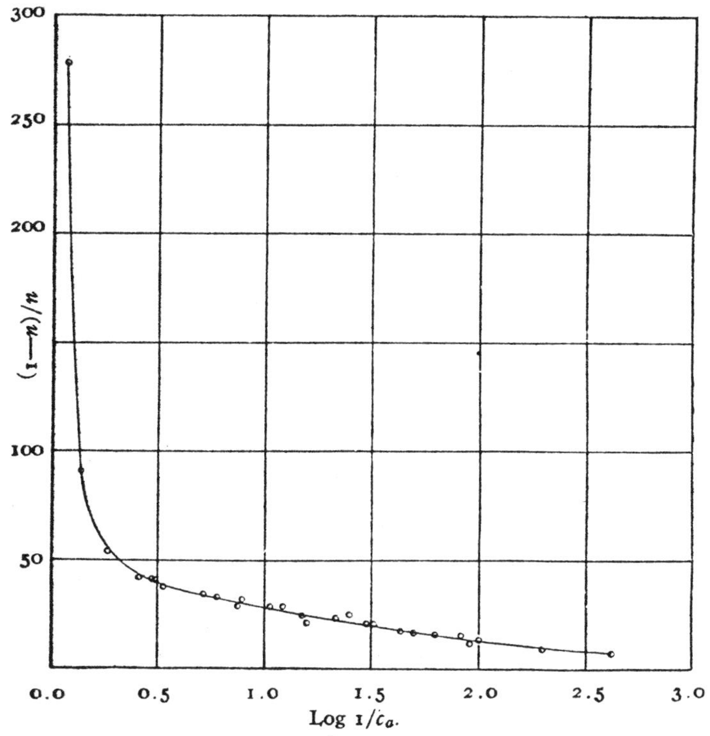

Fig. 2.—Showing change of $(1 - n)/n$ with the concentration.

positive carrier. In other words, the relative amount of current carried by the negative carrier at this concentration is 40 times as great as that at high dilutions.

The values of $(1 - n)/n$ are, of course, affected to some extent by the approximations made in the calculations. The ionization is changing with concentration, so that, even at the highest dilutions, the concentration ratio c_1/c_2 is somewhat greater than the ratio $c_1\gamma_1/c_2\gamma_2$ of the ion concentrations. The error due to this source is negligible in comparison with the observed change in the ratio $(1 - n)/n$. At higher concentrations the ionization of electrolytes does not follow the mass-action law. This error also is small in comparison with the change observed in this ratio. As to order of magnitude, therefore, the values of $(1 - n)/n$ are doubtless correct.

It follows, then, that the fraction of the current carried by the negative carrier increases enormously with increasing concentration. This shows that the mean speed of this carrier increases greatly with increasing concentration.

In dilute solutions the value of $(1 - n)/n$ appears to be approaching a definite limit which is probably not far from 7, the lowest value measured. The ionic conductance due to the sodium ion as deduced from conductance and transference measurements is 130.[1] The conductance of the negative carrier in a sodium solution is, therefore, $7 \times 130 = 910$, and therefore the molecular conductance of sodium at zero concentration is 1040. As will be shown in another paper, this is in excellent agreement with the directly measured value.

Since, in dilute solutions, the speed of the negative carrier is of the same order of magnitude as that of the sodium ion, it follows that the negative carrier must be associated with solvent molecules. In concentrated solutions, where the vapor-pressure lowering due to the solute becomes appreciable, the solvent associated with the negative carrier has an appreciable influence on the electromotive force, and a maximum value of the number of solvent molecules (m) associated with one negative carrier can be calculated. It can thus be shown that a portion of the current is carried by carriers not associated with the solvent. Namely, by placing n equal to zero in Equation III we may calculate a maximum value of m from the measured electromotive force (0.080 volt) of the pair of solutions used in Expt. III, No. 1, where the two concentrations are 1.014 and 0.627 normal, and where the ratio of the two vapor-pressures is 1 : 1.006.[2] The value so calculated is 0.67. The assumption that n is zero implies, of course, that there is *no* transference of the positive ion, and, therefore, that the whole electromotive force corresponds to the work done in the transference of ammonia. Since this assumption, even in the more concentrated solutions, is not strictly true, the value of m is doubtless considerably less than 0.67, but it can not be greater than this. This fact shows that at least one-third of the current is carried by electrons which are not associated with solvent, since those that are so associated must carry at least one molecule of ammonia.

Moreover, it is evident from the table that, at higher concentrations, the electromotive force approaches zero, or at least a very small value, as a limit. If one molecule of ammonia were associated with a single electron, a very considerable electromotive force would necessarily result, since the vapor-pressure of the solution changes rapidly with concentration.[2] The greater portion of the current is, therefore, carried by negative carriers which are wholly unassociated with ordinary matter. Doubtless, except in the case of very concentrated solutions, the larger proportion of the negative electrons is associated with ammonia; but their

[1] Kraus and Bray, THIS JOURNAL, 35, 1368 (1913).

[2] *Ibid.*, 30, 1210 (1908).

speed under a given potential gradient is relatively so small that they contribute but little to the conducting power of the solution.

We are now able to form a very good picture of the process by which conduction takes place in the solutions of metals in ammonia. As we have seen, the metal dissociates in solution according to the equation

$$Me = M^{+} + e^{-},$$

where Me represents the neutral atom of metal which is to be looked upon as a combination between the positive metal ion M^{+} and the negative electron e^{-}. For the alkali metals both ions M^{+} and e^{-} are univalent as will be shown in a later paper. The process of ionic dissociation obeys the mass-action law, so that

$$\frac{(M^{+})(e^{-})}{(Me)} = K,$$

where the symbols in parentheses represent the activities of the different molecular species, which may be replaced by the concentrations themselves in dilute solutions. It is probable that both the metal ions M^{+} and the neutral metal molecules Me are ammoniated to a greater or less extent. That the electrons e^{-} are associated with ammonia in dilute solutions can not be doubted, for the ratio of the mobilities of the two ions Na^{+} and e^{-} approaches a limiting value lying in the neighborhood of 7.[1]

Since the viscosity of the solution does not change materially with the concentration, the change in speed of the negative electron can not be due to viscosity change. But if the electrons are associated with ammonia, then, as soon as the activity of the solvent becomes sensibly constant, the ammoniation of the electron becomes sensibly constant, and the speed being determined by the friction of the ionic envelope with the solvent molecules, the mobility remains constant.

As the concentration increases, the size of the ionic envelope must diminish and since the frictional resistance diminishes with the size of the envelope the mobility of the negative carrier increases. But this is not a complete description of the processes involved as the concentration of the solution increases. The speed of an electron associated with only one molecule would not be greatly different from that of an electron associated with a larger number of molecules. Since the current carried by the negative carrier increases at least 40 times, when the concentration changes from about 0.001 normal to normal, a mere change in the size of the envelope will not suffice to account for the increased carrying capacity of the negative ions. What takes place as the concentration becomes greater is that a portion of the electrons are completely freed from ammonia for a fraction of the time. No individual electron is free for any considerable length of time, but the free electrons, the solvent, and the solvated elec-

[1] Compare Kraus, This Journal, 36, 35 *et seq.* (1914).

trons are in kinetic equilibrium. In concentrated solutions we have, in fact, three carriers present, namely: Na^+, $e^-(NH_3)_x$, and e^-. When free from ammonia, the electrons possess a mobility comparable with that of the electrons in a metal. A very small number of free electrons will, therefore, suffice to produce an enormous increase in the carrying capacity of the negative carriers in solution. The great increase in the proportion of current carried by the negative carrier is thus due to the presence of a small number of free negative electrons.

The free electrons on reaching the electrodes pass into it freely and produce no material change in state in this process. The process of conduction from the solution to the electrode is thus like that in the metal itself in that it consists entirely in a motion of the electrons.

On the basis of these considerations, we are able to predict what will be the properties of very concentrated solutions of the metals in ammonia. The number of electrons, which at any instant possess no envelope of ammonia in a solution of metal at normal concentration, is very small and they possess a mobility many hundred times greater than that of the metal ion. As the solution becomes more concentrated, the number of electrons free from ammonia will finally be comparable with the number of sodium ions present. As a result, such solutions must possess a conducting power many hundred times greater than that of a normal solution of sodium in ammonia. If such is the case, these solutions should begin to approach the metals in conducting power. If the properties of metals are due to the presence of free electrons, it follows that, since concentrated solutions of the metals contain these electrons, these solutions should exhibit those properties which are characteristic of metals. In another paper I shall show that this is the case.

The negative carrier present in dilute solutions of the metals in ammonia possesses considerable interest. The fact that this carrier possesses a speed approximately seven times that of the sodium ion indicates that the solvent envelope surrounding the negative electron is relatively small. It is interesting to note that the speed of the negative electron in ammonia is many times greater than that of the negative ion produced by radiations in hexane.[1] The dielectric constant of ammonia is much greater than that of hexane and it is not unlikely that the greater speed in ammonia is a consequence of this circumstance, for the higher the dielectric constant of the medium, the smaller will be the field of force due to the electric charge.

Let us now consider what takes place around an electrode when a current passes continuously through the solution of a metal in ammonia. At the cathode the concentration of metal is increased under the action

[1] THIS JOURNAL, 36, 59 (1914).

of the current. As soon as such concentration increase reaches an appreciable value, a relatively smaller quantity of metal is carried up for a given current. Consider a number of equipotential surfaces in the neighborhood of the cathode at which the concentration of the metal increases as the cathode is approached. The amount of metal carried across these surfaces decreases in proportion to the decrease of the transference number of the metal ion as the electrode is approached. There is, thus, a distributed accumulation of the metal throughout the volume of the solution in the neighborhood of the electrode. In the end, if the current be sufficiently great, no further change will take place immediately at the electrode surface. At a distance from the electrode the concentration change will increase up to a certain point and then again decrease until the normal concentration of the solution is reached.

The metal, then, is not deposited at the electrode surface, but is distributed throughout the electrode volume. Of course, the phenomenon is modified by diffusion and, under certain conditions, by convection. It is also greatly influenced by concentration. At very high concentrations the phenomena are much less marked than at low concentrations. The phenomena described in the fourth paper of this series are completely accounted for by the above hypothesis.

At the anode, with a non-soluble electrode, the phenomenon differs greatly from that at the cathode, if the solution is not too concentrated. Assume an electrode surrounded by a dilute solution of given concentration. When an electromotive force is applied, the negative carriers move to the electrode while the positive carriers move away from it. Now, the negative carrier moves up to the anode and then enters the metal. The only physical change accompanying this process is that ammonia is left behind at the electrode surface. In a dilute solution the ammonia thus left behind is relatively so small that it has no material influence on the observed phenomena. At the electrode surface there is, therefore, no manner of mechanical disturbance and we are able to observe phenomena which in ordinary solutions would be obscured by secondary effects.

The withdrawal of positive metal ions from the immediate vicinity of the anode surface causes an almost instantaneous removal of all metal from this region. The speed of this process is accelerated owing to the fact that when the concentration change sets in, the potential gradient in this region rises and thus increases the speed of the carriers.

In the immediate neighborhood of the anode there will, therefore, result a region which contains no metal whatever and across which the current is carried entirely by the negative carriers moving freely under the action of the applied potential. The extent of this region will depend on the electromotive force applied, the concentration of the solution, the shape and size of the electrodes, and other specific factors. As was described in the fourth paper of this series, the boundary of the metal-free region is usually very clear and distinct. Up to this boundary the metal is maintained in position by processes of diffusion and convection, the latter factor being the most important if a considerable potential is applied.

It is thus seen that in the case of an anode in a dilute metal solution, where one of the carriers is removed from solution when it reaches the electrode surface, the current is carried entirely by one carrier over an appreciable distance. That a similar phenomenon obtains in solutions of ordinary electrolytes appears highly probable.

Summary.

The electromotive force of concentration cells of solutions of sodium in liquid ammonia has been measured over an extended concentration range.

From these measurements approximate values of the transference numbers have been calculated for different concentrations. The ratio $(1 - n)/n$ of the fraction of the current carried by the negative and positive ions approaches a limiting value of approximately 7 in dilute solutions, and increases to a value of 280 at a mean concentration somewhat less than normal.

Assuming the speed of the positive ion to remain constant, the mean speed of the negative ion increases 40 times between about 0.001 and 1.0 normal. Assuming the equivalent conductance of the positive ion to be 130 (that of the sodium ion), the equivalent conductance of the dilute metal solution is calculated to be 1040.

These results are accounted for on the assumption that the ions Na^+ and e^- exist in solution. At high dilutions the negative electron e^- which serves as negative ion, is surrounded by an envelope of ammonia which determines its speed. At higher concentrations some of the negative electrons are free from the ammonia envelope for a fraction of the time. Under these conditions they move with a speed comparable with that of the negative electrons in metals. This accounts for the rapid increase of the transference number of the negative carrier at higher concentrations, for the number of free electrons increases as the proportion of metal to ammonia increases.

The electrode phenomena described in the fourth paper of this series are in accord with the above hypothesis.

On the basis of the osmotic work involved, it is shown that in the most concentrated solutions a portion of the current is carried by carriers not associated with ordinary matter.

BOSTON, MASS.

The Absorption Spectra of Metal–Ammonia Solutions

Kraus had postulated that the characteristic properties of metal–ammonia solutions were due to the electron, either "free" or in association with ammonia. In the investigation described in the following paper, Gibson and Argo studied the visible absorption spectra of sodium and magnesium[1] solutions. They found that the spectra were the same and that Beer's law was at least approximately obeyed by sodium solutions. The extreme difficulty, in those days, of obtaining quantitative absorption spectra of unstable solutions requiring refrigeration can be appreciated by reading the experimental sections of the paper.

[1]Magnesium is not readily soluble in ammonia; the very slow dissolution which does occur has sometimes been ascribed, perhaps not justifiably, to calcium impurity.

[Reprinted from the PHYSICAL REVIEW, N.S., Vol. VII, No. 1, January, 1916.]

27

THE ABSORPTION SPECTRA OF THE BLUE SOLUTIONS OF SODIUM AND MAGNESIUM IN LIQUID AMMONIA.

BY G. E. GIBSON AND W. L. ARGO.

I. INTRODUCTION.

THE blue color of solutions of metals in liquid ammonia and alkyl amines has been observed and several of their physical properties have been investigated by Kraus and others. In order to account for the electrical properties of these solutions Kraus[1] assumes that in all cases the metal dissociates on dissolving according to the equation

$$Me = Me^{+} + \theta,$$

where Me^{+} is the cation present in solutions of salts of the metal and θ is an electron. Both Me^{+} and θ may be combined to a greater or less extent with the solvent molecules. From measurements of the electromotive force of concentration cells Kraus[2] concludes that at least one third of the current in a normal solution of sodium is carried by the unsolvated electron.

The color of these metallic solutions appears to be independent of the metal and the solvent used. They are intensely blue to transmitted light, and, as far as the eye can judge, the shade of color is the same in all cases. In daylight .001 normal solutions of potassium and sodium in ammonia can be matched against ammoniacal copper solutions. This is also true of a saturated solution of magnesium in liquid ammonia.

Similar blue solutions are obtained when sodium is dissolved in molten sodamide[1] and when calcium is dissolved in a molten mixture of potassium and sodium chloride.[3]

The blue color of these metallic solutions may be due (*a*) to un-ionized molecules (or atoms) of the metal, (*b*) to solvated electrons, (*c*) to unsolvated electrons.

(*a*) We should expect the color of the un-ionized metal atoms to vary with the metal and to be similar to that of the vapor of the metal, provided the metal atoms were not affected by the presence of the solvent.

[1] J. Am. Chem. Soc., *29*, 1557 (1907).

[2] J. Am. Chem. Soc., *36*, 864 (1914).

[3] Titherley, J. Chem. Soc., *65*, 508 (1894).

[4] A description of this experiment will be published shortly.

Sodium vapor has a purple color. The absorption spectrum of the vapor, however, has been investigated by Wood and others, and is in no way similar to that of the solution in ammonia.

If the un-ionized metal atoms combined with the solvent molecules we should expect the color of the compound to vary with the solvent.

(*b*) If the color were due to solvated electrons we should expect an absorption band differing from solvent to solvent. It is unlikely that electrons bound to molecules so different in character as ammonia and fused alkali chlorides should produce the same color.

On the whole, therefore, the experimental evidence hitherto obtained favors the assumption (*c*) that the color is due to unsolvated electrons. In this paper the latter assumption will be tested quantitatively by comparing the theoretical absorption spectrum deduced from this hypothesis with the aid of Drude's theory of metallic dispersion with the actual absorption spectrum determined by experiment.

II. The Relation between Extinction Coefficient and Wave Length According to Drude's Theory.

Let us make the same assumptions for the unsolvated electron in the blue solutions that Drude[1] employs in his electron theory of metallic dispersion and conduction, *i. e.*, that the resistance to the motion of the electron through the medium is proportional to its velocity.

The equation of motion of an unsolvated electron is, therefore,

$$ma = eE - re^2v. \qquad (1)$$

The term on the left is the product of the mass m of the electron and its acceleration a. E is the intensity of the electric field, e the charge on the electron, and r is such a quantity that re^2 is the resistance to the motion of an electron moving with unit velocity through the medium. Electrostatic units will be used throughout in this paper.

Equation (1) combined with the equations of the electromagnetic field, leads to the equations

$$\begin{aligned} n^2(1-\kappa^2) &= 1 + \Sigma\frac{\theta_h}{1-(\tau_h/\tau)^2} - 4\pi\frac{m'\mathfrak{N}}{r^2+(m'/\tau)^2}, \\ n^2\kappa &= 2\pi\tau\frac{r\mathfrak{N}}{r^2+(m'/\tau)^2}, \end{aligned} \qquad (2)$$

where n is the refractive index of the solution. The index of absorption κ is defined by the equation

$$J = J_0 e^{-\frac{4\pi\kappa}{\lambda}d} \qquad (3)$$

[1] Lehrbuch d. Optik, Leipzig, 1912, p. 386.

where J_0 is the intensity of the incident ray of wave-length λ and J is the intensity after a layer of thickness d has been traversed. The symbol m' is the ratio of the mass of the electron to the square of its charge ($m' = m/e^2$). $\mathfrak{N}$ is the number of unsolvated electrons in a cubic centimeter of the solution. The radian time τ is equal to the periodic time divided by 2π. θ_h and τ_h are constants of the bound electrons.

Equations (2) are the general equations of Drude for an absorbing medium containing both bound and unbound electrons. The term under the sign of summation in the first of equations (2) represents the influence of the bound electrons.

For a colorless solvent both κ and τ_h/τ are very small in the visible spectrum. The refractive index of the pure solvent, therefore, varies very slightly with the wave-length, and will be assumed to be constant in the following. In terms of the symbols of equations (2) its value would be given by

$$n_0^2 = 1 + \Sigma\theta_h.[1] \tag{4}$$

Neglecting τ_h/τ in equations (2) and eliminating n we obtain

$$\frac{1-\kappa^2}{\kappa} = \left(\frac{n_0^2 r}{2\pi\mathfrak{N}} - \frac{2m'}{r}\right)\frac{1}{\tau} + \frac{n_0^2 m'^2}{2\pi\mathfrak{N}r} \cdot \frac{1}{\tau^3}. \tag{5}$$

In terms of the wave-length $\lambda = 2\pi c\tau$ and the extinction coefficient $\epsilon = (4\pi\kappa \log_{10} e)/\lambda$ this becomes

$$\frac{a^2}{\epsilon} - \epsilon\lambda^2 = 2\pi ac\left(\frac{n_0^2 r}{2\pi\mathfrak{N}} - \frac{2m'}{r}\right) + \frac{4\pi^2 ac^3 n_0^2 m'^2}{\mathfrak{N}r} \cdot \frac{1}{\lambda^2}, \tag{5a}$$

where a is a contraction for $4\pi \log_{10} e$.

Except for very large values of ϵ (such as occur in pure metals) the term $\epsilon\lambda^2$ may be neglected in the visible spectrum, and (5a) may therefore be written

$$\frac{a^2}{\epsilon} = b_1 + \frac{b_2}{\lambda^2}, \tag{6}$$

where b_1 and b_2 are independent of the wave-length, *i. e.*,

$$b_1 = 2\pi ac\left(\frac{n_0^2 r}{2\pi\mathfrak{N}} - \frac{2m'}{r}\right),$$
$$b_2 = \frac{4\pi^2 ac^3 n_0^2 m'^2}{\mathfrak{N}r}. \tag{7}$$

Equation (6) expresses the theoretical relationship between the extinction

[1] The refractive index of liquid ammonia n_0 has not been determined. In the calculations n_0^2 was taken to be equal to 2 as the refractive index of liquid ammonia cannot be far from that of water.

coefficient ϵ and the wave-length λ in the visible spectrum. According to (6) a^2/ϵ is a linear function of $1/\lambda^2$.

If $2m'/r$ is negligible compared with $(n_0^2 r)/2\pi\mathfrak{N}$ and if λ is small enough to make $\epsilon\lambda^2$ negligible compared with a^2/ϵ the extinction coefficient ϵ for a given wave length will be proportional to the concentration $\mathfrak{N}$ of the unsolvated electrons. These are therefore necessary conditions that Beer's law shall hold.

For sufficiently large wave-lengths we see from ($5a$) that

$$\frac{a^2}{\epsilon} - \epsilon\lambda^2 = 0,$$

or

$$\epsilon = \frac{a}{\lambda}.$$

In other words ϵ is independent of the concentration when λ is large. Hence Beer's law must cease to hold when the wave-length exceeds a certain amount determined by the values of b_1 and b_2. The extinction coefficient vanishes for infinite wave-lengths in all cases.

For very short wave-lengths ($5a$) reduces to

$$\frac{a^2}{\epsilon} = \frac{4\pi^2 a c^3 m'^2 n_0^2}{r\mathfrak{N}} \cdot \frac{1}{\lambda^2},$$

so that in this case ($b_1 = 0$) Beer's law is always obeyed.

If b_1 is negative we must have $2m'/r > (n_0^2 r)/2\pi\mathfrak{N}$ and Beer's law cannot hold, since b_1 is no longer inversely proportional to $\mathfrak{N}$.

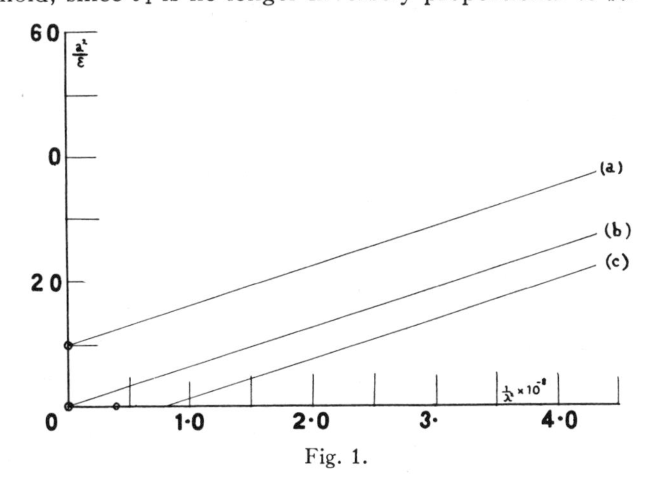

Fig. 1.

The graphs of a^2/ϵ against $1/\lambda^2$ as they follow from the general equation ($5a$) are drawn in Fig. 1. There are three types of curve, a, b, and c, according as b_1 is positive, zero, or negative. The equation of the asymptote in all cases is equation (6) viz.,

$$\frac{a^2}{\epsilon} = b_1 + \frac{b_2}{\lambda^2}.$$

On the scale of Fig. 1 the remainders of the three curves a, b, c are indistinguishable from the axes. The minima of a^2/ϵ are marked with circles.

A more elaborate theory of metallic dispersion has recently been developed by Jaffé[1] in which the frictional resistance r is assumed to be due to collisions between the free electrons and the atoms of the medium. In place of Drude's equations (2) his theory leads in our case to the equations

$$n^2(1 - \kappa^2) = n_0^2 + \frac{2\sqrt{\pi^3}\sigma}{\nu}\psi_1,$$

$$n^2\kappa = \frac{2\pi\sigma}{\nu}\psi_2, \qquad (8)$$

where ψ_1 and ψ_2 are functions of the argument $z = (3\sqrt{\pi}/4)(m'/r)\nu$. For large values of z we obtain from Jaffé's theory

$$\psi_1 = -\frac{3}{2}\cdot\frac{1}{z}$$

$$\psi_2 = \frac{2}{z^2}.$$

Substituting these values of ψ_1 and ψ_2 in (8), and eliminating n, we obtain in place of equation (5a)

$$\frac{a^2}{\epsilon} = -\frac{9\pi^2}{8}\frac{am'c}{r} + \frac{9\pi^3}{8}\frac{n_0^2am'^2c^3}{\mathfrak{N}r}\cdot\frac{1}{\lambda^2}.$$

When $(n_0^2r)/2\pi r$ is negligible this equation is identical in form with (5a) and differs only slightly in the values of the constants. Jaffé's theory therefore leads to the same conclusions as the theory of Drude.

III. The Theories of Drude and Jaffé in View of the Experimental Determinations.

With a view to testing the hypothesis that the blue color is due to unsolvated electrons, the absorption spectra of solutions of sodium and magnesium in liquid ammonia were determined in the visible spectrum. The results of these experiments will be discussed now and compared with the requirements of the theories of Drude and Jaffé while the details of the experiments will be found in Section IV.

[1] Ann. d. Physik [4], 45, 1217, 1914.

In Fig. 2 the values of a^2/ϵ calculated from our determinations (See col. IX. and X., Table II., and col. XI. and XIII., Table VII.) are plotted against $1/\lambda^2$. There are three curves for sodium at different concentrations, Na(1), Na(2) and Na(3) (circles), and one for a nearly saturated solution of magnesium (triangles).

The concentration of sodium is 2.8×10^{-3} normal in series (1) and

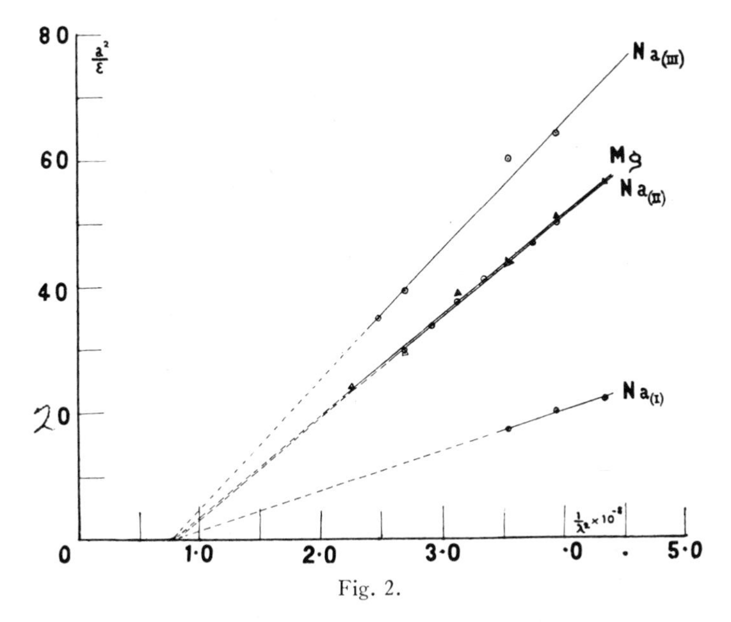

Fig. 2.

1.3×10^{-3} normal in series (2). The concentration in series (3) is less than 1.3×10^{-3} normal, but was not determined exactly. No attempt was made to determine the concentration of the magnesium solution.

The curve for magnesium (Fig. 2) coincides, within the limits of experimental error, with one of the sodium curves. *This is strong evidence in favor of the assumption that the coloring principle is the same for both metals.*

Within the limits of error the graphs are all straight lines and intersect at a common point on the axis of abscissæ. In other words, the ratio of the ordinates of the various curves at a given wave-length is independent of the wave-length. While this does not prove that Beer's law is obeyed, it is one of the necessary conditions for the proportionality of the extinction coefficient and the concentration of the coloring molecules.

If Beer's law were true for the total dissolved sodium, the concentration N of sodium multiplied by the value of a^2/ϵ should be constant for all wave-lengths. This is approximately confirmed by our experiments. At wave-length $\lambda = 502\mu\mu$ we have $(a^2/\epsilon)N = 0.56$ for series (1) and

$(a^2/\epsilon)N = 0.65$ for series (2). The difference between these two values can be accounted for by the loss of sodium due to the amide reaction in the time which elapsed between the two measurements (see Section IV., *d*). Beer's law therefore holds approximately for the total dissolved sodium.

The equation (5 a) obtained from Drude's theory in Section II. may be tested now in the light of these measurements.

The values of b_1 and b_2 obtained by the method of least squares from the experimental data are given in columns II. and III. of Table I.

TABLE I.

I. Series.	II. b_1(Exp.).	III. $b_2 \times 10^8$ (Exp.).	IV. $\mathfrak{N} \times 10^{-20}$ (Calc.).	V. r (Calc.).	VI N(Mols per Liter) (Calc.).	VII. N(Exp.) (Mols. per Liter).	VIII. $K \times 10^{-5}$. (Calc.).	IX. $K \times 10^3$ (Kraus).
Na(1)	−4.57	6.18	1.672	1801	0.271	0.00281	1.03	2.92
Na(2)	−12.83	15.97	1.715	641	0.295	0.00130	3.14	1.35
Na(3)	−15.38	20.28	1.816	535	0.278		3.56	
Mg	−12.21	15.88	1.738	674	0.281		2.86	

Columns IV. and V. contain the values of the number of electrons $\mathfrak{N}$ per cubic centimeter and the frictional resistance constant r as they would follow from Drude's theory. Column VI. contains the concentration of free electrons in mols per liter calculated from the values of $\mathfrak{N}$ in column IV. Column VII. contains the concentration of sodium determined by experiment. Column VIII. contains the values of the conductivity K (ohm^{-1}cm.$^{-1}$) calculated from the values of $\mathfrak{N}$ and r in columns IV. and V., and column IX. the specific conductivity derived from the experiments of Kraus.[1]

Drude's theory, therefore, gives values for the conductivity which are about thirty million times too large. Thus the specific conductivity K of a .0028 normal solution of sodium is $2.92 \times 10^{-3}\omega^{-1}$cm.$^{-1}$ while the conductivity calculated from Drude's theory for the same solution is $1.03 \times 10^5\omega^{-1}$cm.$^{-1}$ (see Table I.).

Drude's theory also leads us to conclude that Beer's law would not be even approximately true for the free electrons. The calculated concentrations of free electron are very great and independent of the concentration of sodium. For each sodium atom in a .001 normal solution Drude's theory would require about one thousand electrons, and we should have to ascribe the increased depth of color on adding more sodium to an increased frictional resistance to the motion of the electrons without any increase in their number.

[1] J. Am. Chem. Soc., *36*, 877 (1914).

If, therefore, we grant the validity of Drude's theory we are forced to conclude that the blue color of these solutions is not wholly due to unsolvated electrons.

It is important to determine the effect on the coloring principle of changing the metal and the solvent. With this end in view we propose to continue our measurements with other metals such as potassium and lithium and with other solvents such as methylamine, ethylamine, hydrazine and possibly with fused salts.

IV (*a*). THE SPECTROPHOTOMETER.

The absorption spectrum was measured by means of a spectrophotometer of the type described by Grünbaum and Martens.[1] A source of light *A* (Fig. 3) illuminates the ground-glass window *B*. The light from

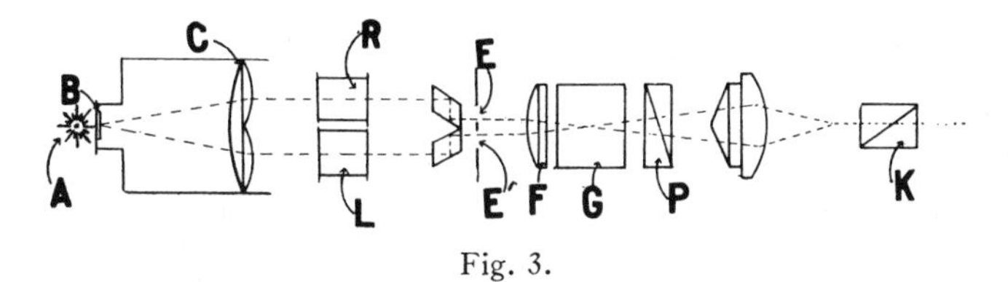

Fig. 3.

B passes through two lenses *C*, which produce an image of *B* on the slits *E*, *E'*. The two beams pass respectively through a cell *R* containing the solution to be examined, and a cell *L* containing water which is used to compensate for the loss of light due to reflection from the windows of the cell. After passing through the slits *E*, *E'* the two beams traverse successively the collimator *F*, the prism *G*, the polarizer *P* and the analyzer *K*. The analyzer is rotated until the fields illuminated by the two beams appear of equal intensity.

The calculation of the extinction coefficient from the angle of rotation is discussed in section IV (*d*).

The source of illumination is an acetylene flame, one inch wide and shielded by a metal chimney three inches in diameter and about two feet high. The flame, when so arranged, burns steadily and illuminates the ground glass more uniformly than a tungsten filament lamp.

The arbitrary wave-length scale of the spectrophotometer had already been calibrated by Adams and Rosenstein[2] in this laboratory. The calibration was repeated for the sodium *D* lines and agreed exactly with their determination.

[1] Ann. der Phys., *12*, 984 (1903).

[2] J. Am. Chem. Soc., *36*, 1452 (1914).

IV (*b*). The Absorption Cell.

The ordinary metal absorption cell with plane glass windows cannot be used for solutions of metals in liquid ammonia. In the first instance, the cell must be perfectly airtight, as the least trace of moisture is sufficient to destroy the blue color. It must also be capable of withstanding a pressure of eight to ten atmospheres, which is the vapor pressure of liquid ammonia at room temperatures. The first cells used were made of copper with glass windows pressed tightly against a thin flange of the metal. Although it was possible to obtain a vacuum-tight joint in this way, the cells were difficult to clean, and slight traces of impurity, such as solder or brass, catalyzed the amide reaction and made it difficult to obtain a permanent blue color.

The cell finally adopted was made entirely of quartz. It is shown in completed form in Fig. 4. The two plates *A*, *A*, one centimeter in diameter, were made of optical quartz free from bubbles. They were

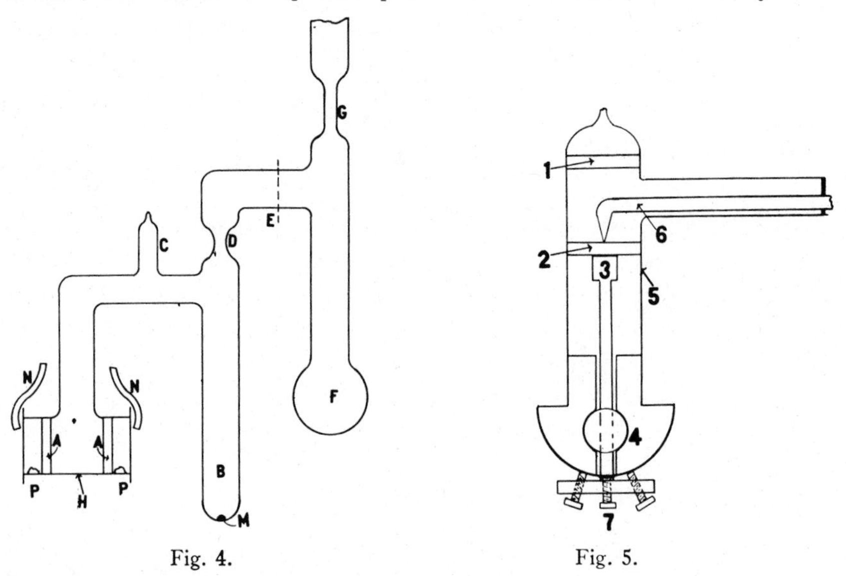

Fig. 4. Fig. 5.

ground plane parallel and were fused into the quartz tube *H*, by means of the oxygen gas flame, in the following manner:

The window 1, Fig. 5, was first sealed into the quartz test tube 5 in a position as nearly perpendicular to the axis as possible. The window 2 was then adjusted parallel to 1 by means of the appliances shown. A cylindrical block of quartz 3 was ground plane on its upper surface and connected by a quartz rod with the universal joint 4. By manipulating the three screws, 7, the upper surface of the block could be adjusted to a position making any desired angle with the axis of the tube. The window

2 was kept in close contact with the block, 3, by means of a quartz pointer, 6, which was depressed by a glass spring not shown in the figure. The window 2 was then adjusted until the image of a point source of light reflected from its surface coincided exactly with the image of the same source reflected from the window, 1. The source, which consisted of a small loop of incandescent platinum wire, was placed in the line of the normal to the two windows at a distance of about 150 cm. from the cell. The reflected images were observed through the incandescent loop. The two windows could be adjusted in this way to within 10′ of arc. Window 2 was then sealed into position by means of the oxygen gas flame.[1]

The two ends of the quartz tube were then cut off, so as to leave an edge of about 3 mm. beyond the windows. The completed cell is shown in Fig. 4.

IV (*c*). The Method of Filling the Cell.

The system *ABCD* in Fig. 4 was made entirely of quartz. Beyond the Khotinsky cement seal *E* glass was used throughout. The whole system was first evacuated and left in contact with phosphorus pentoxide for several hours. After the apparatus had been washed out several times with dry ammonia gas a quantity of ammonia sufficient to fill the cell was distilled into the bulb *F* by immersing it in alcohol cooled to −60° C. The bulb *F* contained a small piece of sodium which dissolved in the ammonia and removed the last traces of moisture. The constriction *G* was then sealed and the ammonia in *F* distilled into the tube *B*, which contained a portion *M* of the metal to be examined. After the cell had been sealed off at the constriction *D* by means of the oxygen-gas flame it was ready for the measurements.

IV (*d*). Measurements with Sodium in Liquid Ammonia at —33.5° C.

The quartz cell was surrounded with a copper jacket *A* (Fig. 6), which could be filled with boiling liquid ammonia through the tube *B*. The ammonia vapor was led out through the tube *C* to a large flask filled with water where it was absorbed. Copper filings were placed between the quartz cell and the copper jacket in order to secure good conduction of heat. The jacket was then fitted into a wooden cylinder *D* which served the double purpose of a heat insulator and a support for the cell in the spectrophotometer.

To prevent the deposition of moisture on the cold quartz windows,

[1] It is difficult to avoid entirely the sublimation of silica on the windows in the process of sealing them into the tube. The slight irregularities resulting from this became invisible on the inner surfaces when the cell was filled with liquid ammonia. The fog on the outside of the windows was made to disappear by cementing thin cover glasses to them with Canada balsam.

cover glasses were cemented with Khotinsky cement to the projecting edges of the cell, in the manner shown in Fig. 4. The air chambers were dried by small pellets of partly deliquesced phosphorus pentoxide, *p*, *p*. Jets of dry air were directed against the outer surfaces of the cover glasses by means of the tubes *N*, *N*. When the air jets were adjusted correctly they prevented entirely the deposition of moisture.

After the cell had been dried by evacuation, pieces of capillary tube of known length and diameter, full of sodium, were introduced through *C* (Fig. 4). The cell was then filled as described in paragraph IV (*c*). The amount of sodium introduced into the cell was more than sufficient for the highest concentration desired. After mixing thoroughly the position of the meniscus of the solution in the tube *B* was measured. A portion of the solution was then transferred to the cell *H*, and the position of the meniscus in *B* was again measured. By cooling the cell the remainder of the solvent was distilled from *B* to *H*. This process was repeated until the desired dilution was obtained. The volumes corresponding to the various positions of the meniscus in *B* were determined after the tube had been opened at the end of the experiment.

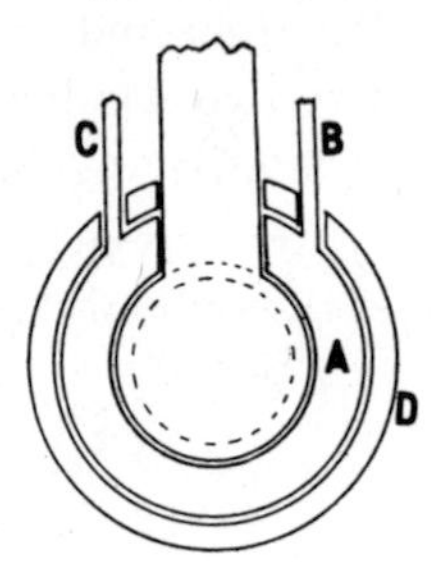

Fig. 6.

The method of reading the spectrophotometer and the method of averaging the readings were essentially the same as those of Grünbaum and Martens.[1] Grünbaum and Martens compare the cell containing the colored solution with a cell of the same dimensions containing the pure solvent. This method would have necessitated the construction of a second quartz cell, and a duplication of the cooling apparatus. To avoid this, the colored solution and the pure solvent were compared in the same cell at different times with a glass cell containing water. The water cell was used to compensate approximately for the reflections from the windows of the quartz cell. Under these conditions the sensibility of the readings is improved.

A complete determination at any wave-length involves the determination of four angles:

1. The angle α_1, with the quartz cell containing liquid ammonia in position *R* (Fig. 3) and the glass water cell in position *L*.

2. The angle α_2, with the quartz cell containing liquid ammonia in position *L* and the glass water cell in position *R*.

3. The angle α_3, with the quartz cell containing sodium solution in position *R* and the glass water cell in position *L*.

[1] Loc. cit.

4. The angle α_4, with the quartz cell containing sodium solution in position L and the glass water cell in position R.

The glass water cell might be replaced by an imaginary quartz cell excactly similar to the cell containing the ammonia solution but filled with a hypothetical liquid of extinction coefficient such that the fraction of the incident light absorbed by the hypothetical cell would be exactly equal to the fraction absorbed by the actual glass water cell. The difference between the extinction coefficient ϵ_{NH_3} of the pure liquid ammonia and the hypothetical extinction coefficient ϵ_h is then given by the relation

$$\epsilon_{NH_3} - \epsilon_h = \frac{\log \tan \alpha_2 - \log \tan \alpha_1}{d},$$

where d is the length of the column of liquid traversed by the light in the quartz cell.

The difference between the extinction coefficient ϵ_{Na} of the blue solution and ϵ_h is

$$\epsilon_{Na} - \epsilon_h = \frac{\log \tan \alpha_4 - \log \tan \alpha_3}{d}.$$

Hence the required difference between the extinction coefficients of the blue solution and the pure solvent (liquid ammonia) is

$$\epsilon_{Na} - \epsilon_{NH_3} = \frac{\log \tan \alpha_4 - \log \tan \alpha_3 - \log \tan \alpha_2 + \log \tan \alpha_1}{d}.$$

Even at $-33.5°$ C. the fading of the blue color due to the amide reaction is sufficiently rapid to necessitate correction. The angles α_3 and α_4 for each wave-length were therefore determined at measured intervals and plotted against the time of reading. The values of α_3 and α_4 at the time of the first reading of the series were obtained by graphical extrapolation. The largest correction applied to any reading amounted approximately to 4 per cent. of the extinction coefficient. The error in the graphical extrapolation is certainly not greater than 25 per cent. of this correction, so that the resultant error in the extinction coefficient due to this cause cannot exceed 1 per cent. This maximum error applies only to reading (6), series III., in the table below. The uncertainty in the fading correction is much less for all other readings (see Column VIII., Table II.).

IV (*e*). Measurements with Magnesium.

Cottrell[1] had observed that solutions of magnesium in ammonia were extraordinarily stable even at room temperature. Magnesium was

[1] J. Phys. Chem., *18*, 85 (1914).

Table II.

Temperature = − 33·5° C. Length of Column d = 1.523 cm.

	I. Reading No.	II. Wave-Length $\mu\mu$.	III. $\epsilon_{NH_3} - \epsilon_h$ (See Table IV.).	IV. Reading α_3.	V. Time of Reading (Minutes).	VI. Reading α_4.	VII. Time of Reading (Minutes).	VIII. Corrected Reading.		IX. $\epsilon_{Na} - \epsilon_{NH_3}$	X. $\frac{a^2}{\epsilon}$	XI. $1/\lambda^2 \times 10^{-8}$ (cm^{-2})
								α_3	α_4			
Series (1)......	(1)	529	.116	87.09°	0	1.87°	6	87.1°	1.75°	1.73	17.2	3.54
Conc. of Na =	(2)	502	.123	85.13	14	2.81	18	85.4	2.5	1.48	20.0	3.94
2.8×10^{-6} mols	(3)	480	.133	83.23	36	3.58	30	84.0	3.0	1.35	22.0	4.33
per cc.	(4)	529		86.14	44	2.58	47					
Series (2)......	(1)	608	.098	79.48	0	6.84	3	79.7	6.7	0.99	29.8	2.70
Conc. of Na =	(2)	585	.107	76.88	8	8.32	5	77.4	8.0	.880	33.8	2.92
1.3×10^{-6} mols	(3)	564	.112	74.62	9	9.69	11	75.3	9.0	.745	37.5	3.13
per cc.	(4)	546	.114	72.66	15	10.87	13	73.7	10.1	.728	40.9	3.34
	(5)	529	.116	70.76	18	12.00	20	72.0	10.6	.682	43.6	3.54
	(6)	515	.119	69.38	25	12.87	22	71.0	11.6	.636	46.7	3.74
	(7)	502	.123	68.32	27	14.00	28	70.1	12.5	.596	49.9	3.94
	(8)	608		77.54	34	7.93	30					
	(9)	608		77.37	41	8.08	44					
	(10)	529		69.17	48	12.99	46					
	(11)	608		76.74	50	9.30	53					
	(12)	529		68.34	58	13.34	56					
	(13)	608		75.22	64	9.09	67					
	(14)	529		67.04	72	13.92	69					
	(15)	608		73.84	88	11.92	91					
	(16)	529		65.75	98	15.63	94					
Series (3)......	(1)	608	.098	74.31	0	10.18	3	74.3	10.2	.753	39.5	2.70
	(2)	635	.084	75.60	13	9.34	7	76.6	8.7	.860	34.9	2.48
	(3)	608	.098	72.83	16	11.45	18					
	(4)	529	.116	65.13	25	16.09	22	66.6	15.1	.496	60.0	3.54
	(5)	502	.123	63.25	32	17.09	39	65.0	15.3	.464	64.1	3.94
	(6)	480	.133	59.28	59	18.13	48	62.0	15.7	.409	72.8	4.33
	(7)	608		68.98	63	13.88	68					
	(8)	608		65.90	102	15.96	99					
	(9)	564		62.24	106	18.65	109					
	(10)	529		59.64	117	20.59	113					
	(11)	502		58.20	121	20.71	127					
	(12)	480		55.31	140	21.08	132					
	(13)	608		62.07	150	19.33	154					
	(14)	564		58.68	159	21.35	158					
	(15)	529		57.00	162	22.44	163					
	(16)	502		55.70	168	23.03	165					
	(17)	491		54.55	171	23.52	177					
	(18)	608		59.70	183	20.85	181					

therefore chosen for our earlier measurements in order to avoid the difficulties involved in cooling the cell. The vapor pressure of liquid ammonia is about eight atmospheres at room temperature, and two cells exploded under this pressure before a measurement was obtained. The strength of the cell was then increased by diminishing its size considerably, but the apparatus with which the determination was finally obtained exploded just as the series of measurements had been completed. The cell, however, was intact, and the explosion may have been caused by the pressure of a metal clamp which was used to mount the apparatus in the spectrophotometer.

As the construction of the quartz cells is somewhat troublesome, our subsequent experiments with sodium were performed at the boiling point of ammonia as described in paragraph IV (*d*), in order to avoid the risk of explosion.

Except that the cell was not cooled, the method of performing the measurements was the same for magnesium as for sodium. The magnesium was introduced in the form of freshly cut turnings. It was found necessary to leave the magnesium in contact with the liquid ammonia for at least a week before the blue color appeared. This delay in the appearance of the color is probably due to the formation of a protecting layer of hydroxide on the surface of the metal.

The fading during the course of the reading was very small. In 100 minutes ϵ for wave-length 6,080 changed from 0.942 to 0.914. This amounted to 0.03 per cent. per minute. The values for the fading correction given in Column IX., Table VII., were calculated from this value.

The explosion of the cell prevented the direct determination of the value $\epsilon_{NH_3} - \epsilon_h$. To obtain this value the following measurements were made:

(1) The "magnesium" cell filled with water compared with the glass water cell, $\epsilon_w - \epsilon_h$ (Table III.).

TABLE III.

Water in Mg Cell—Glass Water Cell.

λ μμ	Cell *R*	Cell *L*	$\epsilon_w-\epsilon_h$
664	39.53	36.54	.0306
608	29.29	35.81	.0358
564	39.09	35.23	.0401
529			.0494
502	38.41	33.70	.0553
480			.0600

TABLE V.

Water in Na Cell—Glass Water Cell

λ μμ	Cell *R*	Cell *L*	$\epsilon_w'-\epsilon$
664	45.72	32.83	
608	44.68	32.68	
564	45.02	33.23	.131
529	45.66	32.34	
502	46.09	32.16	

(2) The " sodium " cell filled with NH_3 compared with the glass water cell, $\epsilon_{NH_3} - \epsilon_h$ (Table IV.).

(3) The " sodium " cell filled with water compared with the glass water cell, $\epsilon_w' - \epsilon_h$ (Table V.).

By subtracting (3) from (2) we obtained $\epsilon_{NH_3} - \epsilon_w$. Adding to this the value (1), $\epsilon_w - \epsilon_h$, we obtained the required value $\epsilon_{NH_3} - \epsilon_h$ (Table VI.).

TABLE IV.

NH_3 in Na Cell—Glass Water Cell.

λ μμ	Cell R	Cell L	$\epsilon_{NH_3}' - \epsilon$
664			.072
635			.084
608	43.37	33.80	.098
585			.107
564	44.03	33.15	.112
546			.114
529	44.27	32.97	.116
515			.119
502	44.35	32.44	.123
491			.128
480			.133

TABLE VI.

NH_3 in Mg Cell—Glass Water Cell (calc.).

λ μμ	$\epsilon_{NH_3} - \epsilon_h$
664	−.028
608	+.003
564	+.021
529	+.034
502	+.047
480	+.062

TABLE VII.

Magnesium at Room Temperature (about 16° C.).

Exp. No.	λ μμ	Reading.		$\epsilon_{Mg} - \epsilon_h$	$\epsilon_{NH_3} - \epsilon_h$	$\epsilon_{Mg} - \epsilon_{NH_3}$	Time in Min.	Per Cent Correction.	$\epsilon_{Mg} - \epsilon_{NH_3}$ (Corrected).	a^2/ϵ	$\frac{1}{\lambda^2} \times 10^{-8}$ (Cm.$^{-2}$).
		a_3	a_4								
1	664	80.50	4.90	1.212	−.028	1.240	0	0	1.241	24.0	2.26
2	608	74.70	7.65	0.942	+.003	0.940	23	0.8	1.014	29.4	2.70
3	564	70.06	10.21	0.779	+.021	0.758	43	1.3	0.769	38.8	3.13
4	529	66.90	11.18	0.705	+.034	0.671	57	1.7	0.684	43.6	3.54
5	502	64.35	13.25	0.622	+.047	0.575	68	2.0	0.587	50.8	3.94
6	480	62.04	13.92	0.579	+.062	0.517	84	2.5	0.529	56.3	4.33
7	608	74.87	8.51				100				

The difference between the extinction coefficients of the blue magnesium solution and of liquid ammonia was then obtained by subtracting the value $\epsilon_{NH_3} - \epsilon_h$ from the value $\epsilon_{Mg} - \epsilon_h$. The measurements made with magnesium are summarized in Table VII., and are plotted in Fig. 2.

SUMMARY.

1. Kraus has shown that the electrical properties of solutions of metals in liquid ammonia can be explained on the assumption that the atoms of

the metal dissociate into electrons partly combined with the solvent and into the cations present in solutions of its salts in ammonia.

In order to test the assumption that the blue color of these solutions is due to the unsolvated portion of the electrons resulting from this dissociation, the absorption spectra of dilute solutions of sodium and magnesium in liquid ammonia were determined in the visible spectrum.

2. A solution of magnesium and a solution of sodium of the same intensity of color were found to have the same absorption spectrum. This is strong evidence in favor of the assumption that the coloring principle is the same in both cases.

3. Approximate measurements of the concentrations of the sodium solutions make it probable that Beer's law holds for the total dissolved sodium.

4. Qualitatively the evidence is in favor of the assumption that the absorption is due to the unsolvated electrons, but our measurements cannot be interpreted quantitatively in terms of Drude's theory of metallic dispersion.

CHEMICAL LABORATORY OF THE UNIVERSITY OF CALIFORNIA,
July 15, 1915.

The Conductivity of Metal–Ammonia Solutions

The following paper, by Kraus, on the conductance of alkali metal–ammonia solutions is truly a classic paper because the conductance-vs.-concentration data were invaluable in the testing of various models of metal–ammonia solutions. The data were unchallenged until 1968, when Dewald and Roberts reported ther conductivity study of the sodium–ammonia system.[1] They obtained good agreement with Kraus in the relatively concentrated range, but differed from Kraus in the region below 10^{-3} *M*. Although Kraus' work is usually considered to be flawless, this is probably a case in which his data are in error. Only in recent years has it been fully appreciated how difficult it is to eliminate all traces of water from liquid ammonia solutions.

[1]R. R. Dewald and J. H. Roberts, *J. Phys. Chem.*, **72**, 4224 (1968).

28

SOLUTIONS OF METALS IN NON-METALLIC SOLVENTS. VI. THE CONDUCTANCE OF THE ALKALI METALS IN LIQUID AMMONIA.[3]

By Charles A. Kraus.

Received January 19, 1921.

I. Introduction.

In the fifth paper of this series it was shown that, for a given concentration interval, the electromotive force of a concentration cell, in which sodium is employed as electrolyte in liquid ammonia between platinum electrodes, decreases rapidly at higher concentrations and apparently approaches a value of zero in very concentrated solutions. From the results of the fourth paper of this series it follows that, in solutions of sodium in ammonia, we have present normal sodium ions and a negative ion which appears to be characteristic of solutions of all metals in ammonia. If the electrode processes in the case of these concentration cells are reversible, and many facts indicate that this is the case, then the relative speeds of the positive and negative carriers may be calculated approximately, assuming the laws of dilute solutions to hold true. While these laws do not hold precisely, it may, nevertheless, be expected that the results of calculations based upon this assumption will be correct as to the order of magnitude.

The results indicate that, as the concentration of the sodium solution increases, the relative speed of the negative carrier increases. This increase is at first relatively slow, but, at the higher concentrations, the

[3] Previous papers of this series have appeared as follows: I, This Journal, **29**, 1557 (1907); II, *ibid.*, **30**, 653 (1908); III, *ibid.*, **30**, 1157 (1908); IV, *ibid.*, **30**, 1323 (1908); V, *ibid.*, **36**, 864 (1914); *Trans. Am. Electrochem. Soc.*, **21**, 119 (1912).

rate of increase with increasing concentration is greatly accelerated. If it be assumed that the positive carrier is in fact the sodium ion and that the speed of this ion does not vary greatly with the concentration, then it follows that the speed of the negative carrier increases with increasing concentration, this increase being the greater the greater the concentration.

If the above conclusion is correct, it follows that the conductance of a solution of sodium in liquid ammonia should increase with the concentration at higher concentrations and that this increase should be the greater, the greater the concentration of the solution. On the other hand, in dilute solutions the speed of the negative carrier is approximately independent of concentration and, since sodium dissociates according to a binary process of ionization, the conductance should increase with increasing dilution and should approach a limiting value in very dilute solutions corresponding to complete ionization.

The only measurements heretofore recorded on the conductance of solutions of the metals in liquid ammonia are due to Cady,[1] who found that solutions of sodium in liquid ammonia are excellent conductors, the conductance being considerably greater than that of typical salts in the same solvent. It is the purpose of the present investigation to supply data relating to the conductance of metal solutions in ammonia as a function of their concentration. At the same time preliminary observations are given on the temperature coefficients of these solutions and on their photo-electric activity.

The chief difficulty met with in measuring the conductance of such solutions is due to the reaction of the metal with ammonia to form the metal amide. Traces of impurities serve to catalyze this reaction to a measurable extent, particularly in dilute solution. By excluding all impurities, particularly oxygen, from the apparatus it is possible to overcome this difficulty in a very large measure. In the case of conductance measurements, however, it is necessary to introduce electrodes which serve to catalyze the reaction; it was not found possible to overcome this difficulty entirely. Various metals, such as gold, were employed in place of platinum as electrodes but no appreciable advantage resulted from such substitution. The only solution of the difficulty lay in reducing the surface of the electrodes to a minimum and in stirring the solutions during the course of the measurements.

II. Apparatus and Manipulation.

The apparatus employed in carrying out the measurements on the conductance of the metals in liquid ammonia at its boiling point is shown diagrammatically in Fig. 1.

The conductivity cell A was provided with a pair of platinum electrodes, B, con-

[1] Cady, *J. Phys. Chem.*, 1, 707 (1897).

sisting of 2 platinum wires having a diameter of 0.7 mm. and projecting into the cell a distance of about 4 mm.; the axes of the wires were separated by a distance of about 1.5 mm. These electrodes were connected externally with the tubes CC into which mercury was introduced. The cell was provided with a third electrode, D, which was intended for the purpose of measuring the conductance of solutions of higher specific conductance. In this case the resistance of the solution was measured between the pair of electrodes B and the electrode D, through the tube E, which had a diameter of approximately 5 mm. It was possible to measure the conductance of solutions up to concentrations as high as 2 *N*. The electrodes were not platinized, since the presence of platinum black greatly accelerates the reaction between metal and solvent, and since, owing to the reversibility of the metal electrodes in these solutions, a perfect minimum is obtained with unplatinized electrodes even at the highest dilutions. One difficulty, however, arises in the case of the more dilute solutions; reaction between the solvent and the dissolved metal takes place immediately in the neighborhood of the electrode surfaces. This results in a diminution of the measured conductance of the solution, even though the effect on the concentration of the solution as a whole is negligible. It was found possible to overcome this difficulty by means of the stirrer F which was suspended by means of a platinum-iridium spring G. A soft iron core H, enclosed in glass, was attached to the lower end of the spring. The tube I, within which the stirrer system was suspended, was provided externally with

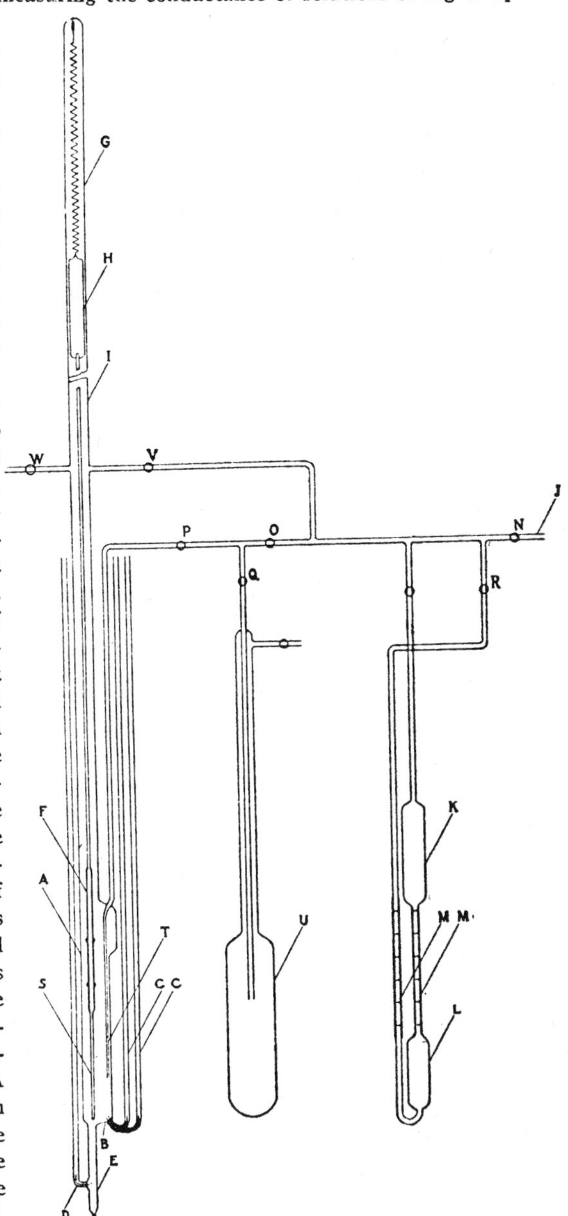

Fig. 1.—Conductance apparatus.

a solenoid (not shown in the figure), by means of which it was possible to actuate the stirrer. A slight motion of the stirrer served to equalize the concentration of the metal in the neighborhood of the electrodes and made it possible to obtain consistent measurements.

The metal was introduced at the bottom of the tube E by a method which has been described in a previous paper.[1] The stopcock W made connection with a vacuum pump. The ammonia supply was contained in a metal cylinder in which it was treated with metallic sodium; this stock cylinder was attached to the tube J. Since it was necessary to determine the concentration of the solutions, the ammonia introduced into the cell was measured before introduction. In order to do this, it was first condensed in the measuring cell K, which consisted of the chamber L and 2 graduated Tubes MM. These tubes were about 10 cm. in length and had a total capacity of 1 cc., graduated to 0.01 cc.; the cell having been calibrated, the volume of the ammonia could be determined within a few thousandths of one cc. The measuring cell was immersed in boiling ammonia during the process of condensation. When the desired amount of ammonia had been condensed in the measuring cell, the stopcocks N, R were closed and the stopcocks O, P and V were opened. It should be stated that, prior to these operations, the entire apparatus was exhausted to a pressure at least as low as 0.001 mm. of mercury. The measuring cell was provided with the enlarged chamber K in order to facilitate the boiling of the ammonia in the cell during the course of its transfer to the conductivity cell A. The solution was stirred while ammonia was being introduced into the cell, the stirrer F being provided with a smaller extension S reaching to the bottom of the tube E.

When the desired amount of ammonia had been measured and introduced into the conductivity cell, conductance measurements were made by means of a bridge and telephone. In the case of the more concentrated solutions, it was possible to employ a direct current in conjunction with a galvanometer, since polarization effects appear to be wanting in these solutions. It was found more convenient, however, to employ a telephone, inasmuch as this instrument had to be employed in the more dilute solutions in any case. The measuring cell had a volume of approximately 5 cc. and the conductivity cell was so designed that its volume up to the end of the small capillary Tube T was also approximately 5 cc. While carrying out the measurements on the conductance of the solution in the conductivity cell, a fresh supply of ammonia was condensed in the measuring cell. After having completed the conductance measurement, this ammonia was transferred to the conductivity cell, as has already been described. The volume added in each case was approximately the same, namely, 5 cc., and thus the concentration of the solution was reduced by one-half on each addition of fresh solvent. After completing the measurements on the conductance of the new solution, it was necessary to remove a known fraction, about $^1/_2$, of the solution, in order to be able to introduce new solvent; this was done by withdrawing a portion of the solution through the tube T. In order to do this, there was provided an auxiliary Tube U, immersed in liquid ammonia boiling below normal pressure; this tube, like the other parts of the apparatus, had been previously exhausted. When it was desired to transfer the solution from the conductivity cell to the tube U, the stopcock O was closed and the stopcocks P and Q were opened. On opening these stopcocks, the solution was forced under its own vapor pressure through the capillary T and the stopcocks P and Q into the cell U. Since the ammonia surrounding this cell was boiling under reduced pressure, sufficient pressure difference was maintained between cell A and the tube U to provide for the transfer of the ammonia from A to U. When the level of the solution in A reached the bottom of the capillary T, the transfer of solution ceased. In this way there was left

[1] THIS JOURNAL, 30, 1206 (1908).

in the cell A a quantity of solution approximately equal in volume to that originally present, and at a concentration approximately $^1/_2$ that of the original concentration. Into this solution a fresh quantity of ammonia was condensed and the conductance of the resulting solution measured. These operations were repeated until the concentration had been carried to as low a value as desired.

It was, of course, necessary to determine exactly the amount of ammonia left behind in the cell T after each operation. This was done by measuring the amount of solution transferred from the conductivity cell to the tube U. For this purpose, the bath surrounding the tube U was removed, and the ammonia evaporated rapidly and absorbed in a weighed flask of water. When the ammonia in U had been completely absorbed in water, the flask was again weighed and the amount of ammonia which had been transferred to the cell U was thus determined. The amount of solution left in the cell was then found by subtracting the weight of the solution transferred from that originally present in the cell. In calculating the weight of ammonia in the measuring cell KL, the density of the liquid was assumed to be 0.674 at the boiling point of liquid ammonia.

When the conductivity measurements were completed, the ammonia remaining in the cell A was evaporated and absorbed in water. The amount of ammonia introduced into the cell, as measured in the measuring cell, could thus be checked by the total of the amounts absorbed in water. In the course of a run, practically all the metal was transferred from the conductivity cell A to the tube U. On the completion of the conductance measurements, a small amount of alcohol was introduced into U, thus converting the metal to an alcoholate. Water was then added, and the contents of the tube were washed out into a beaker and titrated against standard acid. In this way the amount of sodium present in the inital solution was determined. It should be noted in this connection that, prior to these operations, any residual ammonia in the apparatus was pumped out in order to insure the accuracy of the titrations.

The conductance cell was calibrated with respect to standard solutions of potassium chloride. Owing to the small area of the electrode surfaces and their unplatinized condition, it was not possible to determine the cell constant directly. This difficulty was overcome by comparing the constant of this cell with that of an auxiliary cell whose constant had been determined by means of standard potassium chloride solutions. For the purpose of this comparison, a solution of potassium iodide, nearly saturated with iodine, was employed. The electrodes in the case of potassium iodide-iodine solutions do not polarize, and minima can be obtained even with very small electrode surfaces. The relative constants of the pair of electrodes B and of the electrode D, with one of these electrodes, could be obtained in the course of any given series of measurements by measuring the resistance of a given solution across these 2 pairs of electrodes. The constants of the cell were changed slightly from time to time owing to minor alterations which were made in the cell. In the Series 16 to 19 the constants of the 2 pairs of electrodes were 39.51 and 0.9102. In the other series, however, the constants differed somewhat from these values.

III. Conductance of Sodium, Potassium, Lithium, and Mixtures of Sodium and Potassium at the Boiling Point of Ammonia.

The results of the measurements on the conductance of sodium in liquid ammonia are given in Table I. At the head of the table is given the series of the experiment. In the first column is given the number of the experiment, in the second column the dilution, V, in liters per gram atom of sodium, and, in the third, the equivalent conductance H in the customary

TABLE I.—CONDUCTANCE OF SODIUM IN AMMONIA.

Series III.

No.	V.	H.	l.
1	0.5047	82490	163.5
2	0.6005	44100	73.43
3	0.6941	23350	33.63
4	0.7861	12350	15.71
5	0.8778	7224	8.230
6	0.9570	4700	4.872
7	1.038	3228	3.082

Series X.

No.	V.	H.
1	0.8478	10050.0
2	1.694	1118.0
3	3.525	679.5
4	7.023	542.1
5	13.86	481.5
6	27.50	476.7
7	55.35	524.3
8	109.0	613.2
9	203.3	688.7
10	417.8	778.4
11	847.2	844.8
12	1750.0	879.2
13	3621.0	906.6
14	7302.0	916.0
15	14980.0	916.2

Series XV.

No.	V.	H.
1	0.7165	20280.0
2	1.438	1429.0
3	3.411	690.8
4	7.485	535.3
5	16.08	477.5
6	37.04	491.7
7	80.65	565.2
8	174.3	676.6
9	389.1	800.4
10	865.8	888.8
11	1945.0	938.7
12	4369.0	957.9
13	9750.0	957.2
14	21990.0	961.8
15	51340.0	155.2

Series XVI.

No.	V.	H.
1	0.7101	18000.0
2	1.460	1308.0
3	3.007	712.9
4	6.845	538.3
5	14.98	473.6
6	33.25	483.9
7	73.99	555.1
8	162.9	668.8
9	360.8	790.3
10	796.2	886.8
11	1763.0	944.3
12	3873.0	969.9
13	8352.0	981.5
14	19050.0	960.7
15	33230.0	1020.0

Series XVII.

No.	V.	H.
1	0.7332	20430.0
2	1.519	1292.0
3	3.589	670.1
4	7.697	519.5
5	16.42	473.9
6	35.70	488.5
7	78.54	562.0
8	172.1	676.9
9	377.3	801.5
10	850.0	898.2
11	1887.0	954.1
12	4075.0	977.9
13	8954.0	988.8
14	19550.0	1000.0
15	44260.0	1020.0

Series XIX.

No.	V.	H.
1	0.6107	46340.0
2	1.239	2017.0
3	2.798	749.4
4	6.305	554.7
5	13.86	478.3
6	30.40	478.5
7	65.60	540.3
8	146.0	650.3
9	318.6	773.4
10	690.1	869.4
11	1551.0	956.6
12	3479.0	988.6
13	7651.0	1009.0
14	17260.0	1016.0
15	37880	1034.0

Kohlrausch units. In Series III the specific conductance l of the solution in reciprocal ohms is given in the last column. In calculating the dilutions, V, corrections were not made for the densities of the solutions. All values given are based on the value 0.674 for the density of pure ammonia. A saturated solution of sodium in liquid ammonia has a density only a little greater than 0.5. Consequently the true dilutions in the more concentrated solutions are appreciably greater than those given in the table.

Each series of experiments was carried out as rapidly as possible and required from 3 to 5 hours for completion. The results of some of the series carried out were discarded owing to accidents or obvious errors in the course of the experiments and other series have not been included as they add nothing material to the results here given. A number of the earlier series of experiments have not been included owing to the fact that the results in the more dilute solutions are obviously in error.

In the earlier experiments the equivalent conductance in the more dilute solutions passed through a maximum value. This result is unquestionably due to reaction between the metal and the solvent. The sodium amide formed is relatively only a poor conductor and as a consequence the conductance of the solution falls. As the experimental methods were refined, this difficulty was in a large measure overcome. This is illustrated, for example, in Fig. 2, where the points in dilute solutions for each succeeding run from 15 to 19 lie one above the other. In the more concentrated solutions, there was no indication of a change of the conductance with time, but in the most dilute solutions such a change was appreciable. The change, however, in the later series of experiments was not large and it is believed that the error introduced, owing to this effect, is comparatively small, probably not exceeding a few per cent., if as great as this.

TABLE II.—CONDUCTANCE OF POTASSIUM IN AMMONIA.

	Series IV			Series III.	
No.	*V*.	*H*.	No.	*V*.	*H*.
1	1.695	1414.0	1	2.635	848.6
2	3.459	751.9	2	5.384	652.5
3	7.560	596.1	3	11.30	554.2
4	15.87	532.1	4	23.80	526.0
5	34.14	534.8	5	49.56	582.4
6	70.94	594.6	6	105.2	646.0
7	150.6	686.6	7	226.2	752.6
8	324.4	787.2	8	483.5	844.5
9	680.6	860.0	9	1056.0	908.0
10	1434.0	894.3	10	2226.0	933.3
11	2990.0	884.1	11	4695.0	936.5
			12	9759.0	945.0

In Table II are given the results of two series of measurements with potassium. It is evident from the form of the curve, as may be seen by referring to Fig. III, that the solutions of potassium were much less stable than those of sodium. In Series IV the curve exhibits a maximum in dilute solution and in Series III the conductance values are obviously low.

In Table III are given values for the conductance of lithium in ammonia. These measurements were not extended to the more dilute solutions owing to the fact that in these solutions the metal reacted with the solvent to a greater extent than either sodium or potassium.

TABLE III.—CONDUCTANCE OF LITHIUM IN AMMONIA.

No.	*V.*	*H.*
1	0.4229	25600.0
2	0.8323	9642.0
3	1.990	879.4
4	4.250	619.7
5	8.817	501.2
6	18.13	454.6
7	40.17	502.2
8	85.94	569.9
9	181.6	654.4

Finally, in Table IV are given the results of measurements on a mixture of sodium and potassium. The amounts of sodium and potassium present in the mixtures were determined approximately by weighing as chlorides which give the values of 59.4 atomic per cent. of sodium and 40.6 atomic per cent. of potassium present in the mixture.

The mixtures of sodium and potassium were made up by melting these metals together in the course of their introduction into the cell according to the method described in an earlier paper. If ammonia is employed for the purpose of driving the metal over into the conductivity cell, care must be exercised not to allow ammonia vapor to remain in contact with the metal for an appreciable period of time, since this vapor is soluble in the liquid alloy. Reaction takes place between the ammonia and the alloy, resulting in the formation of solid amide. That ammonia is soluble in the liquid metal, is shown by the fact that, when left in contact with the metal, there is a diminution of pressure. At the same time the surface of the metal becomes tarnished and minute gas bubbles are formed within the body of the metal due to an evolution of hydrogen within the metal. In the course of time the liquid becomes spongy, resembling in many respects the product obtained by the action of sodium amalgam on an ammonium salt. Ultimately the pressure again rises, due to the evolution of hydrogen. From these facts it appears that ammonia has a considerable solubility in the liquid metal and that reaction takes place between the ammonia and the metal within the body of the liquid metal phase. If the temperature of the alloy is raised, the amides fuse. The point of fusion depends upon the concentration of the amide, and, under proper conditions, the point of fusion may be brought down below 100°. The alloy is somewhat soluble in the mixtures of fused amides, as is indicated by a blue color which appears to be in every way identical with the color due to the metals dissolved in liquid ammonia[1]. This color, moreover, resembles that obtained when ammonia acts upon metallic sodium at elevated temperatures. Ordinarily the process of reaction consists in an initial solution of the metal in the fused amide and a reaction between the am-

[1] McGee, THIS JOURNAL, 43, 500 (1921).

monia and the metal in this phase. The solutions of the alkali metals in the fusion of mixed amides appear to be entirely stable.

TABLE IV.—CONDUCTANCE OF MIXTURES OF SODIUM AND POTASSIUM IN AMMONIA.

No.	*V*.	*H*.
1	1.158	2493.0
2	2.292	885.1
3	4.983	633.3
4	10.53	526.6
5	21.44	495.0
6	44.78	524.2
7	93.56	604.5
8	193.8	712.0
9	407.5	828.1
10	846.1	917.3
11	1765.0	984.5
12	3645.0	1018.0
13	7556.0	1040.0
14	15250.0	1064.0

Preliminary measurements were also carried out on very concentrated solutions of sodium. For this purpose a special cell was designed in which the conductance of the solution was compared directly with that of mercury. As these measurements are only preliminary, it will be unnecessary to describe the cell in detail. It may be stated, however, that the resistance element consisted of a glass tube having a length of approximately one meter and having a diameter somewhat smaller than one millimeter. The resistance of this cell, when filled with mercury at 0° was 2.65 ohms; and when filled with a saturated solution of sodium in liquid ammonia at the boiling point of liquid ammonia, was 5.49 ohms. The specific conductance of the saturated solutions of sodium in liquid ammonia is 0.479 that of mercury, or 0.51×10^4 reciprocal ohms.

IV. Temperature Coefficient of the Conductance of Sodium in Ammonia.

The temperature coefficients of solutions of sodium in liquid ammonia were likewise measured for several solutions whose concentrations, however, were not determined. The method employed was similar to that employed by Franklin and Kraus.[1] The results are given in Tables V, A and V, B. In Table V, A the temperature is given in the first column and the resistance in the second column, the reciprocal of the resistance in the third column, and in the last column the mean temperature coefficient of the conductance in terms of the conductance at —33°. This series of measurements was carried out with a fairly dilute solution. The results of the series of measurements with a much more concentrated solution are given in Table V, B, in which the temperature is given in the first column and the resistance in the second column.

[1] Franklin and Kraus, *Am. Chem. J.*, 24, 83 (1900).

TABLE V.—TEMPERATURE COEFFICIENT OF THE CONDUCTANCE OF SODIUM AMMONIA.

A. Dilute solution.				Concentrated solution.	
t, °C.	R.	$1/R \times 10^3$.	$\frac{\Delta(1/R)}{\Delta t(1/R)_{33}} \times 100$.	t, °C.	R.
—33	124.3	8.04	....	+18	0.70
—13	85.7	1.17	2.25	—33	2.10
+17	43.4	23.9	4.69	+16	0.71
+48	28.2	35.0	5.34	+50	0.47
+85	15.6	64.0	9.00	+24	0.65
+23	47.7	....	....	—33	2.70

It is apparent from these results that at the higher temperatures the rate of reaction between the metal and the solvent is greatly accelerated. This was quite evident during the course of the experiment, since, when the temperature was maintained constant, the resistance of the solution increased. After having raised the temperature of the cell above that of the room, on bringing it back to room temperature, it was invariably found that the resistance of the solution had increased. This will be seen from a comparison of the results at 17° and 23° in Table V, A, and further from the results of the two measurements at —33° in Table V, B. In the latter case, the resistance was increased from 2.1 to 2.7 ohms, after heating up to 50°. It is obvious that the conductance as measured at the higher temperatures is considerably below the true values, owing to this interaction between solvent and solute.

For the purpose of comparing the temperature coefficient of the metal solutions with those of other substances dissolved in ammonia, the following tables are included. In Table VI are given the results of measurements on the conductance of solutions of a compound of sodium and lead $NaPb_x$[1] at a series of temperatures, and in Table VII are given the results of measurements on the conductance of solutions of potassium amide.

TABLE V.

Temperature coefficient of $NaPb_x$.

t, °C.	R.
—33	134
— 9	150
+ 6	156
+25	155
+38	150

TABLE VII.

Temperature coefficient of potassium amide.

t, °C	R.
—33	69.9
— 6	71.8
+20	63.4
+38	50.5

V. Photo-Electric Properties.

One of the characteristic properties of the alkali metals is their photoelectric activity under the action of visible radiations. This property is found not alone in the case of the elementary metals, but also in the case of their alloys. A preliminary experiment was carried out to determine whether or not the solutions of the metals in ammonia exhibited photoelectric activity. For this purpose 0.25 g. of lithium was introduced

[1] Smyth, THIS JOURNAL, 39, 1299 (1900).

into a tube having a diameter of approximately 3 cm. and a length of 35 cm. A platinum wire was sealed in the bottom of the tube making contact with the metal solution, and an aluminum wire was fixed about 3.5 cm. above the surface of the metal and connected with an electrometer. When ammonia gas was introduced, photo-electric activity was observed as soon as the metal began to go into solution. With an illumination in the room due to one or two 16-candle power lamps, the charge was still observable at the end of 5 minutes. With a 16-candle power lamp held about a foot from the metal, the electroscope was discharged in the course of 5 seconds. It should be noted, however, that the discharge took place with negative, as well as positive charge of the electrometer system. This action was doubtless due to the fact that, owing to the presence of the platinum lead wire in the solution, decomposition took place rapidly and particles of the solution were projected on the aluminum wire with which the electrometer was connected. It is obvious that the photo-electric action was not due to the presence of free metal, since the vapor pressure of the saturated solution of lithium in liquid ammonia is very low. As the pressure due to ammonia in the apparatus was much above the saturation pressure of the solution, it is obvious that free metal could not have existed. It follows, therefore, that, in a solution of an alkali metal in liquid ammonia, the metal is photo-electrically active just as it is in the case of the free element or of an alloy.

VI. Discussion of Results.

The results of the conductance measurements with the sodium solutions at the boiling point of ammonia are shown graphically in Fig. 2, where the equivalent conductances are plotted as ordinates and the logarithms of the dilutions as abscissas. Curve A relates to the more dilute solutions, up to a concentration of about 1 N. In Curve B are plotted

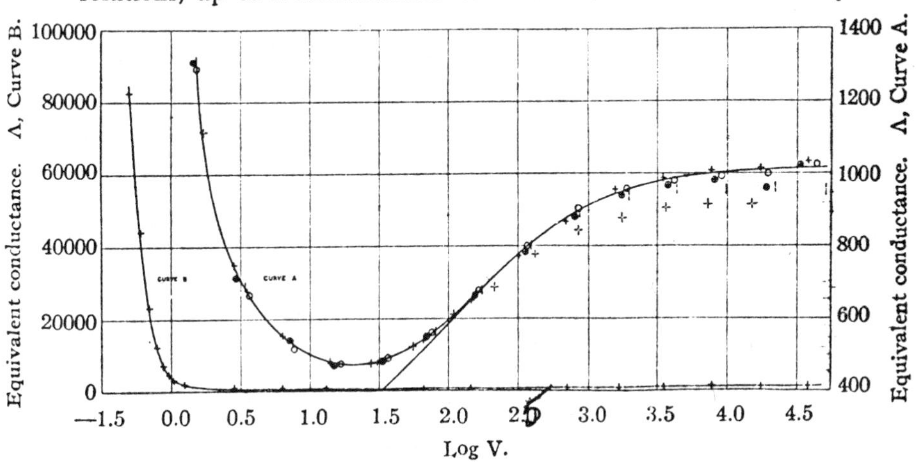

Fig. 2.—Showing conductance of sodium as a function of dilution.

the results up to a concentration of approximately 2 N. In order to represent the results at these higher concentrations, the scale of ordinates has been diminished in the ratio of 100 to 1. The scale for the more dilute solutions appears on the right-hand margin of the figure, and that for the more concentrated on the left. On Curve A, values due to Series 19 are represented by crosses, Series 17 by circles, Series 16 by combined crosses and circles, Series 15 by a broken vertical line, and Series 10 by an open cross. The points for the more concentrated solutions on Curve B are taken from Series 3. The remainder are taken from Series 19.

Considering first the more dilute solutions, it is apparent that as the dilution increases the equivalent conductance increases and approaches a limiting value. The form of the curve in general appears to be very similar to that of ordinary electrolytes dissolved in liquid ammonia. In the dilute solutions the curve appears to conform to the mass-action law. In Fig. 2, the curve as drawn in the more dilute solutions corresponds to the mass-action constant 77.27×10^{-4}, assuming the value of $\Lambda_0' = 1016$. As may be seen from the figure, the points lie upon this calculated curve within the limits of error up to a concentration of approximately 0.01 N, after which the experimental curve begins to diverge very rapidly from the calculated one, which is here continued as a line sloping downward. It is doubtful, however, whether the mass-action law applies up to this concentration in reality. In dilute solutions the chief source of error is due to the formation of amide which tends to reduce the observed value of the conductance of the solution. It is possible, therefore, that the curves in dilute solutions are somewhat flattened out owing to this effect. This might readily lead to an apparent agreement with the mass-action law, which does not exist in fact. In this connection it is to be noted that, as the manipulations were refined, the values observed in the dilute solutions rose continuously as shown in Fig. 2. Furthermore, it will be noted that, in Fig. 3, the curve for a mixture of sodium and potassium rises continuously and apparently considerably more steeply in dilute solutions than does the curve for sodium. Since the sources of error are in a direction to yield too low a value of the conductance, it appears not improbable that the highest values observed in the case of the sodium solutions are lower limits and that the true values should be somewhat higher than these. If this is the case, then the mass-action law does not hold at the higher concentrations; and the value of the mass-action constant would be materially decreased, if it be assumed that the mass-action law applies in the more dilute solutions as it does in the case of typical salts.

In any case, it is evident that the general form of the curve is similar to that of the salts in liquid ammonia solutions. It is remarkable, however, that the limiting value which is approached in the case of the ammonia solutions is very much higher than it is in the case of typical salts. The highest value so far observed in the case of salts is in the neighborhood of 340, which is only about $^1/_3$ that of the metal solutions.

The most striking result obtained, however, is that in the more concentrated solutions. After passing through a minimum at a concentration of about 0.05 *N*, the curve of the equivalent conductance begins to rise rapidly with increasing concentration. The general type of the curve corresponds somewhat to an exponential relation between the conductance and the concentration. For a concentration of approximately normal, the equivalent conductance rises to a value of approximately 3000, and in doubling the concentration the equivalent conductance reaches a value of approximately 80,000. For example, the equivalent conductance of a solution of sodium having a dilution of 0.5047 liter per gram atom is 82490, which corresponds to a specific conductance of 0.01635×10^4. The order of the conductance at this concentration is evidently approaching that of a metal; thus, the specific conductance of mercury is 1.063×10^4 or about 6 times that of the solution in question.

The results for potassium and lithium and for mixtures of sodium and potassium are shown in Fig. 3. Here the uppermost curve is that of potassium, the intermediate one that of a mixture of sodium and potas-

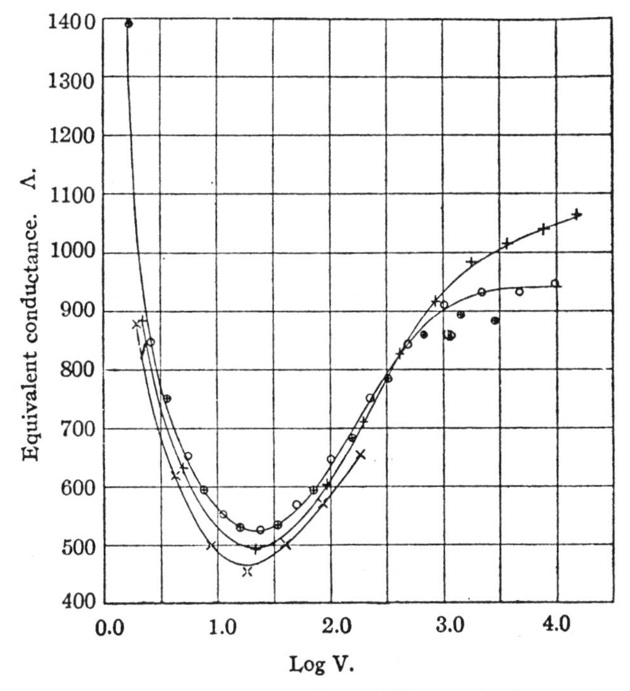

Fig. 3.—Showing conductances of potassium, lithium and mixtures of sodium and potassium as functions of the dilution.

sium, and the lowest one that of lithium. The curves in all cases have approximately the same form. In the more dilute solutions the curve for potassium falls below that of sodium. There can be no question but that this result is due to the formation of amide in reaction between potassium and the solvent; this reaction apparently takes place much more readily in the case of potassium than in the case of sodium. It is safe to assume, therefore, that the actual curve for potassium lies everywhere above that for sodium in the more dilute, as well as in the more concentrated solutions. The curve for lithium lies somewhat below that of sodium, although not so much as might be expected. Since only one run was made with lithium, it is possible that the position of this curve is somewhat in error owing to possible errors in determining the concentration. In all three cases the minimum point of the curve occurs at approximately the same concentration. The forms of the curves in the more concentrated solutions are similar in the cases of the three metals. The same holds true for the mixture of sodium and potassium. This mixture of sodium and potassium appears to have been exceptionally stable and a comparison of this curve with that of pure sodium affords perhaps a better means for determining the relative conductance of sodium and potassium than do the solutions of pure potassium. If the negative carrier is the same for the three metals, then it follows that in the more dilute solutions, the difference in the conducting power of two metals should correspond to the difference in the conducting power of their positive ions. Roughly, this is borne out by the results as shown in the curves for sodium and potassium and in those for the mixture of sodium and potassium, although the results so far are not sufficiently precise to make an exact comparison possible. These results indicate, also, that if an equilibrium exists between the positive and negative ions, this equilibrium is approximately the same for a solution of a mixture of sodium and potassium as it is for solutions of these elements alone. This is what we should expect if the solutions of the metals themselves behaved in a manner similar to that of the salts, or if the solutions of the metals had dissociation functions of approximately the same value.

The solutions of the metals differ markedly from solutions of ordinary electrolytes as regards their temperature coefficients. In general, the temperature coefficients of salts are positive in the neighborhood of the boiling point of ammonia where they have a value of approximately two per cent. per degree. As the temperature rises, the coefficient decreases, passes through zero, and thereafter becomes negative. This behavior is exemplified in the case of the examples given for solutions of potassium amide and $NaPb_x$. In the case of the alkali metal solutions in liquid ammonia, the temperature coefficient in the neighborhood of the boiling point of ammonia is slightly higher than that of typical electrolytes, and, as the temperature increases, the value of the temperature coefficient increases rapidly. For the temperature interval between —33° and —13°, the coefficient has a value of 2.25%, while between 48° and 85° the coefficient has a value of 9.0%. These values, it is to be borne in mind, are lower limits, since the conductance as measured at the higher temperatures is materially reduced, owing to the formation of amide. It will be observed also that the coefficient for the more concentrated solutions differs but little from that of the more dilute. Between —33° and +17° the resistance decreases in the ratio of 124.3: 43.4, or approximately 3:1. For the temperature interval —33° to +16°, the resistance of the more concentrated solution decreases in the ratio of 2.10: 0.71, a value almost identical with that of the more concentrated solution. Similar relations hold for the higher temperature intervals.

In order to determine the value of Λ_0, the limiting value which the equivalent conductance approaches as the concentration decreases, it is necessary to assume or determine the form of the conductance function. As has already been stated, the results up to 0.01 N are in agreement with the law of mass action assuming 1016 for the value of Λ_0. This value is a lower limit since the sources of error tend to give too low a value of the conductance. It seems unlikely that the upper limit can be higher than 1050 if the experimental values in dilute solutions are anywhere near correct. The lower limit 1016 may be assumed as correct for the purpose of this discussion.

If we assume 1016 as the limiting value of the equivalent conductance for a sodium solution in liquid ammonia, and if we assume for the equivalent conductance of the sodium ion the value 130, we obtain for the equivalent conductance of the negative carrier in dilute metal solutions the value 886. This result may be compared with that obtained in the preceding paper of this series. It was there shown that the speed of the negative carrier approaches in dilute solution a value approximately seven times that of the positive carrier. The ratio of the conductance of the negative carrier in a sodium solution to that of the sodium ion, on the basis of the values given above, yields the value 6.8 for the ratio of the speeds of the two ions. This value is, therefore, in excellent agreement with the results of the electromotive force measurements in dilute solutions of sodium in liquid ammonia.

If the carrier in these solutions is the negative electron, it must be assumed to be associated with ammonia molecules in order to account for the low value of its equivalent conductance. It is interesting to compare the speed of the negative electron in ammonia with the speed of the negative electron in other solvents. Jaffe[1] has measured the speed of the negative electron produced by radiations in hexane. He found the

[1] Jaffe, *Ann. Physik.*, **32**, 148 (1910).

value 4.17×10^{-4} cm. per volt-second. On the basis of the above value for the conductance of the negative electron in liquid ammonia solutions, we calculate for the speed of this ion the value 91.7×10^{-4} cm. per second. We thus see that in liquid ammonia the speed of the negative carrier is approximately 20 times as great as that of the negative electron in hexane. The fluidity of ammonia at its boiling point is 376 and that of hexane is 312. It has been found that the speeds of ions of a given type are roughly proportional to the fluidities of the solvents in which they are dissolved—at any rate, this is very nearly true in the case of the larger ions. It may be inferred, therefore, that the negative electron in ammonia is associated with a much smaller number of solvent molecules than it is in hexane, as is to be expected since the dielectric constant of ammonia is 22 and that of hexane is 1.85. It is not possible to say what the exact nature of the combination between the electron and the ammonia molecule is. That the combination is indeed a very loose one is fairly certain from the behavior of the more concentrated solutions. Such a combination is not to be looked upon as a chemical compound in the ordinary sense of the word; such as we have, for example, in the case of the chlorine ion. In the latter case the combination between the negative electron and the chlorine atom is a fixed one, and the formation of such an ion is unquestionably associated with the evolution of considerable energy. This is not the case with the association formed between the negative electrons and the ammonia molecule.

As was shown in the fifth paper of this series, at higher concentrations, the relative speed of the negative carrier increases enormously. If the speed of the positive carrier remains fixed, or approximately fixed, it follows that at higher concentrations the speed of the negative carrier increases rapidly with increasing concentration. As a result, it follows that, unless the number of carriers at higher concentrations decreases very greatly, the conductance of these solutions should increase largely at the higher concentrations. Such is indeed found to be the case.

Between 0.5 and 1 N, the relative speed of the negative carrier begins to increase very rapidly, according to the electromotive force measurements, and over this concentration range the conductance of these solutions likewise begins to increase very rapidly. The fact that, at somewhat greater concentrations, the equivalent conductance reaches values of an entirely different order of magnitude from that in the more dilute solutions, shows clearly that this increase in the conductance cannot be due to an increase in the number of ions. It must, therefore, be due to an increase in the mean speed of the carriers. This may be accounted for on the assumption that at higher concentrations the carriers in part become free from their association with the surrounding ammonia molecules and therefore move with velocities comparable with that of the negative electrons in metals. The current transported by such carriers fulfils the conditions for metallic conduction. It may be surmised, therefore, that conduction in metals is effected by the same carrier.

In dilute solutions, as has been shown above, the conductances of the alkali metals differ in the order of the conductance of their positive ions, and the difference in these conductances corresponds approximately with the difference in the conductances of these ions. It follows, therefore, that the negative carriers in solutions of the different alkali metals have the same conducting power, or in other words are identical.[1] Since the

[1] This is further borne out by the photometric investigation of these solutions by Gibson and Argo. (*Phys. Rev.*, **1**, 33 (1916); THIS JOURNAL, **40**, 1327 (1918)). They found that the absorption curves for the solutions of different metals in liquid ammonia are practically identical. Such a result can only be accounted for on the assumption that there is present in these solutions a charged particle which has the same optical properties in all of them. The solutions in question were comparatively dilute, in which case there were present no carriers which were not associated with ammonia. Apparently, therefore, the absorption spectrum is due to a complex between the negative electron and the solvent. Since it has been shown that it is only at much higher concentrations that the carriers become free, to an appreciable extent, from the ammonia molecules, it follows that evidence as to the presence of these free carriers by means of photometric observations can be gained only by a study of the more concentrated solutions. In the dilute solutions studied Beer's Law apparently holds. Since the ionization in these solutions is relatively high, it follows that this result cannot be interpreted as excluding the possibility that Beer's Law may fail in the more concentrated solutions; that is, the absorption spectrum, due to a negative electron combined with a metal atom, may differ from that of the negative electron combined with an ammonia molecule alone. In the case of ordinary electrolytes the absorption due to the un-ionized fraction appears to be identical with that due to the ions. In order to harmonize this result with our conceptions of the ionization of electrolytes, it is necessary to assume that the optical properties of the ions are not materially affected by the union of the ions to form a molecule. Such an hypothesis seems quite permissible in the case of ordinary electrolytes. In the case of the metal solutions, however, it is scarcely to be assumed that the motion of an electron associated with a metal atom will be identical with the motion of an electron associated with an ammonia molecule. It would appear, therefore, that at higher concentrations deviations from Beer's Law should be found. Should this not prove to be the case, it would be necessary to assume that the neutral molecule consists not merely of the metal atom, but of the metal atom in association with ammonia; and, moreover, it would have to be assumed that the combination of the negative carrier with ammonia is not materially affected by the presence of the metal atom. In view of the fact that at higher concentration the negative carrier is dissociated from the ammonia molecule, it seems improbable that this is the case. It would seem, therefore, that a photometric study of the more concentrated solutions should afford means for establishing the existence of neutral molecules in the solutions of the metals. In this connection it may be noted that the absorption curves for the metals in methyl amine, as observed by Gibson and Argo, are much more complex than they are in the case of ammonia. In this case, Beer's Law does not hold. Although conductance data are not available, it appears probable that the ionization of the metals in methyl amine is much lower than it is in ammonia. The ionization of the metals in ammonia is of the same order of magnitude as that of the salts in this solvent. If, similarly, the

conductance in the more concentrated solutions appears in every way similar to that in ordinary metals, it appears highly probable that all metals owe their power to conduct the current to the presence of this carrier.

These solutions, therefore, constitute a connecting link between metallic and electrolytic conductors. In dilute solutions the process is, at least in part, electrolytic. A portion of the current is carried by the positive carriers which differ in no respect from the positive carriers as they appear in solutions of the common salts. The negative carrier is chemically uncombined, but is associated with one or more molecules of the solvent. These carriers are identical for solutions of all metals, and, when discharge occurs at the anode, the only material process which takes place is that a portion of the solvent is left behind in the immediate neighborhood of this electrode. As the concentration of the solutions increases, the nature of the phenomenon changes only insofar as the combination of the negative carrier with ammonia is affected. At the higher concentrations, the negative carriers are free from association with the ammonia molecules to a greater and greater extent. And, since under these conditions, the negative carrier is associated with no matter of atomic dimensions, it follows that all material effects cease so far as these carriers are concerned. It is not to be understood that a given carrier is free from association with the solvent molecules for any considerable period of time. Obviously, an equilibrium must exist between the free carriers and the combined carriers and ammonia, which results in a constant interchange between the free and the bound carriers. During the interval over which these carriers are free from the solvent molecules, they conduct just as they do in the metals. As the concentration is further increased, the number of these free carriers constantly increases. It is evident that their number in the more dilute solution, for example in the neighborhood of normal, must be relatively small, since at higher concentrations the equivalent conductance reaches values some one hundred times as great as that at normal concentration. It is not possible to determine the actual number of carriers in the more concentrated solutions. In the more dilute solutions, however, it appears that the number of carriers decreases with increasing concentration, just as it does in the case of normal electrolytes in liquid ammonia.

VII. General Summary.

The purpose of the present investigation, the results of which have been given in the preceding series of papers, was to determine the nature of

ionization of the metals in methyl amine were of the same order of magnitude as that of the salts in methyl amine, it follows that at the concentrations at which the photometric measurements were made, appreciable amounts of the un-ionized molecules are present.

the solutions of the metals in liquid ammonia and in particular to determine the nature of the conduction process in these solutions.

The alkali metals do not form compounds with ammonia, and, while it is possible or even probable that ammonia may be associated with the metals in solution, such an association must be a comparatively feeble one, since the energy change accompanying the process of solution is inconsiderable. Calcium forms a hexammoniate and in all likelihood barium and strontium likewise form ammoniates. The compound formed between calcium and ammonia possesses metallic properties. It is evident that the mechanism of the union between calcium and ammonia is very different from that of compounds of calcium in which this metal appears combined with a strongly electronegative element, or group of elements.

From measurements of the vapor pressure of dilute solutions of sodium in liquid ammonia, it has been possible to reach certain conclusions with regard to the molecular state of these solutions. The apparent molecular weight of sodium in liquid ammonia, as calculated on the assumption of the laws of dilute solutions, decreases with decreasing concentration and apparently approaches a value less than the normal atomic weight of sodium. While it has been shown that the solutions of the metals in liquid ammonia are very exceptional in their osmotic behavior, so that the laws of dilute solutions do not appear in general to be applicable, nevertheless, from the general trend of the molecular weight as a function of the concentration, it may be concluded that the true molecular weight of sodium at low concentrations is less than 23. It appears, therefore, that, in solutions of sodium in liquid ammonia, there is present a molecular species other than the sodium atom. Since these solutions are conductors of the electric current, it may be concluded that the low value of the molecular weight is due to the presence of the negative carrier in these solutions.

It has been shown that, in dilute solutions of the alkali metals in liquid ammonia, the conduction process is an ionic one, and that the metallic ion is positively charged and moves toward the cathode in these solutions. It has also been shown that there is present in these solutions a negative carrier which is instrumental in carrying a portion of the current through the solution. When the negative carrier is discharged at the anode, there is no indication of a product at this electrode. If a solution of potassium in liquid ammonia is placed between two solutions of potassium amide in ammonia, then, under the action of an electromotive force, the characteristic blue color moves toward the anode. That portion of the solution from which the color disappears apparently contains nothing but potassium amide after the process has taken place. Since it has been shown that the metallic ion in these solutions moves toward the cathode, it follows that the metallic ion is apparently identical with the positive ion of the amide of this metal. At the same time, it appears that the blue color of the solution, as well as the other properties which are characteristic of this solution, are due primarily, not to the metallic constituent, but to the negative carrier, either free or combined with the metal. It follows, also, that an equilibrium exists between the positive and negative carriers and the un-ionized metal, as well as between the positive ion, the amide ion and the neutral amide. The fundamental metallic constituent, or, in other words, that constituent to which the metals in common owe their characteristic metallic properties, is the negative carrier. So far as the results obtained with these solutions are concerned, there is no evidence indicating that any other ion or molecule is primarily concerned with the metallic properties of metals. In the more concentrated solutions the electrolytic properties diminish and ultimately become inappreciable. Apparently this is due to an increase in the speed of the negative carriers in the more concentrated solutions, as a result of which the fraction of the current carried by the cation gradually decreases with increasing concentration and ultimately becomes so small as not to be recognizable.

That this is the case, follows from the results of measurements on the electromotive force of concentration cells, as well as from measurements of the conductance of these solutions. From a study of the electromotive force of concentration cells, it has been found possible to determine the relative amounts of the current carried by the two ions. In dilute solutions the negative carrier has a carrying power approximately 7 times that of the sodium ion. The greater carrying power of the negative ion is doubtless due to its greater mean speed. As the concentration increases, the relative speed of the negative carrier increases slowly, at first, and then more rapidly. In the neighborhood of normal concentration the increase in the speed of the negative carrier becomes very great, so that, at this concentration, its speed is several hundred times that of the positive carrier. In order to account for the observed speed of the negative carrier in dilute solutions, it must be assumed that this carrier has dimensions of the same order of magnitude as that of the sodium ion. Since there are no other molecules present, save sodium and ammonia, and since the sodium atom is a constituent of the positive ion, it follows that the negative carrier must be associated with one or more molecules of ammonia. At higher concentrations, however, some of these carriers are freed from their combination with the ammonia molecules for a fraction of the time. If these carriers have sub-atomic dimensions, then, as their association with ammonia is lost, their speed should be very great under the action of a given potential gradient. In other words, the conducting power of a solution of a metal in ammonia should increase very greatly with the concentration at the higher concentrations.

In the case of concentrations in the neighborhood of normal, a portion of the negative carriers cannot be associated with the solvent. This follows from the fact that if ammonia is associated with a negative carrier, then, in a concentration cell, it will be carried from a solution having a higher vapor pressure to one having a lower vapor pressure, which process involves work. The vapor pressures of these solutions being known, it is possible to calculate the work and consequently the electromotive force of a cell, on the assumption that the only source of energy is due to this process. Or, rather, given the electromotive force of a cell, it is possible to calculate an upper limit to the mean solvation of the negative carrier. As a result of such a calculation, it was found that the electromotive force of a cell operating between concentrations of approximately 1 N and 0.5 N is such as to yield a value of 0.67 for the mean solvation of the negative carrier. The true value must necessarily be below this. It is not to be understood from this that one-third of the negative carriers are unassociated with the solvent under these conditions, but that at least one-third of the current is carried by carriers which are not associated with the solvent. Since the conductance of these solutions is not much greater than that of the more dilute solutions and since the conductance of the negative carrier is unquestionably very great, it follows that the actual number of negative carriers which are not associated with ammonia at a given instant of time is a very small one indeed. This number, however, increases very greatly at higher concentrations.

The chief results of the foregoing considerations are: first, that an ionic equilibrium exists in solutions of a metal in liquid ammonia; and second, that the negative carrier is identical for all metals and exhibits abnormal conducting power in the case of the more concentrated solutions. It follows as a consequence that the conductance curve for such solutions should exhibit a minimum. In the more dilute solutions the conductance curve should rise and approach a limiting value in virtue of the process of ionization. In dilute solutions, therefore, the conductance curve should correspond approximately with that of ordinary salts in liquid ammonia. In the more concentrated solutions the conductance curve should rise with increasing concentration owing to the increase in the speed of the negative carrier. It has been shown that such is in fact the case. The form of the conductance curve corresponds precisely with that required by these considerations. In dilute solutions the conductance approaches a limiting value according to a curve which is of the same general form as that of ordinary electrolytes. At a concentration of approximately 0.05 N, the equivalent conductance has a minimum value after which it begins to increase rapidly with increasing concentration. This increase is particularly pronounced in the neighborhood of 1 N, corresponding with the increase in the relative speed of the negative carrier as deduced from the results of measurements of the electromotive force of concentration cells. At still higher concentrations, the conductance approaches that of metallic conductors. At approximately 2 N a solution of sodium in liquid ammonia has an equivalent conductance of approximately 83,000 and a specific conductance of 0.0164×10^4. A saturated solution of sodium in liquid ammonia has a specific conductance approximately half that of mercury at 0°.

The conductance curves of sodium, lithium and potassium, as well as of mixtures of sodium and potassium, are similar in form but are displaced as regards the value of the conductance. The difference in the conductance of the more dilute solutions corresponds approximately to the difference in the conductance of the positive ions of these metals. This is in agreement with the conclusion reached above, that the negative carriers in the case of these three solutions are identical.

It appears that the solutions of the metals in liquid ammonia form the connecting link between electrolytic and metallic conductors. It has been definitely shown that the conduction process in the case of these solutions is an ionic one. There is nothing to distinguish the more concentrated solutions from actual metallic substances. It may be concluded, therefore, that the process of conduction in the case of ordinary metals is effected by means of the same negative carrier. Since this carrier is negatively charged and has sub-atomic dimensions, we may conclude that it is identical with the negative electron as it appears in radio-active and other phenomena.

WORCESTER, MASS.

The Magnetic Properties of Sodium–Ammonia Solutions

Huster was the first to measure the magnetic susceptibility of a metal–ammonia solution as a function of concentration. The paper in which he reports the results of such a study for sodium solutions is very long, and we provide just a translation of his summary.

Annalen der Physik, vol. 33, 1938, p. 477

29

Uber die Lösungen von Natrium in Flüssigem Ammoniak: Magnetismus; Thermische Ausdehnung; Zustand des Gelösten Natriums (summary)

E. HUSTER

1. Solutions of alkali metals in liquid ammonia make possible the investigation of typical metallic properties as a function of electron concentration. With regard to magnetic properties, one would expect sodium solutions to show a continuous transition from the weak paramagnetism of a degenerate electron gas to the stronger "normal" paramagnetism of very dilute sodium solutions. These results would be expected whether the dilute solutions contain neutral atoms or the degeneracy of the electrons is lifted. The particular course of this transition should make it clear which of the two possibilities occurs.

2. An apparatus is described that enables the introduction, under vacuum, of the very air-sensitive solutions into tubes that can be sealed so that the susceptibility can be determined by the cylinder-weighing method. Measurements were made at $-35°$ and $-75°$ down to dilutions of about 6×10^4 NH_3/Na. Water served as the reference substance.

3. The density of the solutions, which must be known for the calculation, was determined in pycnometers filled in the same way.

4. The susceptibility of the dissolved sodium was calculated using the rule for mixtures. The results are:

The atomic susceptibility of the dissolved sodium, χ_A(Na), is, at high concentrations (6–7 NH_3/Na), about 6 times as great as that of the compact metal at room temperature. From $-75°$ to $-35°$ χ_A(Na) increases about 7%; the temperature dependence decreases with increasing concentration. After passing through a minimum at 50–100 NH_3/Na, χ_A(Na) attains, at $-35°$ and at high dilution (3–$6 + 10^4$ NH_3/Na), a value corresponding to one Bohr Magneton per atom, although the limits of error are ±25 to 30%. At $-75°$ the onset of the steep increase is shifted to solutions which are about 6 times more dilute. Surprisingly the sodium is strongly diamagnetic at low temperatures and intermediate concentrations (around 100 NH_3/Na). At $-60°$, $\chi_A(Na) = -28 \times 10^{-6}$, compared with $\chi_A(Na) = +27 \times 10^{-6}$ at $-35°$.

5. The magnetic behavior shows that the solutions, up to dilutions of around 2000 NH_3/Na, contain no neutral atoms. At higher dilutions their absence can be deduced from the electrical measurements of Kraus. Thus one concludes that the properties of the solutions are essentially determined by the contained electrons and that the solutions are to be looked upon as dilute metals.

6. The diamagnetism of the dissolved sodium at intermediate concentrations and low temperatures is consistent with the assumption that the solutions contain Na_2 molecules. The molar susceptibility of Na_2 can be theoretically estimated as -100 to -150×10^{-6}; this estimate is in reasonable agreement with the observed minimum value of $-56 + 10^{-6}$.

7. Accordingly, the state of sodium in ammonia solutions which are progressively diluted can be described by the following scheme:

$$(Na;NH_3)\text{-metal} \rightarrow Na_2 \rightarrow \underset{\text{not perceptibly present!}}{(2Na)} \rightarrow 2Na^+ + 2e^-$$

Increase in temperature suppresses the Na_2 molecules out of the series.

8. For the dissociated Na-portion: In a magnetic field the valence electrons of the dissolved sodium behave like free electrons up to the highest dilutions; by mere dilution in a homogeneous phase, a transition from a degenerate to a nondegenerate electron gas is made accessible to measurement. The lifting of the degeneracy occurs in the theoretically predicted region.

9. The above interpretation is not inconsistent with other properties of the solutions. The light absorption of the solutions and the position of the minimum of the susceptibility and the maximum of the temperature coefficient of the electrical conductivity strongly support the proposed interpretation.

The Impulsive Professor Ogg

The following two communications by Ogg startled the scientific community of 1946. In the first he reported a marked photoconductivity for extremely dilute sodium–ammonia solutions and an enormous volume increase for the dissolution of sodium in ammonia. In the second he reported superconductivity for frozen sodium–ammonia solutions at −180°. Subsequent work has shown that none of these observations can be reproduced.[1-3] However the communications spurred intense interest in these solutions, and Ogg's theoretical remarks in these communications were of considerable value. For example, his quantum mechanical calculations for the *F* and *F'* center model[4] served as a sound starting point for other theoretical studies, and his explanation of the liquid phase separation in metal–ammonia systems in terms of electron pairs and Bose–Einstein statistics was very ingenious (and plausible, in view of then-current ideas regarding superconductivity).

---

[1]R. L. Potter, R. G. Shores, and J. L. Dye, *J. Chem. Phys.*, **35**, 1907 (1961).
[2]A. J. Stosick and E. B. Hunt, *J. Amer. Chem. Soc.*, **70**, 2826 (1948).
[3]See the paper by M. J. Sienko, reprinted on p. 274.
[4]Suggested earlier by S. Freed and N. Sugarman [*J. Chem. Phys.*, **11**, 354 (1943)].

30

ELECTRONIC PROCESSES IN LIQUID DIELECTRIC MEDIA. THE PROPERTIES OF METAL-AMMONIA SOLUTIONS

Sir:

New experimental results appear significant in the elucidation of the properties of dilute liquid ammonia solutions of alkali and alkaline earth metals. Extremely dilute sodium solutions (some 10^{-5} molar) observed in the temperature range -35 to $-75°$ displayed a marked increase in electrical conductivity upon irradiation with visible light. It is particularly important that the quantum efficiency of this photoconductivity is an inverse function of the concentration of the solution. This fact renders the effect experimentally observable with reasonable light intensities only at the low concentrations employed.

Dilatometric measurements (at constant temperature), in which sodium was extracted from dilute solutions by metallic mercury, indicated a volume change amounting to some 700 cc. per mole of solute (temperature $-35°$, concentration of solutions some 3×10^{-3} molar). That is, the dissolving of metallic sodium under these conditions results in an expansion nearly thirty times as great as the volume of the solid metal sample.

The close parallelism of the above photoconductivity experiments to those of Hilsch and Pohl[1] dealing with solid solutions of alkali metals in alkali halide crystals suggests a mechanistic explanation similar to that commonly accepted[1] for the latter case. Whereas in ionic crystals the electrons are trapped at vacant negative ion sites, giving "F centers," in the solutions in question they are trapped in cavities which they have "dug" in the solvent. That these cavities are relatively enormous (of the order of 7×10^{-8} cm. in radius) is indicated by the very great expansion attendant upon dissolving the metal to form highly dilute solutions. The bodily mobility of such huge ions would be negligible, but thermal or photoexcitation may raise the trapped electrons to the conduction band. The mobility of a conduction electron would appear to be limited by "redigging" its cavity and by capture in a cavity containing a trapped electron, forming a pair analogous to the "F' centers" of Hilsch and Pohl. The importance of this latter process is indicated by the above inverse dependence of the photoeffect upon concentration.

Considerations of quantum mechanics lend theoretical support to the above model. These considerations, which visualize the electron trapped in a spherical cavity, lead to the following results: (1) the ground state of the system is an "S" state of total energy -0.21 volt (-4800 cal./mole) in a cavity of radius 7.6×10^{-8} cm. (1100 cc./mole); (2) all other states are unstable; (3) photoconductivity follows a transition from the ground state to the lowest "P" state; (4) *two* electrons trapped in the same cavity are appreciably stable with respect to either two electrons in separate cavities or one trapped and one conducting electron. The absorption spectrum and magnetic susceptibility of metal ammonia solutions are in at least qualitative agreement with these calculations which will be described in detail later.

DEPARTMENT OF CHEMISTRY
STANFORD UNIVERSITY
STANFORD UNIV., CALIF. RICHARD A. OGG, JR.

RECEIVED DECEMBER 19, 1945

(1) For literature references, see N. F. Mott and R. W. Gurney, "Electronic Processes in Ionic Crystals," Oxford Press, New York, N. Y., 1940.

31

Bose-Einstein Condensation of Trapped Electron Pairs. Phase Separation and Superconductivity of Metal-Ammonia Solutions

RICHARD A. OGG, JR.
Department of Chemistry, Stanford University, California
March 2, 1946

WITH the invaluable assistance of Drs. Claudio Alvarez-Tostado and William Perkins, the previously reported[1] studies of the behavior resulting from rapid freezing of very dilute solutions of alkali metals in liquid ammonia have been extended (in the case of sodium solutions) over the entire concentration range, up to the saturation point. Sufficiently rapid cooling to temperatures in the range from −90°C to −180°C resulted in the production of apparently homogeneous deep-blue solid solutions, of bronze-like luster when extremely concentrated. All of the solid samples proved to be good electrical conductors, although shrinkage and cracking frequently caused erratic measurements. No abnormal resistance change accompanying solidification was observed, except for solutions in the concentration range (of the order of one molar) characterized by the remarkable phenomenon of separation into two dilute liquid phases[2] at sufficiently low temperatures. Extremely rapid freezing of such solutions (initially at −33°C, i.e., above the upper consolute temperature) caused a relatively enormous decrease of measured resistance. In a representative case, the resistance of the liquid sample at −33°C was some 10,000 ohms, while that of the solid at −95°C was only 16 ohms. Variation of conditions suggested that even such small residual resistances might be due to "end effects" and faulty contact with the platinum electrodes. That the solutions in this special concentration range actually became superconducting was demonstrated by adaptation of the classical Kammerlingh Onnes "ring experiment." Thin-walled glass cells having the shape of an annular disk were filled with the proper solution at −33°C and then rapidly plunged into a vessel of liquid air between the poles of an electromagnet (field strength some 1500 gauss). After removal from the magnetic field, the existence of persistent currents was shown by tests with a sensitive magnetometer. Numerous control experiments obviated any other possible explanation. The magnetometer tests were conducted at −180°C, but intervening warming of the ring samples to much higher temperatures did not destroy the persistent currents. In all probability such solid solutions remain superconducting up to the melting point, i.e., to absolute temperatures of the order of 180 to 190 degrees.

The probable explanation of the above phenomena is to be found in the behavior of trapped electron pairs, recently demonstrated[3] to be a stable constituent of fairly dilute metal-ammonia solutions. In the concentration range characterized by liquid-liquid phase separation, experimental studies[4] show the solute to be diamagnetic at temperatures just above the consolute point. This suggests the electron constituent to be almost exclusively in the trapped electron pair configuration. Because of their zero angular momentum, such pairs must obey Bose-Einstein statistics. If the effective mass does not exceed twice the electron mass by an extremely large factor, then the calculated degeneration temperature[5] at the concentrations in question is relatively high—of the order of a few hundred degrees absolute. It is postulated that the liquid-liquid phase separation which occurs on slow cooling (upper consolute temperature 232°K) is the device adopted by the systems to avoid the Bose-Einstein condensation, with its unfavorable free energy change. In the more dilute phase the electron constituent is still predominantly the trapped pair, but at a concentration low enough to raise the degeneracy temperature to just above the prevailing temperature. In the more concentrated phase, the trapped electron pairs have become unstable because of the greater interionic forces, and one has essentially a liquid metal, the trapped single electrons being below the Fermi-Dirac degeneration temperature. The small, temperature independent paramagnetism[4] of very concentrated solutions would appear to support this latter model.

By sufficiently rapid cooling, it appears that the liquid-liquid phase separation is prevented, and that the system becomes frozen and hence metastable in the "forbidden" concentration region, which is thus characterized by the Bose-Einstein condensation of trapped electron pairs. From the discussion of London,[5] apparently such a state must display the phenomenon of electrical superconductivity, in agreement with the above experimental observations.

The extension of the above model to explain previously observed superconductivity is apparent, and is the more plausible in view of the essentially only quasi-metallic character of the large number of alloys and compounds which display the phenomenon.[6]

[1] R. A. Ogg, Jr., J. Chem. Phys. 13, 533 (1945).
[2] For literature references, see W. C. Johnson and A. W. Meyer, Chem. Rev. 8, 273 (1931).
[3] R. A. Ogg, Jr., J. Chem. Phys. 14, 114 (1946).
[4] E. Huster, Ann. d. Physik [5], 33, 477 (1938); S. Freed and N. Sugarman, J. Chem. Phys. 11, 354 (1943).
[5] F. London, Phys. Rev. 54, 947 (1938).
[6] H. G. Smith and J. V. Wilhelm, Rev. Mod. Phys. 7, 237 (1935).

Erratum: Lines 21 to 23 of the second paragraph should read: "but at a concentration small enough to lower the degeneracy temperature to the prevailing temperature."

Paramagnetic Resonance Studies

Quantitative paramagnetic resonance measurements are, in contrast to static magnetic susceptibility measurements, relatively unaffected by diamagnetism. Hutchison and Pastor applied the former technique to potassium–ammonia solutions and, by comparison of their data with the static susceptibilities of Freed and Sugarman,[1] concluded that no abnormally large diamagnetism exists in these solutions. A reprint of their work follows.

---

[1]S. Freed and N. Sugarman, *J. Chem. Phys.*, **11**, 354 (1943).

32

Paramagnetic Resonance Absorption in Solutions of K in Liquid NH_3*

Clyde A. Hutchison, Jr., and Ricardo C. Pastor

Institute for Nuclear Studies and Department of Chemistry, University of Chicago, Chicago, Illinois

SOLUTIONS OF METALS IN NH_3

The alkali metals and the alkaline earth metals dissolve readily in liquid NH_3.[1-5] The solubilities[4] of Na and K at the boiling point of NH_3, $-33.4°C$, are about 5.4 and 4.9 mol l^{-1}, respectively, and the solubility changes very little with temperature. The concentrated solutions have the color of bronze and a metallic luster. The more dilute solutions are blue, the absorption spectrum of the dilute solutions of all the alkali metals being identical.[6] When the solvent is evaporated, the pure alkali metals remain as residues. (Reaction of the metal with the solvent will occur at varying rates depending upon temperature, impurities, etc., but in the absence of impurities the solutions contained in Pyrex glass at room temperature are quite stable enough to make possible a variety of measurements of physical properties.) A study of the vapor pressure[7] as a function of concentration shows the nonexistence of compounds in the freshly prepared solutions of the alkali metals.

One of the striking properties of such solutions is their electrical conductivity.[8-11] The conductivity is shown in Fig. 1. At all concentrations the equivalent conductivity is greater than that of any known salt in any known solvent. In the most dilute solutions the equivalent conductance reaches a limiting value of approximately 1000 ohm^{-1} cm^2 mol^{-1} which is three times that of the best conducting salts in liquid NH_3. Even more remarkable is the behavior of the conductivity at the higher concentrations. A saturated solution of K in NH_3 has a conductivity of approximately 4500 ohm^{-1} cm^{-1}. This may be compared with the conductance of Hg at 20°C, which is 10 400, and it is seen that these solutions have a conductivity of the same order as that of metals. The conductances of the solutions of the various alkali metals in liquid NH_3 agree very closely at all concentrations. Both electromotive force and conductance studies lead to the conclusion that in the most dilute solutions of sodium the speed of the negative charge carrier is about 7 times that of the positive charge carrier. When the concentration is increased, the equivalent conductance decreases, passes through a minimum at about 0.04 *M*, and then increases to the enormous values already mentioned. In the concentrated solutions the speeds of the negative carriers approach those of the electrons in metals.

One is led to the conclusion that when any of the alkali metals is dissolved in NH_3 there is a dissociation into positive alkali metal ions and electrons. The negative charge carrier is apparently the electron in each case. In the dilute solutions the electrons are evidently bound not too tightly in some way to the solvent molecules. In the concentrated solutions the electrons are much less tightly bound and move in much the same manner as in metals. In such solutions we have the possibility of examining the properties of electrons over a range of concentrations varying by a factor of 10^4 to 10^5.

The behavior of the conductivity with concentration was interpreted by Kraus and his co-workers in terms of equilibria between sodium atoms, sodium ions, free electrons, and solvated electrons. Several lines of evidence lead to the discarding of such models. One of the clearest reasons for rejecting models involving the existence of large concentrations of alkali metal atoms or solvated electrons is a consideration of the results of studies of the static magnetic susceptibilities of such

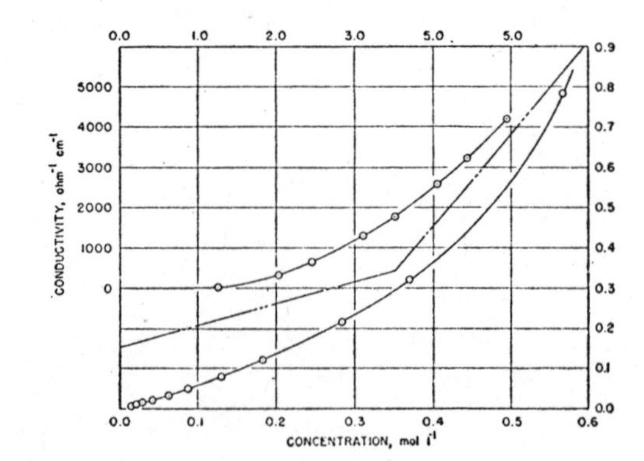

Fig. 1. Conductivity of K in NH_3 at NH_3 B. P. [C. A. Kraus, J. Am. Chem. Soc. 43, 755 (1921)].

* This research was assisted by the U.S. Office of Naval Research.

[1] C. A. Kraus, J. Franklin Inst. 212, 537 (1931).
[2] C. A. Kraus, *The Properties of Electrically Conducting Systems* (Chemical Catalog Company, New York, 1922.)
[3] D. M. Yost and H. Russell, *Systematic Inorganic Chemistry* (Prentice-Hall, Inc., New York, 1944), p. 136.
[4] W. C. Johnson and A. W. Meyer, Chem. Rev. 8, 273 (1931).
[5] W. C. Fernelius and G. W. Watt, Chem. Rev. 20, 195 (1931).
[6] Gibson and Argo, J. Am. Chem. Soc. 40, 1327 (1918).
[7] Kraus, Carney, and Johnson, J. Am. Chem. Soc. 49, 2206 (1927).
[8] C. A. Kraus, J. Am. Chem. Soc. 437, 49 (1921).
[9] C. A. Kraus and W. W. Lucasse, J. Am. Chem. Soc. 43, 2529 (1921).
[10] C. A. Kraus and W. W. Lucasse, J. Am. Chem. Soc. 45, 2551 (1923).
[11] G. E. Gibson and T. E. Phipps, J. Am. Chem. Soc. 48, 312 (1926).

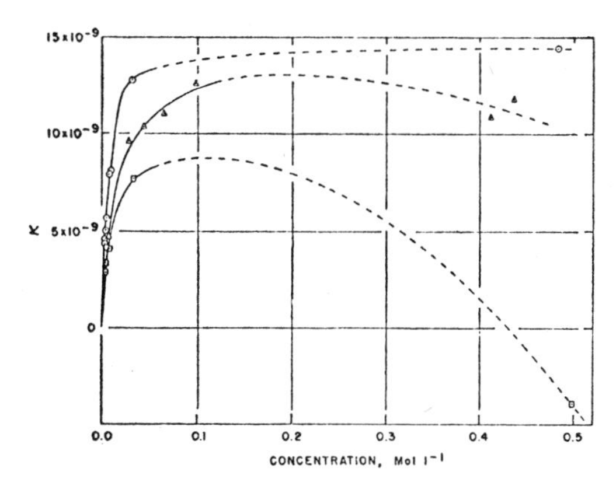

FIG. 2. Magnetic susceptibilities of solutions of alkali metals in liquid NH_3 (see reference 12). ⊙ K at 240°K. ⊡ K at 220°K (see reference 13). △ Na at 238°K.

solutions.[12,13] If such species existed in large numbers, the solutions would be paramagnetic and their susceptibilities would obey Curie's law with respect to variation of temperature and concentration. This type of behavior is not observed. The experimental measurements of the static susceptibilities are shown in Figs. 2, 3, and 4. In the very concentrated solutions and down to a few tenths normal, the susceptibiltiy is quite small of the order found in the metal and has a very small temperature coefficient. As the concentration is lowered, the molar susceptibility increases rapidly and approaches in the most dilute solutions the value for one mol of spins. This is the general qualitative type of behavior to be expected if the electrons are free or in a periodic potential as in a metal. Under such conditions they would at high concentrations form a degenerate Fermi gas with most of the electrons paired and not contributing to the paramagnetism. At the lower concentrations, however, the degeneracy would be removed and the susceptibility would be of the order of that of a mol of spins. Hence, in this particular respect the solutions behave much as would be expected for a metal in which the concentration of electrons could be varied over a very wide range.

With respect to both conductivity and magnetic susceptibility the behavior is not in detail like that of metals. For example, the temperature coefficient of the conductivity is positive at all concentrations.[10,11] Also the magnetic susceptibility is considerably lower than that calculated for a free electron gas,[13a] and at all concentrations investigated the susceptibility increases with increasing temperature.[12] Various attempts have been made to account for this behavior of the susceptibility. Huster[13] suggested an equilibrium between electrons and Na ions, on the one hand, and diatomic molecules on the other. The necessary molecular concentrations are, however, in disagreement with the apparent molecular weight measurements of Kraus.[14] Freed and Sugarman[12] have proposed a decrease of density of energy levels near the top of the Fermi distribution because of the binding of the electron in a variety of resonance structures involving NH_3 molecules and solvated metal ions. This would result in a reduction of the magnitude of the susceptibility. They also proposed that there are pairwise interactions between electrons similar to those which lead to the F' centers in crystals in which two electrons are trapped in a single vacancy.[15] These pairs would be expected to be diamagnetic and would thus lead to lowered susceptibilities. In particular they might be responsible for the increase of susceptibility with increasing temperature. Ogg[16–18] has discussed in some detail the formation of F'-like centers. He proposes that individual electrons are trapped in cavities which they create in the solution and are in equilibrium with pairs of electrons similarly trapped. He believes that he can account for the behavior of several physical properties, including the magnetic susceptibility on the basis of such an equilibrium. Moreover, he believes that the cavities are very large, and that as a consequence of this there is an extraordinarily large diamagnetic contribution to the susceptibility.

EXPERIMENTAL METHODS

Some time ago we observed paramagnetic resonance absorption in solutions of K in liquid NH_3 at microwave frequencies.[19] Observations of the resonance have also been made by Garstens and Ryan[20] and by Levinthal,

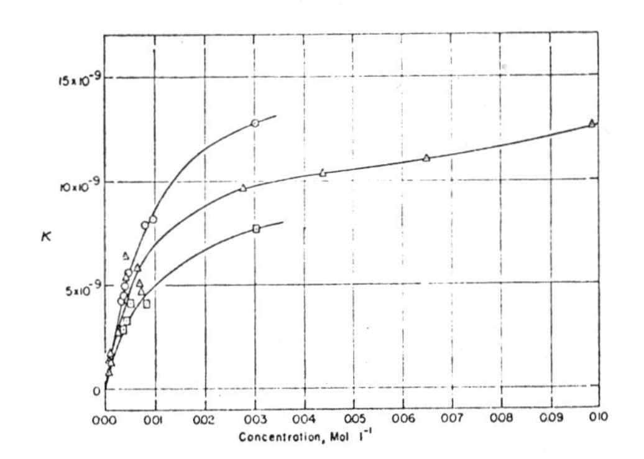

FIG. 3. Magnetic susceptibilities of solutions of alkali metals in liquid NH_3 (see reference 12). ⊙ K at 240°K. ⊡ K at 220°K (see reference 13). △ Na at 238°K.

[12] S. Freed and N. Sugarman, J. Chem. Phys. 11, 354 (1943).
[13] E. Huster, Ann. Physik 33, 477 (1938).
[13a] N. F. Mott and H. Jones, *Theory of Properties of Metals and Alloys* (The Clarendon Press, Oxford, 1936), p. 184.
[14] C. A. Kraus, J. Am. Chem. Soc. 30, 1197 (1908).
[15] F. Seitz, Revs. Modern Phys. 18, 384 (1946).
[16] R. A. Ogg, J. Chem. Phys. 14, 114 (1946).
[17] R. A. Ogg, J. Chem. Phys. 14, 295 (1946).
[18] R. A. Ogg, J. Am. Chem. Soc. 68, 155 (1946).
[19] C. A. Hutchison, Jr., and R. C. Pastor, Phys. Rev. 81, 282 (1951).
[20] Martin A. Garstens and Alden H. Ryan, Phys. Rev. 81, 888 (1951).

Rogers, and Ogg.[21] The width of 0.1 gauss, which we initially reported, turned out to be caused largely by magnetic field inhomogeneities. Whereas the resonances in paramagnetic salts are generally several thousands of times this broad and must be studied at microwave frequencies, we have investigated this extremely sharp and consequently enormously intense resonance at low frequencies and field strengths where it was much simpler for us to produce a homogeneous field and measure the actual widths.

We have studied the paramagnetic resonance absorption in solutions of K in liquid NH_3 at frequencies in the range 5.5 to 8.2 mc sec^{-1} corresponding to field strengths of 2.0 to 2.9 gauss. The static magnetic field was supplied by a solenoid which could be accurately aligned with the earth's field. The sample of K in liquid NH_3 was placed at the center of the solenoid in the coil of a regenerative oscillator-detector circuit. The magnetic field was modulated at 40 cycle sec^{-1} by means of a small auxiliary coil. A phase sensitive detector was employed and its output was recorded on a recording potentiometer. The modulation amplitude was sufficiently small to insure a signal proportional to the first derivative of the absorption. The static magnetic field was varied linearly with time by means of a motor-driven potentiometer in the current supply circuit. The current through the solenoid was recorded on the same chart as the magnetic signal. A Watkins-Pound calibrator[22] circuit was employed for determining signal intensities. With this device one may simulate the change in coil resistance produced by the magnetic signal by means of a changing plate resistance of a thermionic tube. One then employs the regenerative oscillator-detector-amplifier-recorder system to compare the magnetic signal with the calibrator signal produced when a 40 cycle sec^{-1} signal of known magnitude is placed on the grid of the calibrator tube at known fixed plate current. The recorder chart was

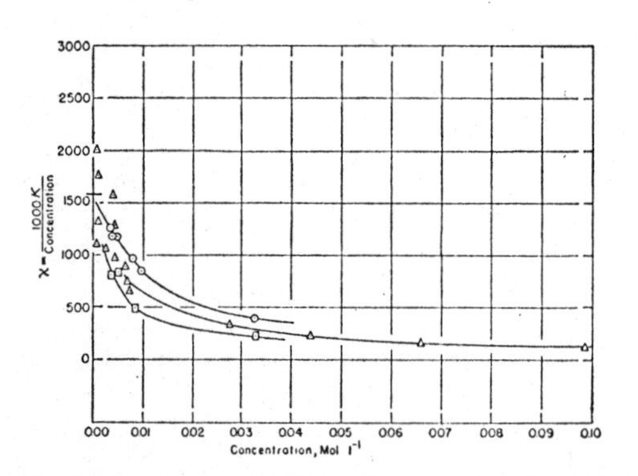

FIG. 4. Magnetic susceptibilities of solutions of alkali metals in liquid NH_3 (see reference 12). ⊙ K at 240°K. ⊡ K at 220°K (see reference 13). △ Na at 238°K. The bar at 1564×10^{-6} denotes the value of the susceptibility of a mol of electron spins at 240°K.

[21] Levinthal, Rogers, and Ogg, Phys. Rev. 83, 182 (1951).
[22] G. D. Watkins and R. V. Pound, Phys. Rev. 82, 343 (1951).

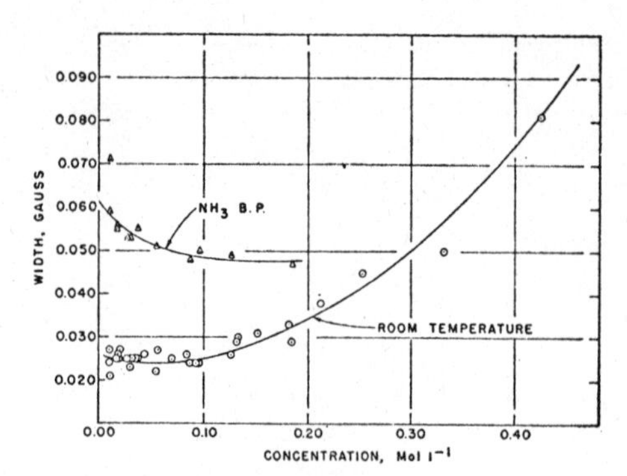

FIG. 5. Width of resonance at maximum slope *vs* concentration.

calibrated in terms of calibrator grid voltage immediately after each scanning of the magnetic resonance signal.

The samples of K in liquid NH_3 were prepared by distilling K several times in high vacuum and finally distilling it into Pyrex glass capsules. Then NH_3 dried by addition of K and subsequent distillation was added and the capsule sealed. The amount of K in these solutions was determined after completion of magnetic investigation by dissolving in H_2O and boiling to remove the NH_3. The resulting solution was titrated with standard acid. The weight of NH_3+K was determined by weighing the capsule before and after opening it.

Measurements were made at a room temperature of 28°C and at the boiling point of liquid NH_3 which is −33°C. For the low temperature runs the sample was immersed in a Dewar filled with liquid NH_3.

SPECTROSCOPIC SPLITTING FACTOR

The spectroscopic splitting factor for the solutions of K in liquid NH_3 was compared with the factor for 2,2-diphenyl-1-picrylhydrazyl[19,23] at approximately 8.2 mc sec^{-1}. In the case of the solution of K in NH_3 the rf coil field strength was 0.0023 gauss rms. The shift of the center of the resonance under these conditions due to the Bloch-Siegert[24] effect is by the factor

$$1-[(2H_1)^2/16H_0^2]=1-1.6\times10^{-7},$$

and is consequently small enough to be neglected. Assuming the splitting factor for 2,2-diphenyl-1-picrylhydrazyl to be 2.0037[19,23] (the microwave value), we find the factor at room temperature for K in NH_3 to be 2.0010 ± 0.0002 at 0.425 mol l^{-1}, and 2.0011 ±0.0002 at 0.0365 mol l^{-1}. This is the same as the value 2.0012 ± 0.0002 found at 23 500 mc sec^{-1}.[19] This factor is considerably lower than the value of the *g* factor, 2.0023, for a free electron spin. However, it is not nearly so low as the value 1.995,[25] which we have observed for additively colored crystals of KCl.

[23] Holden, Kittel, Merritt, and Yager, Phys. Rev. 77, 147 (1950).
[24] F. Bloch and A. Siegert, Phys. Rev. 57, 522 (1940).
[25] C. A. Hutchison, Jr., and G. A. Noble, Phys. Rev. 87, 1125 (1952).

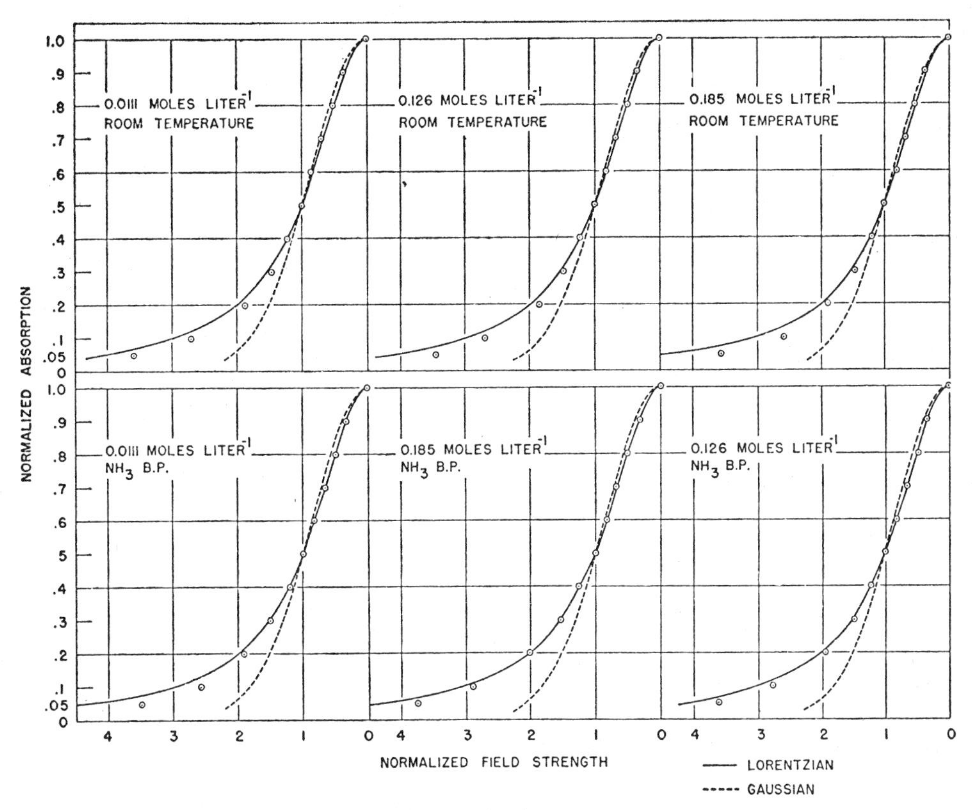

FIG. 6. Line shape.

WIDTH

The width of the resonance between the points of maximum slope was determined by measuring the distance between the maximum and minimum on the recorder chart which, when combined with the current record on the same chart, gave the width. The results are shown in Fig. 5. The width was measured at room temperature from a concentration of 0.425 down to 0.011 mol l^{-1}, and at the boiling point of NH_3 from 0.185 to 0.011 mol l^{-1}. It is clear that this resonance is by far the sharpest electronic resonance so far observed in a condensed phase. In the case of a resonance as sharp as this, one must consider the Rabi[26] width associated with the flipping of the spin by the perturbing rf magnetic field. This is approximated by

$$\Delta H/H = 4\sin(\theta/2),$$

where ΔH is the width at half-height and where θ is given by

$$\theta = \tan^{-1}\frac{\text{rf field strength}}{\sqrt{2}H}.\dagger$$

[26] I. I. Rabi, Phys. Rev. 51, 652 (1937).
† This is the rms field strength.

When θ is very small, as in our case,

$$\Delta H = \sqrt{2}\text{ rf field strength.}$$

For a sample whose concentration was 0.152 mol l^{-1} the variation of width with rf field strength was investigated with the following results:

$\sqrt{2}$ rf field strength gauss	width at max slope gauss
0.0032	0.0311
0.0081	0.0327
0.0163	0.0335
0.0324	0.0388

It is clear that at the higher coil voltages the Rabi width is seriously interfering with observation of the true width. Consequently, all scannings of the resonance absorption were made at coil voltages corresponding approximately to the smallest rf field listed above, or about 0.05 volt. In the case of an organic free radical which is much wider than the K resonance there is, of course, no observable change of width with coil voltage.

The width of this resonance is several orders of magnitude less than that observed in aqueous solution of paramagnetic salts of about the same concentrations. There is evidently an extraordinarily large exchange narrowing associated with the great mobility of the electrons in these solutions. The average value of β/r^3 (r is the distance between electrons) ranges from 0.06 to 2.6 gauss. The broadening by magnetic nuclei would be expected to be several gauss. At room temperature there is perhaps a slight increase in the width at the lowest concentrations, and the increase is marked at the lower temperature. This increase becomes pronounced in the vicinity of the region in which the susceptibility begins to rise rapidly with dilution. The room temperature measurements were extended to considerably higher concentrations than at the lower temperature, and there is apparently quite a large increase of width at the higher concentrations.

LINE SHAPES

The areas under the curves drawn by the recording potentiometer were obtained by numerical integration. The scale on the field strength axis was quite linear as determined from the current recording. The scale on the magnetic signal axis was quite linear in calibrator grid voltage. Moreover, it was found that the grid voltage at constant plate current was very close to being proportional to magnetic signal over the range employed in the experiments. This last was determined by comparing calibrator signals with magnetic signals of various sizes produced with an organic free radical by field modulation amplitudes of various sizes. It was consequently a simple matter to obtain the absorption curves with errors not greater than probably 5 percent. Some absorption curves were also determined using the Watkins-Pound RF Spectrometer.[22] The results were indistinguishable from those obtained with the regenerative oscillator-detector. The regenerative oscillator-detector was employed in our investigations because of its much more favorable signal-to-noise ratio. The results of the investigation of the line shapes are shown in Fig. 6. The points are taken from the experimental curves. The line shapes of three samples over a range of concentrations and at both room temperature and the boiling point of NH_3 were compared with Gaussians and Lorentzians. It is clear that these absorptions correspond much more closely to the Lorentizian than to the Gaussian over most of the range of the absorption, but drop considerably below the Lorentzian in the wings of the curve.

INTENSITIES

A second numerical integration yielded the areas under the absorption curves. By means of the current and signal calibrations the relative areas under the absorption curves for the various samples could be obtained.

Since the static susceptibilities of the samples are related to the rf susceptibilities by the Kronig-Kramers relation,

$$\chi = \frac{2}{\pi}\int_0^\infty \frac{\chi''(\nu)}{\nu} d\nu,$$

these second integrals afford a means of comparing the static susceptibilities of the spin systems. Our experiments are carried out at constant frequency and varying field strength. We may, however, for a sharp symmetrical absorption make the transformation of the above relation from the one type of experiment to the other and hence obtain static susceptibilities from our measurements of the rf susceptibilities. We must, of course, consider the changes in Q of the coil and the filling factors from sample to sample. We determined Q, ω, and C during each run. Filling factors were determined for each sample. The desired relationship is

$$\chi = \frac{2\beta}{\pi\xi\omega^2 Ch}\int_0^\infty \frac{\Delta R}{R^2} dH,$$

where χ is the static susceptibility, β is the Bohr magneton, ξ is the filling factor, ω is the angular frequency, C is the capacity across the coil, h is Planck's constant, R is the rf resistance of the coil, and H is the field strength. The grid voltage of the calibrator is proportional to $\Delta R/R^2$. We replaced $\int_0^\infty(\Delta R/R^2)dH$ with the second integral of the recorder curves and employed a standard substance to eliminate the proportionality constant. The standard substance employed was tris-p-nitrophenylmethyl, a completely dissociated free radical in which resonance was first observed by Pake, Weissman, and Townsend.[27] A sample of this material was prepared and later assayed for us by Professor Weissman. We carried out scannings of the resonance of this material. The comparison of the second integrals of the curves so obtained with the similar ones for the K in liquid NH_3 enabled us to

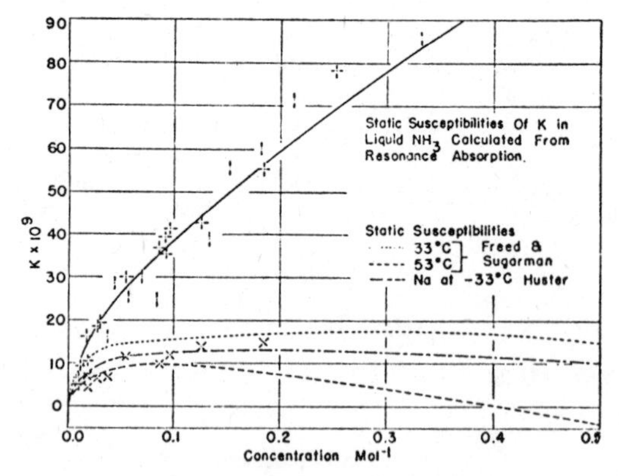

FIG. 7. Static susceptibilities of K in liquid NH_3 calculated from resonance absorption. + Room T. Series I, 28°C; | Room T. Series II, 26°C; × bp of NH_3, −33°C.

[27] Pake, Weissman, and Townsend, personal communication.

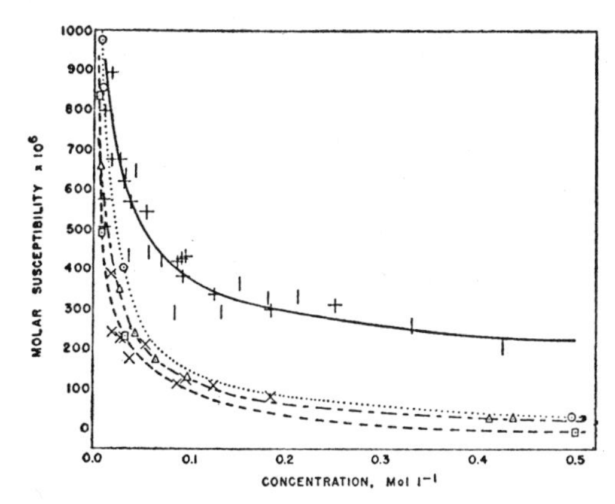

FIG. 8. Static susceptibilities of K in liquid NH_3 calculated from resonance absorption. + Room T. Series I, 28°C. | Room T. Series II, 26°C. × B.P. of NH_3 −33°C. ······−33°C.; ----- −53°C, Freed and Sugarman; —·—·— Na at −33°C, Huster.

calculate static susceptibilities of the solutions relative to that of the organic compound. The free radical was assumed to have the static susceptibility of 1 mol of spins per mol of free radical at 28°C. The number of mols in the sample of organic free radical was known to approximately 5 percent.

The results of the susceptibility determinations are shown in Figs. 7 and 8. At room temperature the calculated static susceptibility is seen to follow the same sort of pattern as determined in static field experiments at low temperatures. The susceptibilities are considerably higher than those of the static experiments at the boiling point. Like the latter, they rise sharply with dilution at about the same concentration. The low temperature runs were made some time subsequent to the high temperature runs, and two points have been omitted for the two lowest concentrations where some decomposition is known to have occurred. These two samples also give obviously low results at the higher temperatures. Some of the other points at the lower concentrations are probably suspect for the same reason. The points for the molar susceptibility at the higher concentrations at the lower temperature are in reasonably good agreement with the static field measurements. The resonance phenomenon is peculiar to the paramagnetic spins and is relatively unaffected by the diamagnetism. Consequently, such a comparison of static field measurements with static susceptibilities calculated from rf measurements enables one to detect the presence of any abnormal diamagnetic susceptibilities. It is clear that no abnormally large diamagnetism exists in these solutions at the boiling point of NH_3 at the concentrations investigated by both methods. Such determinations of static susceptibilities from rf measurements afford a powerful tool for a number of problems of chemical and physical interest where large amounts of diamagnetic material or abnormally large diamagnetisms of the paramagnetic species itself may obscure the primary effect. We have discussed this problem recently for the case of the detection of organic free radicals in solution.[28]

ACKNOWLEDGMENT

We wish to thank Clarence Arnow for the design and construction of the electronic equipment used in these experiments; Jack Boardman for the numerical integrations and other computations; and Edward Bartal for the construction of the solenoid, low temperature apparatus, and other machine work.

[28] Hutchison, Kowalsky, Pastor, and Wheland, J. Chem. Phys. 20, 1485 (1952).

Near-Infrared and Visible Absorption Spectra

Blades and Hodgins were the first, after Gibson and Argo (see p. 158), to report, with experimental details, a detailed study of the absorption spectra of metal–ammonia and metal–amine solutions.[1] Their most significant finding with respect to metal–ammonia solutions was that a single absorption band at 6650 cm^{-1} appeared in the spectra for all the metals that were studied.

---

[1] A tantalizing reference to unpublished work of Hilsch and Vogt is given in E. Vogt, *Z. Elektrochem.*, **45**, 597 (1939). Vogt reported that the absorption peak for both Li and Na solutions lies at approximately 1.8 μ (5600 cm^{-1}).

Reprinted from Canadian Journal of Chemistry, 33: 411–425. 1955

ABSORPTION SPECTRA OF METALS IN SOLUTION[1]

33

BY H. BLADES[2] AND J. W. HODGINS[3]

ABSTRACT

Measurements of the absorption spectra were made on solutions of alkali and alkaline earth metals in ammonia, methylamine, ethylamine, and mixed solvents. In ammonia, a single absorption band was measured which is common to all the metals examined. In the amines and in mixtures of ammonia with methylamine, however, bands were found which were characteristic of the metal employed. A hypothesis has been advanced to explain the existence of the different types of energy traps responsible for the variations in spectra.

INTRODUCTION

The alkali metals and some of the alkaline earth metals dissolve in liquid ammonia without reaction; it is thought that simple dissolution occurs, yielding metal ions and electrons which are trapped or solvated in the liquid (5). Various physical properties have been examined which are consistent with this postulate, although a concise model has not yet been formulated for the electron traps. For example, the expansion exhibited on dissolving the metal has been attributed to the formation of holes in the liquid which represent energy barriers for the escape of electrons (8). The electrical conductivity in dilute solution is greater than can be explained by simple ionic transport, but less than would be expected by electronic conduction (5). Magnetic susceptibility measurements indicate that the metal is ionized, but some difficulty is experienced in explaining the variation of the susceptibility with temperature (3). The measurements of paramagnetic resonance absorption are consistent with the existence of trapped electrons (4). Finally, the absorption spectra of a number of the solutions have been measured, and the observation made that in ammonia there is a single absorption maximum whose position is independent of the particular metal in solution (1, 9). This maximum is in a region where none appears in the spectrum of bulk sodium metal; and in the spectrum of atomic sodium, very strong absorption occurs at 5889 Å (*D* line), which does not appear in the solution spectrum. Since the maximum observed is also absent from the spectrum of sodium ion and of liquid ammonia, it has been ascribed to the trapped electrons.

Primary amines will also dissolve some of the metals; sodium, cesium, potassium, lithium, and calcium are soluble in methylamine, while ethylamine dissolves at least lithium and potassium. Although these solutions have been examined less exhaustively than the ammonia solutions, their behavior is similar. Thus the variation of electrical conductivity with concentration in metal solutions in methylamine is similar to that in anhydrous ammonia (2), and the absorption spectra, although exhibiting maxima at a different wave

[1]*Manuscript received October 18, 1954.*
Contribution from Department of Chemistry and Chemical Engineering, Royal Military College of Canada, Kingston, Ontario. Issued as D.R.B. Report No. SW-2.
[2]*Present address: DuPont de Nemours Company, Wilmington, Delaware.*
[3]*Present address: Royal Military College of Canada, Kingston, Ontario.*

length, were shown to be independent of the metal for a limited number of solutions (1).

The experimental observations of the absorption spectra are limited to quite early work done with a visual spectrophotometer (1) and to some work by Vogt (9). Because the solutions in methylamine had maxima in the absorption spectrum at visible wave lengths, these were observed by the early workers, but the peaks for ammonia solutions were missed as they occur in the near infrared. Vogt describes the spectra of lithium and sodium in ammonia solutions of unspecified concentration. The dearth of spectrophotometric data is probably explained by the difficulty of manipulating the solutions, their very high absorbance, and the awkward spectral region in which the absorption maximum for ammonia is located.

In the present paper, absorption spectra are presented for solutions of lithium, sodium, potassium, and calcium in ammonia and in methylamine, lithium and potassium in ethylamine, sodium and potassium in mixed ammonia and methylamine, and sodium in mixed ammonia and ethylamine.

EXPERIMENTAL PROCEDURE

The main experimental problems associated with the measurement of liquid ammonia and amine spectra arise from the instability of the solutions, their low boiling points (which necessitate constant refrigeration), and the very high molar absorbancy index which make dilute solutions and very thin cells necessary.

The spectra were determined in two parts, the region from 0.3 to 1.0μ on a Beckman D.U. ultraviolet spectrophotometer and from 1.0 to 2.5μ on a Beckman IR-2 infrared spectrophotometer. Quartz cells of thickness 10, 1.0, and 0.1 mm. as supplied by the American Instrument Company were used. In both instruments the effective band width is considerably smaller than the half-height width of the maxima being investigated so that the shape of any spectrum was not affected by the dispersion of the instruments.

The cell was contained within a thick-walled Lucite box with quartz windows on either side for the passage of the light beam. The absorption cell was removed from the light path by moving it upwards. It was kept at the desired temperature by a stream of cold dry air, which entered the box by the ducts shown in Fig. 1 (at *G*). The air was dried with silica gel and cooled by liquid air. The outside windows of the box were kept clear of frost by jets of warm dry air.

The cold box made a light-tight fit into either spectrophotometer, and fogging of the instrument optics was prevented by the warm dry air which continually swept the volume outside the box. Temperature was controlled manually both by regulating the flow of air through the box and by regulating the temperature of the air entering it. A thermocouple cemented on the cell wall near the face showed fluctuations of two to three degrees.

The apparatus for manipulating the solvents and preparing the solutions is shown schematically in Fig. 1. Solvents were stored in a detachable trap *A*. Solutions were prepared in trap *B* by various methods and were transferred to

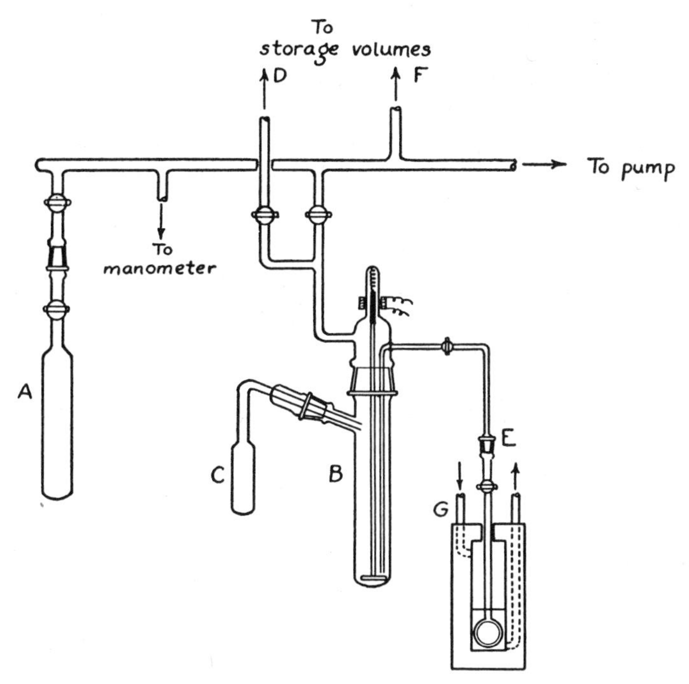

FIG. 1. Apparatus for preparing solutions and filling absorption cell.

the cell by admitting nitrogen from D thus forcing the liquid through the siphon into the cell. Excess liquid in the neck of the cell was removed by pumping off the nitrogen and distilling the solvent back into B.

The cell was disconnected at the ground glass joint E for measurements in the spectrophotometer. At the completion of these measurements the cell could be reconnected at E and the solvent distilled off. Another solvent could then be distilled into the cell thus avoiding the necessity of forcing over-all combinations of metals and solvents.

A saturated solution of potassium in methylamine was found to be quite stable, could be successfully forced over, and was just the correct concentration for spectral measurements in the 0.1 mm. cell. Solutions were thus prepared simply by placing a piece of potassium in trap B and distilling in a suitable quantity of methylamine. It thus became possible by the procedure described above to obtain spectra for any solvent in which potassium was more soluble than in methylamine.

A similar procedure could be used for sodium solutions although for this metal the combination of a saturated solution in methylamine and the 1.0 mm. cell was most convenient.

For more soluble metals it was necessary to resort to a dilution technique where a piece of the metal was placed in trap C and solvent added by distillation. Trap C could then be rotated on its ground glass joint until drops of solution poured from C into B where a relatively large volume of solvent diluted the drop to a suitable concentration.

Since lithium and calcium metal are relatively stable in the air, solutions of these metals were prepared either by the above dilution procedure or by placing a small piece of the metal in trap *B*. However, solutions of these metals are much less stable than those of sodium and potassium and hence greater difficulty was experienced in forcing them over.

When a solvent was distilled into the thin cells containing metal residue the first few drops of solution were sufficient to cover the whole cell face since the liquid rose by capillary action. Thus the concentrations of solutions prepared in this manner were not reproducible. Mixed solvents were condensed from a large gas storage bulb *F* in order to prevent preferential condensation. Table I

TABLE I

METHODS OF PREPARATION OF THE SOLUTION

Metal	Solvent	Method of preparation
Li	Methylamine	Small piece of metal placed in trap, or by diluton
Li	Ammonia	Residue from methylamine
Li	Ethylamine	Dilution
Na	Methylamine	Saturated solution
Na	Ammonia	Residue from methylamine
Na	Mixed solvent	Residue from methylamine
K	Methylamine	Saturated solution and dilution of saturated solution
K	Ammonia	Residue from methylamine
K	Ethylamine	Saturated solution
Ca	Ammonia	Small piece of metal placed in cell
Ca	Methylamine	Saturated solution

shows the specific fashion in which the various solutions were prepared. Solvents were distilled, before use, in a Podbielniak type of column and at a pressure of 300 to 400 mm. Hg. All the amines as received had ammonia fractions in them but the separation of this fraction was quite sharp on the still used. Only a middle fraction was used for the experiment. The solvents were dried over one of the alkali metals but ultimately it was found that lithium was the best drying agent probably because it is the most soluble.

No attempt was made to purify the metals used. It was observed, however, in the case of sodium and calcium that these metals did contain something which made it difficult to produce a solution in methylamine. This material could be removed in one of two ways. The metal could be fused in trap *B* under vacuum (feasible only for sodium) or a solution could be first made in ammonia and the ammonia subsequently evaporated. The latter method was used for calcium, but either method left the metal in a finely divided state which caused some trouble because metal particles tended to remain suspended in the liquid.

The absorption spectra of liquid ammonia and methylamine are shown in Fig. 2. From these spectra it can be seen that little difficulty will be encountered from solvent contribution to solution spectra, down to about 7000 cm^{-1}. At frequencies above this almost any cell thickness is acceptable while below this a cell 0.1 mm. in thickness is desirable. Spectra of sodium solutions in the mixed solvent and in ammonia have a band at about 7350 $cm.^{-1}$ which for ease of preparation were measured in the 1.0 mm. cell. Consequently, measurements of this band are more uncertain than most of the spectra reported.

Considerable variation of absorbance was observed from solution to solution even in the case of those which were originally saturated. This was probably due to change in concentration which occurred when the liquid was forced over. It is possible that sufficient water and oxygen remained adsorbed on the surface of the vessels to affect the dilute solutions slightly, particularly in the region between the windows.

Variations in absorbance were also observed when the temperature of a specific solution was changed, particularly in the case of the saturated solutions. This variation followed no regular pattern and if the solubility changed with temperature and some suspended particles of metal were present such behavior would be predicted. However, when the temperature was held steady, absorbance measurements on solutions in pure solvents were reproducible throughout the time required for determination of the spectrum.

No detectable fading was observed in sodium or potassium solutions in methylamine up to −20°C. Lithium and calcium solutions in methylamine tended to fade at temperatures as low as −40°C. Solutions in ammonia were stable without exception but in this case the temperature was not allowed to rise above −40°C.

MATERIALS USED

Ammonia	Matheson	Anhydrous
Methylamine	Matheson	Anhydrous
Ethylamine	Matheson	Anhydrous
Sodium	Merck	Reagent grade
Potassium	Baker and Adamson	Reagent grade
Lithium	Metalloy	Low sodium, reagent grade
Calcium	Purest grade	Dominion Magnesium Limited

EXPERIMENTAL RESULTS

The spectra of lithium, sodium, potassium, and calcium in ammonia are identical, having a single broad band with a maximum at 6650 $cm.^{-1}$ (−60°C.)* and a half-height width of 2950 cm^{-1}. The specific observations are shown in

**When a band was not determined at −60°C. a value has been estimated using the temperature coefficients given below.*

TABLE II

SPECTRA OF Li, Na, K, AND Ca IN AMMONIA, METHYLAMINE, ETHYLAMINE, AND MIXED SOLVENT

Solution	Temp., °C.	Position of maximum, cm.$^{-1}$	Half-height width, cm.$^{-1}$	Density at maximum	Cell thickness, mm.
Li in ammonia	−70	6700	3200	2.43	0.1
Na in ammonia	−62	6800	3370	2.57	1.0
K in ammonia	−42	6500	3250	2.36	0.1
	−52	6550	3150	2.47	0.1
	−60	6700	3100	2.58	0.1
	−61	6600	3150	2.91	0.1
	−63	6650	3050	2.60	0.1
	−71	6850	2900	2.70	0.1
	−78	6750	2900	2.82	0.1
Li in methylamine	−55	7350	4350	0.76	0.1
	−55	7600	4350	0.73	0.1
	−75	8000	4350	0.71	0.1
Na in methylamine	−14	14,700	3450	1.83	1.0
	−23	15,550	3000	0.42	1.0
	−25	15,000	3300	1.61	1.0
	−30	15,300	2950	0.45	1.0
	−48	15,450	3250	1.20	1.0
	−50	15,550	3000	0.57	1.0
	−66	15,750	2750	0.53	1.0
K in methylamine	−62	12,200 15,300			0.1
	−67	12,100 15,100			1.0
	−73	12,500 15,900			10
Ca in methylamine	−60	7800	4550	0.815	0.1
K in ethylamine	−64	15,500	3600	1.66	10.0
Li in ethylamine	−40	7050	5900	2.14	0.1
Na in 1:1 ammonia–methylamine	−81	7500*	3150	0.54	1.0
Same	−63	7500*	3750	0.61	1.0
Same	−79	7500*	3350	0.51	1.0
K in 1:1 ammonia–	−72	7500	3350	1.11	0.1
methylamine	−81	7600	3200	1.08	0.1
Same	−52	7000	3550	0.88	0.1
Same	−43	7200	3600	0.75	0.1

**Low frequency maximum only.*

Table II. Calcium does not appear there because the solution used was too concentrated, so that it was possible only to show that no additional bands appeared and that the maximum was located similarly to the other three metals.

In methylamine different bands were observed for the various metals. This is illustrated in Fig. 4. Lithium and calcium have a single band each, which is the same for both metals and appears at 7680 cm.$^{-1}$ (−60°C.). Sodium has a single band at 15,300 cm^{-1}. Potassium has two bands close together with the

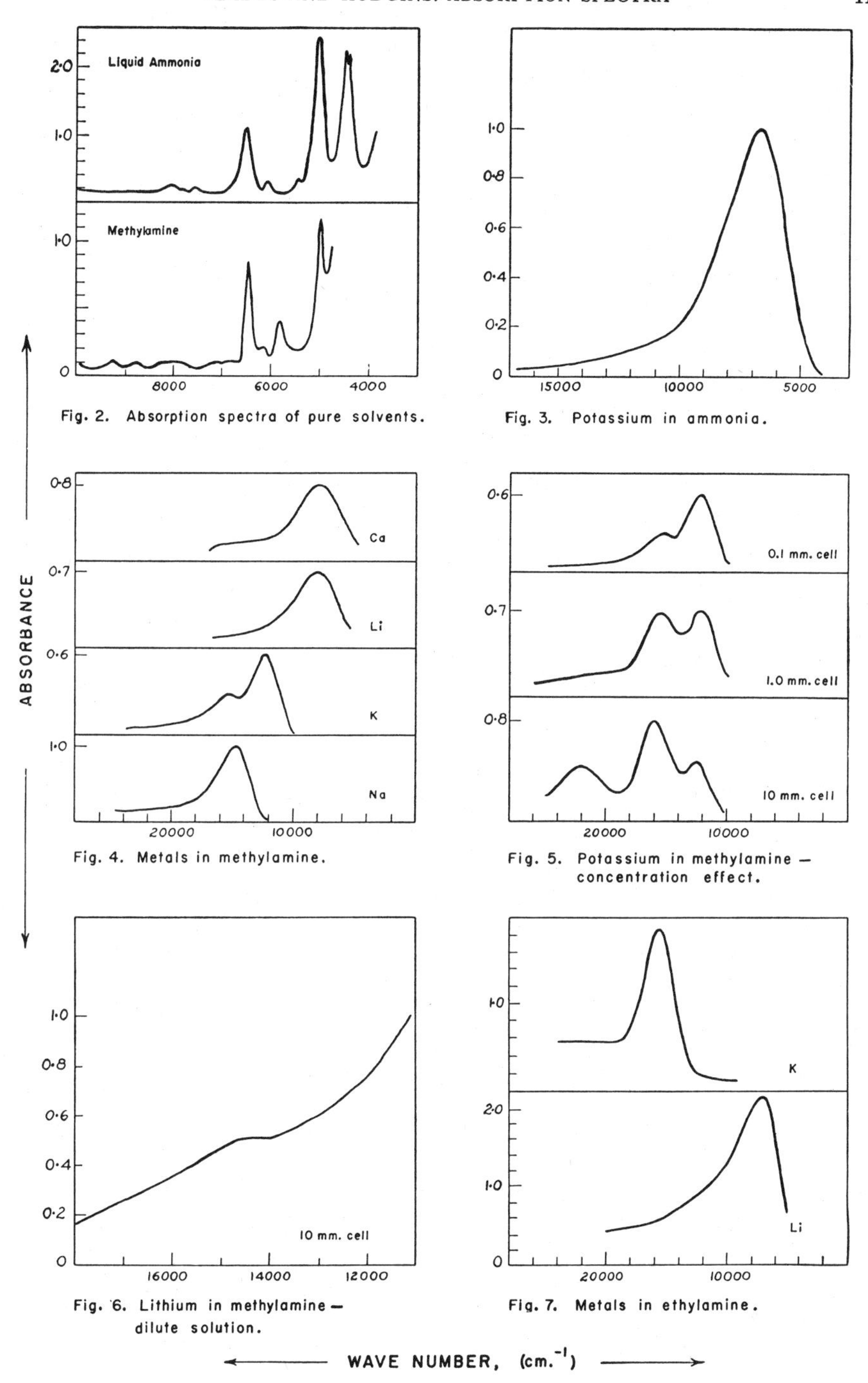

Fig. 2. Absorption spectra of pure solvents.

Fig. 3. Potassium in ammonia.

Fig. 4. Metals in methylamine.

Fig. 5. Potassium in methylamine — concentration effect.

Fig. 6. Lithium in methylamine — dilute solution.

Fig. 7. Metals in ethylamine.

result that the observed position of either band is somewhat affected by the other. It seems likely, however, that the high frequency band is identical with that found for sodium.

The effect of concentration on the potassium spectrum was investigated by adjusting the concentration, in three separate cells of 10, 1.0, and 0.1 mm. thickness, to give nearly constant absorbance. This will produce a change in concentration of about 100-fold. The resulting spectra are shown in Fig. 5. It will be noted here that the relative height of the two maxima varies with concentration and this indicates that absorption occurs by two independent processes.

In ethylamine only lithium and potassium were successfully dissolved. Their spectra are shown in Fig. 7. Two different bands are observed, the high frequency band being very similar to the high frequency band produced in methylamine. The low frequency band as observed for lithium, however, has a much greater half-height width than the comparable bands observed in methylamine.

The spectrum of sodium in a mixed solvent of methylamine and ammonia (molar ratio 1 : 1), shown in Fig. 8, exhibits two main absorption bands. The position of the high frequency band was established at 14,700 cm.$^{-1}$; its

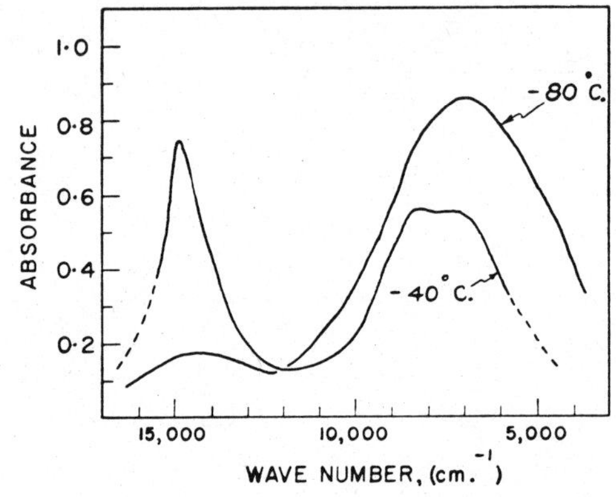

FIG. 8. Sodium in mixed ammonia–methylamine.

precise location was made difficult by the large variation of absorbance with temperature. The position and half-height width of the low frequency band are shown in Table II. From these it can be seen that this band is essentially the same as that observed for pure ammonia. Although its position is slightly shifted, its half-height width indicates that it does not contain a new hidden band. The temperature dependence of the absorbance at the maximum is shown for both bands in Table III.

The absorption spectrum of potassium in mixed methylamine and ammonia (molar ratio 1 : 1) reveals one band. The location and half-height width of this band serves to identify it with the low frequency band observed for sodium in the mixed solvent. This spectrum was obtained in the 0.1 mm. cell so that the concentration of potassium is probably about 10 times that of the corresponding sodium solution.

A solution of sodium in a mixture of ethylamine and ammonia (molar ratio 1 : 1) was examined for its absorption spectrum. This showed two bands, one at 7500 cm.$^{-1}$ and one at 14,800 cm^{-1}. The high frequency maximum can

TABLE III

EFFECT OF TEMPERATURE ON THE ABSORBANCE AT THE MAXIMA FOR THE MIXED SOLVENT

Temperature, °C.	Absorbance of the low frequency maximum	Absorbance of the high frequency maximum
−38	0.72	0.11
−49		0.13
−57		0.13
−59	0.68	0.16
−62		0.18
−63	0.61	
−68	0.62	0.21
−68	0.60	
−69		0.24
−71		0.25
−74	0.58	
−76		0.29
−78	0.51	
−79	0.52	
−80		0.33
−80		0.38
−81	0.44	
−82	0.43	0.44
−83	0.44	
−83	0.47	

probably be ascribed to sodium effectively dissolved in ethylamine although no solution is formed in the pure solvent.

From the data in Table II it might be expected that the position of the band in the spectrum of sodium in methylamine would be affected by both temperature and concentration. The effect of temperature on a specific solution for which the absorbance remained fairly constant is shown in Table IV. A plot of these data gives a temperature coefficient of −13 cm.$^{-1}$/degree. If the data in Table II for sodium in methylamine are now adjusted to a single temperature, using the above temperature coefficient it can be demonstrated that a change in position of the maximum is produced as the absorbance (i.e. concentration) is changed. The variation is 330 cm.$^{-1}$/unit of absorbance.

TABLE IV

THE EFFECT OF TEMPERATURE ON THE POSITION OF THE MAXIMUM FOR A SOLUTION OF SODIUM IN METHYLAMINE

Temperature, °C.	Position of maximum	Absorbance at maximum
−14	14,710	1.89
−25	14,810	1.80
−28	14,930	1.73
−50	15,150	1.53
−50	15,250	1.55
−58	15,270	1.45
−73	15,490	1.39
−73	15,380	1.65
−74	15,490	1.48

The position of the maximum for potassium in ammonia varies with temperature in a similar fashion, and the temperature coefficient for this system is -9 cm.$^{-1}$/degree.

The temperature coefficients for the half-height widths are 12 cm.$^{-1}$/°C. for potassium in ammonia and 14 cm.$^{-1}$/°C. for sodium in methylamine. As would be expected these coefficients are positive but the data are not precise enough to test the fit of a plot of $\sqrt{T}$ against band width.

The spectra of sodium and potassium in methylamine were observed for saturated solutions in a 1.0 mm. and a 0.1 mm. cell respectively. An attempt was made to determine the solubility of these two metals in methylamine. This was done by preparing saturated solutions at -80°C. and filling a small bulb with the solution in a manner similar to that used for filling the absorption cells. The solvent was then evaporated and the residual metal determined by measuring the hydrogen evolved on reacting it with water. The method of preparation of sodium solutions was such that the production of small particles of suspended metal was very likely. While this is not important in the measurement of spectra it does make analysis for the metal in solution difficult. Solutions were allowed to stand for some time before a sample was drawn off for analysis but this does not preclude the possibility of particles being swept off the delivery tube. With potassium the problem was not so severe because the metal was not in a finely divided state and particles were much less likely to be formed. Thus figures obtained for potassium are more likely to be accurate. The best estimates of concentration were 8×10^{-4} moles/liter and 1.9×10^{-3} moles/liter for saturated solutions of sodium and potassium respectively. It is suspected that the figure for sodium is somewhat high. A value of about 10^4 is thus observed for the molar absorbancy index of sodium in methylamine. The molar ratio of potassium to methylamine as used in the 0.1 mm. cell was thus about 5×10^{-5}; that for sodium to methylamine as used in the 1.0 mm. cell was probably 10^{-5} or less.

DISCUSSION

Originally it was hoped that information about the electron traps in liquid ammonia and amines could be obtained by observing spectra in mixtures of ammonia and an amine. It had been established that the spectrum in ammonia had a band at about 6000 cm.$^{-1}$ (9) and that a corresponding band occurred for methylamine at about 15,000 cm.$^{-1}$ (1). If traps exist in the structure of the liquid rather than in the form of a molecular orbital about a given molecule of solvent then it was reasoned that, in mixtures, intermediate bands between the two parent bands would appear. Indeed there was the possibility of only the intermediate band appearing.

The present observations confirm previous conclusions regarding the spectra in ammonia but the spectra for potassium, lithium, and calcium in methylamine are new and unexpected. At first glance it would appear that it is necessary to postulate a trap which will involve the metal ion as well as the solvent. But since identical bands appear for more than one metal (i.e. 7680 cm.$^{-1}$ band for lithium and calcium and the 15,500 cm.$^{-1}$ band for sodium and potassium) it is evident that a given metal ion does not produce a unique trap. This is adequate evidence for rejecting a metal ammoniate or solvent metal complex as the location of the trap. It will be shown below that it is possible to fit all the observed facts into the presently held theory that the electron trap is formed by a hole in the liquid which has a more or less regular structure.

It is suggested by Lipscomb (6) that the density of metal ammonia solutions is consistent with the hypothesis of a hole in the liquid of 3 Å radius. The solvent molecules at the edge of such a hole will have their more positive ends, in this case the three hydrogen atoms, directed inwards thus forming a region of low potential in which the electron is trapped. By analogy with *F* centers and impurity phenomena in solids such a trap might be expected to have an excited state with an attendant energy transition which is manifested by the absorption spectrum.

The positive end of the ammonia molecule is made up of three hydrogen atoms and is symmetrical so that the array of molecules at the edge of any hole will be identical. This will make the configurational co-ordinates and consequently the transition levels for each hole identical, with the result that a single maximum is observed for all the solutions in liquid ammonia. As the temperature is changed the intermolecular association will be disturbed with the result that the configuration of the hole is slightly altered. This leads to a slight change in the transition levels in the trap and is manifested by a shift in the position of the absorption maximum.

The methylamine molecule is not symmetrical about the nitrogen atom and hence it is possible that two different aspects of the molecule may be presented at the boundary of a trap. These two orientations may or may not be mixed depending on how readily they fit around the hole, the size of the hole, and on what other forces are at work in the rest of the liquid. On the basis of this model the methylamine spectra can be broken into two groups, called for convenience, amine bands and aliphatic bands. The former represents the condition where the methylamine molecules are oriented so that an ammonia-like trap is produced. This case is probably represented by the spectra of calcium and lithium where a band is observed at about 7680 cm^{-1}. The small difference in position of the maximum from that observed in ammonia is accounted for by the effect of the methyl group on the intermolecular forces and hence on the configurational co-ordinates. Aliphatic bands arise from traps in which the methyl group takes part in the formation of the trap boundary. Such a trap would be expected to have quite different configurational co-ordinates and a different set of transition levels. Such a situation is in fact observed for the sodium and potassium spectra in methylamine where a band appears at about 15,500 cm^{-1}.

In potassium solutions in methylamine an intermediate band is produced which must arise from a third trap because the relative height of the two bands observed varies with concentration and temperature. Such behavior would not be observed if two transition levels in a single oscillator were being represented. The fact that there is a discrete additional band and not a diffuse broadening of the original implies that a fixed combination of the two possible orientations exists.

Aliphatic bands in ethylamine are represented by the spectrum of potassium in solution. A mixture of the orientations producing aliphatic and amine traps is probably represented by the lithium spectrum in ethylamine. The single band is very much broadened and close inspection shows that the broadening is largely on the high frequency side which is what would be expected for mixed traps where the amine orientation predominated.

The particular trap which is formed in the amines depends on the metal ion which is present in the solution. This implies that the metal ions have a relatively long range effect on the orientation of solvent molecules so that one trap or the other is formed depending on the strength of this influence. Thus as the potassium solutions were made more dilute the number of pure aliphatic traps tended to increase at the expense of the other mixed trap. In very dilute lithium solutions in methylamine (Fig. 6) there is definite evidence that a weaker band at about 15,000 cm.$^{-1}$ exists which was not observed in the more concentrated solutions. In even more dilute solutions this band presumably would become relatively stronger and in fact Gibson and Argo (1) report a definite band in that position. It would appear that aliphatic traps would normally form in methylamine but the influence of certain metal ions favors the amine type trap to a degree depending on their concentration.

Further evidence for the long range effect of the metal ions is presented by the concentration effect on the position of the maximum in sodium solutions in methylamine. Here it was observed that even in very dilute solutions a variation in concentration shifted the position of the maximum to a significant degree.

It will be shown below that the difference in depth of traps of the two kinds is only about 0.1 ev., so that even a very weak influence from the metal ion may be sufficient to favor the orientation producing one trap or the other.

In mixed ammonia and methylamine, the following six types of trap may be expected:

Amine type	Mixed type	Aliphatic type
1. Pure ammonia	3. Ammonia+methylamine (amine oriented)	6. Pure methylamine
2. Pure methylamine		
	4. Ammonia+methylamine (aliphatic oriented)	
	5. Methylamine (amine oriented) +methylamine (aliphatic oriented)	

In pure methylamine, a trap of type 5 above is represented only in the potassium spectrum; from this it can be predicted that traps of type 4 and 5 will be rare in the mixed solvents. In fact no corresponding bands have been observed.

Traps of types 1 and 2 have maxima which lie quite close together (Table V). In the mixed solvent an intermediate band is observed, probably due to traps of type 3. A similar band could of course be produced from the sum of two

TABLE V

AVERAGE VALUES, INDEPENDENT OF METAL, FOR BAND POSITION AND HALF-HEIGHT WIDTH FOR THE LOW FREQUENCY BANDS

Measurement	Solvent		
	Ammonia	1:1 Ammonia methylamine	Methylamine
Band maximum	6650	7310	7680
Half-height width, cm.$^{-1}$	3080	3460	4420

separate bands each representing the spectrum observed in the pure solvents, but such a band would be wider than either of the two parent bands. This, however, is not observed since the band width also is intermediate as shown in Table V. It is therefore concluded that the intermediate band observed results from the formation of combination traps involving both solvents.

The band of the pure aliphatic type is observed only for sodium in the mixed solvents. A corresponding band does not appear in the potassium solution. This is consistent with the pattern in pure methylamine where it is observed that potassium has a stronger directing influence towards the formation of amine type bands than does sodium.

The relative depths of traps of the two kinds, i.e. those involving the amine group only and those involving only the aliphatic group, can be estimated from the effect of temperature on the height of the two maxima appearing in the sodium solution in mixed methylamine and ammonia. Heights of these two maxima as a function of temperature are given in Table III. By plotting the absorbance of the two maxima against temperature it can be shown that the one maximum gains at the expense of the other when the temperature is changed.

If it is assumed that the activation energy for the formation of traps is zero, we may write:

$$B/A = a\, e^{-(E_b - E_a)/RT} \quad [1]$$

where A = concentration of traps of the amine type,
B = concentration of traps of the aliphatic type,
a = constant,
E_a = energy of formation of Trap A,
E_b = energy of formation of Trap B.

If it is further assumed that Beer's Law holds for the maxima, it follows that

$$B^1/A^1 \propto B/A$$

where B^1 = absorbance at maximum of high frequency band when the concentration of aliphatic traps is B,
A^1 = absorbance at maximum of low frequency band when the concentration of amine traps is A.

Thus the plot of $\log B^1/A^1$ against $1/T$ should permit an estimation of $E_b - E_a$.

Such a plot is shown in Fig. 10 and from this a value of −0.17 ev. is obtained for $E_b - E_a$. Traps of mixed amine and ammonia types are thus slightly deeper than the aliphatic type. Because of the great similarity of these bands it is suggested that ammonia traps, methylamine traps of the amine type, and mixtures of the two will have approximately the same depth. The difference in depth between traps in ammonia and methylamine for a metal such as sodium then is about 0.2 ev.

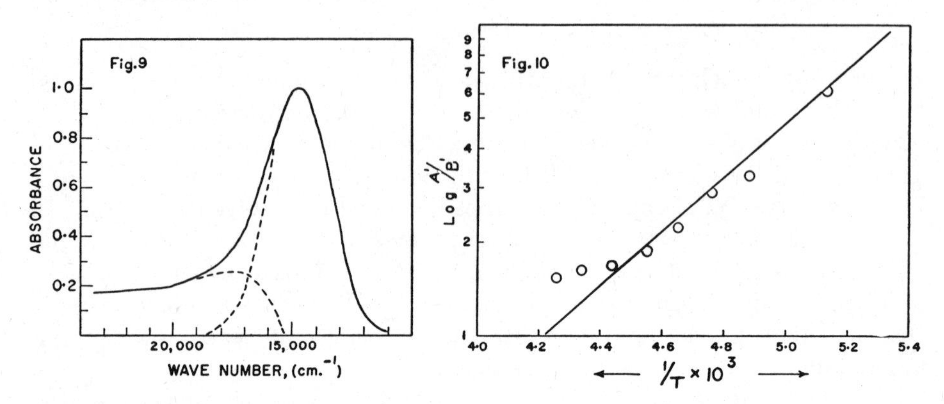

FIG. 9. Sodium in methylamine.
FIG. 10. Effect of temperature on peak height in spectrum of sodium in mixed ammonia–methylamine.

The separation of the two bands for spectra in methylamine and ammonia is about 1 ev. This value is much larger than the difference between the potential energies of the two traps so that it must be concluded at least for the low frequency bands that the absorption process involved in producing the bands is not one resulting in dissociation of the traps but rather is a transition between two energy levels of the trap.

Fig. 9 shows a more detailed representation of the absorption spectrum of sodium in methylamine. If this spectrum be compared with that found for potassium in ammonia (Fig. 3) it will be noted that for the high frequency band in methylamine there is a rather high level of continuous absorption on the high frequency side. In ammonia the level of absorption falls off much more quickly in the corresponding region.

By analogy with *F* centers in alkali halide crystals (7) it is suggested that the continuous absorption in the solution spectrum in methylamine represents a dissociation continuum which begins in the neighborhood of the main band. This implies a dissociation energy of about 2 ev. for the high frequency trap.

SUMMARY

1. A single band in the absorption spectra at 6650 cm.$^{-1}$ was found for all the solutions of metals in liquid ammonia which were studied.

2. The absorption spectra of solutions of metals in methylamine have a single band with the exception of potassium which has two. The position of the single band may be at one of two points, 15,300 cm.$^{-1}$ or 7680 cm.$^{-1}$ (−60°C.) depending on the metal.

3. The absorption spectrum of potassium in ethylamine has a single maximum in the 15,300 $cm.^{-1}$ region.

4. A solution of sodium in mixed methylamine–ammonia solvent has an absorption spectrum with two bands, one in the 15,300 $cm.^{-1}$ and one in the 7300 $cm.^{-1}$ region.

5. From the sodium in methylamine absorption spectrum it is argued that a dissociation continuum begins in the neighborhood of the band, which permits the estimation of 2 ev. for the depth of traps represented by bands in the 15,300 $cm.^{-1}$ region. From temperature effects in the mixed solvent it was shown that the trap represented by the band in the 7300 $cm.^{-1}$ region is about 0.2 ev. deeper.

6. Since the separation in trap depth is much less than the separation of the two bands it is argued that the absorption bands result from energy transitions and not from dissociation of the trap.

ACKNOWLEDGMENT

The foregoing investigation was part of the work performed at the Royal Military College under D.R.B. Research Grant No. 357. This opportunity is welcomed for thanking the Defence Research Board for its fine cooperation and financial assistance.

REFERENCES

1. Gibson, G. E. and Argo, W. L. J. Am. Chem. Soc. 40: 1327. 1918.
2. Gibson, G. E. and Phipps, T. E. J. Am. Chem. Soc. 48: 312. 1926.
3. Husten, E. Ann. Physik, 33: 477. 1938.
4. Hutchison, C. A., Jr. and Pastor, R. C. J. Chem. Phys. 21: 1959. 1953.
5. Kraus, C. A. J. Am. Chem. Soc. 43: 749. 1921.
6. Lipscomb, W. N. J. Chem. Phys. 21: 52. 1953.
7. Mott, N. F. and Gurney, R. W. Electronic processes in ionic crystals. The Clarendon Press, Oxford. 1940. p. 114.
8. Ogg, R. A., Jr. J. Am. Chem. Soc. 68: 155. 1946.
9. Vogt, E. von. Naturwissenschaften, 35: 298. 1948.

A Provocative Model

The following paper by Becker, Lindquist, and Alder was important in the development of our understanding of metal–ammonia solutions. These authors showed that the conductivity data of Kraus (see p. 175) and the magnetic susceptibility data of Hutchison and Pastor (see p. 204) could be reasonably well accounted for by assuming the two dilute-solution equilibria which had been proposed by Huster[1] (see p. 198):

$$M \rightleftharpoons M^+ + e^-_{am}$$

$$2M \rightleftharpoons M_2$$

The species e^-_{am} and M were assumed to be paramagnetic, and the species M^+ and e^-_{am} were assumed to conrribute to the electrical conductivity of the solution.

Becker, Lindquist, and Alder proposed imaginative yet plausible structures for the species M and M_2, for which they gave theoretical justification. Although these proposed structures are now believed to be incorrect for metal–ammonia solutions and perhaps delayed our detailed understanding of dilute metal–ammonia solutions, they provided the impetus for many physical-chemical studies of metal–ammonia solutions.

---

[1]Huster's statement that the Na monomer species in sodium solutions is not present in significant amounts is essentially consistent with the equilibrium constants calculated by Becker, Lindquist, and Alder. For example, according to their constants for 240°K, the Kmonomer concentration never exceeds 6% of the total metal concentration.

Reprinted from THE JOURNAL OF CHEMICAL PHYSICS, Vol. 25, No. 5, 971–975, November, 1956
Printed in U. S. A.

34

Model for Metal Ammonia Solutions

E. BECKER,* R. H. LINDQUIST,† AND B. J. ALDER
Department of Chemistry and Chemical Engineering, University of California, Berkeley, California
(Received December 30, 1955)

A model is proposed for dilute metal ammonia solutions in which the neutral species is a metal ion solvated by ammonia molecules, with the valence electron from the metal ion localized on the protons of these ammonia molecules. This species can dissociate in a manner similar to an ion pair, and the equilibrium constant for this dissociation as estimated agrees with the one derived from electrical conductivity data. Also, two neutral species can associate to form a dimer species, and the calculated energy of formation of this dimer is in satisfactory agreement with the binding energy derived from magnetic susceptibility data. This model is consistent with the large volume increase on mixing and with some aspects of the absorption spectrum.

INTRODUCTION

THE physical and chemical properties of solutions of alkali metals in liquid ammonia have been studied extensively for the last fifty years. Several models have been proposed to explain one or more of the unusual properties of these solutions.

The currently accepted model for these solutions is the "electron cavity" model.[1,2] It is used to interpret the results of the recent spin resonance experiments, and seems adequate to explain the general behavior of the magnetic susceptibility data as well as the unusual volume increase on mixing. However, there are several inconsistencies that led us to evaluate the data on the basis of an entirely different model. The basic assumption of the electron cavity model, that the electrons are completely independent of the metal atoms, is inconsistent with the differences found in the magnetic susceptibility and in the electrical conductivity for the different alkali metals. The argument from the absorption experiments that the electrons are independent of the metal ion is misleading, since it is based on measurements in a very narrow spectral region[3] where only the same slope of a tail of an absorption band was observed.

Other shortcomings of the model are the difficulty to explain: (a) the nature of the energy of pairing these electrons,[4] (b) the source of energy sufficient to form large cavities (the size of 2 to 4 ammonia molecules) for the electrons, (c) the unusually high limiting conductance (about 900 $ohms^{-1}$) for the electron in such a cavity, and (d) the absorption spectra.

An attempt has been made earlier to associate an electron with the metal atom.[5] The paired electron species was thought to be Na_2, which is known to exist in the vapor. This model, however, was abandoned for reasons[6] which are invalid when ionization equilibria are considered.

PRESENTATION OF MODEL

The proposed model for metal ammonia solutions pictures the solute as four species in equilibrium. First, a neutral species, called a monomer, consisting of an alkali metal ion surrounded by approximately six oriented ammonia molecules with an electron circulating

* Predoctorate Fellow, National Science Foundation.
† Predoctorate Fellow, Allied Chemical and Dye Corporation. Present address: California Research Corporation, Richmond, California.
[1] J. Kaplan and C. Kittel, J. Chem. Phys. **21**, 1429 (1953).
[2] W. N. Lipscomb, J. Chem. Phys. **21**, 52 (1953).
[3] G. E. Gibson and W. L. Argo, J. Am. Chem. Soc. **40**, 1339 (1918).
[4] T. L. Hill, J. Chem. Phys. **16**, 394 (1948).
[5] E. Huster, Ann. Physik **33**, 477 (1938).
[6] E. Huster, Ann. Physik **43**, 183 (1948).

around the metal ion on the protons of the ammonia molecules. Secondly and thirdly, a solvated metal ion and an electron arising from the dissociation of a monomer. Lastly, a dimer consisting of two monomers bound principally by exchange forces.

Coulter[7] postulated this model in essence previously, but used it only to explain the unusual dependence of the compressibility on concentration and the large decrease of viscosity with concentration.

In the proposed model the electron is removed from the immediate vicinity of the metal atom and is localized to a large extent on the positive end of the dipole of ammonia because these protons still have considerable attraction for negative charge. This is reasonable in view of the formation of hydrogen bonds for protons with atoms which are electron donors. That the presence of such protons is essential for the formation of these unusual solutions can be seen when the protons of ammonia are replaced by alkyl groups. Trimethyl amine, for example, is incapable of forming these solutions while methyl amine still can.

The metal ion is solvated by about 6 ammonia molecules, which are oriented so that, on the average, the nitrogen end of the molecule will point toward the positive charge. The ion polarizes the ammonia molecules,[8] thereby increasing further the electron affinity of the protons.[9] The electron then preferentially circulates among the protons of these ammonia molecules around the central positive ion on a roughly spherical surface. This entire unit will henceforth be called the monomer.

This proposal is quite reasonable in view of the metallic behavior found for solid $Ca(NH_3)_6$ and $Li(NH_3)_4$, where these solvated metal atoms act as an entity having an electron available for metallic bonding. Furthermore, the miscibility gap of Na and K in liquid NH_3 at temperatures near the freezing point can be explained by a second phase consisting of a complexed alkali metal species. The phase diagram of Na[10] indicates that the metallic phase has a stoichiometric ratio of $Na \cdot 5\,NH_3$.

The electron motion around the monomer is governed by the attraction of the protons and the coulombic attraction of the metal ion, and is restricted on both sides by the repulsion of the electronegative nitrogens. The preceding implies that the ammonia molecules adjacent to the monomer should still, on the average, have their nitrogen atoms oriented toward the center of the monomer. That is, the distribution of an electronic charge over 6 ammonia molecules still leaves these molecules with dipoles in the same direction as before. To distribute the charge in this manner[4] is legitimate since the electron moves with such a speed that the dipoles are unable to follow its motion. The possibility exists, however, that the electron can escape from the immediate vicinity of the ion. When the monomer dissociates, the electron overcomes the coulombic attraction of the cation, and becomes associated with other protons of bulk solvent ammonia molecules. This dissociation of the monomer is analogous to the behavior of an ion pair of a weak electrolyte.

As the concentration of metal in solution increases, these monomers can dimerize and as the concentration increases still further higher clusters may form. The energy of binding arises both from exchange energy, as in the binding calculation of a hydrogen molecule, and from van der Waals attraction.

CALCULATIONS

The following calculations deal with the dilute region, where no higher species than dimers have to be considered. Due to the complicated nature of the phenomenon, these calculations are merely intended to show that right orders of magnitude for the various experimental quantities can be obtained from the model proposed above.

A. Volume Expansion

The volume expansion of sodium in liquid ammonia over and above that of sodium ions in liquid ammonia will be calculated. The expansion experimentally observed (in 1 M to 5 M solutions) corresponds to about 72 A^3 per sodium atom. For the expansion per sodium ion in solution,[2] a term must be added which corresponds to the difference in size between the metal and the ion amounting to approximately 36 A^3. Furthermore, an electrostrictive term of about 35 A^3 should be added, so that the difference in volume of about 140 A^3 has to be accounted for.

The size of our monomer is estimated as follows. The radius of a sodium ion is 0.95 A, the distance to the nitrogen 1.2 A (which has been shortened 0.3 A from the van der Waals radius due to the electrostrictive effect), and the resultant N-H distance 0.77 A, so that the radial distance to a hydrogen atom from the central sodium ion is about 3.0 A.

The electron decreases the attractive forces between bulk ammonia molecules and the monomer ammonia molecules. To quantitatively account for this repulsion, the electron is presumed to increase the hydrogen-nitrogen distance between two hydrogen bonded ammonia molecules from 2.37 A[11] to 2.7 A, which corresponds to the van der Waals distance. The electron, in effect, breaks these hydrogen bonds and the size of the monomer is thus increased by a spherical shell of radius 0.33 A. Therefore, the radius of the solvated ion is increased from 4.2 A to 4.53 A, where the size of the cluster ends at the middle of the nitrogen-hydrogen

[7] L. V. Coulter, J. Chem. Phys. 19, 1326 (1951).
[8] J. N. Shoolery and B. J. Alder, J. Chem. Phys. 23, 805 (1955).
[9] W. M. Latimer and C. Slansky, J. Am. Chem. Soc. 62, 2019 (1940).
[10] O. Ruff and J. Zedner, Ber. deut. chem. Ges. 41, 1948 (1908).
[11] L. Pauling, *The Nature of the Chemical Bond* (Cornell University Press, Ithaca, 1948), second edition, p. 306.

distance $[3.0+\frac{1}{2}(2.37)]$. This spherical shell has a volume of 80 A^3, in approximate agreement with the experimental value.

The volume expansion for the dimer is expected to be of this same order of magnitude, but this large volume effect should disappear when the electron dissociates from the monomer. Data on cesium solutions[12] indicate quite strongly that the volume expansion per alkali atom decreases upon dilution.

B. Magnetic Susceptibility

The concentration dependence of the magnetic susceptibility indicates there is a dimer species formed in which the electrons are paired, and the temperature dependence indicates that this dimer is energetically favorable. The proposed model calls for a consideration of the competition of two equilibria involving solvated electrons, e^-, solvated ions, M^+, monomers, M, and dimers, M_2.

$$M=M^+ + e^- \qquad K_1=\frac{[M^+][e^-]}{[M]}, \qquad (1)$$

$$M+M=M_2 \qquad K_2=\frac{[M_2]}{[M]^2}. \qquad (2)$$

The magnetic susceptibility is a measure of the total number of paramagnetic species (e^- plus M), and knowing the total amount of alkali metal, the amount of the diamagnetic species, M_2, can be evaluated. K_1 and K_2 can be expressed in terms of these two experimentally known quantities as follows:

$$\frac{(e^-+M)}{(M_2)^{\frac{1}{2}}}=\frac{K_1^{\frac{1}{2}}}{K_2^{\frac{1}{2}}}+\frac{(M_2)^{\frac{1}{2}}}{K_2^{\frac{1}{2}}}. \qquad (3)$$

These equilibrium constants were determined from the variation of the susceptibility with concentration for potassium at three different temperatures at concentrations up to 0.5 M[13] (Table I).

K_1 affects the susceptibility principally in the dilute region where the uncertainty in the experimental data is quite large. Consequently, no temperature dependence could be established for K_1, but probably the ionization reaction should be slightly endothermic. The values of K_2 show the expected temperature dependence for an exothermic reaction and the heat of formation of the dimer can be evaluated to be 0.4 ev. The values of K_2 might be somewhat in error due to the assumption of an activity coefficient of unity. However, if these values for the equilibrium constants are used, the magnetic susceptibility for potassium can be reproduced up to 0.5 M within experimental error.

The experimental fact that the susceptibility of potassium is greater than that of sodium at lower temperatures, while above −33°C the situation is reversed, can be explained by the different energies involved for reactions (1) and (2). For reactions (1) and (2) one expects that the smaller the alkali metal the higher the dissociation and dimerization energy, respectively, which is the same trend as that for the alkali metals in the gaseous state. The data for sodium permit only an approximate determination of the two equilibrium constants, and confirm that the sodium dimer is energetically more favorable than the potassium dimer. Since reaction (1) is probably endothermic and reaction (2) is exothermic, a crossover in susceptibility is possible.

TABLE I. Equilibrium constants for potassium.

T	298°	274°	240°
K_1	0.02	0.05	0.03
K_2	100	900	9800

It is also quite easy, on the basis of our model, to account for the lower magnetic susceptibility experimentally found for the alkaline earth metals (Ca and Ba)[14] than for the alkali metals. The basic difference is that the corresponding monomer species is diamagnetic.

The constant spectroscopic splitting factor with concentration in paramagnetic resonance experiments is again an indication that the electron is associated with protons at all concentrations. The explanation for the extremely narrow spin resonance line remains virtually unchanged since it depends on interaction of the electron with the protons of the ammonia molecules.[1]

The difference between the total magnetic susceptibility and the paramagnetic susceptibility in dilute solutions allows a diamagnetism which corresponds to an electron confined to a maximum radius of about 4 A.[13] Thus the dimensions of the ion pair are not inconsistent with this measurement.

C. Conductivity

The equivalent conductivity of dilute metal ammonia solutions decreases rapidly with increasing concentration. It appears to be similar to the behavior of strong electrolytes dissolved in liquid ammonia, which are incompletely dissociated due to the relatively low dielectric constant of the solvent (22). Kraus[15] interprets the conductivity of dilute metal ammonia solutions as being of an electrolytic nature and involving the formation of ion pairs. He calculates the dissociation constant for Na to be 0.05 at −33°C, which compares quite well with the value of about 0.03 calculated for potassium from susceptibility data.

In order to estimate the temperature dependence for K_1, the available conductivity data of Na[16] was analyzed

[12] J. W. Hodgins, Can. J. Research **27B**, 861 (1949).
[13] C. A. Hutchison, Jr., and R. C. Pastor, J. Chem. Phys. **21**, 1959 (1953).
[14] S. Freed and N. Sugarman, J. Chem. Phys. **11**, 357 (1943).
[15] C. A. Kraus, J. Chem. Educ. **30**, 83 (1953).
[16] G. E. Gibson and T. E. Phipps, J. Am. Chem. Soc. **48**, 312 (1926).

by the Fuoss method.[17] The uncertainty in the data permits one only to conclude that the temperature dependence is small, and not inconsistent with the calculations quoted in the next section.

The model adequately accounts for the conductivity in relatively dilute solutions and for the variation observed with alkali metal. The smaller the metal ion, the lower the conductivity is observed to be, which is explained by a more tightly bound ion pair. It seems likely that the concentration (0.6 M/l) at which the conductivity rises sharply and eventually reaches a magnitude comparable to metallic conduction corresponds to the point at which tunneling can take place through a distance of the order of 10 A. It should be pointed out here that electron transfer over such a distance is favored in our model since the van der Waals energy between two monomers at that distance is quite appreciable (about 0.14 ev, as calculated in Sec. E).

D. Apparent Molecular Weight

From the decrease in vapor pressure at 15°C of Na solutions ranging from 1.2 to 0.5 M, Kraus[18] has evaluated the effective molecular weight of sodium as decreasing from 30 to 21, respectively. With the aid of the two equilibria, the evaluated molecular weights are roughly in accord with Kraus' data, that is, the molecular weight decreases from 38 to 23. The effects of activity coefficients have been neglected in these concentrated solutions.

E. Energetic Calculations

With a reasonably reliable value for the equilibrium constant for the dissociation of the monomer (Kraus' ion pair) thus established, it is possible, by estimating the entropy change accompanying this process, to get a value for the energy necessary to dissociate the monomer. The entropy change, ΔS, can be estimated by considering the electrons as a perfect gas and applying the Sackur-Tetrode equation, which yields a ΔS of 3.5 eu. In this crude approximation the entropy change of the solvent is entirely neglected, since the first shell of ammonia molecules is already oriented by the ion in the monomer. The ΔS of 3.5 eu is therefore an upper limit, since additional ordering of the solvent could occur in the dissociated monomer, and since in the perfect gas calculation it was assumed that the entire volume of the solution was accessible to the electron. From the entropy the energy required to dissociate the ion pair is calculated to be 0.1 ev.

A rough quantum-mechanical calculation of the dissociation energy of the monomer can be made by considering the electron moving in the screened coulombic field of the central charge. Therefore, a hydrogen-like $1s$ wave function is used for the electron, scaled so that it has a maximum at 3.5 A. This means that an effective dielectric constant of 7 must be used, and that the ionization energy is 0.28 ev, again neglecting all solvent changes.

To calculate the dimerization energy, the dimer was considered as a very large hydrogen molecule in which the nuclei are forced apart to about 7 A. The exchange energy, using the scaled $1s$ hydrogen wave function, at that distance amounts to about 0.35 ev. This value is not unreasonable if it is considered that Cs_2 has a binding energy of 0.45 ev at an internuclear distance of about 5 A.

The van der Waals energy between 2 isolated monomers can be estimated by approximate formulas for the polarizability,[19] which turns out to be very large (5.8×10^{-22} cc), resulting in a van der Waals energy of $(0.25\times10^6)/(R^6(\text{in A}))$ ev. In the solution this attractive energy should be reduced by the dielectric constant (the index of refraction squared).

Thermodynamic calculations for the solvation energy of the electrons were carried out using the following data for sodium: the heat of solution (0.06 ev),[20] the heat of vaporization (1.06 ev), the ionization energy (5.0 ev), and the solvation energy of the sodium ion (−4.4 ev).[21] The solvation energy for the electron is then −1.6 ev.

F. Absorption Spectrum

The spectrum has been investigated incompletely and it is thus only possible to speculate as to the origin of the absorption curves so far reported. On the basis of the model presented, the spectrum might contain the following electronic transitions:

1. The solvated, ionized electron to a free electron.
2. (a) Radial transitions of the electron associated with the monomer. (b) Change in angular momentum of the electron associated with the monomer.
3. (a) Angular momentum and radial transitions of the dimer. (b) Singlet to triplet transition of the dimer.

Experimentally, an absorption has been observed by several investigators[22,23] with a broad maximum at about 0.7 ev. The tail end of this absorption extends into the visible and causes the blue color of these solutions. The outer photoelectric threshold has been determined in these solutions to be approximately 1.5 ev,[24,25] essentially independent of the type of alkali

[17] H. Harned and B. Owen, *The Physical Chemistry of Electrolytic Solutions* (Reinhold Publishing Corporation, New York, 1950), p. 188.

[18] C. A. Kraus. J. Am. Chem. Soc. **30**, 1197 (1908).

[19] L. Pauling and E. B. Wilson, Jr., *Introduction to Quantum Mechanics* (McGraw-Hill Book Company, Inc., New York, 1935), p. 387.

[20] C. A. Kraus and F. C. Schmidt, J. Am. Chem. Soc. **56**, 2298 (1934).

[21] W. M. Latimer, J. Chem. Phys. **23**, 90 (1955).

[22] E. Vogt, Z. Elektrochem. **45**, 597 (1939).

[23] W. L. Jolly, Rept. No. UCRL-2008, University of California Radiation Laboratory, (1952) (unpublished).

[24] G. K. Teal, Phys. Rev. **71**, 138 (1947).

[25] J. Häsing, Ann. Physik, **37**, 509 (1940).

metal. As for the inner photoelectric threshold, the only reliable evidence is that it is less than visible light energies.

Previous calculations showed that the energy necessary to remove an electron from its solvating environment amounted to 1.6 ev, in reasonable agreement with the outer photoelectric effect. A radial transition of an electron associated with the monomer will result in ionization, and hence, an inner photoelectric threshold. Our calculations indicate that this threshold should be less than 0.2 ev.

The energy of the angular momentum transition can be estimated from a free electron model where the electron is restricted to move on the surface of the sphere, that is, the interaction of the angular and radial parts of the wave function has been neglected. The energy levels of such a model correspond to $E = l(l+1)\hbar^2/2mr^2$ and thus the $l=0$ to 1 transition occurs at 0.6 ev for $r=3.5$ A.

The striking blue color of these solutions in the dilute region can then be attributed to the electrons being excited in the potential well around an alkali metal atom, created by the attraction of the protons of ammonia. This proposal is supported by the evidence that this color is nearly independent of the nature of the alkali metal but dependent on the nature of the solvent.[3] Nevertheless, the metal ion appears to play an essential role in this absorption since the introduction of electrons into liquid ammonia, by x-rays or by high potential gradients causing electrolysis, fails to yield the blue color. In our model the presence of metal ions is necessary to provide a center for periodic motion of the electron.

The protons of the ammonia molecules surrounding an alkali atom form a spherical attractive potential well of about 1.5 ev in depth, therefore angular momentum transitions of 0.6 ev are possible without ionization. Although it requires less than 0.2 ev to ionize the electron, in order for this to occur the ammonia molecules surrounding the monomer must be in a suitable configuration. However, the configuration of the ammonia molecules influences the shape and depth of the potential well of the electron so that large broadening of the spectral transition must be expected.

For the dimer, radial and angular momentum transitions similar to that of the monomer can occur and are roughly of the same energy. The singlet-triplet transition can be estimated from the proposed model as 0.7 ev. The detection of this transition might be difficult due to the interference of the angular momentum transition.

ACKNOWLEDGMENTS

The partial financial support of this investigation by the U. S. Atomic Energy Commission is gratefully acknowledged.

One of the authors, B. J. Alder, wishes to acknowledge helpful discussions with the members of the theoretical chemistry department at the University of Cambridge, while a visiting Guggenheim fellow there.

Solutions of Europium and Ytterbium

Until Warf and Korst carried out the work described in the following paper, the only metals known to be soluble in liquid ammonia were the alkali metals and the alkaline earth metals heavier than magnesium. These workers showed that europium and ytterbium are soluble in ammonia. Their presumption that these metals form hexaammines was later verified by Thompson, Stone, and Waugh.[1]

---

[1]D. S. Thompson, J. J. Stone, and J. S. Waugh, *J. Phys. Chem.*, **70**, 934 (1966).

[Reprinted from the Journal of Physical Chemistry, **60**, 1590 (1956).]

35

SOLUTIONS OF EUROPIUM AND YTTERBIUM METALS IN LIQUID AMMONIA

By James C. Warf and William L. Korst

Department of Chemistry, University of Southern California, Los Angeles 7, Calif.

Received June 25, 1956

It is well known that samarium, europium and ytterbium, unlike the other rare earth elements, form a number of compounds in which the oxidation number of the metal is two, and that in this state there are certain resemblances to the alkaline earth elements. We have recently shown, for example, that europium and ytterbium form dihydrides isostructural with the alkaline earth hydrides.[1] On the further basis of structure, density and the volatility of europium and ytterbium, in which respects a certain similarity to calcium, strontium and barium exists, it was considered that, like the alkaline earth metals, these rare earth metals may dissolve in liquid ammonia.

Calcium, strontium, barium, europium and ytterbium are all either face-centered or body-centered cubic, but density considerations are more pertinent, as europium and ytterbium are distinctly less dense than their neighbors. Pauling[2] has shown the relation of the metallic radii of europium and ytterbium to that of barium. Europium and ytterbium are much more volatile than the other rare earth metals.[3] The evidence suggests that in metallic europium and ytterbium, each atom contributes two electrons to the mobile lattice electrons, as in the alkaline earth metals, while in the other rare earth metals, three electrons per atom are contributed.

Experimental

Solutions of Europium and Ytterbium in Liquid Ammonia.—A simple apparatus attached to the vacuum line was used, in which ammonia could be distilled from sodium and condensed on the metal. The ytterbium and samarium metals were kindly supplied by Dr. F. H. Spedding, who also reduced europium salts, obtained through the courtesy of Mrs. Herbert N. McCoy; this metal had been cast in a tantalum crucible.

Both europium and ytterbium were found to dissolve in liquid ammonia at −78°, forming solutions having the characteristic deep blue color of metals in ammonia. Europium appeared to be the more soluble. While no solubility measurements were made, europium solutions 1.14 *f* and ytterbium solutions 0.25 *f* were prepared. Samarium was apparently insoluble. Evaporation of ammonia from europium and ytterbium solutions left golden metallic crystals, presumably the metal hexammoniates.

Ytterbium in Ammonia *via* Electrolysis.—Efforts were made to show that ytterbium(II) salts in ammonia could be electrolyzed, yielding solutions of ytterbium. A solution of ytterbium(II) iodide in liquid ammonia, approximately 0.005 *f*, was prepared by adding the stoichiometric amount of ammonium iodide to ytterbium in ammonia. An electrolytic cell with platinum electrodes was employed. The yellow YbI_2 solution was electrolyzed, and after an initial period of gas evolution, a deep blue color appeared around the cathode. The color was not stable; on diffusing outward, it disappeared with the formation of an unidentified precipitate, conceivably $Yb(NH_2)_2$.

(1) W. L. Korst and J. C. Warf, *Acta Cryst.*, **9**, 452 (1956).

(2) L. Pauling, *J. Am. Chem. Soc.*, **69**, 542 (1947).

(3) (a) A. H. Daane, D. H. Dennison and F. H. Spedding, *ibid.*, **75**, 2272 (1953); (b) E. I. Onstott, *ibid.*, **77**, 812 (1955).

Europium(II) Amide.—An attempt was made to prepare europium amide by catalytic decomposition of the ammonia solution. Europium was leached with liquid ammonia from a fragment of the tantalum crucible in which it was cast, and filtered through a sintered glass disk into a vessel holding a few milligrams of iron(III) oxide. After standing approximately an hour, the ammonia was removed, leaving the amide and the lustrous bronze-colored metal ammoniate; the latter was decomposed, as judged by the disappearance of the golden color, by evacuation at room temperature. The gray-brown residue was analyzed for nitrogen and for europium. This gave an N/Eu ratio of 0.484, corresponding to 27.9% $Eu(NH_2)_2$. Presumably the pure amide could be prepared by more effective decomposition of the ammonia solution of the metal.

Attempted Preparation of Europium Solutions in Ammonia by Leaching Europium Oxide Reduction Products.—Daane, *et al.*,[3a] have shown that ytterbium and samarium oxides may be reduced by lanthanum at 1450°. We thought that it may not be necessary actually to distil the ytterbium (or europium) metal away, but that mere leaching of the reaction product would serve to yield liquid ammonia solutions. Accordingly, a mixture of europium oxide and lanthanum, the former in 10% excess over the amount demanded by the equation $Eu_2O_3 + 2La = 2Eu + La_2O_3$, was fired in a sealed molybdenum crucible under 300 mm. of helium pressure in an induction furnace. The temperature was maintained at 1450–1460° for 25 minutes. After cooling, the reaction product was leached with liquid ammonia, but not a trace of blue color formed. The experiment was repeated under much the same conditions, except that aluminum was substituted for lanthanum; the result was the same. The reduction reaction is evidently rapidly reversible, the reactants being favored at lower temperatures.

Discussion

The heat of solution of europium in liquid ammonia was estimated at −26 kcal./g. atom by the difference of two energy-cycle equations for europium and calcium. This was based on the heats of sublimation of the metals,[4,5] the first two ionization potentials of the metals,[6] and the difference between the heat of ammoniation of the gaseous ions, interpolating Coulter's[7] data at the ionic radius of Eu^{++} (1.10 Å.).

Finally, the possibility that americium metal is soluble in liquid ammonia should be noted, since the position of this element in the actinide series corresponds to that of europium in the lanthanide series. Evidence supporting such a conjecture lies in the distinctly low density of americium[8] compared to neighboring elements and in its somewhat low heat of vaporization.[9] But lack of definite evidence of americium(II) compounds and the existence of a hydride in which the H/Am ratio is approximately 2.7[8] would indicate insolubility of the metal in ammonia. The conclusion of Graf, *et al.*,[10] that in metallic americium there are three valence electrons per atom also indicates insolubility in ammonia.

This work was carried out in part under the auspices of the Office of Naval Research.

(4) A tentative value of 40 kcal./g.-atom for europium was supplied by Dr. Adrian Daane.

(5) National Bureau of Standards, Circular 500, "Selected Values of Chemical Thermodynamic Properties," U. S. Govt. Printing Office, Washington, D. C., 1952.

(6) C. E. Moore, "Atomic Energy Levels," Vol. I, Natl. Bur. Standards, Circular 467, 1949; H. N. Russell, W. Albertson and D. N. Davis, *Phys. Rev.*, **60**, 641 (1941).

(7) L. V. Coulter, This Journal, **57**, 553 (1953).

(8) E. F. Westrum, Jr., and L. Eyring, *J. Am. Chem. Soc.*, **73**, 3396 (1951).

(9) S. C. Carniglia and B. B. Cunningham, *ibid.*, **77**, 1502 (1955).

(10) P. Graf, *et al.*, *ibid.*, **78**, 2340 (1956).

The Mechanism of Electron Migration in Ammonia

Although the electron occupies a large volume in liquid ammonia, its ionic conductance is very high. This fact in itself is enough to suggest that the electron conduction process is not similar to that of an ordinary atom such as I^-. In the following paper, Dewald and Lepoutre show how their thermoelectric data for metal–ammonia solutions indicate a negative heat of transport for the electrons, corresponding to a quantum tunnel process.

[CONTRIBUTION FROM BELL TELEPHONE LABORATORIES AND FACULTÉ LIBRE DES SCIENCES, LILLE, NORD, FRANCE]

The Thermoelectric Properties of Metal–Ammonia Solutions. III. Theory and Interpretation of Results

BY J. F. DEWALD AND GERARD LEPOUTRE

36

RECEIVED OCTOBER 7, 1955

Thermodynamic equations are derived for the thermoelectric power of metal–ammonia solutions which include the effects of electron–electron and electron–ion interactions. The previously reported anomalies in thermoelectric behavior of metal and metal–salt solutions are shown *not* to arise from these interactions, but rather from a large *negative* heat-of-transport of the electrons in these solutions ($Q_e^* \approx -.7$ e.v.). This large negative heat-of-transport is accounted for on the assumption that electrons move through the solutions, even at high dilution, by a quantum tunnel process, rather than by the previously considered ionic, or conduction-band processes.

I. Introduction

In the first two papers of this series[1,2] we have presented data on the thermoelectric properties of metal–ammonia solutions which are in many respects quite puzzling. This is perhaps not surprising for our knowledge of the thermostatic properties of these metal–ammonia solutions, to say nothing of their structure or transport mechanisms, is only partially complete. Furthermore, we do not yet have a complete thermodynamic derivation of the thermoelectric power of thermocells involving weak electrolytes with chemical reaction. Since the solutions in question certainly involve at least one and probably several association reactions, a knowledge of how these reactions affect the thermoelectric properties of the system would appear to be essential.

The organization of this paper may be briefly stated. First we derive equations for the thermoelectric power based on a fairly general model of the solutions. Then, using data obtained from other experiments on the solutions, we show that the anomalies in the thermoelectric data are orders of magnitude larger than could arise from any conceivable thermostatic association effects. Then completely neglecting association effects, we "force" the data to fit the thermoelectric equations for an ideal strong electrolyte. We find that all the anomalies in the data may be understood, in at least a semi-quantitative fashion, if one assumes that the electrons move in the solution with a *negative* heat of transport. Finally, a rational model for the conduction process is presented, not very different from the currently popular model, which could give rise to such a phenomenon.

II. The Thermoelectric Power of a Metal–Ammonia Solution with the Inclusion of Electronic and Ionic Association Effects

We wish here to derive equations for the thermoelectric properties of metal–ammonia solutions (and also mixed salt–metal solutions) which will include at least the major contributions of possible association effects and yet which will not be too general or too complex for comparison with the experimental data.

For concreteness let us consider the diagram of the thermocell shown in Fig. 1. The validity of the Thomson relation for such a system may be shown, using the Onsager reciprocal relations, to be quite general so long as Soret diffusion is restricted and there are no concentration gradients. The Thomson relation may be written in the form

$$\frac{d\varphi}{dT} = \frac{1}{T}\frac{(J_H)}{(J_e)_T} \quad (1)$$

where J_H is the flow of enthalpy from the heat reservoir A which accompanies the isothermal flow of current (J_e) through the thermocell.

Consider now the sub-system within the dotted line in Fig. 1. We assume the electrodes to be inert and reversible to electrons but put no restriction on the mechanism of charge transfer at the electrode or in the solution itself. Now, enthalpy is conserved in any region at constant pressure so we may express J_H as the difference between the enthalpy accumulation rate within the sub-system, ($\dot{H}$), and the enthalpy flows into and out of the sub-system other than that from the heat reservoir. We obtain the equation

$$\frac{(J_H)}{(J_e)_T} = \frac{\dot{H}}{J_e} - (\bar{H}_{el} + Q^*_{el}) - \sum_{\substack{\text{all ions in}\\ \text{soln.}}} [t_i/z_i(\bar{H}_i + Q_i^*)] \quad (2)$$

where

$\dot{H}$ = enthalpy accumulation rate in the sub-system
$\bar{H}_i$ = partial molar enthalpy of ion i
Q_i^* = molar heat-of-transport (or excess enthalpy) of the ion i (DeGroot's "reduced" heat of transfer)[3]
t_i = fraction of the current carried by ion i
z_i = charge of ion i

H_{el} and Q^*_{el} are, respectively, the enthalpy and heat-of-transport of electrons in the metal wires.

Now it can be shown that, if all of the species in the solution were completely dissociated, the various thermostatic terms ($\dot{H}$, $\bar{H}_{el}$ and H_i) in (2) would, by virtue of the equilibrium across the electrode–solution interface, combine into a single entropy term $T(\bar{S}_e - \bar{S}_{el})$, where the $\bar{S}$'s are the entropies of electrons in the solution and in the metal electrode. The result would be exactly that of Holtan, Mazur and DeGroot.[4] However, if the solutes are weak electroytes we have additional thermostatic terms which must be considered. For the three component system, salt–metal–solvent, we may write

$$\frac{\dot{H}}{J_e} = -t_x\bar{H}_{salt} + (1 - t_e)\bar{H}_{metal} \quad (3)$$

(1) J. F. Dewald and G. Lepoutre, THIS JOURNAL, **76**, 3369 (1954).

(2) G. Lepoutre and J. F. Dewald, *ibid.*, **78**, 2953 (1956).

(3) S. R. DeGroot, "Thermodynamics of Irreversible Processes," North Holland Publ. Co., Amsterdam, 1951.

(4) (a) H. Holtan, Jr., P. Mazur and S. R. DeGroot, *Physica*, **XIX**, 1109 (1953); (b) there appears to be a minus sign missing from the second half of their equation 48.

where t_X and t_e are the Hittorf transport numbers, respectively, of X^- and electrons. Equation 3 expresses the fact that since the temperature and pressure of the sub-system remain constant, the enthalpy of the sub-system can change only to the extent that there is an over-all change in the material content of the sub-system. We assume that only the ions are mobile, the solvent and any neutral molecules remaining fixed. $\bar{H}_{salt}$ and $\bar{H}_{metal}$ are the partial molar enthalpies of salt and metal considered as thermostatic components.

Our next problem is to evaluate $\bar{H}_{el}$ in equation 2 in terms of the properties of the solution and the metal electrode. Since the electrodes are assumed reversible to electrons, the electrochemical potential of the electrons in the metal must be equal to that of the electrons in the solution. This may be expressed in the form

$$\bar{H}_{el} = \bar{H}_s - T(\bar{S}_s - \bar{S}_{el}) \tag{4}$$

where the $\bar{H}$'s contain the electrostatic potential and where the subscript "el" refers to electrons in the metal lead wires, and the subscript "s" refers to the single electrons in the solution. The use of the single electrons in this connection is purely for convenience, since the electrochemical potentials of the electrons in all forms are equal by virtue of the equilibrium condition.

We now substitute (3) and (4) into (2), obtaining

$$\frac{(J_H)}{(J_e)_T} = -t_X\bar{H}_{salt} + (1 - t_e)\bar{H}_{metal} - H_s + T(\bar{S}_s - \bar{S}_{el}) - \sum_{\substack{\text{all ons in}\\ \text{soln.}}} t_i/z_i\bar{H}_i - Q^*_{el} - \sum_{\text{all ions}} t_i/z_iQ^*_i \tag{5}$$

Equation 5 involves no assumptions about the detailed constitution of the solutions. To this extent it is both exact and unenlightening. Now we follow Bingel[5] and assume the presence of metal ions (M^+), single electrons (s), paired electrons (p), metallide ions (M^-), in addition to X^- ions and neutral salt molecules (MX) arising from the presence of salt. Expressing the sum over the enthalpies, $\bar{H}_i$ explicitly, and then combining terms appropriately, we obtain equation 6 below. The entropy and heat-of-transport terms of the electrons in the metal have been discarded, since these are orders of magnitude smaller than the terms arising from the solution.[6]

$$T\frac{d\varphi}{dT} = \frac{(J_H)}{(J_e)_T} = T\bar{S}_s - \sum_{\text{all ions}} t_i/z_iQ^*_i - t_pW_p - 2t_mW_m - (1 - t_s - t_p - 2t_m)(\alpha_pW_p + \alpha_mW_m) + t_XW_X(1 - \alpha_X) \tag{6}$$

The terms have the following meanings

t_X = fraction of current carried by X^-
t_s = fraction of current carried by e^-
t_p = fraction of current carried by $e_2^=$
t_m = fraction of current carried by M^-
α_X = fraction of the salt existing as X^-
α_p = fraction of the electrons existing as $e_2^=$
α_m = fraction of the electrons existing as M^-
$W_X = \bar{H}_{M^+} + \bar{H}_{X^-} - \bar{H}_{MX}$ = enthalpy of ionization of MX
$2W_p = 2\bar{H}_s - \bar{H}_p$ = enthalpy of pair dissociation
$2W_m = 2\bar{H}_s + \bar{H}_{M^+} - \bar{H}_{M^-}$ = enthalpy of metallide dissociation

The first two terms in (6) constitute the complete solution for the strong electrolyte case, or for weak electrolyte systems at infinite dilution. This can be seen from the fact that in these two cases $\alpha_X = 1$, $t_p = t_m = 0$, $\alpha_p = 0$, $\alpha_m = 0$, and the last four terms are zero. The last four terms represent the effects of association.

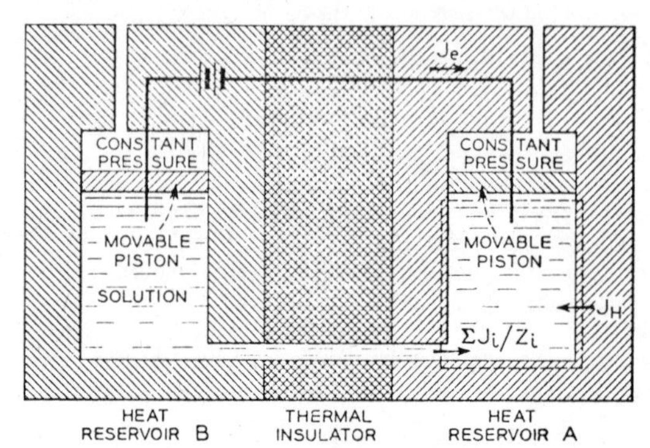

Fig. 1.—Thermodynamic representation of a constant pressure thermocell.

III. Comparison with Experiment

Three major anomalies in the thermoelectric behavior of these solutions have been observed and reported by us. These are: (1) the large dilution effects; (2) the large temperature effects; (3) the very large effect of added salt. Since it is both the largest and the least ambiguous of the three effects we concentrate our attention first on the salt effect.

A. Demonstration of the Second Order Nature of the Thermostatic Contributions to the Salt Effect.—Equation 6 shows that there are two possible ways in which the addition of salt might affect the thermoelectric power of a metal–ammonia solution. It could, by a common ion effect, alter the concentrations of the various electronic and ionic species and thus alter the α's and $\bar{S}_s$ in (6). More generally, the addition of the salt would change the transference numbers of the various electronic and ionic species, thus affecting the thermoelectric power to the extent that the various enthalpies of association and heats-of-transport are finite. Each of the terms in (6) is thus seen to be a possible source of a salt effect. Our problem is to eliminate those terms which play only a minor role.

1. The Entropy Term.—The entropy term in (6) may be shown to be roughly independent of the presence of salt by a simple mass-action argument. The entropy of single electrons, $\bar{S}_s$, may be approximated as

$$\bar{S}_s = S_0 - k \ln [e^-]$$

where $[e^-]$ is the concentration of single electrons and $\bar{S}_0$ is a constant. Thus the change in $\bar{S}_s$ on adding salt may be written

$$\Delta S = k \ln \frac{[e^-]_{R=0}}{[e^-]_{R=R}} \tag{1'}$$

(5) W. Bingel, *Ann. Physik*, **12**, 57 (1953). To our knowledge Bingel's is the most general model considered so far in the literature, all others being special cases of Bingel's.

(6) M. I. Temkin and A. V. Khorochin, *J. Phys. Chem.*, (*U.S.S.R.*), **26**, 500 (1952).

Now we may write the mass action expression for the dissociation of the M^- ion as

$$K = \frac{[e^-][M^+]}{[M^-]} = [e^-]^2 \frac{\left(1 + \frac{2[X^-]}{C} + \frac{[e^-]}{C} + \frac{2[e_2^-]}{C}\right)}{\left(1 - \frac{[e^-]}{C} - \frac{2[e_2^-]}{C}\right)} \quad (2')$$

Since the addition of salt will, if anything, decrease the concentration of both the single and the paired electrons, (2′) may be expressed as an inequality

$$\frac{[e^-]_{R=0}}{[e^-]_{R=R}} \lesssim \sqrt{1 + \frac{2[X^-]}{C}} \lesssim \sqrt{1 + 2R} \quad (3')$$

From (3′) and (1′) we then obtain

$$\Delta S \lesssim \frac{k}{2} \ln (1 + 2R) \quad (4')$$

For the largest value of R employed in our experiments, (R = 28.5), equation 4′ predicts a maximum variation $\bar{S}_s$ of about 175 μv./°C. which is only a small fraction of the experimentally observed change in thermoelectric power on adding sodium chloride at this salt/metal ratio. At 0.002 molar metal concentration the change is ~1600 μv./°C. and at higher dilutions it is even greater than this. Thus the entropy term in (6) makes only a minor contribution, if any, to the salt effect. This is not to say that the term itself is negligible, only that it does not vary much on addition of salt.

2. The t_mW_m and t_pW_p Terms.—There are a number of ways to show that these terms cannot account for more than a small fraction of the experimentally observed salt effect. Both terms will decrease on addition of salt through the decrease in t_m and t_p. Thus the maximum possible contribution to the salt effect would be the value of the terms in the pure metal solution. Since, as Bingel[5a] has calculated, and Hutchison's[7] magnetic data confirmed, $W_p < W_m$, we may write

$$t_pW_p + 2t_mW_m \lesssim (t_p + 2t_m)W_m$$

Now the magnetic data indicate that at 0.002 molar about 80% of the electrons are single electrons. Thus since the mobilities of the pairs and metallide ions should be, if anything, smaller than the single electron mobility, we may write

$$t_p + t_m \lesssim 0.2$$

from which

$$t_p + 2t_m \lesssim 0.4$$

The magnetic data indicate a value of $W_m \approx 0.15$ e.v. Thus using this figure we find that at −33°

$$\frac{1}{T}(t_pW_p + 2t_mW_m)_{\text{pure metal}} < 250 \ \mu\text{v./°C.}$$

This is to be compared again with the experimental value of 1600 μv./°C. Since even with the extreme assumptions used here we cannot account for more than a small fraction of the effects observed, the origin of the large salt effect must lie elsewhere.

3. The Last Two Terms.—The last two terms in (6) may be shown to operate in the wrong *direction* to account for a positive salt effect. W_x, which appears in the last term, may be shown from conductance measurements[8] to be both small and negative; thus, since addition of salt causes an increase in t_x, the last term can only give rise to a negative salt effect.

As shown above, $W_m > W_p > 0$. Thus, since any common ion effect on the α's must increase α_m by more than it decreases α_p, the factor $(\alpha_pW_p + \alpha_mW_m)$ in the next-to-last term is positive and can only increase on salt addition. The factor $(1 - t_s - t_p - 2t_m)$ is quite readily shown to be positive and can only become more positive on salt addition. Thus the next-to-last term can only give a negative salt effect.

We conclude from the reasoning above that the very large effect of salt addition on the thermoelectric power must arise from sum over the heats-of-transport in equation 6.

B. Interpretation with Neglect of Association Effects—the Salt Effect.—In this section we attempt an explanation of the experimental data by completely neglecting the association effects and assuming that the only species in solution are metal ions, X^- ions and single electrons. Equation 6 may be written for this case as (7)

$$\frac{d\varphi}{dT} = \bar{S}_e - \frac{1}{T}\sum_{\text{all ions}} \frac{t_iQ_i^*}{z_i} = \bar{S}_e + \frac{1}{T}(t_eQ_e^* + t_xQ_x^* - t_+Q_+^*) \quad (7)$$

Rather little is known about the absolute values of the heats-of-transport of even the simplest electrolyte systems. However, studies dating back to Richards[9] in the 19th Century, and continuing in the recent work of Bonnemay,[10] Goodrich and coworkers,[11] Tyrrell, and Hollis[12] and Chanu,[13] indicate that, in aqueous solution at least, the heats-of-transport of "normal" ions do not vary greatly from one ion to another; the only "abnormal" ions in this respect seem to be the hydrogen and hydroxide ions, and even these do not differ from the "normal" ions by much more than 0.1 e.v. These results imply that the heats-of-transport of normal ions in aqueous solution are at most a few kT and it seems quite reasonable to postulate a similar behavior for "normal" ions like Na^+ and Cl^- in liquid ammonia solution. Thus, since the transference numbers of the sodium and chloride ions are either quite small or roughly equal (depending on the salt/metal ratio) in our experiments, we may further simplify equation 6 by cancelling the $t_xQ_x^*$ term against the $t_+Q_+^*$ term. Equation 7 then becomes

$$\frac{d\varphi}{dT} = \bar{S}_e + \frac{t_eQ_e^*}{T} \quad (8)$$

Equation 8 is in a form suitable for direct comparison with the experimental data. We may plot

(7) C. A. Hutchison, Jr., and R. C. Pastor, *J. Chem. Phys.*, **21**, 1959 (1953).

(8) (a) E. C. Franklin and C. A. Kraus, THIS JOURNAL, **27**, 191 (1905); (b) A. I. Shatenshtein, *J. Phys. Chem.* (*U.S.S.R.*), **15**, 974 (1941).

(9) T. W. Richards, *Z. physik. Chem.*, **24** 39 (1897).

(10) M. Bonnemay, *J. Chem. Phys.*, **46**, 176 (1949).

(11) J. C Goodrich, *et al.*, THIS JOURNAL, **72**, 4411 (1950).

(12) H. J. V. Tyrrell and G. L. Hollis, *Trans. Faraday Soc.*, **45**, 411 (1949).

(13) J. Chanu, *J. chim. phys.*, **51**, 390 (1954).

the thermoelectric power at any given concentration of metal as a function of the transference number of the electrons in the mixed salt–metal solutions. If (8) is a valid approximation, we may expect an essentially linear plot, the slope of which will yield the heat of transport of electrons in the solution. Such a plot is shown in Fig. 2. The thermoelectric data, taken from the second paper in this series,[2] are for solutions 0.01 molar in metal content.[14] The transference number of the electrons is calculated, for each value of the salt/metal ratio, from the conductivity and transference data of Franklin and Kraus,[8a] and of Kraus,[15] the assumption being made that there are no interactions between the salt and the metal. The qualitative result is unaffected by other more complex assumptions. Details of the calculation are shown in Appendix I.

As seen from Fig. 2, the thermoelectric data conform in not unreasonable fashion to the linear behavior predicted by equation 8. The slope of the thermoelectric power *vs.* t_e curve multiplied by the absolute temperature yields a value for the heat-of-transport of the electrons

$$Q_e^* = -0.70 \text{ e.v.}$$

This large negative value for the heat-of-transport is a surprising result; at first glance it seems to contradict many of the ideas which have been presented in the past regarding the structure of these metal–ammonia solutions, and if the salt effect stood by itself as the only anomalous feature of the data one would be strongly tempted to discard such an explanation. In this connection it should be mentioned that one could quite readily understand a *positive* heat of transport of 0.7 e.v. This would be roughly the value to be expected if the electrons moved *via* the conduction band which, according to Jolly,[16] is located ~0.8 e.v. above the electron pair level. The result that the mobile electrons appear to carry considerably *less* than the thermostatic enthalpy requires extensive confirmation from other sources.

C. The Concentration and Temperature Dependences in the Pure Metal Solutions.—We obtain at least a partial confirmation of the negative heat-of-transport from the anomalous concentration and temperature dependences. Consider first the concentration dependence of the pure metal solutions. There are two major sources of a concentration dependence of thermoelectric power exhibited by equation 8, the entropy term ($\bar{S}_e$) and the transference number (t_e). For an ideal solution the entropy term is of the form

$$\bar{S}_e = \bar{S}_0 - 2.303\, k \log_{10} C$$

For a real solution the entropy variation will be somewhat less than this, but still roughly of this form. The transference number of the electrons may also be empirically approximated as a logarithmic function of concentration in the range from

(14) The result is very nearly independent of the metal concentration in the range of concentration studied here. See below under "concentration dependence."

(15) C. A. Kraus, "Properties of Electrically Conducting Systems, ACS Monograph, 1922.

(16) See the reference cited by Kaplan and Kittel.[17]

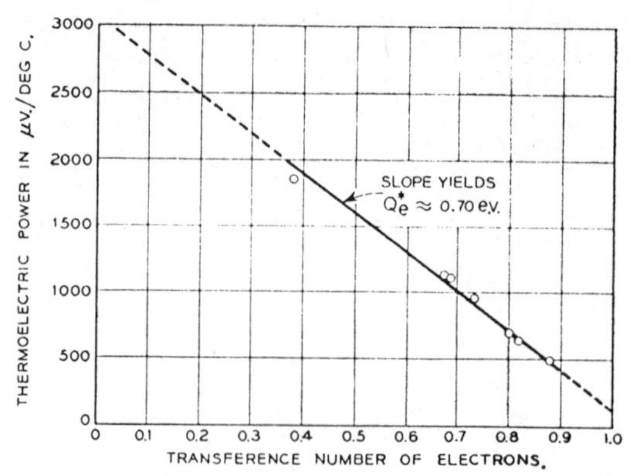

Fig. 2.—The thermoelectric power of mixed sodium–sodium chloride solutions as a function of the transference number of the electrons (calculated). Data are for solutions 0.01 molar in metal at −33°.

approximately 0.002 to 0.05 molar. From Kraus' transference data[15] we obtain

$$t_e \approx 1.048 + 0.065 \log C$$

Inserting these values and the value of Q_e^* derived above into (8) and differentiating, we obtain at −33°

$$\frac{d(d\varphi)/dT)}{d \log_{10} C} \approx -2.303k - 190 = -375 \ \mu\text{v.}/^\circ\text{C.}$$

This value is to be compared with the experimental value of −340 μv./°C. for sodium and potassium solutions at −33°. The agreement is good, probably fortuitously good, in view of the many approximations involved. Regardless of this, at least a major portion of the anomalous concentration dependence of the pure solutions can be accounted for if the heat of transport is negative.

A second confirmation of the large negative heat of transport of electrons is obtained from the fact that the large Thomson coefficients observed for the pure metal solutions are of the order predicted by equation 8. The Thomson coefficient, (σ), may be obtained from (8) by differentiating with respect to T.

$$\sigma = \frac{(d^2\varphi)}{(dT^2)} = \frac{d\bar{S}_e}{dT} - \frac{Q_e^*}{T_2} t_e + \frac{Q_e^*}{T}\frac{dt_e}{dT} \quad (9)$$

We expect that (9) will be valid only in very dilute solution since the Thomson coefficient is a second-order term. Where the discard of association effects at finite concentration might be justified in treating the first-order effects, this cannot be done for the second-order terms. Now, unfortunately, we do not have any precise data to evaluate the first and last terms in (9). However, orders of magnitude for these terms may be obtained, from the Sackur equation for the first, and from the temperature dependence of conductivity of solutions in ammonia for the second. The two terms act in opposite directions and are both appreciably smaller than the observed Thomson coefficients at high dilution. The second term may then be taken to be the major factor in the Thomson coefficient at high dilution. Using the value −0.7 e.v. for Q_e^* and a value of 7/8 for t_e we find

$$-\frac{(Q_e^* t_e)}{(T^2)_{Na}} = 10.6 \ \mu\text{v.}/^\circ\text{C.}^2 \text{ (at } -33^\circ)$$

This is to be compared with the value of the Thomson coefficient obtained at the lowest concentration studied by us (0.001 M)

$$\sigma_{Na}(0.001) = 6.5\ \mu v./°C.^2$$

The discrepancy between the two values is relatively small and since the experimental value is still rising with dilution at 0.001 molar, the real discrepancy may well be even smaller than this.

The difference between the Thomson coefficients observed for sodium and for potassium solutions is also understandable, in terms of the differences in t_e for the two solutions, if the heat of transport is large and negative. Using the ionic conductances at infinite dilution of Kraus and Q_e = −0.7 e.v., we predict that at infinite dilution the difference between the Thomson coefficients of the two metals should be

$$\sigma_{Na} - \sigma_K \approx -\frac{Q_e^*}{T^2}(t_e^{Na} - t_e^{K}) = 0.4\ \mu v./°C.^2$$

The experimental coefficients for the two metals differ by about 0.7 μv./°C.^{2} at 0.001 molar, in order of magnitude agreement with the prediction.

One final correlation of the experimental data can be made with the assumption of a negative heat of transport. It was shown in paper 2 of this series that the concentration dependence of the mixed salt–metal solutions could be expressed as a linear function of the quantity $1/(1 + 0.3R)$ where R was the salt/metal ratio.

The concentration dependence in the mixed solutions was defined and empirically expressed as

$$\text{"Concn. dependence"} \equiv -\left[\frac{\partial(\mathrm{d}\varphi/\mathrm{d}T)_{R,C}}{\partial \log C_{metal}}\right]_R \approx A - B/(1 + 0.3R) \quad (10)$$

The experimental values of A and B were found to be, respectively, 670 μv./°C. and 330 μv./°C. We have already shown that the absolute value of the concentration dependence in the pure metal solutions, (R = 0) can be understood in terms of the negative heat of transport. This is also the case for the mixed solutions as can be seen by differentiating equation 8 with respect to concentration at constant salt/metal ratio. Using only conductivity data and the value derived above for Q_e^* we may calculate the value of B in (10). The calculated value is +370 μv./°C. which is in surprisingly good agreement with the experimental value of 330 μv./°C. Physically what this means is that dilution at constant salt/metal ratio has two effects; it changes the entropy and also the transference number of electrons. At finite concentration the latter effect is appreciably larger in the presence of added salt than in its absence because the sodium chloride is more strongly associated than the sodium metal. Because of this the equivalent conductance of the salt will increase on dilution by a larger amount, percentagewise, than does that of the metal. Thus the fraction of the current carried by the salt will increase with dilution at constant salt/metal ratio.

The one effect which is not understandable in terms of this qualitative picture is the concentration dependence of the Thomson coefficient. Since this is, in effect, a third-order term it could very well arise from one or more of the discarded association terms of equation 6. More extensive knowledge of the thermostatic properties of the solutions is required before consideration of this effect would be profitable.

IV. Physical Basis for a Large Negative Heat-of-transport

A. General Discussion.—The preceding semiquantitative interpretation of all the major anomalies in the thermoelectric data seems to establish the validity of large negative heat-transport fairly unambiguously. We attempt now to understand this thermodynamic result in terms of a physical conductance mechanism.

Several conduction mechanisms have been considered in the past. Perhaps the earliest view with any claim to present-day consideration was that of Kraus, that the electrons were "solvated" and moved, essentially as ions, in the form of $(NH_3)_n^-$ ions. More recently the tendency has been to view the dilute solutions as semi-conductors, with most of the electrons trapped in solvent cavities, and conduction taking place by thermal, or photoactivation to a "conduction band," located some 0.7 to 1.0 volt above the trapping level. Kittel and Kaplan[17] have recently questioned this view,—at least for the thermal process—and have in effect returned to the earlier model of Kraus. They picture that the electron-cavity complex moves as a unit. They make an order of magnitude calculation of the mobility of such an aggregate, obtaining a value of the order of typical ionic mobilities, and then ascribe the extra mobility to other causes.

The reasoning used by Kittel and Kaplan to discard the conduction band process appears to be simply that the electron mobility is much too low. This argument is unconvincing, for the equivalent conductance of the electrons which they use in their calculation, (1000 mhos), represents the number average mobility of *all* the electrons, trapped and untrapped; we would normally expect that the number of *conduction* electrons would be very small compared to the total number of electrons. Without some estimate of the population of the conduction band, calculation of the mobility of the *conduction electrons* from the average mobility is clearly not possible. This is not to say that the mechanism of Kaplan and Kittel is incorrect, merely that it was not demonstrated.

A much more convincing argument for abandonment of the conduction band hypothesis is the relatively small temperature coefficient of conductivity observed in these solutions. If the conduction band were located 0.8 e.v. above the trapped level, as is indicated by the photo-absorption and photo-conductivity experiments, and the dark conductivity were by excitation to this level, we would expect a temperature coefficient of dark conductivity given roughly by[18]

$$\frac{\partial \ln\Lambda}{\partial 1/T} \approx \frac{\Delta\epsilon}{2k} \approx 4600°K. \quad (11)$$

The experimental value of the temperature co-

(17) J. Kaplan and C. Kittel, *J. Chem. Phys.*, **21**, 1429 (1953).

(18) The factor of 2 is used in (11) to take at least order-of-magnitude account of the positive polarization energy of the solvent around a trapped electron. (See below for discussion.)

efficient of dark conductivity is only $\sim$600°K. Thus the conduction band model for the dark conductivity seems highly unlikely.

To our knowledge no other mechanisms have been advanced for the conduction process in dilute solutions. This is rather surprising for we can see no reason to exclude a quantum "tunnel" mechanism. If such a process were in fact operative in the dilute range of concentration one could readily understand both the sign and order of magnitude of the observed heat-of-transport of electrons. Neither the "ionic" type of flow, pictured by Kraus and by Kaplan and Kittel, nor the conduction band model can predict such behavior.

To see how a tunnel conduction process could give rise to a large negative heat-of-transport we need to consider the nature of the electron trapping process. In these liquid solutions the trap may best be visualized, as Ogg[19] has done, as a center of dipole polarization. An electron, finding itself in a region where by chance a few of the solvent dipoles are favorably oriented, will tend to remain in that region and remaining there, will tend to polarize the permanent dipoles of the solvent further, until eventually a fairly stable configuration is achieved. The solvent dipoles will not be able to follow the detailed motion of the electron and thus the electron will "see" an effective charge located at the polarization center, and the electron wave function will be hydrogen-like around this center. Now if another polarization center approaches a trapped electron, the trapped electron may well make a quantum transition to this new center. If it does so, it will leave the energy of dipole polarization behind and since this *positive* energy is included in the thermostatic energy (and enthalpy), a negative heat-of-transport would result, *i.e.*, the mobile species would carry less than the thermostatic energy (or enthalpy).

The details of the process described above are doubtless quite complicated. However we may make an order of magnitude calculation of the size of the heat-of-transport by assuming that the entire polarization energy is left behind when an electron tunnels out of a solvent trap. The repulsive dipolar energy remaining in the dielectric once the electron is removed may be approximated by the Born expression

$$E_D \approx \frac{e^2}{2a}\left(1 - \frac{1}{k}\right) \approx \frac{e^2}{2a} \qquad (12)$$

We may take the result of Lipscomb,[20] that a trapped electron in these solutions is confined very largely in a sphere of radius $\sim$ 3.5 Å. and, assuming that the solvent polarization is largely dipolar, *i.e.*, cannot follow the detailed electronic motion, use it in (12) to find

$$E_D \approx 1.0 \text{ e.v.} \approx -Q_e^*$$

The result is certainly in order-of-magnitude agreement with the experimental value derived from the thermoelectric data, and in the absence of any other plausible mechanism is taken to imply the operation of a tunnel effect in the dark conduction process in these solutions.

(19) R. A. Ogg, Jr., *J. Chem. Phys.*, **14**, 295 (1946); **14**, 114 (1946); THIS JOURNAL, **68**, 155 (1946).

(20) W. N. Lipscomb, *J. Chem. Phys.*, **21**, 52 (1953).

V. The Standard Molar Entropy of Ammoniated Electrons

One of the original purposes of this work was to study the entropy of the electrons in these solutions. Now as Holtan, Mazur and DeGroot[4] have shown, and as will be elaborated by one of us in a subsequent paper, one cannot determine either the thermostatic or the transport quantities for individual ions by strictly thermodynamic measurements. Absolute ionic *entropies*-of-transport may appear to be measurable, however in a strictly thermodynamic sense they are no different from the thermostatic entropies since they involve assuming some value for the entropy-of-transport of electrons in a metal, and while this may well be quite small, only a mechanistic argument can demonstrate this.

Realizing the validity of the above reasoning, we may still use equation 8 above (this equation contains mechanistic as well as thermodynamic elements) and, to the extent that the mechanistic arguments used are valid, obtain a value for the absolute "ionic" entropy of ammoniated electrons in the standard state. Using the value of Q_e^* derived above, we find for the entropy at 0.01 molar metal concentration and $-33°$

$$\bar{S}_e = \frac{d\varphi}{dT} - \frac{t_e Q_e^*}{T} \approx +74 \text{ e.u.}$$

or for the standard molar entropy at $-33°$

$$\bar{S}_e^0 = 65 \pm \sim 5 \text{ e.u.}$$

The value above is in sizable disagreement with the value implied by the recent paper of Latimer and Jolly.[21] Using Hutchison's[7] magnetic data we find that if the absolute standard equivalent entropy of electron pairs is 25 e.u., as given by Latimer and Jolly, the absolute standard entropy of single electrons must be 30 e.u., roughly one-half the value calculated above. The discrepancy is not understood at present.

Appendix I

The Calculation of the Transference Number of Electrons in the Mixed Metal–Salt Solutions.—As a crude first approximation to the transference number of electrons in the mixed solutions, we might assume that both salt and metal were completely dissociated, and that the relative mobilities were given, at finite concentration, by the relative ion conductances at infinite dilution. In this case, using the values of λ_0 given by Kraus,[15] we would find

$$t_e = \frac{1}{1 + \frac{\lambda_{Na^+}}{\lambda_{e^-}}(1 + R) + \frac{\lambda_{Cl^-}}{\lambda_{e^-}} R} = \frac{0.875}{1 + 0.3R} \qquad (1')$$

This gives the transference number of electrons in the mixed solutions at infinite dilution and is the origin of the abscissa of Fig. 3 in our previous paper.[2]

An appreciably more realistic approximation than this may be obtained for finite concentrations by allowing for the incomplete dissociation of both the salt and the metal. The only assumption which we make is that there are no interactions be-

(21) W. M. Latimer and W. L. Jolly, THIS JOURNAL, **75**, 4147 (1953).

tween salt and metal in the solution. We may then write

$$t_e(c,m) = \frac{t_e(c,O)}{\left(1 + R\,\frac{\Lambda_{NaCl}(m)}{\Lambda_{Na}(c)}\right)} \quad (2'')$$

where

c = concentration of metal
m = concentration of salt
R = salt/metal ratio = m/c

The Λ's are the equivalent conductances of the salt and metal at the concentrations m and c, respectively, and the t_e's are the transference numbers of electrons in the pure metal and mixed solutions.

The transference numbers shown in Fig. 2 of this paper were calculated using equation 2″ with $t_e(c, 0)$ calculated from the ratio of ion conductances at infinite dilution. Other methods of approximating t_e yield comparable values.

MURRAY HILL, N. J.

Spectroscopic and Magnetic Studies of Metal–Amine Solutions

In the following paper, Fowles, McGregor, and Symons show that, in solutions of metals in amines, a diamagnetic species is responsible for the absorption band in the visible, and a paramagnetic species for the absorption band in the infrared. Although their description of the diamagnetic species as solvated electron pairs is subject to revision in the light of more recent studies, their observations were fairly accurate and appreciably advanced our knowledge of these solutions.

37

665. *Alkali-metal–Amine Solutions. Spectroscopic and Magnetic Studies.*

By G. W. A. Fowles, W. R. McGregor, and M. C. R. Symons.

One or both of two intense electronic absorption bands in the visible and the infrared region, respectively, are invariably found in dilute solutions of alkali metals in amine solvents. Which band predominates and the precise position of the absorption maximum depend upon the solvent, the metal, and the temperature. Solutions showing only the visible band are not paramagnetic, whereas paramagnetism is found whenever the infrared band is detectable. In agreement with current theory, it is postulated that electrons are solvated, either singly or as pairs, and that the intense absorption band is caused by the transition of an electron to a discrete excited state defined by the solvent shell. Allocation of the visible and the infrared bands to excitations from paired and unpaired electrons respectively, and the factors affecting the pairing and unpairing of electrons, are discussed, together with the effect of changes in environment on the absorption maxima.

It has been established that alkali metals dissolve reversibly in liquid ammonia to give true, rather than colloidal, solutions. The precise nature of these solutions has been investigated in several ways. Conductivity studies [1] have shown that these solutions have an equivalent conductance greater than that found for any salt in liquid ammonia; in very dilute solutions the conductance is about three times that of a salt in liquid ammonia, while in concentrated solutions the conductance closely approaches that of a metal. These results have been interpreted as a dissociation of the metal atoms into metal ions and solvated electrons, $M \rightleftharpoons M^+ + e$, the high conductivity of the dilute solution being attributed to the abnormal mobility of solvated electrons. Further information on the nature of solvated electrons is provided by density measurements,[2] which show that solutions of metals are less dense than pure ammonia to the extent of at least 70 $Å^3$ per alkali-metal atom dissolved, and by measurements of magnetic susceptibility [3] and paramagnetic resonance absorption.[4] Magnetic measurements show that paramagnetic species are present at low concentrations, but that the solutions become increasingly diamagnetic as they become more concentrated, and that, further, the paramagnetism increases with increase in temperature. Since the characteristic spectra of the metal atoms cannot be detected, these changes in paramagnetism are generally attributed to the presence of electrons solvated either singly or in pairs. The metal atoms are completely ionised and the resulting cations and electrons are stabilised by solvation. If the electrons are separately solvated, they will be paramagnetic, whereas they will be diamagnetic if they are solvated in pairs.

In the model described by Kaplan and Kittel,[5] based on the investigations by Hutchinson *et al.*,[4] the solvated electrons exist in delocalised molecular orbitals on all the solvent protons surrounding cavities which the electrons create in the solvent. The molecular orbitals are described by a wave function of the type

$$\Psi = p^{-\frac{1}{2}} \sum_{i=1}^{p} \psi_i$$

where p is of the order of 50, and ψ_i represents the $1s$ atomic orbital on the ith proton together with a small contribution from higher-energy states. In an alternative

[1] Kraus, *J. Amer. Chem. Soc.*, 1921, **43**, 749.
[2] Lipscomb, *J. Chem. Phys.*, 1953, **21**, 52.
[3] Freed and Sugarman, *ibid.*, 1943, **11**, 354.
[4] Hutchinson and Pastor, *Rev. Mod. Phys.*, 1953, **25**, 285; *J. Chem. Phys.*, 1953, **21**, 1959; Hutchinson, *J. Phys. Chem.*, 1953, **57**, 546.
[5] Kaplan and Kittel, *J. Chem. Phys.*, 1953, **21**, 1429.

description, Platzman [6] considers the electron to be bound in the lowest discrete state of the potential created by the oriented solvent molecules. In both these models, the solvated electron is quite independent of the cation.

Recently, Becker *et al.*[7] have suggested that no cavities exist, but that the electron is associated only with the protons of the solvent molecules orientated around the cation. Our results suggest that, although the cation influences the solvation of the electron, the extreme position proposed by Becker is unlikely in dilute solutions.

Spectra.—Examination of the available spectroscopic data [8-13] for dilute solutions of the alkali and alkaline-earth metals in ammonia, amines, and mixed solvents shows the presence of one or both of two characteristic and very intense electronic absorption bands, one in the visible (~6500 Å) and the other in the infrared (~15,000 Å) region. A less intense, but continuous, absorption is also found, starting at about 5500 Å and extending into the ultraviolet. Apparently, the only exception to this generalisation is found in the spectra of some solutions of potassium, where there is an absorption peak at about 8000 Å; this exception is discussed below. The existing data, together with the results obtained in the present study, are summarised in Table 1.

Nature of the Absorption.—The concept that one species is responsible for both bands can be discarded, since often only one band is detectable, whilst in other cases a change of temperature increases one band at the expense of the other—a fact which implies the presence of two different absorbing species in equilibrium. Thus three problems present themselves: (*a*) What is the nature of the absorbing species? (*b*) What is the change induced when light is absorbed? (*c*) What are the reasons for the marked effect of slight changes in environment on the relative amounts of each absorbing species and on the precise positions of the band maxima?

(*a*) Blade and Hodgins [10] attribute the existence of two bands to the variation in physical arrangement of the solvent molecules around the periphery of cavities containing single electrons, and suggest that with amines there are traps or cavities containing single electrons in which the hydrogen atoms of either amino-groups only or both amino- and hydrocarbon groups are oriented inwards. These are termed amine traps and aliphatic traps respectively. The 15,000 Å peak is then attributed to amine traps and the 6500 Å peak to aliphatic traps. This theory clearly cannot account for the appearance of the visible peak in ammonia solutions at low temperatures and high concentrations, and it is also difficult to see why all possible intermediate arrangements of N-H and N-Me groups around the cavity are not just as likely as the two extremes postulated. In any case, on the basis of Kaplan and Kittel's interpretation,[5] the orientation of N-Me bonds around the cavity could only mean that fewer protons are available to contribute to the delocalised molecular orbital, since C-H bonds would not be sufficiently polar for the hydrogen atoms to contribute appreciably.

Platzman and Frank [14] also only consider single solvated electrons, and propose that the observed absorption bands are part of a series of transitions. This theory, again, does not explain why when both bands are observed in one solution a change in temperature favours one at the expense of the other.

We suggest that, since magnetic studies [4] show that the energies of paired (e_2) or unpaired (e_1) solvated electrons are very similar, the most reasonable explanation of the two main bands is that they are caused by e_2 and e_1 species severally. The paramagnetic-resonance results obtained by Hutchinson and Pastor [4] for liquid-ammonia solutions, and

[6] Platzman, unpublished results, quoted in footnote to p. 423 of ref. 14.
[7] Becker, Landquist, and Alder, *J. Chem. Phys.*, 1956, **25**, 971.
[8] Gibson and Argo, *J. Amer. Chem. Soc.*, 1918, **40**, 1327.
[9] Vogt, *Naturwiss.*, 1948, **35**, 298.
[10] Blade and Hodgins, *Canad. J. Chem.*, 1955, **33**, 411.
[11] Jolly, U.S. Atomic Energy Comm. Nat. Sci. Foundation, Washington D.C., 1952, U.C.R.L. 2008, 3.
[12] Bosch, *Z. Physik*, 1954, **137**, 89.
[13] Hohlstein and Wannagat, *Z. anorg. Chem.*, 1956, **288**, 193.
[14] Platzman and Frank, *Z. Physik*, 1954, **138**, 411.

those obtained by us for solutions in methylamine, ethylenediamine, and propylenediamine (see Table 2) strongly suggest that the infrared band is a property of the e_1 species and the visible band of the e_2 species. Thus solutions which show only the visible band are not detectably paramagnetic, although the total metal concentration (*ca.* 10^{-3}M) would give an unpaired-electron concentration many hundreds of times greater than the minimum which

TABLE 1. *Absorption spectra of dilute solutions of metals in amine solvents.*

Solvent	Metal	Temp.*	Range investigated (Å)	Absorption max. (Å)	Comments †	Ref.
NH_3	Li	−65° to RT	4000—7500	—	A	8
	Li	—	4000—20,000	18,000	—	9
	Li	− 70	4000—20,000	14,900	T.	10
	Li	−253	4000—25,000	5850; 12,500	B	12
	Na	−65 to RT	4000—7500	—	A	8
	Na	—	4000—20,000	18,000	—	9
	Na	—	10,000—25,000	15,000	—	11
	Na	− 60	4000—20,000	14,700	T.	10
	Na	−253	4000—25,000	5900; 12,500	B	12
	K	−65 to RT	4000—7500	—	A	8
	K	−70	4000—20,000	15,000	—	10
	Mg	−65 to RT	4000—7500	—	A	8
	Ca	−65 to RT	4000—7500	—	A	8
$MeNH_2$	Li	−60	4000—20,000	13,500	—	10
	Li	—	4000—16,000	7100; 13,000	—	13
	Na	−60	4000—20,000	6400	—	10
	Na	−70 to RT	4000—7500	6500	—	8
	Na	—	4000—16,000	6900	—	13
	K	−60	4000—20,000	6540; 8200	—	10
	K	−70 to RT	4000—7500	6500	—	8
	Ca	−60	4000—20,000	12,800	—	10
$EtNH_2$	Li	−40	4000—20,000	14,200	—	10
	Li	—	4000—16,000	6800	—	13
	Na	—	4000—16,000	6800	—	13
	K	−60	4000—20,000	6450	—	10
$(\cdot CH_2\cdot NH_2)_2$	Na	RT	4000—10,000	6700	—	‡
	K	RT	4000—10,000	6700	—	‡
$NH_2\cdot CHMe\cdot CH_2\cdot NH_2$	Na	RT	4000—10,000	6600	—	‡
	K	RT	4000—10,000	6700	Also small shoulder at 8200 Å	‡
NH_3–$MeNH_2$						
(90—80% $MeNH_2$)	Li	—	4000—16,000	7400—7500; 13,000	—	13
(70% $MeNH_2$)	Li	—	,,	13,000	—	13
(90% $MeNH_2$)	Na	—	,,	7000	—	13
(80—70% $MeNH_2$)	Na	—	,,	7200; 13,200—12,500	—	13
(60—0% $MeNH_2$)	Na	—	,,	12,500—12,000	—	13
(50% $MeNH_2$)	Na	−70	4000—20,000	13,300	—	10
NH_3–$MeNH_2$ (50%)	K	−60	4000—20,000	14,000	—	10
NH_3–$EtNH_2$						
(90—70% $EtNH_2$)	Li	—	4000—16,000	7000—7200; 13,000	—	13
(60% $EtNH_2$)	Li	—	,,	12,500	—	13
(90—80% $EtNH_2$)	Na	—	,,	7000	—	13
(70—60% $EtNH_2$)	Na	—	,,	7000—7500; 13,000	—	13
(50% $EtNH_2$)	Na	—	,,	12,500	—	13

* RT = room temperature. † A = absorption increases with temperature; B = thin films, concentrated; T = temperature-dependent. ‡ This paper.

can be detected. When both bands are observed, or only that in the infrared region, then paramagnetic resonance absorption is always found.

Support for this allocation of the visible and infrared bands to the e_2 and e_1 species is provided by the Hutchinson and Pastor's magnetic measurements of liquid-ammonia solutions, which show that e_2 is favoured as the temperature is lowered, while Blade and Hodgins [10] find that the visible band of a solution in a mixed ammonia–methylamine solvent is favoured at the expense of the infrared band at lower temperatures (thus agreeing with the qualitative observations by Ogg [15] on analogous solutions in liquid ammonia).

[15] Ogg, *J. Chem. Phys.*, 1946, **14**, 114.

(*b*) It is often stated (see, *e.g.*, ref. 5) that the intense absorption associated with solvated electrons corresponds to the total ejection of an electron from the cavity composed of orientated solvent molecules into bulk, unorientated solvent (described as a conduction band). Similar theories have been invoked to describe the intense ultraviolet absorption of halide ions. However, Platzman and Frank [14] recently suggested that the electron in the excited state of a halide ion is still held by the orientated solvent shell, though more weakly. Indeed, one would expect an absorption continuum rather than a discrete band for a process involving total electron ejection.

Evidence for the nature of the act of light absorption in metal solutions is provided by the irradiation experiments of Linschitz, Berry, and Schweitzer,[16] who found that, when lithium in a glass of mixed ether, *iso*pentane, trimethylamine, and methylamine was irradiated, the strong visible absorption band was lost. These authors found that the band was discharged only when light of wavelength less than 5400 Å was used for irradiation, although the maximum absorption was at 6000 Å. A general increase in absorption in the near-infrared region, with a broad maximum at 12,500 Å was also observed. Our qualitative observation that when a blue glass of propylenediamine containing dissolved sodium is irradiated the blue colour is greatly diminished but never quite lost is in agreement with the work of Linschitz *et al.*

We can hence conclude that the two main absorption bands do not correspond to the ejection of an electron (paired or unpaired) from its cavity into a conduction band, but rather to its excitation into a higher-energy orbital still within the same cavity. We further suppose that the continuum observed below about 5500 Å corresponds to the total ejection of the electron into a conduction band. When this electron is forced out of its cavity completely, it is unable to return to its original state because of the rigidity of the solvent, but once the glass is allowed to soften, the electron can create a new cavity, so that the original absorption band returns. When one electron has been ejected from an e_2 cavity, then an e_1 cavity in effect remains, thus accounting in part for the increase in the infrared band observed by Linschitz.

(*c*) *Effect of Environmental Changes on the* $2e_1 \rightleftharpoons e_2$ *Equilibrium.*—It is apparent from Table 1 and from the results on mixed solvents reported by Blade and Hodgins,[10] that quite small changes in temperature, concentration, and the nature of the metal and solvent readily affect the equilibrium $2e_1 \rightleftharpoons e_2$, so that e_1 and e_2 must have very similar energies. Lowering the temperature or raising the concentration clearly favours the e_2 form. Solvents with relatively high dielectric constants (*e.g.*, NH_3) appear to favour e_1 formation irrespective of the metal, but in amine solvents e_1 formation occurs only when the metal ions have a high surface-charge density. It is perhaps significant that lithium is much more soluble in amines than either sodium or potassium, and that lithium alone of these metals forms appreciable amounts of e_1 cavities in such solvents. These trends are consistent and suggest that in amine solvents ion-pair formation may take place, so that the metal ion is close to the solvated electron and has a noticeable influence upon the type of cavity formed.

We are unable to explain the absorption band at 8500 Å which appears in relatively concentrated solutions of potassium in methylamine and propylenediamine. Since the band is reproducible but occurs only in these two solvents, it is unlikely to be caused by impurities in the metal. The band appears at such short wavelengths that it can hardly be caused by e_1 transitions (shifted by the effect of the metal ion) and in any case paramagnetic-resonance measurements show that the concentration of e_1 species is extremely small.

Effect of Environmental Changes upon the Position of Maximum Absorption and Band Width.—The results reported by Blade and Hodgins [10] show that subtle changes in environment have a marked effect, not only on the pairing of electrons, but also upon the

[16] Linschitz, Berry, and Schweitzer, *J. Amer. Chem. Soc.*, 1954, **76**, 5833.

exact positions of maximum absorption and widths of both bands. The magnitude of the shift of maximum absorption does not give a direct measure of the change in energy of either the ground or the excited state, but of the change in energy of one in relation to the other. The various trends may be summarised: (*a*) Increase in temperature broadens the band and shifts it to longer wavelengths. For a given solvent, this effect is more pronounced for e_1 than e_2 bands. (*b*) Increase in concentration shifts the band to shorter wavelengths. This effect is more marked for e_2 than e_1 bands, and is greater in amine solvents than in ammonia. (*c*) Replacement of hydrogen atoms by methyl groups (*i.e.*, in amines) gives a shift to shorter wavelengths, and in mixed solvents the band lies between the two extremes. This gradual change indicates that in mixed solvents the hydrogen atoms of the N–H bonds of both amines and ammonia molecules can contribute directly to cavity formation.

The effects closely parallel those reported for the same changes in environment on the first absorption band of the solvated iodide ion.[17] If either the e_1 or the e_2 bands represented the complete ejection of an electron from a cavity then one might have expected a reversal in the direction of the spectral shifts compared with those for iodide, since the ground state of the solvated electron resembles the excited state of the iodide ion. Since, however, the effects are very similar, the postulate that the excited electron is still held within the e_1 or the e_2 cavity is given further support.

It has been noted that the e_2 band in amine solvents is far more sensitive to changes in concentration than is the e_1 band in ammonia. This can probably be attributed to the combined effect of the higher charge density per cavity and the lower dielectric constant of the amine solvent which makes ion-pair formation significant despite the low concentrations.

Experimental

Purification of Materials.—Sodium and potassium were purified by distillation *in vacuo* into tubes sealed at one end, a modified version of the technique described by Dostrovsky and Llewellyn[18] being used. The samples of the metals taken for the distillation were cut under light petroleum from the centre of larger pieces.

Anhydrous ethylenediamine and propylenediamine were placed over potassium hydroxide pellets for some days, and then distilled through a 3 ft. column packed with glass helices. The middle fraction from the distillation was treated with sodium wire, and then distilled *in vacuo* on to freshly distilled alkali metal (sodium for ethylenediamine, potassium for propylenediamine); a permanent blue coloration was taken as a criterion of dryness.

Methylamine. An aqueous solution (30% w/w) was run on to pellets of sodium hydroxide; the liberated methylamine was passed through a column of glass wool mixed with moist, freshly precipitated mercuric oxide to remove traces of ammonia[19] and collected in a trap cooled with liquid oxygen. The amine was dried *in vacuo* with freshly crushed barium oxide and finally metallic sodium.

Preparation and Stability of Alkali-metal Solutions in Ethylenediamine and Propylenediamine.—The apparatus shown in Fig. 1 was used for the preparation of the metal solutions. A tube containing purified sodium or potassium was placed in *D*, which was then sealed off, and the system was evacuated through the B14 joint, *J*, and well flamed to remove traces of adsorbed moisture on the walls. When gently heated, the metal ran through the constrictions c_1 and c_2 and solidified in *B*; a thin film of the metal was then distilled on to the walls of *A* and the constriction c_3 sealed off. The diamine was condensed into *A*, *via J*, and the apparatus was sealed off at constriction c_4, leaving the metal and amine in an all-glass, highly evacuated system containing no taps or joints liable to introduce impurities. The amine was allowed to warm to room temperature in contact with the metal, and the resulting solution was freed from excess of metal by filtration through the sinter *S* into *E*; the filtration was assisted by gently warming tube *A* (with the hand), thus slightly increasing the vapour pressure of the amine in this section.

[17] Smith and Symons, *J. Chem. Phys.*, 1956, **25**, 1074.
[18] Dostrovsky and Llewellyn, *J. Soc. Chem. Ind.*, 1949, **68**, 208.
[19] Hohlstein and Wannagat, *Z. anorg. Chem.*, 1956, **284**, 191.

Since lithium could not be melted through glass constrictions, a small portion of the metal was cut from the centre of a larger piece under light petroleum and while still wet with solvent was inserted into tube *A* against a counter-current of nitrogen; for this preparation, tube *D*

FIG. 1. *Apparatus for preparation of metal–diamine solutions.*

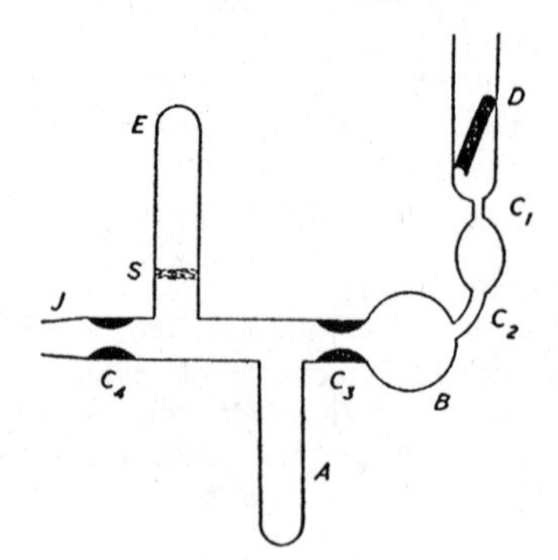

FIG. 2. *Apparatus for preparation of metal–diamine solutions for spectral studies.*

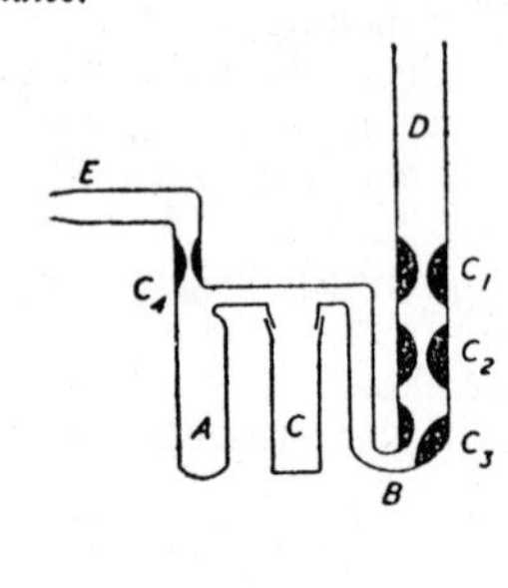

FIG. 3. *Absorption spectra.*

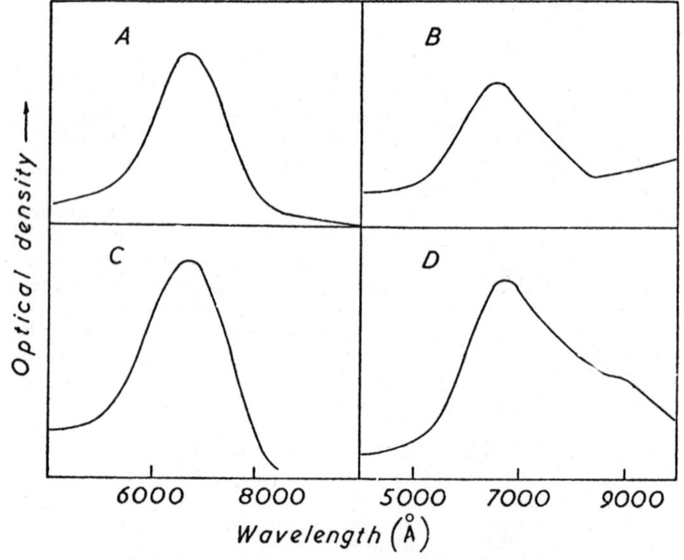

A, Sodium and, *B*, potassium in ethylenediamine.
C, Sodium and, *D*, potassium in propylenediamine.

and constrictions c_1, c_2, and c_3 were omitted from the apparatus. After evacuation, the solution was made up as before.

Sodium, potassium, and lithium gave blue solutions in both diamines. Several independent studies for each metal–amine combination showed that in both solvents sodium and potassium formed very much more stable solutions than did lithium. Thus sodium and potassium solutions generally decomposed over a period of 6—12 hr. while the lithium solutions were stable for only 15—30 min. The rapid decomposition of the lithium solutions can probably be attributed in part to impurities in the metal.

Effect of Metals on the Stability of the Solutions.—In the hope that conductivity measurements could be made over a range of concentration, the effect of possible electrode metals on the stability of the solutions was investigated. A small strip of electrode metal was sealed into a side arm on tube *E*, and after evacuation and flaming of the apparatus the blue solutions were prepared as before and tipped on to the metal strip. With molybdenum, tungsten, and platinum the colour decayed within a minute or two; in view of this, conductivity studies were not attempted.

It seems that these transition metals catalyse the decomposition of the metal–amine solutions because of their ability to accept electrons.

Strength of the Metal–Amine Solutions.—Precise measurement of the concentration of the metal in the solutions is difficult, since some decomposition to amide appears to be inevitable, and any determination of the metal concentration will give only the sum of the metal in solution and that present as amide. Any metal determination thus represents the maximum upper limit. The total concentration of sodium in a freshly prepared saturated solution in ethylenediamine was estimated on a calibrated flame-photometer, and found to be 0·02 g.-atom/per l.

A rough estimate of the concentration of metal in other solutions was then made by a comparison of the relative intensity of colour. On this basis, lithium is by far the most soluble of the alkali metals in either solvent; sodium and potassium in ethylenediamine and potassium in propylenediamine have similar solubilities, but sodium is only sparingly soluble in propylenediamine.

Absorption Spectra of Solutions of the Alkali Metals in Ethylenediamine and Propylenediamine.—Absorption spectra (3700—10,000 Å) were measured with a Unicam S.P. 600 spectrophotometer, the solutions being contained in fused silica cells (1 cm., equipped with standard C10 ground joints) which could be evacuated without distortion or breakage. Measurements below 3700 or above 10,000 Å were not reliable because of the strong absorption of the amine solvents. Unfortunately, cells thinner than 1 cm. could not be used, since they could not be evacuated and sealed.

The metal solutions were made in the apparatus shown in Fig. 2. The whole apparatus (except the cell *C*) was cleaned with hydrogen fluoride cleaning mixture [20] and dried for 12 hr. at 120°. The cell was cleaned with warm "Teepol," washed with pure acetone, and dried with warm air. The cell *C* was attached to the remainder of the apparatus with a trace of "Silicone" grease, and a tube of purified metal was placed in tube *D* which was then sealed. The whole apparatus was evacuated through joint *E*, and gently flamed (except cell *C*). After the apparatus had been pumped out for several hours, the metal was melted through constrictions c_1, c_2, and c_3 until a small globule appeared in *B*; the apparatus was then sealed at c_3. Sufficient solvent to fill cell *C* was then condensed in *A*, and the apparatus sealed at c_4; solvent was then carefully tipped into *C*, and its absorption compared with a "blank" cell containing boiled-out distilled water. A small portion of the solvent was tipped on to the globule of alkali metal in *B*, and the blue solution formed was carefully tipped back again into *C*. Gentle shaking gave a uniform dilute solution, whose absorption was again compared with the "blank" cell. The whole system, *A*, *B*, and *C* was designed to fit into a standard Unicam cell holder. All measurements were made at room temperature. The results for sodium and potassium in ethylenediamine and propylenediamine, shown graphically in Fig. 3, were obtained by subtracting the solvent absorption from that of the solution. Optical densities rather than extinction coefficients are recorded because of the difficulty of precisely estimating the concentration.

Paramagnetic Resonance Measurements on the Metal–Diamine Solutions.—The metal–diamine solutions were made in the usual manner, and tipped into a thin-walled Pyrex tube (3 mm.) so

TABLE 2. *Paramagnetic-resonance absorption in metal–amine solutions at room temperature (metal concentration about* 2×10^{-3}M*).*

Metal	Amine	Magnitude of absorption	Metal	Amine	Magnitude of absorption
Na ...	$(\cdot CH_2\cdot NH_2)_2$	Nil	Li	$MeNH_2$	Strong
K	,,	Very weak	K	,,	Weak
Na ...	$NH_2\cdot CHMe\cdot CH_2NH_2$	Nil			
K	,,	Nil			

attached to the preparative apparatus that it could be placed directly into an H_{012} 3 cm. wavelength rectangular resonant cavity for paramagnetic-resonance measurements. The results

[20] Crawley, *Chem. and Ind.*, 1953, 1205.

obtained are summarised in Table 2. Concentrations were varied between 10^{-2} and 10^{-4}M, as judged from the intensity of the colour, and any absorption was found to increase with concentration. Experiments with lithium in methylamine showed that absorption was readily detected even in 10^{-4}M-solutions.

Irradiation Experiments.—Rapid cooling of a fairly deep blue solution of sodium in propylenediamine gave a clear, uncracked glass, the blue colour being undiminished in intensity. When this glass, immersed in liquid oxygen contained in an unsilvered Dewar flask was irradiated with light from a tungsten-filament lamp, the colour was slowly bleached, until, after about 4 hours' irradiation, only a faint blue colour remained. The original, deeper blue colour reappeared immediately when the glass softened. Repetition gave similar results, showing that no overall decomposition was occurring.

Grateful acknowledgment is made to the Esso Petroleum Company for the award of a maintenance grant to W. R. M., to Mr. D. Austen and Dr. D. Ingram for making the paramagnetic-resonance measurements, and to Mr. M. Smith for valuable discussions.

THE UNIVERSITY, SOUTHAMPTON. [*Received, February 5th,* 1957.]

The Calculation of Activity Coefficients

Kraus calculated transference numbers from his early emf data on sodium–ammonia concentration cells by assuming that the activities of the solutions were proportional to their concentrations (see p. 143). In the following paper, Dye, Smith, and Sankuer combined Kraus' emf data with their own transference number data from moving-boundary measurements to calculate the activity of sodium in dilute solutions. The vapor pressure data of Kraus, Carney, and Johnson[1] were used to compute activities in the concentrated range.

---

[1]C. A. Kraus, E. S. Carney; and W. C. Johnson, *J. Amer. Chem. Soc.*, **49**, 2206 (1927).

38

[Contribution from Kedzie Chemical Laboratory, Michigan State University, East Lansing, Michigan[1b]]

The Activity Coefficient of Sodium in Liquid Ammonia

By J. L. Dye,[1a] G. E. Smith and R. F. Sankuer

Received February 26, 1960

Transference number data from moving-boundary measurements and data from the e.m.f. of cells with transference were used to calculate the activity of sodium in liquid ammonia as a function of concentration. Consideration of ion-pairing and dimerization equilibria allowed extrapolation of the data to infinite dilution and resulted in values of 9.6×10^{-3} and 23, respectively, for the equilibrium constants of the two reactions

$$Na^+\cdot e^- \rightleftarrows Na^+ + e^- \quad (1) \qquad\qquad 2Na^+\cdot e^- \rightleftarrows Na_2\cdot \quad (2)$$

The results were combined with published vapor pressure data for the high concentration region and with calorimetric data to give a partial molar entropy of 16.9 ± 1.8 cal. deg.$^{-1}$ mole^{-1} for the solvated electron in the hypothetical ideal one molal solution.

Introduction

The activity coefficients of alkali metals in liquid ammonia can be determined in the high concentration range from data on the vapor pressure of ammonia above such solutions. Since the precision in this method is poor at low concentrations, it is necessary to use pure sodium as the standard state or simply to compute relative activity coefficients.[2a] Such data are useful for comparing the behavior of different metals and for studying phase equilibria but do not help to determine the nature of the ionized species in dilute solutions. Calorimetric data are available[2b,3] for the heat of solution of the metal to form dilute solutions, so that ΔH^0 for the process

$$NH_{3(l)} + Na_{(s)} \longrightarrow Na_{(am)}^+ + e_{(am)}^- \quad (3)$$

can be calculated. The data of Pleskov[4] on the e.m.f. of cells were used by Jolly[5] and by Coulter[2b] to calculate ionic enthalpies and entropies for the alkali ions and to estimate the heat of solvation of

(1) (a) To whom correspondence should be addressed. (b) This work was supported in part by the U. S. Atomic Energy Commission under Contract AT(11-1)-312.

(2) (a) P. R. Marshall and H. Hunt, *J. Phys. Chem.*, **60**, 732 (1956). (b) L. V. Coulter, *ibid.*, **57**, 553 (1953).

(3) G. A. Candela, M.S. Thesis, Boston University, 1952.

(4) V. A. Pleskov and A. M. Monoszon, *Acta Physicochim. U.R.S.S.*, **2**, 615 (1935); V. A. Pleskov, *ibid.*, **6**, 1 (1937).

(5) W. L. Jolly, U. S. Atomic Energy Comm., *U.C.R.L.-2201*, pp. 23 (1953).

the electron. However, it was not possible to obtain ΔF^0 for the process given by (3). Because of the solubility of the metals in ammonia, one cannot set up suitable cells without transference using the metals as electrodes. The use of dropping amalgam bridge electrodes similar to those used in aqueous solutions[6] would be complicated by the partition equilibrium of the metal between the ammonia and the mercury phases.

Almost fifty years ago, Kraus,[7] in the course of his classic studies of metal–ammonia solutions, measured the e.m.f. of cells with transference employing dilute solutions of sodium in ammonia. The type of cell which he studied was

$$Pt|Na(C_1 \text{ in } NH_3)\vdots Na(C_2 \text{ in } NH_3)|Pt \quad (4)$$

in which the platinum electrodes appeared to behave reversibly towards the solvated electron. These data were used by Kraus to demonstrate the high value of the anionic transference number and its change with concentration. To make this calculation, Kraus assumed the activity ratio to be equal to the concentration ratio. Klein[8] has made e.m.f. measurement on metal–methylamine solutions and Fristrom[9] has extended the e.m.f. data of Kraus to other metals in ammonia and to temperatures other than $-33.5°$.

In view of the modern interpretations of the structure of dilute solution, activity coefficients undoubtedly play an important role. The measurement of transference numbers reported in a companion paper[10] makes possible the calculation of activity coefficient ratios using the data of Kraus. The equilibria postulated by modern theory can be used to extrapolate the data to the infinite dilution reference state. Use of vapor pressure data at higher concentrations allows calculation of the activity coefficient over the entire range of concentration from infinitely dilute solution to the saturated solution. These calculations and the interpretation of the results are the subjects of this paper.

Calculations and Results

The calculation of molar activity coefficient ratios requires a knowledge of the cation transference number, T_+, as a function of concentration. This can be obtained by subtracting from unity the measured value of T_-. Since the transference data covered the range 0.02 to 0.14 molar while the e.m.f. data of Kraus were obtained from 0.002 to 0.82 molar, it was necessary to extrapolate the transference data in both directions. Because of the independent value available for the intercept at $C = 0$,[10] the extrapolation to low concentrations should be quite reliable. Direct extrapolation to concentrations above 0.14 molar is hazardous because of the expected drop of T_+ to very low values. The total conductance passes through a minimum and then rapidly rises to a high value. A rapid drop of T_+ is expected because of an increase in λ_- rather than a rapid decrease in the cation conductance. The sodium ion conductance should decrease slowly and uniformly with increasing concentration. Because of this, the cation transference number at concentrations above the experimental range was calculated by dividing the extrapolated value of λ_+ by the total conductance Λ obtained from the data of Kraus[11] and Fristrom.[9]

Electromotive Force Data.—The e.m.f. data of Kraus[7] were obtained for a number of pairs of solutions, while the calculation of relative activity coefficients requires e.m.f. data relative to a particular reference solution. Thus it was necessary to treat the data of Kraus in a different manner than is normally done. If $\Delta\epsilon$ represents the e.m.f. observed for a cell having metal concentrations C_1 and C_2 and if C_1 and C_2 are not widely different, one can make the approximation

$$\frac{\Delta\epsilon}{\Delta \log C} = \frac{\Delta\epsilon}{\log \frac{C_2}{C_1}} \approx \frac{d\epsilon}{d \log C} \quad (5)$$

A graph of $\frac{\Delta\epsilon}{\Delta \log C}$ *versus* log C (average) was made, and graphical integration was used to obtain ϵ *versus* log C as shown in Fig. 1. The smoothed values of ϵ are given in Table I. A reference concentration corresponding to $\log C_{ref} = -2.20$ was chosen for calculation of activity coefficients. This concentration is high enough to be fairly insensitive to errors in the e.m.f. measurements at the lowest concentrations and yet low enough so that the laws of dilute solution behavior can be applied to the reference solution.

The calculation of $\frac{y_\pm}{y_{\pm ref}}$ followed standard procedure[12] and made use of the equation

$$\log \frac{y_\pm}{y_{\pm ref}} = \log \frac{C_{ref}}{C} + \frac{F}{4.606\,RT}\left[\frac{\epsilon}{T_{+ref}} + \int_0^\epsilon \delta d\epsilon\right] \quad (6)$$

in which

$$\delta = \frac{1}{T_+} - \frac{1}{T_{+ref}} \quad (7)$$

and F is the Faraday.

The value of $\int \delta d\epsilon$ was obtained by graphical integration and the procedure was checked by graphical differentiation of the integral curve. The resulting values of $y_\pm/y_{\pm ref}$ are given in Table I along with the values of T_+, ϵ and $\int \delta d\epsilon$ used. The symbol $y_\pm$ refers to the stoichiometric mean molar activity coefficient. In the discussion of the extrapolation procedure which follows, $f_\pm$ represents the mean molar activity coefficient *of the ions present* and is presumed to obey the Debye–Hückel equation in dilute solutions.

Direct extrapolation of $y_\pm/y_{\pm ref}$ to infinite dilution using the Debye–Hückel equation gives an approximate value of $y_{\pm ref}$, but the curvature is too great for a reliable extrapolation to be made, and to evaluate this parameter the equilibria involved must be considered. The equilibria considered

(6) H. S. Harned and B. B. Owen, "The Physical Chemistry of Electrolytic Solutions," 3rd Ed., Reinhold Publishing Corp., New York, N. Y., 1958, p. 489.

(7) C. A. Kraus, THIS JOURNAL, **36**, 864 (1914).

(8) H. M. Klein, Ph.D. Thesis, Pennsylvania University, 1957; *Dissertation Abstr.*, **17**, 990 (1957).

(9) R. M. Fristrom, Ph.D. Thesis, Stanford University, 1949; *D.D.*, **16**, 32 (1948–1949).

(10) J. L. Dye, R. F. Sankuer and G. E. Smith, THIS JOURNAL, **82**, 4797 (1960).

(11) C. A. Kraus, *ibid.*, **43**, 749 (1921).

(12) R. A. Robinson and R. H. Stokes, "Electrolyte Solutions," Butterworths Scientific Publications, London, 1955, p. 200.

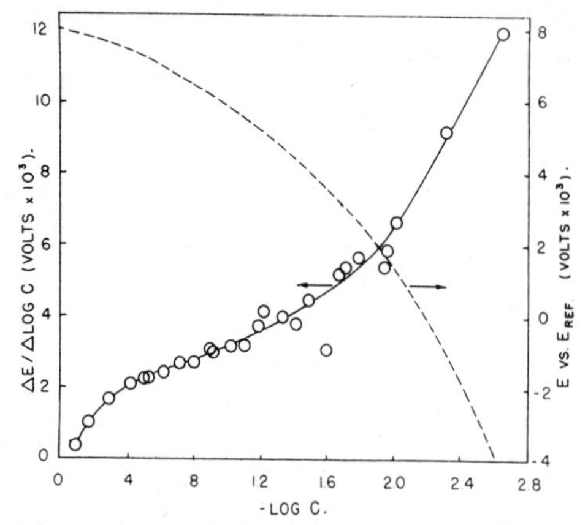

Fig. 1.—The e.m.f. data of Kraus[7] plotted in differential form and the resulting integral curve. Data are for sodium in ammonia at −33.5°.

were those postulated by Becker, Lindquist and Alder[13]

$$M^+{:}e^- \underset{}{\overset{K_1}{\rightleftharpoons}} M^+ + e^- \quad (1)$$

$$2M^+{\cdot}e^- \overset{K_2}{\rightleftharpoons} M_2 \quad (2)$$

It was assumed that the ionic species obeyed the Debye–Hückel equation[14]

$$\log f_\pm = \frac{-A\sqrt{C_i}}{1 + B\,\mathring{a}\sqrt{C_i}} \quad (8)$$

The numerical values of A and B used were 4.75 and 0.693, respectively, for ammonia at −33.5°. The parameter $\mathring{a}$ was arbitrarily chosen to be 5.5 Å. and was *not* adjusted for best fit. The constants K_1 and K_2 were "coupled"; that is, the data could be reproduced reasonably well over a range of values of K_1 provided that K_2 and $y_{\pm\text{ref}}$ were suitably varied.

TABLE I

SMOOTHED DATA FOR ACTIVITY CALCULATIONS

$M \times 10^3$	T_+	ϵ (volt × 10³)	$\int \delta d\epsilon$ (volt × 10³)	$y_\pm/(y_\pm)_{\text{ref}}$
0.178	0.133	−5.64	0.768	1.231
.398	.131	−1.69	.107	1.174
.631	.129	0.00	.000	1.000
1.00	.126	1.297	.115	0.8105
2.00	.118	2.987	.830	.5710
3.98	.104	4.328	2.350	.3839
7.94	.085	5.419	5.143	.2539
15.9	.062	6.309	10.12	.1705
31.6	.042	7.052	18.17	.1200
50.1	.028	7.450	26.47	.1002
79.4	.011	7.717	37.54	.0874

The value of $y_{\pm\text{ref}}$ is sensitive to the value of K_1 used, and so an independent measure of $y_{\pm\text{ref}}$ was used before fitting the two constants to the data. If α represents the fraction of the total sodium present as Na^+, then

$$\alpha = \frac{C_i}{C} \quad (9)$$

(13) E. Becker, R. H. Lindquist and B. J. Alder, *J. Chem. Phys.*, **25**, 971 (1956).

(14) Ref. 6, p. 66.

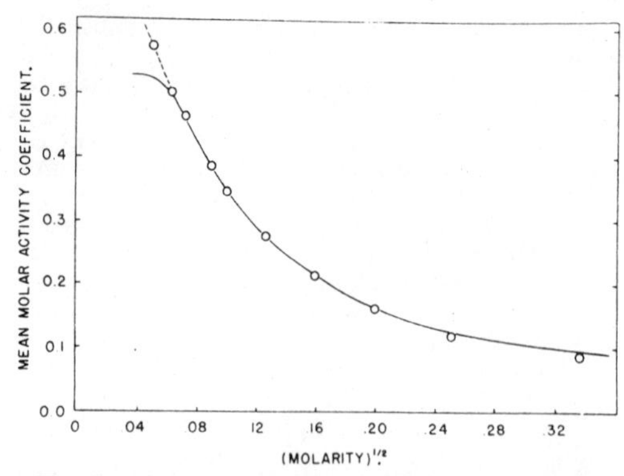

Fig. 2.—Mean molar activity coefficient of sodium in ammonia. Solid line calculated from e.m.f. and T-data; circles represent calculations from equilibrium constants.

and

$$y_\pm = \alpha f_\pm \quad (10)$$

with $f_\pm$ evaluated by (8). *For the reference concentration only*, α was estimated from conductivity data for Na^+ [10] to be 0.79. Since conductivity is very sensitive to the absolute value of α, while $\frac{y_\pm}{y_{\pm\text{ref}}}$ is sensitive primarily to the ratio $\frac{\alpha}{\alpha_{\text{ref}}}$, it was felt that this procedure would give a better value of $y_{\pm\text{ref}}$ than the adjustment of K_1 and K_2 to give the best fit.

Having thus fixed $y_{\pm\text{ref}}$, K_1 and K_2 were evaluated from the relative activity coefficients by a method of successive approximations. Equation 1 gives

$$K_1 = \frac{\alpha^2 C^2 f_\pm^2}{[M^+{\cdot}e^-]} \quad (11)$$

To get an approximate value of K_1, the apparent constant $K_1^{(a)}$ was evaluated at each concentration as

$$K_1^{(a)} = \frac{\alpha^2 C f_\pm^2}{(1-\alpha)} \quad (12)$$

using the measured $y_\pm/y_{\pm\text{ref}}$ and equation 10. Equation 8 was used to obtain $f_\pm$ and the entire procedure was repeated until consistent values of α and $f_\pm$ were obtained.

The parameter $K_1^{(a)}$ varied with concentration because of the presence of the second equilibrium. An extrapolation to infinite dilution gave an approximate value of $K_1^{(t)}$, the true equilibrium constant. K_2 was calculated from $K_1^{(t)}$ and $K_1^{(a)}$ using the relationship

$$K_2^2 = \frac{(K_1^{(t)} - K_1^{(a)})K_1^{(t)}}{2(C\alpha f_\pm)^2 K_1^{(a)}} \quad (13)$$

After adjusting $K_1^{(t)}$ to give the most constant set of K_2 values, the two constants were combined with $f_\pm$ from equation 8, and $y_\pm$ was computed for comparison with the experimental values. The results are given in Table II and are shown graphically in Fig. 2. In this figure, the solid line represents the smoothed activity coefficient curve obtained from e.m.f. and transference number data, and the circles represent the calculated values using $K_1 = 9.6 \times 10^{-3}$, $K_2 = 23$. These compare very favorably with the values of 9.2×10^{-3} and 18.5

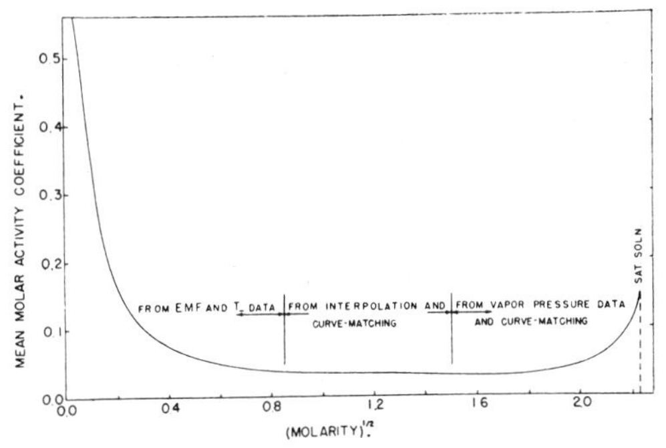

Fig. 3.—Mean molar activity coefficient of sodium in ammonia over the entire range of solubility.

obtained from the sodium ion conductance.[10] The deviation of the experimental curve from the calculated curve at low concentrations is probably due to an error in the e.m.f. measurements. The low concentration region of the e.m.f. curve (Fig. 1) is based upon only two points. Since a dilution method was used by Kraus, decomposition of the dilute solutions could easily account for the deviations.

TABLE II

COMPARISON OF OBSERVED AND CALCULATED ACTIVITY COEFFICIENTS FOR DILUTE SOLUTIONS

$M \times 10^2$	$y_\pm$ (l. mole^{-1}) Calcd.	Obsd.
0.251	0.576	0.527
.398	.504	.504
.502	.466	.464
.794	.388	.387
1.000	.350	.348
1.59	.277	.276
2.51	.213	.216
3.98	.161	.166
6.31	.119	.125
10.0	.087	.095

The vapor pressure data of Kraus, Carney and Johnson[15] were used to compute activity coefficients in the concentrated range based upon pure sodium metal as the standard state. The data of Marshall and Hunt,[2a] obtained at $-45.4°$, could not be used because of the presence of a two-phase region from mole fraction, $X_{Na} = 0.018$ to $X_{Na} = 0.067$ at this temperature.

Integration of the Gibbs–Duhem equation was performed graphically to obtain the activity of sodium in the concentrated region from the measured activity of ammonia. The method was checked by a second integration of the Gibbs–Duhem equation using the calculated sodium activity. The results agreed with the original values of the ammonia activity.

It was desirable to combine the dilute solution results and the results obtained from vapor pressures even though a considerable concentration gap remained between the two sets of measurements. Since two different standard states are involved, the relationship between activities is

$$(y_\pm C)^2 = k a_{Na} \quad (14)$$

(15) C. A. Kraus, E. S. Carney and W. C. Johnson, THIS JOURNAL, **49**, 2206 (1927).

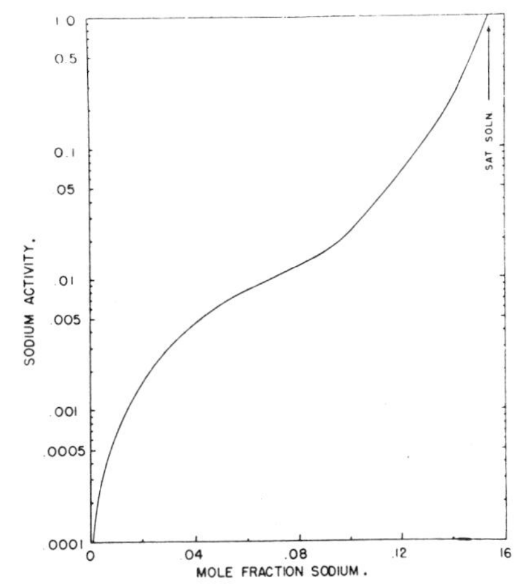

Fig. 4.—Activity of sodium (log scale) in ammonia *versus* mole fraction; standard state is pure solid sodium.

in which a_{Na} is the activity of the sodium based upon pure, solid sodium as the standard state and k is an unknown constant determined by the relative free energies of the two standard states. For the process

$$Na_{(s)} \longrightarrow Na^+_{(am)} + e^-_{(am)}$$

hypothetical, ideal one molar solution

$$\Delta F^0_\pm = -RT \ln k \quad (15)$$

From (14) we can write

$$\log (y_\pm C)^2 = \log k + \log a_{Na} \quad (16)$$

This suggested a method of evaluation of k.[16] Separate sheets of semi-log paper were used with $(y_\pm C)^2$ plotted *versus* C on one sheet and a_{Na} *versus* C on the other. The two sheets of graph paper were then moved vertically until the two sets of data could be connected by a smooth curve. The value of k was given by the ratio of the readings on the two semi-log axes for any concentration. The results are given in Table III and are shown graphically in Figs. 3 and 4. The value of k is 0.54.

TABLE III

ACTIVITY OF SODIUM IN AMMONIA OVER THE ENTIRE RANGE OF SOLUBILITY

X_{Na}	Molarity	a_{Na}[a]	$y_\pm$[b]
0.154	4.97	1.000	0.148
.143	4.65	0.294	.0860
.125	4.16	.0845	.0516
.100	3.43	.0234	.0328
.0833	2.93	.0139	.0296
.0625	2.27	.00872	.0304
.0476	1.77	.00596	.0322
.0278	1.07	.00262	.0354
.0196	0.762	.00153	.0379
.0164	.641	.00118	.0396
.0100	.399	.000642	.0468
.00505	.200	.000317	.0647

[a] Based upon pure sodium as the standard state. [b] Based upon the hypothetical, ideal, one molar standard state.

(16) We are indebted to Professor R. H. Schwendeman for pointing out the feasibility of this method.

Discussion

The activity coefficient data in dilute solutions are certainly compatible with the ion-pairing and dimerization equilibria. The close agreement of the constants with those obtained from λ_+ data lends added support to the assumption that these two equilibria are involved. As with the conductance data the results reported here cannot distinguish between an ion-pair and a monomer unit.

Figure 3 shows that the mean molar activity coefficient is nearly constant over a very large concentration region. The extremes are only 25% apart, from 0.64 to 3.76 molar. Over this range, the mean ionic activity is nearly proportional to the total concentration, and the solution behaves as a nearly "ideal" binary electrolyte.

Figure 4 shows the variation of total activity (on a semi-log scale) with concentration. The inflection in the curve reflects the fact that in this concentration range at a temperature only 8.5 degrees below this, the system separates into two phases.

These results permit calculation of the standard free energy change for the solution process. For the process, $Na_{(s)} \rightarrow Na^+_{(am)} + e^-_{(am)}$, the values are:

(1) To form the hypothetical ideal solution, $C = 1$; $\Delta F_1^0 = +0.3$ kcal. mole^{-1}. (2) To form the hypothetical ideal solution, $m = 1$; $\Delta F_2^0 = -0.1$ kcal. mole^{-1}. (3) To form the hypothetical ideal solution, $X_{Na} = 1$; $\Delta F_3^0 = +3.8$ kcal. mole^{-1}.

These data can be combined with heat of solution data,[2b] which give $\Delta H^0 = +4.4$ kcal. mole^{-1}, to yield $\Delta S_1^0 = 17.1$ cal. deg.$^{-1}$ mole^{-1}; $\Delta S_2^0 = 18.7$ cal. deg.$^{-1}$ mole^{-1}; $\Delta S_3^0 = 2.5$ cal. deg.$^{-1}$ mole^{-1}. Using the value of 12.6 cal. deg.$^{-1}$ mole^{-1} for the partial molal entropy of sodium ion[2] and 10.8 cal. deg.$^{-1}$ mole^{-1} for the entropy of sodium at $-33.5°$ [17] gives $\bar{S}^0 = 16.9$ cal. deg.$^{-1}$ mole^{-1} (hypothetical ideal, $m = 1$) for the solvated electron.

Since extrapolation and interpolation were used extensively and also because smoothed curves were used for ϵ and T_+, we attempted to estimate the error in the final result. The estimate of the probable error in the smooth T_- curve is $\pm 0.6\%$.[10] Since T_- varies from 0.86 to 0.94 over the range studied, the probable error in T_+ is about $\pm 7\%$. The error in ϵ is not easy to estimate, but it is certainly less than the error in T_+ at high concentrations. The root mean square deviation of $\frac{\Delta \epsilon}{\Delta \log C}$ from a smooth curve is $\pm 4.6\%$ (probable error $= \pm 3\%$). Combination of all effects leads to an estimated probable error in $y_\pm / y_{\pm\,ref}$ ranging from 0.8% at 0.01 molar to 9.6% for a 0.63 molar solution. Because of extrapolation errors, this increases to about 16% at the point of tie-in with the vapor pressure data. Another uncertainty is introduced by choosing α_{ref} from conductance data, with the magnitude of this error unknown because of the dependence upon the conductance function chosen to represent the behavior of the ions. However, a rough estimate may be made based upon the experimental error. The estimated probable error in α_{ref} (6%) leads to an error of 4.2% in $y_{\pm ref}$. The combined effects give a probable error in $y_\pm$ varying from 4.3% at 0.01 molar to 10.5% at 0.63 molar. Consideration of all of these effects leads to an estimate of the probable error of ΔF^0 of ± 0.36 kcal. mole^{-1} which gives an uncertainty of ± 1.8 cal. deg.$^{-1}$ mole^{-1} for ΔS^0 assuming ΔH^0 to be accurate to ± 0.1 kcal mole^{-1}.

Acknowledgment.—The authors express appreciation to the U. S. Atomic Energy Commission for financial support of this research.

(17) F. Simon and W. Zeidler, *Z. physik. Chem. (Liepzig)*, **123**, 383 (1926); National Bureau of Standards, *Circular 500*, "Selected Values of Chemical Thermodynamic Properties," 1952, p. 447.

Erratum: The equation and numbers listed for K_2 should refer to the equation

$$M^+ \cdot e^- \rightleftharpoons \tfrac{1}{2} M_2$$

rather than

$$2M^+ \cdot e^- \rightleftharpoons M_2$$

A Simple Model for Dilute Metal–Ammonia Solutions

In the following communication, Gold, Jolly, and Pitzer describe a simple model for dilute metal–ammonia solutions that is consistent with the observed properties of these solutions. The spectral data of Gold and Jolly[1] were the principal experimental basis for this model.

[1] M. Gold and W. L. Jolly, *Inorg. Chem.,* **1**, 717 (1962).

39

[Reprinted from the Journal of the American Chemical Society, **84**, 2264 (1962).]

A REVISED MODEL FOR AMMONIA SOLUTIONS OF ALKALI METALS

Sir:

It is well established that, in extremely dilute solutions of alkali metals in liquid ammonia, the metal is completely dissociated into ammoniated metal ions and ammoniated electrons. Each ammoniated electron is believed to exist in a large spherical cavity in the solvent and to be stabilized by the orientation of the ammonia dipoles on the periphery of the cavity. The decrease in the equivalent electrical conductance of these solutions as the concentration is increased to about 0.05 *M* indicates that ion pairing or assembly into larger aggregates takes place. The decrease of the molar paramagnetic susceptibility of the solutions with increasing concentration indicates that the ammoniated electrons associate to form species containing electron pairs. Both of these effects may be reproduced by assuming appropriate constants for these equilibria

$$M^+ + e^- = M$$

$$2M^+ + 2e^- = M_2$$

At $-33°$ these equilibria shift in the vicinity of 0.01 *M* and the associated species predominate at higher concentrations.

It has been proposed by Becker, Lindquist, and Alder[1] that the monomer, M, consists of an ammoniated M^+ ion with the electron located in an expanded orbital on the protons of the coördinated ammonia molecules. Likewise, it has been proposed that the dimer, M_2, consists of two ammoniated M^+ ions held together by a pair of electrons in a bonding molecular orbital located principally between the two ions. These descriptions of the monomer and dimer are inconsistent with certain properties, in part newly measured, of alkali metal–ammonia solutions which are summarized below. We propose a revised description of the monomer and dimer species which more adequately accounts for all the properties of the solutions.

(1) The absorption spectra of sodium–ammonia solutions which are less concentrated than 0.03 *M* follow Beer's law within an experimental uncertainty of ± 2–5% at all wave lengths between 4000 and 25,000 Å. In fact, Beer's law is obeyed with similar precision for wave lengths between 4000 Å. and the absorption maximum (15,000 Å.) for concentrations at least as high as 0.1 *M*.[2] There is no indication of a separate absorption band near 6700 Å. Such a band has been observed in amines by Fowles, McGregor and Symons[3] and was attributed to a diamagnetic species. We have been unable to substantiate the claimed[4] appearance of shoulders at 6700 and 8000 Å. in the absorption spectra of ammonia solutions of sodium and sodium iodide. Thus the three species e^-, Na, and Na_2 have almost identical absorption spectra.

(2) The absorption spectra of solutions of lithium, potassium[5] and cesium are identical with those for sodium solutions of similar concentration within the limits of experimental uncertainty, which for the former solutions was ± 10%. Hence the spectra of the M and M_2 species (in the visible and infrared) do not depend upon the particular metal.

(3) Gunn and Green[7] have found that the apparent molar volume of lithium in ammonia 0° is practically constant at 50 ml./mole from 0.02 to 0.99 *M*. For sodium, the apparent molar volume changes only from 56.3 to 58.6 ml./mole from 0.0093 to 0.34 *M*. The latter data are inconsistent with some preliminary data of Evers and Filbert[8] which indicate a marked minimum in the volume–concentration curve for sodium at 0.03 *M*. However, until the latter findings are confirmed, we shall favor the former data and therefore conclude that the partial molal volumes of $M^+ + e^-$, of M, and of $^1/_2$ M_2 are nearly the same.

(4) Nuclear magnetic resonance data[9] are available only above 0.05 *M*. In the range 0.05 to 0.2 *M* the unpaired electron spin density on the sodium nuclei is only about 0.1% of that expected for isolated sodium atoms. The n.m.r. shift for nitrogen is rather large and indicates that an appreciable fraction of the unpaired electron spin density extends into the ammonia solvent regions. Since in this concentration range the concentration ratio of M to e^- is not small, the electron spin density on the metal atom nucleus must be small for the species M.

This evidence, taken as a whole,[10] is quite inconsistent with the Becker, *et al.*, model where the volumetric and spectral properties of $(e^- + M^+)$, M, and $^1/_2$ M_2 would be expected to differ widely. Douthit and Dye[6] and Evers[11] have pointed out that if the monomer were simply an ion-pair consisting of an ammoniated metal ion and an ammoniated electron, the constancy of the spectra in the dilute range could be explained. They make no such proposal for the dimer. However, our spectral data require a new picture for M_2 as well. We picture the M_2 species as a quadrupolar ionic assembly of $2e^- + 2M^+$ in which there is little distortion of either the ammoniated electrons or ammoniated metal ions. Presumably the electrons and ions are held in a square or rhombic configuration. The probability density for the electron in the solvated e^- species extends with decreasing intensity through several solvent layers.[12] Thus the wave functions for the two e^- in M_2 will overlap significantly and it is reasonable that the singlet

(1) E. Becker, R. H. Lindquist and B. J. Alder, *J. Chem. Phys.*, **25**, 971 (1956).

(2) M. Gold and W. L. Jolly, unpublished data.

(3) G. W. A. Fowles, W. R. McGregor and M. C. R. Symons, *J. Chem. Soc.*, 3329 (1957).

(4) H. C. Clark, A. Horsfield and M. C. R. Symons, *ibid.*, 2478 (1959).

(5) We find no evidence for a negative deviation from Beer's law for potassium as reported by Douthit and Dye.[6]

(6) R. C. Douthit and J. L. Dye, *J. Am. Chem. Soc.*, **82**, 4472 (1960).

(7) S. R. Gunn and L. G. Green, *J. Chem. Phys.*, **36**, 363 (1962).

(8) E. C. Evers and A. M. Filbert, *J. Am. Chem. Soc.*, **83**, 3337 (1961).

(9) H. M. McConnell and C. H. Holm, *J. Chem. Phys.*, **26**, 1517 (1957); J. Acrivos and K. S. Pitzer, unpublished data.

(10) W. E. Blumberg and T. P. Das (*J. Chem. Phys.*, **30**, 251 (1959)) treat the n.m.r. data on the basis of the Becker, *et al.*, model with reasonable agreement. The fit is strained, however, and would be easier on our model.

(11) E. C. Evers, *J. Chem. Ed.*, **38**, 590 (1961).

(12) J. Jortner, *J. Chem. Phys.*, **30**, 839 (1959).

state should be lower in energy than the triplet by more than kT. Also a small electron density at the sodium nucleus in the ion pair (M) species is to be expected. Both the volume and the 1s–2p spectral frequency depend primarily on the cavity size for the solvated electron. Thus, if this cavity retains its size through ion pair and quadruplet formation, the results cited in (1), (2), and (3) above become understandable.

Presumably the solvated electron retains its structure in further polymeric species (*e.g.*, M_4) which probably form in more concentrated solutions. However, in highly concentrated solutions, one should expect that the nature of the ammoniated electrons would change if for no other reason than that there are insufficient ammonia molecules to properly coördinate both the metal ions and the electrons. We believe the deviations from Beer's law which occur in sodium solutions more concentrated than 0.03 M indicate the formation of high polymers with incipient metallic bonding.

DEPARTMENT OF CHEMISTRY AND
LAWRENCE RADIATION LABORATORY
UNIVERSITY OF CALIFORNIA
BERKELEY 4, CALIFORNIA
MARVIN GOLD
WILLIAM L. JOLLY
DEPARTMENT OF CHEMISTRY
RICE UNIVERSITY
HOUSTON 1, TEXAS
KENNETH S. PITZER

RECEIVED FEBRUARY 12, 1962

High-Precision Calorimetry

Although heats of solution of metals in ammonia at −33° had been determined by other investigators using calorimeters based on the measurement of vaporized ammonia, the data were not accurate enough to permit estimation of heats of dilution. Gunn and Green devised a calorimeter containing a glass reaction vessel with which they could measure heats of solution at 25° with relatively high accuracy. The data were found to be consistent with the equilibria that had been proposed by Becker, Lindquist, and Alder (p. 227).

Reprinted from the JOURNAL OF CHEMICAL PHYSICS, Vol. 36, No. 2, 368–370, January 15, 1962
Printed in U. S. A.

Heats of Solution of Alkali Metals in Liquid Ammonia at 25° *

STUART R. GUNN AND LEROY G. GREEN
Lawrence Radiation Laboratory, University of California, Livermore, California

(Received August 9, 1961)

40

Heats of solution in ammonia at 25° have been measured for Li, Na, and K at concentrations down to 0.005 M, and for Na in the presence of NaI. Dilution is endothermic, and the heat of dilution is greater in the presence of added Na ion; these results support the Becker-Lindquist-Alder theory of the solutions.

HEATS of solution and heats of dilution of metals in liquid ammonia are of interest for two reasons: first, in contributing to an understanding of the nature of these unusual solutions, and second, because the solutions might have some application in thermochemical studies due to their powerful reducing properties.

Heats of solution of all the alkali metals and the readily soluble alkaline earth metals have been reported.[1–5] These data were all obtained at −33° using vaporization calorimeters, and are not sufficiently accurate to permit estimation of heats of dilution except at the highest concentrations; in some cases the agreement between independent investigations is poor. Ogg[6] has referred to then-current calorimetric studies of alkali metal solutions in the 0.1 to 0.01M range as indicating an endothermic heat of dilution, but no further information has appeared in the literature. Coulter[7] has referred to unpublished work of G. Candela as indicating $\Delta H \approx +3$ kcal mole^{-1} for the process

$$\tfrac{1}{2}e_2^{=} \rightarrow e^{-}. \qquad (1)$$

One of the difficulties in these measurements is the instability of the solutions; the rate of ammonolysis increases with temperature and is catalyzed by metallic surfaces. We attempted to measure heats of solution of sodium and potassium in the rocking-bomb calorimeter previously described[8]; however, with both a coinage-gold bomb and a tantalum bomb, fairly rapid heat evolution continued following the initial heat effect due to dissolution. The rate was greater in the tantalum than in the coinage gold. In the coinage-gold bomb, a black solid was found after the run, the amount being larger with potassium than with sodium. Ignition of this solid in a porcelain boat produced a small button which appeared to be metallic gold. Evidently, the metallic surfaces catalyzed the ammonolysis and in addition an intermetallic compound with gold was formed. A calorimeter was finally developed in which the solution contacted only Pyrex glass and a very small area of Apiezon W wax; this proved to give satisfactorily low ammonolysis rates at 25°.

EXPERIMENTAL

The calorimeter, laboratory designation IF, consists essentially of a Pyrex bottle mounted within a copper shell. The bottle is made of tubing 1.76-in. o.d., 0.14-in. wall, flat bottomed, and with a $\frac{19}{38}$ female standard taper at the top, 8.8-in. over-all length. A 5-mm tube attached to the shoulder of the bottle was used for evacuating and filling. The shell consisted of copper tubing 2.00-in. o.d. and 0.083-in. wall, closed at the ends by copper plates 0.125 in. thick. The body of this was 7.50-in. long and the cap 2.50-in.; the two sections were joined by screws at a flange sealed with an O-ring. A Manganin heater was wound bifilarly on the bottle, which was set into the shell with Apiezon W wax completely filling the gap. The heater leads were brought out through a vacuum-tight seal.

The bottle was closed by a plug made from a $\frac{19}{38}$ male standard taper with through tip; this is cut off short on the outer end and is sealed off closely on the inner end with a hook attached at this point. A bulb breaker consisting of a copper rod encased in Pyrex is attached to the hook by an eye at its top and is fitted at the bottom with a socket into which the sample bulb is wedged with a small wad of glass wool. The plug is sealed in place with Apiezon W wax.

Temperature was measured with a bead-type thermistor mounted in a small well soldered to the outside of the copper shell. The shell was chrome plated to decrease radiative heat transfer; the thermal leakage modulus was about 1.5×10^{-3} min^{-1}. The calorimeter was mounted by means of nylon loops in the evacuated rocking submarine[7] and the contents were stirred by oscillation at about 18 cycles/min between a clockwise extreme with the capped end 10° above the horizontal and a counterclockwise extreme with the same end 50° below the horizontal. At the desired time, the sample was broken by continuing clockwise rotation to the vertical position, the hook being so designed that the bulb holder fell off at this point. Further details of the

* This work was performed under the auspices of the U. S. Atomic Energy Commission.
1 C. A. Kraus and F. C. Schmidt, J. Am. Chem. Soc. **56**, 2297 (1934).
2 F. C. Schmidt, F. J. Studer, and J. Sottysiak, J. Am. Chem. Soc. **60**, 2780 (1938).
3 L. V. Coulter and R. H. Maybury, J. Am. Chem. Soc. **71**, 3394 (1949).
4 L. V. Coulter and L. Monchick, J. Am. Chem. Soc. **73**, 5867 (1951).
5 S. P. Wolsky, E. J. Zdanuk, and L. V. Coulter, J. Am. Chem. Soc. **74**, 6196 (1952).
6 R. A. Ogg, Jr., J. Chem. Phys. **14**, 295 (1946).
7 L. V. Coulter, J. Phys. Chem. **57**, 553 (1953).
8 S. R. Gunn, Rev. Sci. Instr. **29**, 377 (1958).

auxiliary equipment, sample bulbs, loading procedure, and operation were substantially as described elsewhere.[8,9]

The internal volume of the bottle, with the bulb breaker inserted, was 175 ml. About 5 moles (141 ml at 25°) of ammonia was used. The total heat capacity of the system was about 270 cal deg^{-1}; thus the heat capacity of the solid parts was about twice that of the solution. Because of the large heat capacity of the solid parts and the low thermal conductivity of glass, the thermal lag was quite large. It is estimated that the limiting useful sensitivity is about 1×10^{-4} deg for small heat effects, and limit of accuracy is about 0.1% for large heat effects.

Metals were purified by distillation and handled in the dry box[10]; sodium iodide was reagent grade, oven-dried, and stored in a desiccator. Before evacuating the calorimeter bottle, about 5×10^{-5} gram atom of the metal being measured was dropped in the bottle; this formed a faint blue solution which served to reduce any impurities which might have reacted with the metal solution at the time of dissolution.

The corrections necessary to calculate the heat of solution ΔH from the observed heat effect q_{obs} are the same as those for salts in ammonia with the exceptions that the energy of the falling weight (about 0.005 cal) is neglected and that the apparent molal volume V_m of the metal cannot be neglected

$$-\Delta H=(q_{obs}+1.960V_b-1.960V_m-0.00289\Delta pV_v)/n. \tag{2}$$

n is the number of gram atoms of metal, V_b is the internal volume of the sample bulb, V_v is the volume of the vapor phase, and Δp is the reduction of vapor pressure, in centimeters of mercury, of the ammonia over the solution. For V_m, we use our values[11] at 0° without correction for temperature change: Li, 49; Na, 57; and K, 65 ml mole^{-1}. To calculate the vapor pressure depression term we use the data of Marshall and Hunt[12]; a plot of $P/P°$ vs C from 0.067 to 0.217 is approximately linear with a slope of -0.018 liter mole^{-1} for both lithium and sodium. Assuming the same coefficient at 25°, a liquid volume of 140 ml, and a vapor volume of 35 ml, the correction is 10 cal mole^{-1}. independent of concentration. Thus the heats of solution can be calculated from the simplified expression

$$\Delta H=-[(q_{obs}+1.960V_b)/n]+C, \tag{3}$$

where the values of C are: Li, 106; Na, 122; and K, 137 cal mole^{-1}.

Table I. Heats of solution.

C (moles liter^{-1})	ΔH (kcal mole^{-1})	C	ΔH
Li		K	
0.161	−12.74	0.228	−3.15
0.063	−12.46	0.213	−3.16
0.0312	−12.21	0.074	−2.77
0.0066	−12.04	0.0377	−2.47
0.0065	−11.90	0.0112	−2.25
		0.0051	−2.15
Na		Na into 0.29 M NaI	
0.81	−1.48	0.166	−0.65
0.346	−1.47	0.070	−0.17
0.238	−1.44	0.046	+0.06
0.090	−1.19	0.0164	+0.57
0.066	−1.08	0.0093	+0.83
0.0347	−0.85		
0.0095	−0.57		
0.0073	−0.56		
0.0052	−0.68		
0.0042	−0.65		

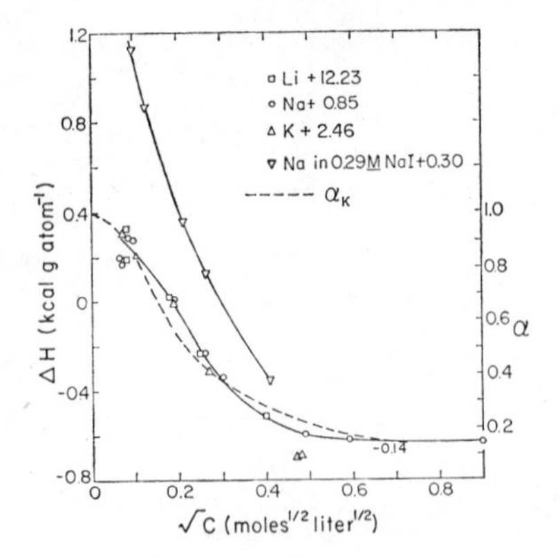

Fig. 1. Heat of solution in ammonia at 25°.

RESULTS

Results of the measurements are given in Table I and plotted in Fig. 1. The data for lithium, sodium, and potassium are normalized at 0.0354 M (mole ratio 1000). The heats of dilution are quite similar, although that of potassium is perhaps slightly greater. Also plotted against an arbitrary scale on the right are values of α, the degree of dissociation of diamagnetic species into paramagnetic species, for potassium solutions at 25° given by Hutchison and Pastor.[13] The shape of the two curves is quite similar; it may be concluded that the heat of the dissociation process is about +1.2 kcal g atom^{-1}. Also plotted, on an arbitrarily displaced scale, is the heat of solution of sodium in ammonia containing 0.29 M sodium iodide. The shape of this curve is significantly different, the heat of dilution being much greater.

[9] S. R. Gunn and L. G. Green, J. Phys. Chem. **64**, 1066 (1960).
[10] S. R. Gunn and L. G. Green, J. Am. Chem. Soc. **80**, 4782 (1958).
[11] S. R. Gunn and L. G. Green, J. Chem. Phys. **36**, 363 (1962).
[12] P. R. Marshall and H. Hunt, J. Phys. Chem. **60**, 732 (1956).
[13] C. A. Hutchison, Jr., and R. C. Pastor, J. Chem. Phys. **21**, 1959 (1953).

DISCUSSION

Theories of metal-ammonia solutions generally consider up to five species: paramagnetic monomeric atoms M and diamagnetic dimeric molecules M_2, or "expanded metals" in which the valence electron is localized on the metal or nearby solvent molecules, and monomeric and dimeric solvated electrons e^- and $e_2^=$ ("cavity model") where the electrons are trapped in holes in the solvent and are independent of the metal ions, M^+. Trapped electrons e^- and $e_2^=$ may be considered formally similar to F and F' centers of crystals.[14] Ogg,[6,15] Hill,[16] Lipscomb,[17] Kaplan and Kittel,[18] and Stairs[19] have discussed the cavity model and the equilibrium

$$\tfrac{1}{2}e_2^= = e^-. \qquad (4)$$

Symons[20] has recently argued that the optical properties of the solutions are best explained by the equilibrium

$$\tfrac{1}{2}M_2 = M^+ + e^- \qquad (5)$$

originally suggested by Huster.[21] Becker *et al.*[22] have proposed that four species and two equilibria are involved

$$\tfrac{1}{2}M_2 = M \qquad K_6 = (M)/(M_2)^{\frac{1}{2}}, \qquad (6)$$

$$M = M^+ + e^- \qquad K^7 = (M^+)(e^-)/(M). \qquad (7)$$

It is to be noted that reaction (5) is the sum of reactions (6) and (7) and if K_7 is much larger than K_6 the the concentration of M will be small and the system approximately described by reaction (5). Furthermore, magnetic methods do not readily distinguish between the systems of (4) and (5).

Becker *et al.*[22] used the magnetic resonance data of Hutchison and Pastor[14] on potassium to calculate K_6 and K_7, finding for K_6, 0.010, 0.03, and 0.1, and for K_7, 0.03, 0.05, and 0.02 at −33°, 0°, and +25°. Evers and Frank[23] analyzed the conductance data of Kraus[24] for sodium at −33° and calculated 0.037 for K_6 and 0.0072 for K_7. Dye *et al.*[25] measured transport numbers and, with conductance data, calculated 0.054 for K_6 and 0.0092 for K_7; with improved activity coefficient calculations, Dye *et al.*[26] obtained 0.043 for K_6 and 0.0096 for K_7. Evers,[27] in supporting the BLA model, has criticized Symon's[19] analysis of the optical evidence and argued that monomeric species cannot be ignored.

Our observation of differences in the heats of dilution of sodium in the presence and absence of excess sodium iodide bears on this problem. Equilibrium (4) would be unaffected (to a first approximation) by added Na^+. Equilibrium (5) would be repressed; the dilution heat should occur at a lower concentration but be of the same magnitude. On the other hand, our observations are consistent with the BLA model; reaction (7) is substantially repressed but reaction (6) is unaffected. Thus, reaction (6) is strongly endothermic, reaction (7) is exothermic (as are heats of dilution of salts[9] in the concentration range where ion-pair dissociation occurs) and their sum, reaction (5), corresponds to the +1.2 kcal effect we observe.

The heat of solution data in 0.29*M* NaI are now used to estimate ΔH_6 and ΔH_7 independently. For the ion-pair dissociation equilibrium of NaI we estimate[11]

$$NaI = Na^+ + I^- \qquad K_8 = (Na^+)(I^-)/(NaI), \qquad (8)$$

6.7×10^{-4} for K_8 and −3.5 kcal mole^{-1} for ΔK_8. We neglect activity coefficients throughout. Taking trial values of ΔH_6 and $1.2-\Delta H_6$ for ΔH_7, and starting with the values of Evers and Frank[23] at −33°, we use the van't Hoff equation to calculate K_6 and K_7 at 25° (this assumes the heat of reaction at 25° to be equal to the effective average in the range −33 to +25°). Then by a method of successive approximations, shifts in the equilibria and associated heat effects for reactions (6), (7), and (8) can be calculated and compared with the four measured dilution heats in 0.29 *M* NaI given in Table I. A good fit is obtained with +3.5 for ΔH_6, −2.3 for ΔH_7, 0.154 for K_6, and 0.00028 for K_7. However, for the pure sodium solutions, these values give a heat of dilution about 60% larger than that observed over the measured concentration range, and predict that at lower concentrations the heat of dilution would be exothermic due to dominance of dissociation (7). The calculations are quite uncertain due to neglect of activity coefficients, particularly for solutions with salt added, and to the uncertainty of conversion of equilibrium constants from −33° to +25°. However, it would appear that ΔH_6 and ΔH_7 are approximately +3 and −2 kcal (g atom)$^{-1}$, respectively.

[14] M. F. Deigen, Trudy Inst. Fiz. Akad. Nauk Ukr. S.S.R. **5**, 105 (1954); Izvest. Akad. Nauk S.S.S.R. Ser. Fiz. **18**, 716 (1954) Zuhr. Ekseptl. i. Teoret. Fiz. **26**, 293 (1954).
[15] R. A. Ogg, Jr., J. Am. Chem. Soc. **68**, 155 (1946); J. Chem. Phys. **14**, 114 (1946); Phys. Rev. **69**, 668 (1946).
[16] T. L. Hill, J. Chem. Phys. **16**, 394 (1948).
[17] W. N. Lipscomb, J. Chem. Phys. **21**, 52 (1953).
[18] J. Kaplan and C. Kittel, J. Chem. Phys. **21**, 1429 (1953).
[19] R. A. Stairs, J. Chem. Phys. **27**, 1431 (1957).
[20] M. C. R. Symons, J. Chem. Phys. **30**, 1028 (1959).
[21] E. Huster, Ann. Physik **33**, 477 (1938).
[22] E. Becker, R. H. Lindquist, and B. J. Alder, J. Chem. Phys. **25**, 971 (1956).
[23] E. C. Evers and P. W. Frank, Jr., J. Chem. Phys. **30**, 61 (1959).
[24] C. A. Kraus, J. Am. Chem. Soc. **43**, 749 (1921).
[25] J. L. Dye, R. F. Sankuer, and G. E. Smith, J. Am. Chem. Soc. **82**, 4797 (1960).
[26] J. L. Dye, G. E. Smith, and R. F. Sankuer, J. Am. Chem. Soc. **82**, 4803 (1960).
[27] E. C. Evers, J. Chem. Phys. **33**, 618 (1960).

Knight Shift Studies

Acrivos and Pitzer measured the Knight shift of ^{23}Na and ^{14}N in sodium–ammonia solutions over fairly wide intervals of concentration and temperature. The change in the electron densities at these nuclei on going from concentrated to dilute solutions was consistent with a change from metallike solutions to solutions in which the unpaired electrons are not closely associated with the sodium nuclei. The data confirmed the importance of the Becker, Lindquist, and Alder equilibria (p. 227); it was possible to calculate equilibrium constants and heats of reaction which were in fair agreement with values evaluated from other physicial measurements.

[Reprinted from the Journal of Physical Chemistry, **66**, 1693 (1962).]

41

TEMPERATURE DEPENDENCE OF THE KNIGHT SHIFT OF THE SODIUM-AMMONIA SYSTEM

BY J. V. ACRIVOS

Lawrence Radiation Laboratory, Berkeley 4, California

AND K. S. PITZER

Department of Chemistry, Rice University, Houston 1, Texas

Received March 29, 1962

The Knight shift of Na^{23} and N^{14} in sodium–ammonia solution was measured over the temperature interval -33 to $+22°$ and in the concentration range corresponding to mole ratio 5.7 to 700 (NH_3/Na). The results in the dilute region, $R \geqq 300$, were interpreted in terms of the equilibrium constants K_1 and K_2 for the reactions Na(am) = Na^+(am) + e^-(am) and Na(am) = $^1/_2Na_2$(am). The effective Knight shifts, k_0 for Na^{23} in Na(am) and k_1' for N^{14} in e^-(am) were found to be $k_0 = (0.034 \pm 0.005)T^{-1}$ and $k_1' = (13.5 \pm 1)T^{-1}$. The measured standard enthalpy and entropy of reaction for the dissociation and the dimerization equilibria are, respectively, ΔH_1^0 (298°) $\cong -6.6$ and $\Delta H_2^0 = -7.3 \pm 1$ kcal./mole and ΔS_1^0 (298°) $\cong -34$ and $\Delta S_2^0 = -24.1 \pm 3$ cal./deg. mole. The change in enthalpy for the dissociation equilibrium was temperature dependent and indicated that a large negative change in heat capacity accompanied the reaction. The electron densities at the Na^{23} nucleus were $\rho_1(Na^{23}) = 0.071\ a_0^{-3}$ and $0.00098\ a_0^{-3}$ for the concentrated ($R \cong 5.7$) and dilute solutions ($R \geqq 300$), respectively.

The study of the electromagnetic properties of the alkali and alkaline earth metals in liquid ammonia has supplied a great deal of information about the chemical nature of these solutions. As a result of conductivity measurements, Kraus[1] has proposed that there exist, present in solution, solvated atoms, positive ions, and electrons. The presence of paramagnetic species was indeed verified by Huster[2] and Freed and Sugarman[3] from the static magnetic susceptibility, χ, and by Hutchison and Pastor[4] from the paramagnetic absorption by means of e.s.r. The main conclusions to be drawn from these measurements are: (a) the magnetic susceptibility of the metal in ammonia solution always lies below the expected Curie value, which is approached asymptotically only as the dilution increases to infinity, and (b) the $1/T$ temperature dependence of χ is not obeyed. Becker, Lindquist, and Alder[5] explained these results by assuming the existence of four different species, solvated metal dimers, atoms, and positive ions and electrons in the dilute solutions, and then proceeded to evaluate the chemical equilibrium constants for the dissociation and dimerization of the solvated metal atoms or monomers from the e.s.r. data.[4] These reactions may be written

$$\text{Na(am)} = \text{Na}^+\text{(am)} + \text{e}^-\text{(am)} \quad (1)$$

$$\text{Na(am)} = {}^1/_2\text{Na}_2\text{(am)} \quad (2)$$

with the combined reaction

$$^1/_2\text{Na}_2\text{(am)} = \text{Na}^+\text{(am)} + \text{e}^-\text{(am)} \quad (3)$$

where the respective equilibrium constants are K_1, K_2, and K_3. The Knight shift (KS) data of McConnell and Holm[6] for the Na^{23}-$N^{14}H_3$ solutions at room temperature in the concentration range R = 5–500 supported these views. Pitzer[7] and Blumberg and Das[8] were able to explain some of the features of the KS data[6] by calculating the distribution of electron densities in the solvated paramagnetic species. Moreover, both Dye, Smith, and Sankuer[9] and Evers and Frank[10] derived sets of equilibrium constants from conductance measurements of Kraus[1] at $t = -33°$ together with transference number data. Their respective results are in fair agreement.

(1) C. A. Kraus, *J. Am. Chem. Soc.*, **43**, 749 (1921); for review work also see *J. Chem. Educ.*, **30**, 83 (1953).
(2) E. Huster, *Ann. Physik*, **33**, 477 (1938).
(3) S. Freed and N. Sugarman, *J. Chem. Phys.*, **11**, 354 (1943).
(4) C. A. Hutchison, Jr., and R. C. Pastor, *ibid.*, **21**, 1959 (1953).
(5) E. Becker, R. H. Lindquist, and B. J. Alder, *ibid.*, **25**, 971 (1956).
(6) H. M. McConnell and C. H. Holm, *ibid.*, **26**, 1517 (1957).
(7) K. S. Pitzer, *ibid.*, **29**, 453 (1958).
(8) W. D. Blumberg and T. P. Das, *ibid.*, **30**, 251 (1959).
(9) J. L. Dye, R. F. Sankuer, and G. E. Smith, *J. Am. Chem. Soc.*, **82**, 4797, 4803 (1960).
(10) E. C. Evers and P. W. Frank, *J. Chem. Phys.*, **30**, 61 (1959).

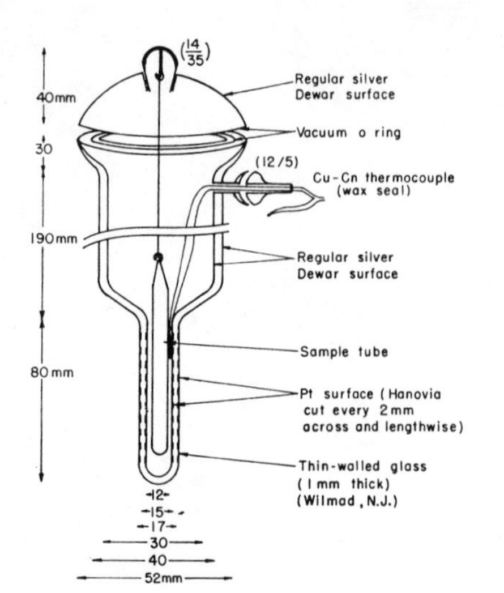

Fig. 1.—Rf. permeable dewar flask.

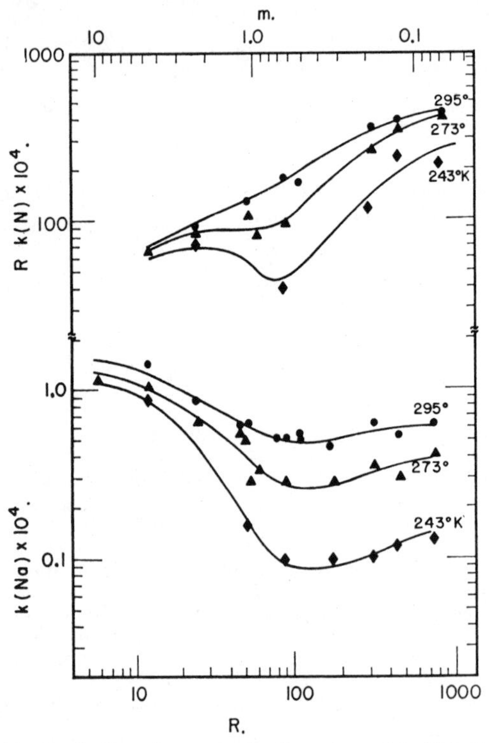

Fig. 2.—Isothermal concentration dependence of the Knight shift in the $Na-NH_3$ solutions.

The vapor pressure studies of Dewald[11] and the calorimetric results of Gunn and Green[12] also support the proposed equilibria of Becker, *et al.*[5] However, the equilibrium constants K_1 and K_2 as obtained from e.s.r.[5] are larger, by a factor of three, than those computed from the conductance experiments.[9,10] However, K_3 has substantially the same value from either source. In this work, the KS of Na^{23} and N^{14} in the $Na-NH_3$ solutions has been determined in the temperature interval -40 to 22°. In the concentration range where the chemical equilibria given in eq. 1 and 2 are valid, the constants K_1 and K_2 are evaluated from the KS data together with the activity coefficient of the charged species obtained from the Debye–Hückel theory. Although the relative accuracy of the measurements is very low at $-33°$, fair agreement is obtained with the values obtained from the conductance measurements.[1,9,10]

Experimental Results

The KS's of the $Na-NH_3$ solutions with respect to a standard of 0.5 m NaCl in NH_3 were measured with a Varian V-4200 wide line n.m.r. spectrometer operating at 10.0 and 2.77 Mc./sec. for Na^{23} and N^{14} resonance, respectively, in a constant magnetic field of 8881 oersteds.

The stabilities of the magnetic field and radiofrequency were the determining factors in the experimental accuracy. The radiofrequency was determined with a Hewlett–Packard counter No. 524B to ±2 c.p.s. while the field, stable to ±5 p.p.m., was swept with a linear potentiometer. The spectra were recorded by means of the sideband technique[13] with a modulation frequency ν_M = 412 c.p.s. The separation between the sidebands of 824 c.p.s. was then used to calibrate the potentiometer reading.

The temperature of the samples was measured to ±0.5°, by means of a copper–constantan thermocouple in contact with the sample tube. The sample tubes were immersed in a freezing mixture and contained in a dewar flask which allowed the rf. to penetrate. The constant temperatures were obtained as follows: 0° with an ice–water mixture, $-15.3°$ with a solid–liquid mixture of benzyl alcohol, $-30.6°$ with a solid–liquid mixture of bromobenzene, and below $-33°$ with acetone–Dry Ice mixtures. The samples then were allowed to attain the equilibrium temperature inside the closed dewar. The dewar flask is shown in Fig. 1. The tip which contained the samples was not silvered but was covered with a thin layer of Pt (Hanovia) in which a cross and lengthwise grill was cut at 2 mm. intervals in such a manner that the inner and outer surfaces would be concentric so as to allow perfect rf. penetration.

The $Na-NH_3$ samples were prepared *in vacuo* by first distilling a known weight of Na into the side arm of a sample tube and then distilling the required volume of NH_3 from a $Na-NH_3$ solution. The sample tubes were first aged in dilute HCl, then passed through hot cleaning solution, and finally steamed and dried in the absence of dust. No decomposition was noticed when warming the samples to room temperature for long periods, as shown by the reproducibility of the n.m.r. measurements within the expected accuracy. The samples were stored in liquid nitrogen when not in use. The concentration of the $Na-NH_3$ solution is reported in terms of the mole ratio, R, or the sodium molality, m.

$$R = \frac{[NH_3]}{[Na]} = \frac{58.7}{m} = 1.35\,\frac{V_{NH_3} \times \rho_{NH_3}}{w_{Na}}$$

where V is the volume of ammonia determined before the solution was made, ρ its density at that temperature (see, for instance, Yost and Russell[14]), and w is the weight of sodium. For the more dilute samples, a chemical analysis for total sodium was carried out after the measurements were finished. For R = 730, the nuclear resonance signal to noise ratio at room temperature for Na^{23} was barely unity but rose to 10 at $-33°$. The KS data are given in Table I. Figure 2 shows the isothermal concentration dependence of the KS, according to the data given in Table I. The

(11) J. F. Dewald, Ph.D. Thesis, California Institute of Technology, 1948.

(12) R. S. Gunn and L. R. Green, *J. Chem. Phys.*, **36**, 363, 368 (1962).

(13) J. V. Acrivos, *ibid.*, **36**, 1097 (1962).

(14) D. M. Yost and H. Russell, Jr., "Systematic Inorganic Chemistry," Prentice-Hall, New York, N. Y., 1948, p. 138.

TABLE I

KNIGHT SHIFT OF Na^{23} AND N^{14} IN $Na-NH_3$ SOLUTIONS WITH RESPECT TO 0.5 MOLAL NaCl IN NH_3[a]

Sample no.	R	m	t (°C.)	k(Na) × 10⁴ ±0.10	k(N) × 10⁴ ±0.10	Rk(N) × 10⁴
1	5.7	10	9.5	1.40	[a]	[a]
			0	1.25		
20	12.2	4.8	22	1.45	..	..
			0	1.06	5.58	68.07
			−27	1.00	..	..
			−46	0.94	..	..
18	25	2.4	22	.87	3.62	91
			0	.68	3.46	87
			−27	.63	2.78	70
			−30	.35[b]	..	..
				.78		
			−31	..	2.82	71
			−60		0.26[d]	7[d]
10	46	1.3	22.5	.61	[a]	[a]
			0	.56		
15	49	1.2	0	.51	[a]	[a]
25	52	1.1	26.3	.68	..	..
			22	..	2.55	133
			14	.48	..	..
			5	..	2.23	116
			0	.29	2.05	107
			−15	.16	1.61	84
			−31.2	.06	..	..
			−47	..	0.28	15
9	60	0.98	0	.34	1.54	80
			−19.5	..	1.40	84
			−24	.27	..	..
			−65	..	.40	24
			0	..	1.20[c]	72
					1.73	104
12	78	.75	22	.52	[a]	[a]
			− 7	.31	..	..
26	90	.66	25.1	.55	..	..
			23	..	1.87	168
			15.5	.50	..	..
			4	..	1.14	103
			0	.27	1.06	95
			−12	..	0.82	74
			−22	.13	.47	42
			−39	.07	..	..
			−56	.03	.20	18
11	110	.53	22	.56	[a]	[a]
17	111	.53	22	.51	1.66	184
			− 7	.26	..	..
			−19.5	..	.89	99
			−24	.21	..	..
7	118	.50	22	.46	[a]	[a]
			10	.37	..	..
			0	.29	..	..
			−46	.03	..	..
13	210	.28	− 7	.28	..	..
31	323	.18	22.8	.64	..	..
			21	..	1.11	358
			13.5	..	1.14	368
			9	.42	..	..
			4.5	..	0.90	291
			0	.35	.81	262
			− 9	..	.46	149
			−22	.15	..	..
			−27.5	..	.44	142
			−31	..	.37	120
			−52	..	.29	94
14	450	.13	22	.53	.83	374
			0	.31	.78	351
			− 7	.30	..	..
			−17	..	.48	216
			−24	.20	..	..
					..	..
			−29	..	.55	247
			−44	.07	..	..
			−46	.07	..	..
			−63	..	.10	45
37	730	.080	18.5	.62	.53	387
			11.5	.53	.60	438
			4.2	.49	.57	416
			0	.42	..	..
			−20	.18	..	..
			−22	..	.43	314
			−28.5	.18	..	..
			−30.5	..	.30	219
			−34	..	.13	95
			−43	..	.02	15
			−51	.02	..	..

[a] Poor annealing of the glass led to cracks at liquid nitrogen temperature and not all the runs could be completed. [b] Here the Na^{23} resonance shows two absorption lines. Although the temperature is above the critical value, the cause may be a different phase adsorbed on the surface of the sample tube. The N^{14} absorption gave a single line at this temperature. [c] Here the N^{14} resonance shows two absorption lines. Shaking the sample produced no effect; however, upon refreezing and warming up to 0° only one absorption was observed. The Na^{23} absorption was a single line at this temperature. [d] Measured for the dilute phases in the two-phase region.

isotherm at 243°K. was obtained by interpolating or extrapolating the data in Table I. The values at room temperature are in agreement, within the experimental accuracy, with the results of McConnell and Holm,[6] when a correction is made for the chemical shift of $Na^+(am)$ with respect to $Na^+(aq)$ of 17 p.p.m.

Knight Shift

The value of the effective field at the nucleus under observation, $\mathbf{H}_e$, is in general different from that of the externally applied one, $\mathbf{H}_0$. Thus,

$$\mathbf{H}_e = \mathbf{H}_0 + \mathbf{H}_s + \mathbf{H}_d + \mathbf{H}_c \quad (4)$$

where $\mathbf{H}_s$ is the contribution due to the magnetization, $\mathbf{M}$[15,16]

$$\mathbf{H}_s = \left(\frac{4}{3}\pi - \alpha\right)\mathbf{M} + q\mathbf{M}$$

Here the Lorentz cavity field, $(4\pi/3)\mathbf{M}$, arises from the induced dipoles on the surface of a microscopic hypothetical sphere which contains the nucleus, α is the bulk diamagnetic correction factor, $\alpha = 4\pi/3$ or 2π, respectively, for a sample shaped as a sphere or an infinite cylinder. For $\chi \approx 10^{-7}$,[4] and infinite cylinder sample shape, the first term is of the order of 0.1 p.p.m. q is a steric factor which depends on the anisotropy of the electronic g factor of any paramagnetic species present in solution.

$$q = (16\pi/45)(b/a)^3 \frac{(g_\perp^2 - g_\parallel^2)}{g^2}$$

(15) N. Bloembergen and W. C. Dickinson, *Phys. Rev.*, **79**, 179 (1950).

(16) W. C. Dickinson, *ibid.*, **81**, 717 (1951).

where b is the radius of the paramagnetic species and a is its distance from the nucleus under observation, $g_{\parallel}$ and $g_{\perp}$ are, respectively, the g-factors in the directions parallel and perpendicular to the applied field. The g-factor for the Na–NH_3 solutions is $g = 2.0012$, which is lower than the free electron value of 2.0025. Since $g^2 = (1/3)(2g_{\perp}^2 + g_{\parallel}^2)$ in solution, if one assumes $g_{\perp} = 2.0025$, the anisotropy in the g factor gives a negligible contribution, $q\mathbf{M} \sim 0.003$ p.p.m. $\mathbf{H}_d$ is the field due to the orbital motion of the electrons in the individual species. For Na^+(am) it may be assumed that $\mathbf{H}_d$ is constant and independent of the anion as is the case for aqueous solutions of different sodium salts.[17] However, the diamagnetic correction for Na_2(am) and Na(am) is likely to be different from that for Na^+(am) and the estimate of this value is probably the largest source of error. The diamagnetic contribution from the electrons in expanded orbitals in the Na(am) and Na_2(am) species is thus (see for instance Pople, Bernstein, and Schneider[18])

$$-\frac{H_d}{H_0} = \frac{e^2}{3mc^2} \int \frac{\rho_1}{\epsilon r} d\tau$$

where ρ_1 is the electron density and ϵ the dielectric constant, which is larger than unity but smaller than the value for the bulk material. If $\langle(1/\epsilon r)\rangle \sim 0.1a_0^{-1}$, the diamagnetic shift at the central atom is of the order 2 p.p.m. for each electron, and zero at the coördinated NH_3 molecules. $\mathbf{H}_c$ is the field at the nucleus due to the Fermi contact term[19] in the paramagnetic species. Thus

$$\mathbf{H}_c = (8\pi/3)g\beta\rho_1(\mathbf{r}_n)\langle\mathbf{S}^z\rangle, \{\rho_1(\mathbf{r}_n) = \langle|\psi_e(\mathbf{r}_n)|^2\rangle\}$$

where

$$\langle S^z\rangle = \frac{1}{2}\left[\exp\left(\frac{g\beta H_0 S^z}{kT}\right) - \exp\left(-\frac{g\beta H_0 S^z}{kT}\right)\right]$$

$$= \chi^{\mathrm{mol}}\frac{H_0}{N_0}(g\beta)$$

ψ_e is the unpaired electron wave function and N_0 is Avogadro's number. This term is now assumed to give the leading contribution to $\mathbf{H}_e$ in addition to $\mathbf{H}_0$. Thus, the field shift with respect to NaCl in NH_3

$$k = \frac{H_e - H_0}{H_0} \cong \frac{H_c}{H_0} = \frac{8\pi}{3}\frac{\chi^{\mathrm{mol}}}{N_0}\rho_1(\mathbf{r}_n) \quad (5)$$

is the KS[20] and in the case where the chemical species containing the nucleus, n, under observation undergoes fast chemical exchange it can be shown that[21]

$$k(\mathrm{n}) = \sum k_i x_i \quad (6)$$

where k_i and x_i are the mole fraction and KS of the corresponding species.

Chemical Equilibria

In the concentration range where the chemical equilibria 1 and 2 are valid, $R \geqq 150$, the observed KS obey the relationships

$$k(\mathrm{Na}^{23}) = x_0 k_0$$

$$R \times k(\mathrm{N}^{14}) = (x_0 k_0' + x_1 k_1')$$

$$= k_1'((k_0'/k_1')x_0 + x_1) \quad (7)$$

where k_0 is the KS for Na^{23} in Na(am), and k_0' and k_1' are the KS for N^{14} in the individual species Na(am) and e^-(am), respectively. The mole fractions of Na(am), e^-(am), and Na_2(am) within the solute are, respectively, x_0, x_1, and x_2. The volumetric and optical spectral properties of Na(am) and of Na^+(am) + e^-(am) are practically identical. Gold, Jolly, and Pitzer[22] concluded from these facts that Na(am) probably consisted of ion pairs of solvated sodium ions and electrons. From this model one would expect the KS for N^{14} to be substantially unchanged by this ion pair association and we shall hereafter assume $k_0' = k_1'$.

The room temperature value of k_1' is obtained by extrapolating $[R \times k(\mathrm{N})]$ to infinite dilution: see Fig. 3.

$$k_1' = \lim\,[R \times k(\mathrm{N})], \text{ as } R \rightarrow \infty$$

$$\cong 500 \times 10^{-4}$$

or

$$k_1' = 14.6/T \quad (8)$$

when the Curie temperature dependence of the susceptibility is introduced. In order to determine k_0, the functional dependence of k(Na) with respect to m must be known. Thus, if the solutions obey the equilibrium relationships given by eq. 1 and 2

$$K_1' = \frac{K_1}{\gamma_\pm^2} = \frac{m x_1^2}{x_0}$$

$$K_2^2 = \frac{x_2}{m x_0^2}$$

$$x_0 + x_1 + 2x_2 = 1 \quad (9)$$

it follows that

$$\left(\frac{\partial \ln x_0}{\partial \ln m}\right)_{\gamma_\pm, T} = \frac{x_1 - 4x_2}{1 + x_0 + 6x_2}$$

$$\left(\frac{\partial \ln (x_1 + x_0)}{\partial \ln m}\right)_{\gamma_\pm, T} = -x_2\frac{\left(4 + \dfrac{2x_1}{x_1 + x_0}\right)}{1 + x_0 + 6x_2} \quad (10)$$

where $\gamma_\pm$ is the activity coefficient for the charged species. Hence, if one assumes that the value of

(17) J. E. Wertz and O. Jardetzky, *J. Chem. Phys.*, **25**, 357 (1956).
(18) J. A. Pople, W. G. Schneider, and H. J. Bernstein, "High Resolution Nuclear Magnetic Resonance," McGraw-Hill Book Co., Inc., New York, N. Y., 1959, p. 175.
(19) E. Fermi, *Z. Physik*, **60**, 320 (1930).
(20) C. H. Townes, C. Herring, and W. D. Knight, *Phys. Rev.*, **72**, 852 (1950).
(21) H. S. Gutowsky, D. W. McCall, and C. P. Slichter, *J. Chem. Phys.*, **21**, 279 (1953).
(22) M. Gold, W. L. Jolly, and K. S. Pitzer, *J. Am. Chem. Soc.*, **84**, 2264 (1962).

the activity coefficient will not vary appreciably with concentration, the value of k_0 may be determined when $k(\mathrm{Na})$ attains a maximum value. Thus, from eq. 9 and 10

$$k_0 = \left(\frac{[k(\mathrm{Na})] \times k_1'}{3[R \times k(\mathrm{N})] - 2k_1'}\right)_{k(\mathrm{Na})_{\max}}$$

$$\left(\frac{\partial \ln [R \times k(\mathrm{N})]}{\partial \ln m}\right)_{\gamma\pm, T, k(\mathrm{Na})_{\max}} = -2x_2\left(1 + \frac{2x_2}{1 - 2x_2}\right) \quad (11)$$

At room temperature, $(dk(\mathrm{Na})/dm) = 0$, for $R \cong 700$; see, for instance, Fig. 2. Thus according to eq. 11

$$k_0 = 1.2 \pm 0.2 \times 10^{-4} = \frac{0.035 \pm 0.005}{T} \quad (12)$$

and

$$\left(\frac{\partial \ln [R \times k(\mathrm{N})]}{\partial \ln m}\right)_{\gamma\pm, T, k(\mathrm{Na})_{\max}} = -0.19$$

is to be compared with the mean slope

$$\frac{\Delta \ln [R \times k(\mathrm{N})]}{\Delta \ln m} = -0.16$$

between $R = 450$ and 730.

Here the error in k_0 is due to the uncertainty of the value of $[Rk(\mathrm{N})]$ when $k(\mathrm{Na})$ attains its maximum value. In addition, the values of k_0 and k_1' were determined by comparison of the KS data interpolated or extrapolated to $-33°$ with the average values of the equilibrium constants at this temperature[9,10]

$$K_1 = 0.013 \text{ (moles/kg.)}$$

$$K_2 = 19.0 \text{ (moles/kg.)}^{-1/2}$$

Thus, at $t = -33°$

$$k_0 = 1.4 \times 10^{-4}$$

$$k_1' = 508 \times 10^{-4}$$

or

$$k_0 = 0.034/T$$

$$k_1' = 12.3/T \quad (13)$$

Here the activity coefficients were evaluated from the Debye–Hückel theory, making use of the dielectric constant[23] and the density of pure ammonia[14] together with the distance of nearest approach of 5.5 Å., as chosen by Dye, *et al.*[9] Hence, the agreement of the room temperature and $-33°$ values of k_0 and k_1' justifies the assumption that $k_0' = k_1'$. The equilibrium constants for eq. 1 and 2 in the temperature interval 22 to $-33°$ are now obtained from the KS data together with the average values

(23) "Table of Dielectric Constants of Pure Liquids," NBS Circular 514 (1951).

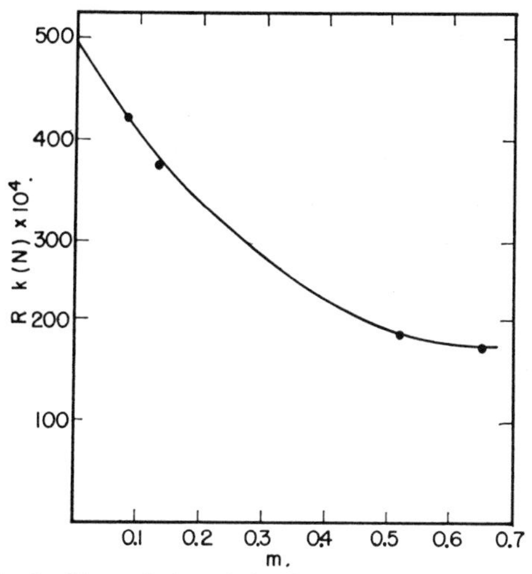

Fig. 3.—Extrapolation of the N^{14} Knight shift to infinite dilution.

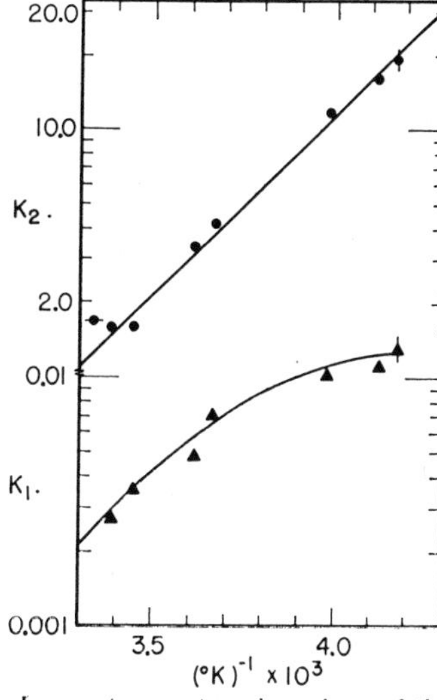

Fig. 4.—Inverse temperature dependence of the dissociation and dimerization equilibrium constants of Na(am) in the $Na-NH_3$ solutions. Values of K_2 and K_1 obtained by other authors are indicated as follows: (-●-) vapor pressure measurements of Dewald[11]; (▲, ●) conductance and other measurements by Dye, *et al.*,[9] and Evers, *et al.*[10]

$$k_1' = (13.5 \pm 1)/T$$

$$k_0 = (0.034 \pm 0.005)/T \quad (14)$$

They are given in Table II. Figure 4 shows the temperature dependence of the equilibrium constants.

The chemical equilibria 1, 2, and 3 are not satisfied for $R \leqq 111$; however, in this concentration range higher order sodium polymers, leading to a minimum in the KS, start to appear. As the concentration increases, $R \leqq 90$, the appearance of a metallic state is evidenced by an increase in the KS, which for Na^{23} tends to the limiting value of $k =$

TABLE II

EQUILIBRIUM CONSTANTS FOR THE DISSOCIATION AND DIMERIZATION OF Na(am) IN $Na-NH_3$

T	K_1 (moles/kg.)	K_2 (moles/kg.)$^{-1/2}$	$K_3 \times 10^3$ (moles/kg.)$^{1/2}$
295	0.0027 ± 0.001	1.6 ± 0.2	1.7
290	.0035	1.6	2.2
277	.0049	3.4	1.4
273	.0070 ± .0005	4.2	1.7
252	.010	12.2 ± 0.5	0.84
244	.011	15.8	0.67
240[a]	.013 (0.045)[b]	19 (67)[b]	0.69 (0.65)[b]

[a] The values of the equilibrium constants at this temperature are taken from the analysis of the conductance data of Kraus,[1] by Dye, Smith, and Sankuer,[9] and Evers and Frank.[10] [b] The equilibrium constants obtained from the e.s.r. data[4] by Becker, Lindquist, and Alder[5] are given for comparison.

1.5×10^{-4} for the saturated solution at room temperature.

Energy Calculations

The standard heats of reaction for the chemical equilibria 1, 2, and 3 are obtained from the temperature dependence of the respective equilibrium constant; thus, from Fig. 4.

$$\Delta H_1^0 \ (298) \cong -6.6 \text{ kcal./mole}$$

$$\Delta H_2^0 = -7.3 \pm 1 \text{ kcal./mole}$$

$$\Delta H_3^0 \ (298) = \Delta H_1^0 \ (298) - \Delta H_2^0 \cong 0.7 \text{ kcal./mole} \quad (15)$$

Here ΔH_2^0 is found to be constant throughout the temperature interval whereas ΔH_1^0 and consequently ΔH_3^0 are not. The value of ΔH_3^0 at room temperature is in agreement with the heat of dilution determined by Gunn and Green,[12] $\Delta H_3 = 1.2$ kcal./mole. Also, the temperature dependence of ΔH_1^0 indicates that a large negative change of heat capacity ΔC_p^0,[24,25] of the order of -100 cal./mole deg., accompanies the dissociation reaction.

The changes in entropy for the dissociation and dimerization reactions are found to be

$$\Delta S_1^0 \cong -34 \text{ cal./deg. mole at } T = 298°\text{K.}$$

$$\Delta S_2^0 = -24.1 \pm 3 \text{ cal./deg. mole}$$

Electron Densities

The parameters k_0 and k_1' in eq. 7 are, respectively, measures of the KS for Na^{23} and N^{14} in the single species Na(am) and e^-(am). Thus, the electron densities at the respective nuclei may be evaluated according to eq. 5 if one assumes χ^{mol} to be the theoretical susceptibility for $S = 1/2$. In the concentrated region, the density at the sodium nucleus was obtained from the KS making use of the combined Pauli spin susceptibility for a free electron gas with the Fermi energy evaluated from the metal electronic heat capacity.[26] The measured electron densities are given in Table III together with the values for the sodium-free atom[27] and the metal.[26]

TABLE III

ELECTRON DENSITIES IN THE $Na-NH_3$ SOLUTIONS IN UNITS OF a_0^{-3}

	$\rho_1(Na^{23})$					$\rho_1(N^{14})$
	$Na-NH_3$: R		Na	Na	Na^+	$Na-NH_3$
	5.7	≧300	(atom)	(metal)	(F-ctr.)	$R \geqq 300$
Exptl. value	0.071	0.00098	0.7525[c]	0.54[d]	0.014 to 0.009[e]	0.39
Theor. value		.066[a]				2.1[a]
		.014[a]				1.0[a]
		.016[b]				0.02[f]

[a] From two different solutions of the wave equation for the free electron in $Na \cdot 6NH_3$ (Blumberg and Das[8]). [b] From the solution of the wave equation for e^-(am) + Na^+(am) (Jortner[30]) where the distance between the ions is 7.2 a_0. [c] From atomic beam measurements (Kusch and Taub[27]). [d] From KS measurements (Knight[26]). [e] From the observed line widths in NaF and NaCl, F-centers (Lord[28]). [f] From the solution of the wave equation for the electron in $e^- \cdot NH_3$ (Pitzer[7]).

In conclusion, the variation of the electron density at the Na^{23} nucleus, from 9% of the value for the free atom in the concentrated region, R = 5.7, to 0.13% of the value for the free atom in the dilute region, $R \geqq 300$, indicates in the same manner as the conductance measurements of Kraus[1] the existence of two different types of $Na-NH_3$ solutions. The concentrated solutions possess properties characteristic of a metal, but in the dilute solutions the unpaired electrons are not closely associated with the sodium nuclei and therefore must occupy expanded orbitals in the dielectric medium. This also is evidenced by the high electron density observed for $N^{14}H_3$ in the dilute solutions. Here $\langle \rho_1(N^{14}) \rangle$ is the sum of the densities at the nitrogen nuclei of each of the ammonia molecules coördinated to the paramagnetic species, and since the unpaired electron moves in an expanded orbital, it will contribute to the density at several layers of coördination shells in the same manner that the electron in an F-center contributes to the density at nuclei removed several lattice distances from the vacancy.[28,29] The electron densities cannot be explained on the basis of the $Na \cdot 6NH_3$ species, assumed by Blumberg and Das,[7] since the values of $\rho_1(Na^{23})$ obtained from this model are larger than the observed ones by more than an order of magnitude; see Table III. On the other hand, although the models proposed by Pitzer,[7] Jortner,[30] and Gold, *et al.*,[22] are more plausible, the solution of the wave equation for the unpaired electron has to be refined to determine the extension of the expanded orbitals in the dielectric medium, in order to obtain more accurate values of the electron density.

Acknowledgment.—This work was performed under the auspices of the U. S. Atomic Energy Commission.

(24) K. S. Pitzer, *J. Am. Chem. Soc.*, **59**, 2365 (1937).
(25) H. S. Harned and N. D. Embree, *ibid.*, **56**, 1042, 1050 (1934).
(26) W. D. Knight, "Solid State Physics," Vol. 2, Academic Press, Inc., New York, N. Y., 1956, p. 93.
(27) P. Kusch and H. Taub, *Phys. Rev.*, **75**, 1477 (1949).
(28) N. W. Lord, *Phys. Rev.*, **105**, 756 (1957).
(29) G. Feher, *ibid.*, **103**, 834 (1956); **105**, 1122 (1957).
(30) J. Jortner, *J. Chem. Phys.*, **30**, 839 (1959); **34**, 678 (1961).

Hyperfine Splitting of ESR Spectra

In the following note, Vos and Dye report the first observation of nuclear hyperfine splitting in a metal–amine solution. The fact that such interaction has never been observed in dilute metal–ammonia solutions is significant; in the latter solutions there are no appreciable concentrations of metal-centered monomer species as there are in amine solutions.

Reprinted from THE JOURNAL OF CHEMICAL PHYSICS, Vol. 38, No. 8, 2033–2034, 15 April 1963
Printed in U. S. A.

42

Hyperfine Interactions in Solutions of Cs and Rb in Methylamine*

KENNETH D. VOS† AND JAMES L. DYE
Kedzie Chemical Laboratory, Michigan State University
East Lansing, Michigan

(Received 5 February 1963)

ESR spectra have been reported for a number of different metal–ammonia and metal–amine solutions, but to our knowledge, this is the first observation of nuclear hyperfine splitting in these systems. We have examined the ESR spectra of solutions of all of the alkali metals in purified methylamine and ethylenediamine at various temperatures, but only solutions of Rb and Cs in methylamine showed hyperfine splitting.

Monomethylamine was purified by fractional distillation in a stream of nitrogen. The metals were purified by successive distillations *in vacuo*. Samples in 3-mm Pyrex tubes were examined using a Varian Model 4500 spectrometer with 100-kc/sec field modulation. The variable sample temperature was measured with a thermocouple, and the field strength was measured using the frequency of proton absorption.

Typical first-derivative spectra for Rb and Cs in methylamine are shown in Fig. 1. The first-derivative curves of the six-line Rb spectrum and the eight-line Cs spectrum are asymmetric about the base line. The low-field side is farther from the base line than the high-field side, independent of the direction of sweep. In addition, the spectra generally show the presence of an extra absorption line which overlaps the hyperfine absorption and varies in relative intensity with concentration and temperature. The effect of temperature was investigated for Cs solutions. The intensity of the extra absorption line increases markedly as the temperature is lowered and its linewidth decreases from about 50 G at 5°C to about 2.7 G at −95°C. Over this same temperature interval, the total hyperfine separation (distance between the extreme lines of the spectrum) decreased from 320 to 90 G.

The presence of this nuclear hyperfine splitting is direct evidence for appreciable electron density at the metal nucleus. Knight shift measurements on solutions of Na in liquid ammonia at relatively high concentrations[1,2] indicate electron densities at the Na^{23} nucleus varying from 9% of the value for the free atom in concentrated solutions to 0.13% in dilute solutions. The electron density values for Cs and Rb calculated from ESR hyperfine separations can be compared with those of the free atoms determined from atomic beam experiments.[3] These separations are: Rb^{85}, 0.1013 cm^{-1}; Rb^{87}, 0.2280 cm^{-1}; Cs^{133}, 0.3066 cm^{-1}. Using these values and the Fermi equation,[4] the values of Ψ_0^2, the electron densities at the nuclei, in units of a_0^{-3} are: Rb^{85}, 2.35; Rb^{87}, 2.34; Cs^{133}, 3.92. The corresponding electron densities in methylamine solutions

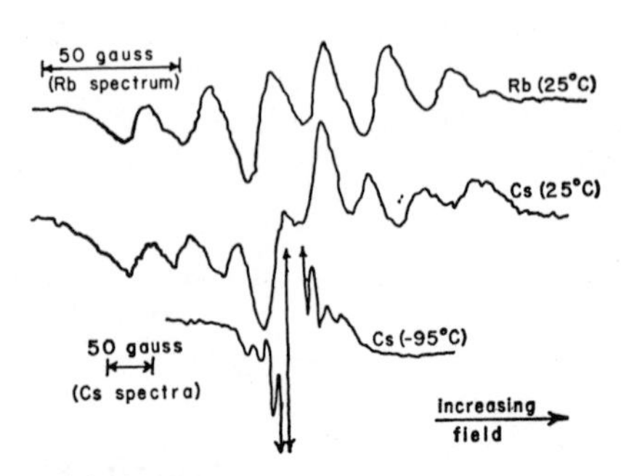

FIG. 1. Typical ESR first-derivative curves for solutions of Rb and Cs in monomethylamine.

can be calculated[5] from the total hyperfine separations: Rb, 128 G at 25°C; Cs, 394 G at 25°C, 90G at −95°C. These calculations give for Cs at 25°C, $\Psi_0^2=0.269a_0^{-3}$ which is 6.9% of the value for the free atom, decreasing to 1.6% at −95°C. It is interesting to note that the measurements of Acrivos and Pitzer[2] show a large decrease of the Knight shift with decreasing temperature, which indicates a lowering of the unpaired electron density at the nucleus.

Calculation of the electron density for Rb solutions is complicated by the presence of two isotopes. According to the abundances and nuclear moments involved, the four-line Rb^{87} spectrum should have a peak-to-peak separation twice that of Rb^{85} and an intensity 0.76 times as large. We may therefore assign the outside peaks in the Rb spectrum to Rb^{87}. This gives an electron density at the Rb^{87} nucleus at 25°C which is 3.5% of the value for the free atom.

Because of the asymmetry of the absorption, it is difficult to assign g values. The extra absorption line has a g value of 2.0018 ± 0.0002. Within experimental error, this value is the same as that found for the single-line spectra of Cs and Rb in ethylenediamine and Li, Na, and K in methylamine and ethylenediamine.

We wish to thank Professor H. S. Gutowsky for helpful discussions about these spectra. We plan to publish a more complete account of these and other ESR studies in ammonia and amines at a later date.

* This work was supported by a Research Contract with the U. S. Atomic Energy Commission.

† Present address: General Atomic, Inc., San Diego, California.

[1] H. M. McConnell and C. H. Holm, J. Chem. Phys. **26,** 1517 (1957).
[2] J. V. Acrivos and K. S. Pitzer, J. Phys. Chem. **66,** 1693 (1962).
[3] P. Kusch and H. Taub, Phys. Rev. **75,** 1477 (1949).
[4] E. Fermi, Z. Physik **60,** 320 (1931).
[5] W. Kohn and J. M. Luttinger, Phys. Rev. **97,** 883 (1955).

The Miscibility Gaps in Metal–Ammonia Systems

The liquid–liquid miscibility gaps that are found in most metal–ammonia systems are among the more fascinating and puzzling phenomena in this generally fascinating area of chemistry. In the following review paper, Sienko describes these phase separations and discusses various theoretical interpretations. The paper by Chieux and Sienko[1] should be consulted for a more recent statement of the theory of the phenomena.

---

[1]P. Chieux and M. J. Sienko, *J. Chem. Phys.*, **53**, 566 (1970).

43

ON THE COEXISTENCE OF LIQUID PHASES IN METAL-AMMONIA SYSTEMS AND SOME SURFACE TENSION STUDIES ON THESE SOLUTIONS ABOVE THEIR CONSOLUTE POINTS

by

M. J. SIENKO
(Baker Laboratory of Chemistry, Cornell University, Ithaca, New-York)

SUMMARY. — *The phase diagrams of metal-NH_3 systems are reviewed with particular attention to the miscibility gap in the Li, Na, and K systems. The unique character of the parabolic coexistence curves and their implication of long range forces are considered, together with a simple model of electron delocalization. The latter leads to predicted consolute concentrations of 3.9 atomic % M in NH_3, in remarkable agreement with observed values. The course of surface tension versus concentration has been investigated above the consolute temperature. Behavior is salt-like at low concentrations of M in NH_3 and metallic at high. Vestigial traces of the critical phenomena appear in the surface tension curves but, as expected, diminish with increased departure from the critical temperature.*

INTRODUCTION

I should like to dedicate this paper to the memory of Professor Richard Andrew Ogg, Jr., whose zeal in the study of liquid ammonia chemistry did so much to reawaken interest in the metal-ammonia solutions. I was fortunate in being able to spend 15 months working with Professor Ogg at Stanford University, most of which time was spent in unsuccessful attempts to repeat those famous experiments in which " superconductivity was discovered " in quenched Na-NH_3 solutions. Like much of liquid ammonia experimentation, the supraleitung effects were not reproducible, but in the process of studying them we learned an enormous amount about the liquid-liquid phase separation in metal-ammonia systems. Much of the experimental data reported below were gathered by Donald E. Loeffler, a graduate student at Stanford University at the time. Because his thesis results have never been published in the journals and therefore remain relatively inaccessible, it is particularly gratifying to be able to call attention to that excellent and interesting piece of work. The remainder of the data reported below, which has also not been published, is largely the work of a recent doctoral student of mine at Cornell University, Frank Holly. I should also like to acknowledge here the invaluable discussions I have had with my colleague, Professor Benjamin Widom, who first called my attention to the truly unique character of the critical solution behavior in the Na-NH_3

system. Finally, the generous support by the Air Force Office of Scientific Research and by the Advanced Research Projects Agency of our basic research program is gratefully recognized.

The following discussion is divided into two main sections :

Part I. — Phase Equilibria in Metal-Ammonia Solutions.

Part II. — Surface Tension of Metal-Ammonia Solutions.

I. Phase Equilibria in Metal-Ammonia Solutions.

The temperature-composition diagrams for metal-ammonia systems show in general three principally distinctive features : A very steep solubility curve for the equilibrium between saturated solution and metal ; a sharp

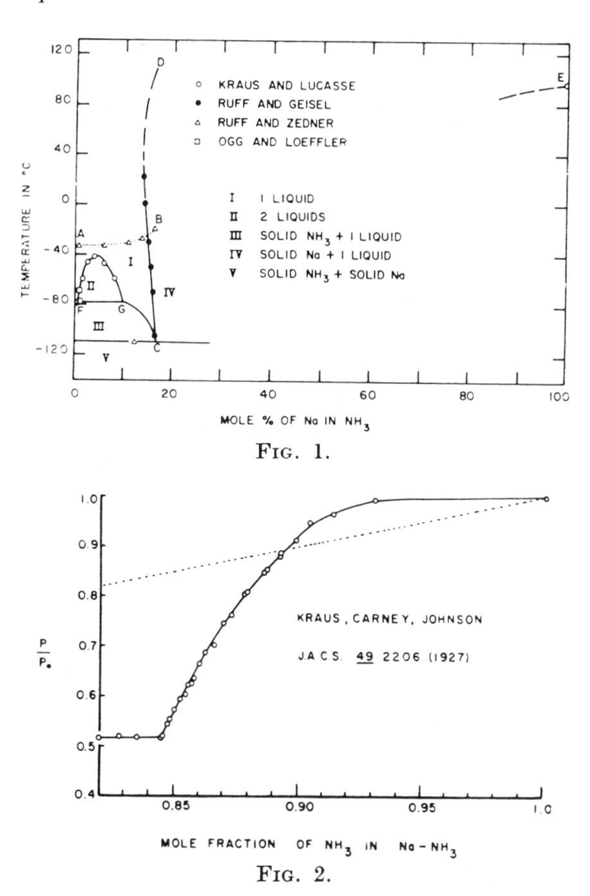

FIG. 1.

FIG. 2.

eutectic occurring in most cases at temperatures much lower than other equilibrium temperatures in the system ; and, finally a miscibility gap in the liquid region corresponding to coexistence of two liquid solutions of metal in ammonia. The most complete data have been accumulated for the sodium-ammonia system, the results for which are shown in Figure 1 as a conventional temperature-mole fraction (x) phase diagram. The solubility line, which rises almost vertically on the right side of the diagram, appears to be least ambiguously determined, since the points on it correspond to easily established equilibrium situations. As shown in Figure 2, the saturation composition of a sodium-ammonia solution at a given temperature can be fixed by distilling ammonia onto a fixed weight of sodium and measuring the ammonia pressure as a function of NH_3/Na ratio. The pressure of ammonia remains constant so long as any solid sodium remains undissolved but then rises abruptly when the saturated solution commences to be diluted. (The dotted line in the figure is a reference line for ideal solution behavior.) The fact that the solubility of sodium in liquid ammonia does not change appreciably with temperature, going from 16.7 at % Na at — 105 °C to 14.0 at % Na at + 22 °C [O. Ruff and E. Geisel, *Berichte* **39**, 831 (1906); W. C. Johnson and A. W. Meyer, *Chem. Rev.*, **8**, 273 (1931)], is consistent with rather low heat of solution, at least for the differential heat of solution into the saturated solution. This retrograde solubility with increasing temperature is unusual; the other metals in NH_3 show no solubility change with temperature or increased solubility at higher temperatures. I believe it was Kraus who first suggested that at much higher temperatures even the retrograde solubility of Na in NH_3 would have to give way to a positive increase with temperature, since the phase line needs to curve over to meet the Na axis. Unfortunately, chemical decomposition due to reactions such as $Na + NH_3 \rightarrow NaNH_2 + 1/2\ H_2$ are likely to preclude attainment of these higher temperature equilibria except under pressures in the kilobar range. Kraus and Schmidt [*J. Am. Chem. Soc.*, **56**, 2297 (1934)] have determined directly an integrated heat of solution of metals in ammonia in more dilute solutions by a clever method based on measuring the amount of NH_3 condensed or evaporated in the solution process. With this method they find + 1.4 kcal of heat absorbed per mole of Na dissolved in NH_3 at — 35 °C and 0.0 kcal for K. Coulter and Monchick (*J. Am. Chem. Soc.*, **73**, 5867 (1951)] find — 9.7 kcal for Li. The relatively large negative heat found for lithium suggests a special interaction, which may be connected with the formation of $Li(NH_3)_4$ mentioned below.

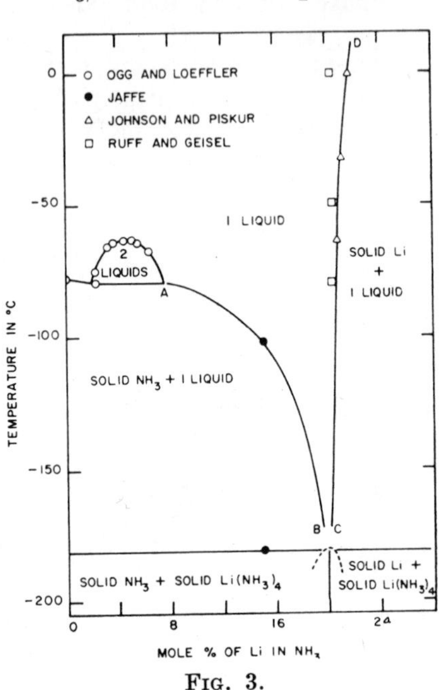

FIG. 3.

The phase diagrams for the other well-investigated systems Li-NH_3,

$K-NH_3$, and $Ca-NH_3$ are represented in Figures 3, 4, and 5, respectively. In the case of calcium and presumably also that of the other alkaline earth elements the solid phase in equilibrium with the " saturated solution " is $M(NH_3)_6$ instead of M.

The least adequately investigated feature of the $M-NH_3$ diagrams is

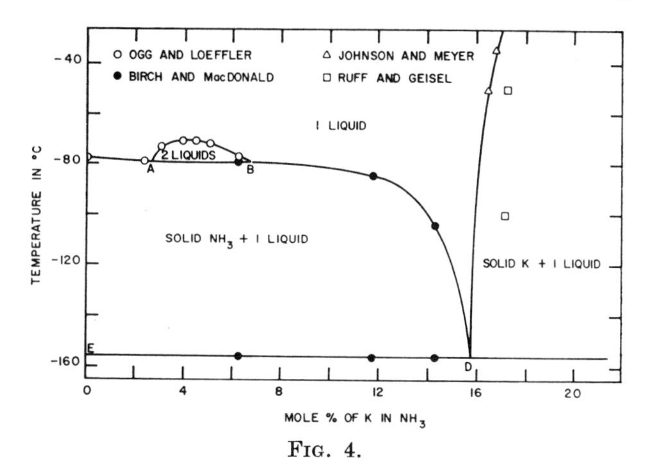

FIG. 4.

the region near the low temperature eutectic. A. J. Birch and D. K. C. MacDonald [*Trans. Farad. Soc.*, **43**, 792 (1947) ; **44**, 735 (1948)] investigated these eutectics in the course of their search for superconductivity, proposing that the formation of a highly conducting eutectic and supercooling of this eutectic mixture to give a highly conducting liquid accounted for the extremely low electrical resistances reported by R. A. Ogg [*Phys. Rev.*, **69**, 243 (1946) ; **69**, 544 (1946) ; **70**, 93 (1946)] for metal-NH_3 solutions quenched in liquid nitrogen. The suggested eutectic temperatures (and compositions, where known) are — 185 °C for Li (22 at %), — 110° for Na (17 at %), — 157° for K (15 at %), — 118° for Cs, — 87° for Ca, — 89° for Sr, and — 89° for Ba. The nature of the eutectics is not at all unambiguously established, which is particularly unfortunate in the case of lithium where the composition data are such as to indicate possible formation of the compound Li $NH_3)_4$. It is interesting to note that the simplest probable

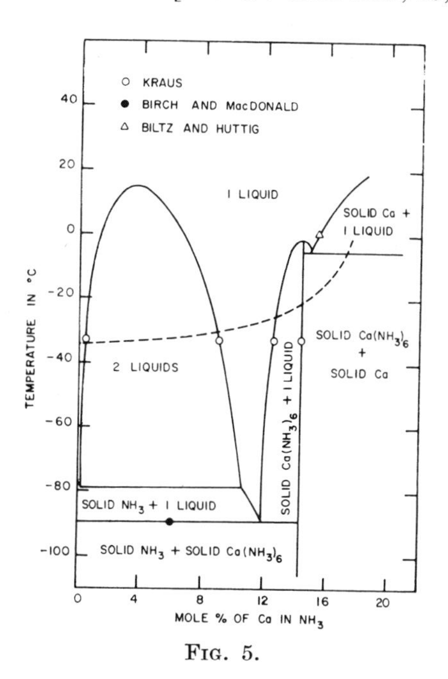

FIG. 5.

model for such a tetrammine lithium accounts equally well for the observation that the Hall voltage of Li $(NH_3)_4$ is consistent with one negative carrier per lithium atom [H. von R. Jaffe, *Zeitschrift für Physik*, **93**, 741 (1935)] and also with measurements of the nuclear magnetic resonance in metal-NH_3 solutions [H. M. McConnell and C. H. Holm, *J. Chem. Phys.*, **26**, 1517 (1957)]. If we assume in Li $(NH_3)_4$ that the 2*s* electron of the lithium has been promoted to a 3*s* orbital, thereby making available four sp^3 hybrids on the lithium to accommodate the lone pair electrons of the NH_3 groups, then conventional radii and bond distances lead to the conclusion that the outer node of the 3*s* orbital coincides with the proton positions. This is similar to the picture that Pitzer has postulated for explaining the lack of any observable Knight shift for protons in Na-NH_3 solutions. [K. S. Pitzer, *J. Chem. Phys.*, **29**, 453 (1958)]. Furthermore, the orbital extension of the 3*s* electron, particularly after correction for the appreciable dielectric constant of the medium, should be very large, leading to extensive overlap with the 3*s* orbitals of neighboring Li $(NH_3)_4$ groups. Resulting electron delocalization would lead to formation of a conduction band, in which carrier mobility would likely be quite high — possibly even greater than the 10^3 cm^2 per volt second characteristic of very good conductors. However, existence of Li $(NH_3)_4$ as a true compound necessitates modification of the Li-NH_3 phase diagram in the eutectic region to allow for a maximum melting hump (presumably very narrow) between two eutectics instead of a single eutectic. It would be interesting to investigate the effect of precise stoichiometry control in the neighborhood of $Li(NH_3)_4$ on the electrical properties observed for quenched lithium-ammonia solutions. If the compound $Li(NH_3)_4$ is definitely established, it might not be unreasonable to expect that the comparable " eutectics " for Na and K are in fact $Na(NH_3)_6$ and $K(NH_3)_6$, both of which would be expected to have higher carrier mobility than $Li(NH_3)_4$ due to increased orbital extension of 4*s* and 5*s* distributions.

Probably the best known and certainly the most easily observed feature of metal-ammonia solutions is the liquid-liquid phase separation. Visually, it is most striking. For example, a solution containing 4.2 atomic % Na in NH_3 on cooling below — 41.6 °C separates into two distinct layers — a more dense, les concentrated, dark blue phase at the bottom and a light, mobile, more concentrated, bronze phase at the top. The precise temperature at which the single phase solution splits into two liquid layers can readily be determined by measuring the electrical resistance of the solution as a function of decreasing temperature. A typical cell for such measurements is shown on the right side of Figure 6. Metal M is introduced through the side arm, which can then be sealed off and NH_3 condensed into the cell. Alternate closing of the two stopcocks in the side arms forces the liquid to mix back and forth through the capillary portion. Figure 7 shows representative behavior for a typical solution as its resistance is measured with decreasing temperature. The straight line portion to the left is characteristic of a single phase system. At the break at about — 67° there occurs an abrupt rise in the electrical resistance of the capillary portion of the circuit, this being the place where the dilute, significantly-less-conducting phase collects as phase separation proceeds. Figure 8 summarizes the phase separation data for Li, Na, and K as obtained in this way. It might be noted that the concentrated side of the two-liquid phase region is the more precisely defined, since the discontinuities in the resistance-temperature curves are easily determinable to 0.1 °C.

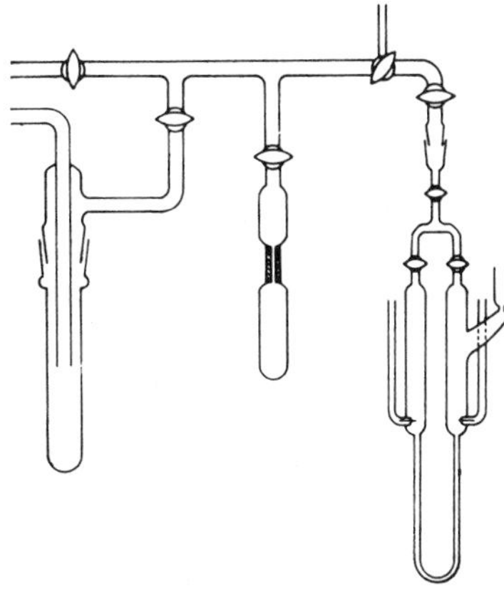

Fig. 6.

The general form of the phase separation curve is reminiscent of that found for miscibility gaps in other systems — e. g., water-phenol—with the major differencethat in the metal-ammonia system both layers are dilute solutions of one component in the other. Speculation as to the nature and cause of the phase separation has generally treated it as a condensation phenomenon. Birch and MacDonald [*Trans. Farad. Soc.*, **44**, 740 (1948)] ascribe the positive departure from Raoult's law [see Fig. 2] at low concentration of

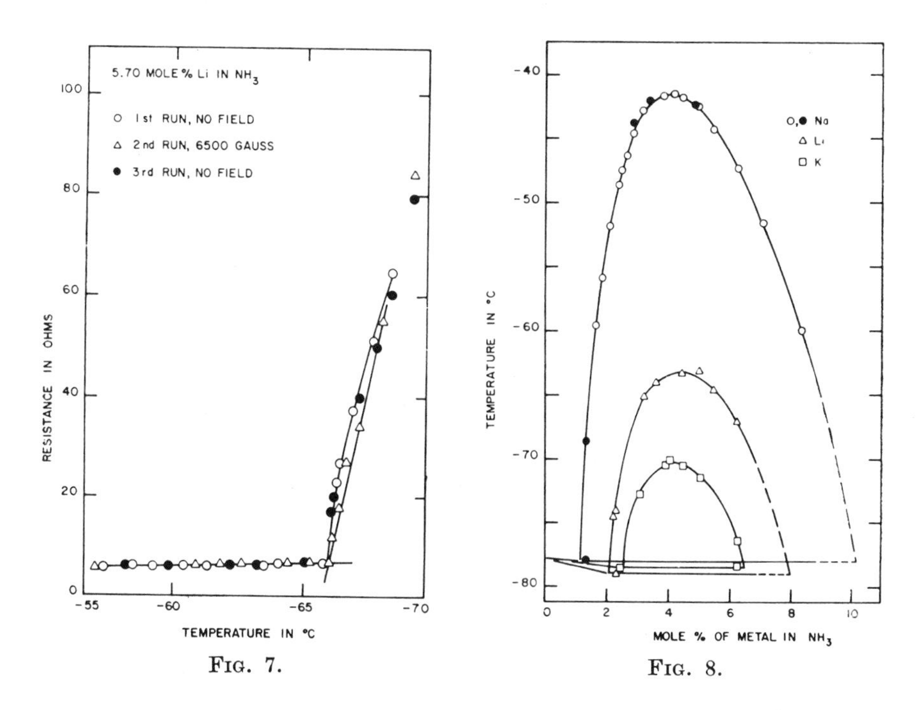

Fig. 7. Fig. 8.

M in NH_3 to association of the solute and the negative deviation at high concentration to strong association of the solvent and solute molecules.

In their view " the phase separation is to be regarded as a direct consequence of the aggregation vapor-pressure phenomenon ". They claim, in fact, that a semi-quantitative approximation for the phase separation curve may be derived by use of van der Waals equation in van't Hoff's law of osmotic pressure.

Pitzer [*J. Am. Chem. Soc.*, **80**, 5046 (1958)] suggests that the ammonia be regarded as a dielectric medium in which the sodium is considered to undergo liquid-vapor separation when the temperature drops below the critical temperature. In effect, the dilute side of the two phase region corresponds to gaseous sodium (no metallic properties observed) and the concentrated side, to liquid sodium (possessing all properties characteristic of liquid metals). Proceeding, as has been done in semiconductor theory, to allow for expanded electron distributions because of the dielectric constant of ammonia, Pitzer makes semi quantitative predictions of the critical temperature and critical volume which are surprisingly good considering the approximations introduced.

Ogg, on the other hand, proposed a much more drastic departure from ideal reasoning, stating that the observation of two liquid phases in equilibrium both of which are *dilute* solutions of metal in ammonia strongly suggested that this phenomenon had an origin different from that of other critical liquid phase separations. His ingenious explanation (*Phys. Rev.*, **69**, 243 (1946)] was that electron pairs, postulated to account for the observed diamagnetic nature of the solute in the intermediate concentration range, were responsible for the phenomenon. Because the electron pairs would have zero angular momentum, they would if independent obey Bose-Einstein instead of Fermi-Dirac statistics and would be expected to show significant condensation in momentum space to a lowest linear momentum state at quite high temperatures. Ogg postulated that " the liquid-liquid phase separation which occurs on slow cooling [as distinct from quenching] is the device adopted by the systems to avoid the Bose-Einstein condensation, with its unfavorable free energy change.... By sufficiently rapid cooling, it appears that the liquid-liquid phase separation is prevented, and that the system becomes frozen and hence metastable in the ' forbidden ' concentration region, which is thus characterized by the Bose-Einstein condensation of trapped electron pairs ". At the time Ogg proposed his model it was believed that the onset of superconductivity and the helium lambda point could be described as Bose-Einstein condensation phenomena, a view which has since been considerably modified. Also serious as an objection to Ogg's model is the probability that trapped or quasi-free electron pairs in a dielectric medium are not describable as Bose-Einstein particles. However, working on the premise that any influence, such as presence of a strong magnetic field, which would decrease the concentration of electron pairs would also depress the critical solution temperature, Loeffler [Ph. D. thesis, Stanford University, 1949] explored the temperature at which phase separation occurs in a magnetic field. As indicated in Figure 7 for Li-NH_3, the depression of the critical temperature, if it occurs at all, is considerably less than 0.1 °C. Even smaller effects were observed for $Na\text{-}NH_3$ and for $K\text{-}NH_3$.

That an unconventional approach is needed in uncovering the explanation of the liquid-liquid phase separation becomes clear from a close examination of the detailed shape of the coexistence curve. Unlike other carefully examined liquid-liquid separations [B. H. Zimm, *J. Chem. Phys*. **20**, 538 (1952)] and unlike the simple gas-liquid coexistence curves [E. A. Guggenheim, *J. Chem. Phys.*, **13**, 253 (1945)], which are cubic, the coexistence curves in metal-ammonia systems are parabolic. Figure 9, which was drawn at the suggestion of Professor Widom, shows how the data for $K\text{-}NH_3$ have been analyzed. Circled points represent temperatures at which phase separation occurs on cooling the solutions of different composition.

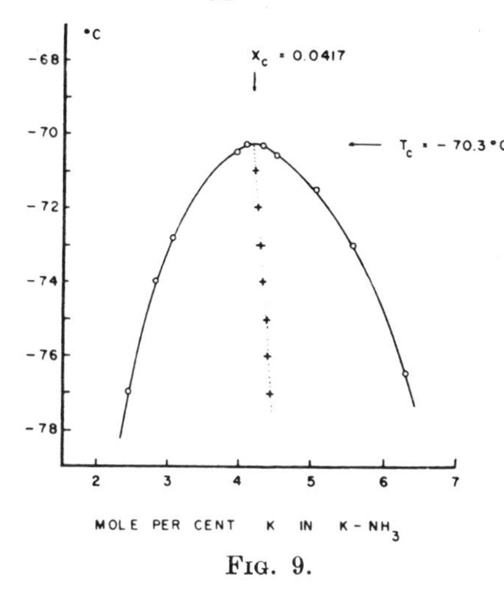

FIG. 9.

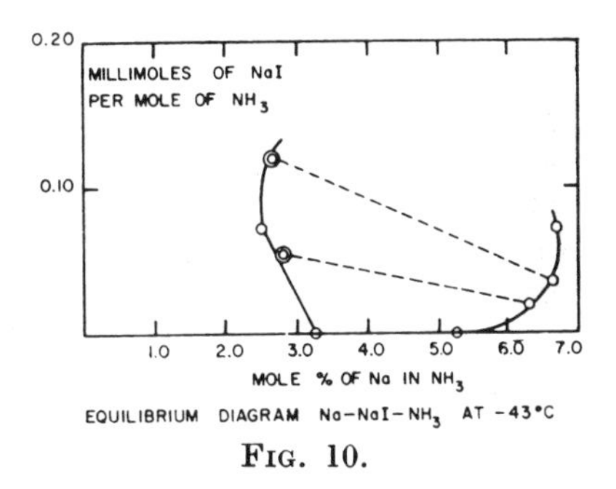

FIG. 10.

From the smooth curve drawn through these points, the average composition of each pair of coexisting solutions is calculated at various temperatures. As expected, the average compositions so defined are a linear function of the temperature, consistent with the law of the rectilinear diameter. The functional dependence of the concentration difference between each solution and the critical concentration for that temperature as defined by the rectilinear diameter is then obtained as a function of the temperature difference from the maximum critical temperature. Expressed in the form

$$T_c - T = a(x_1 - x_2)^n,$$

where T_c is the maximum critical temperature, a is a constant, x_1 and x_2 are mole fractions of metal in NH_3 in the coexisting layers, the exponent n comes out to be 2.1 ± 0.05 for the case of $Na\text{-}NH_3$, 2.1 ± 0.2 for the case of $Li\text{-}NH_3$, and 2.0 ± 0.1 for the case of $K\text{-}NH_3$. With the exception of the well-examined system $Na\text{-}NH_3$, the data from which these dependences are derived are extremely sketchy ; still, it is clear that the coexistence curves are more nearly parabolic than cubic. Because of the apparent fundamental significance of this difference in functional dependences, it would be most desirable to have independent determinations of the metal-NH_3 coexistence curves by techniques which did not involve the discontinuity in the resistance-temperature behavior. At present, we cannot exclude the possibility that the parabolic nature of the coexistence curve is only an artifact of the particular experimental technique employed.

However, there is additional evidence that the liquid-liquid separation in metal-NH_3 solutions is indeed quite different from other condensation phenomena. Some years ago we had occasion to study the effect of added salt, NaI, on the phase separation in Na-NH_3 systems [Sienko, *J. Am. Chem. Soc.*, **71**, 2707 (1949)]. The results are summarized in Figure 10, which shows a constant temperature slice of a portion of the ternary system Na-NaI-NH_3. The most obvious effect is a broadening of the miscibility gap due to the first addition of NaI. When recently I showed these results to Professor Widom he expressed regret that I did not have more data on the course of the maximum critical temperature and composition as a function of added NaI. For coexistence curves which are cubic, the consolute concentration is a quadratic function of added salt concentration, but theory predicts that for a coexistence curve which is parabolic the consolute point should be a linear function of added salt concentration. Though pressed by other commitments and faced at the time by apparent failure in our attempts to make the surface tension measurements to be described below, Frank Holly carried out another set of coexistence measurements to supplement the two complete sets we had. The three sets of results are summarized in Figure 11. If one analyzes the three sets of results by means of the rectilinear diameter intersection, which is the most precise way of fixing the consolute point, one finds the following interesting results :

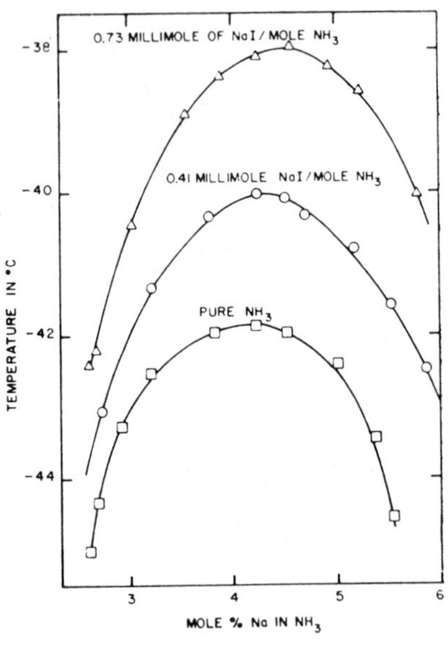

FIG. 11.

	CONSOLUTE POINT
Pure NH_3 + Na	4.12 At% Na
Added 0.41 millimole NaI	4.32
Added 0.73 millimole NaI	4.50

The dependence of consolute concentration on NaI concentration is apparently linear, as predicted by the theory.

The immediate question, of course, is what theoretical significance can be attached to a parabolic coexistence curve rather than a cubic one.

Strangely enough, the simple theory of van der Waals actually predicts coexistence curves that are parabolic — in clear-cut discrepancy with the experimental observations of cubic dependence, not only for liquid-vapor coexistence of pure substances but also for similar critical phenomena such as remanent magnetization of ferromagnetic materials below their Curie temperatures and the liquid-liquid phase separations in two-component systems. Agreement with experiment can be obtained by treating the molecular interactions through the Bragg-Williams approximation for a " lattice gas " [Refer, for example, to Terrell L. Hill, *Introduction to Statistical Thermodanymics*, Addison-Wesley (1960), particularly Chapter 14]. For a one-dimensional gas, no phase separation is predicted. For a two-dimensional gas, only a partial solution has been worked out, but it predicts phase separation with an eight-power coexistence curve. For a three-dimensional gas — that is, an ordinary real gas — only a numerical solution has been achieved, it indicates phase separation with a 16/5 power coexistence curve, close enough to a cubic dependence to be considered in agreement with experiment. To get a parabolic coexistence curve, the dimensionality of the lattice-gas *would have to be extended to infinity.* In other words, the van der Waals parabolic type of coexistence curve is what is expected as the exact solution for a system in which short range forces are considered to be operative to an infinite number of neighbors. Alternatively, we could have one in which the intermolecular forces extend to infinity [See B. Widom, *J. Chem. Phys.*, **37**, 2703 (1962), for references and further discussion]. Since we are dealing with a real system, it must be that the liquid-liquid phase separation observed in the metal-ammonia systems is due to the existence of long-range forces. Any interaction that falls off less rapidly than r^{-3} would suffice; it is not unlikely that electrical interactions, possibly associated with the delocalization of electrons, are the ones involved.

Simple semiconductor theory appears to support the above interpretation. If we regard metal atom M as an electron donor in the liquid ammonia dielectric, then the binding of the electron to a localized center may be treated in terms of the hydrogen atom problem corrected for departure of the dielectric constant from unity. More important, the orbital extension of the trapped electron r will be given by

$$r = \varkappa(m/m^*)\, a_0,$$

where $\varkappa$ is a dielectric constant, m is the electron rest mass, m^* is the effective mass of the electron, and a_0 is the first Bohr radius. Assuming that the effective mass equals the rest mass and that the dielectric constant $\varkappa$ can be calculated [as done for M in WO_3 — see B. L. Crowder and M. J. Sienko, *J. Chem. Phys.*, **38**, 1579 (1963)],

$$\varkappa^{-1} = \varepsilon^{-1} + \frac{5}{16}\,(\varepsilon_0{}^{-1} - \varepsilon^{-1}),$$

where ε, the static dielectric constant of liquid ammonia, is 22 and ε_0, the optical frequency dielectric constant, is estimated from the square of the refractive index to be 1.87, then with $\varkappa = 5$ we calculate r to be about 2.6 Å. Mott [*Nuovo Cimento*, **7**, 312 (1958)], in treating the problem of transition from non-metallic state to the metallic state, estimates that the cross-over from Heitler-London description to one using Bloch functions should occur when the mean distance between atoms is 4.5 times the mean radius of any one of them. Carrying this analysis over to our case we expect delocalization of the electrons to set in when the M atoms are no more than 11.7 Å apart. For a close-packed array of atoms, this corresponds to 8.85×10^{20} M atoms per cc. For metal in liquid NH_3, the equivalent concentration is 3.87 atomic % M, in remarkable agreement with the observed consolute concentrations : 4.12 atom % for Na in NH_3, 4.35 for Li in NH_3, and 4.17 for K in NH_3.

If, as indicated above, the onset of delocalization of electrons leads to long-range interactions, then similar behavior is to be expected in other systems where metal atoms are dissolved in a dielectric medium. Specifically, we expect such long range interactions to appear in the well-known solutions of metals in molten salts. Also, there is the corollary prediction that the coexistence curves for their miscibility gaps should be parabolic and not cubic. In the one case that we have examined in detail, Bi in $BiCl_3$ [S. J. Yosim, A. J. Darnell, W. G. Gehman, and S. W. Mayer, *J. Phys. Chem.*, **63**, 230 (1959)], the liquid-liquid phase coexistence curve is indeed parabolic and not cubic !

II. Surface Tension of Metal-Ammonia Solutions.

In very dilute solutions the metal-NH_3 systems are best considered as salt-like in which the metal M is considered to be more or less dissociated into solvated cation and anion. In concentration solutions, the behavior is obviously metallic. It would seem, then, that these solutions would be ideally suited for investigating the influences that operate in passage from salt-like to metallic behavior. In particular, we undertook a study of the surface tension in order to provide data for testing the various theories of surface tension, all of which included a functional dependence on electron density but all of which remained essentially untested since there is very little one can do to vary significantly the electron concentration in a liquid metal.

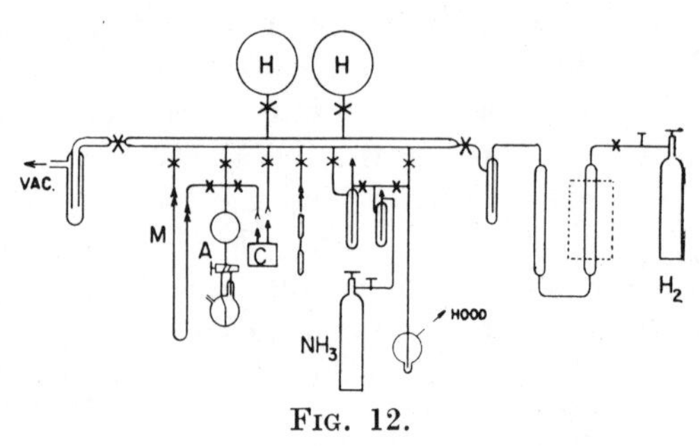

Fig. 12.

As might be expected, the experimental problem turns out to be enormous not only because the solutions are sensitive to attack by air and moisture but also because they are inherently unstable to decomposition via the reaction $M + NH_3 \rightarrow MNH_2 + 1/2\ H_2$. The method we chose for our

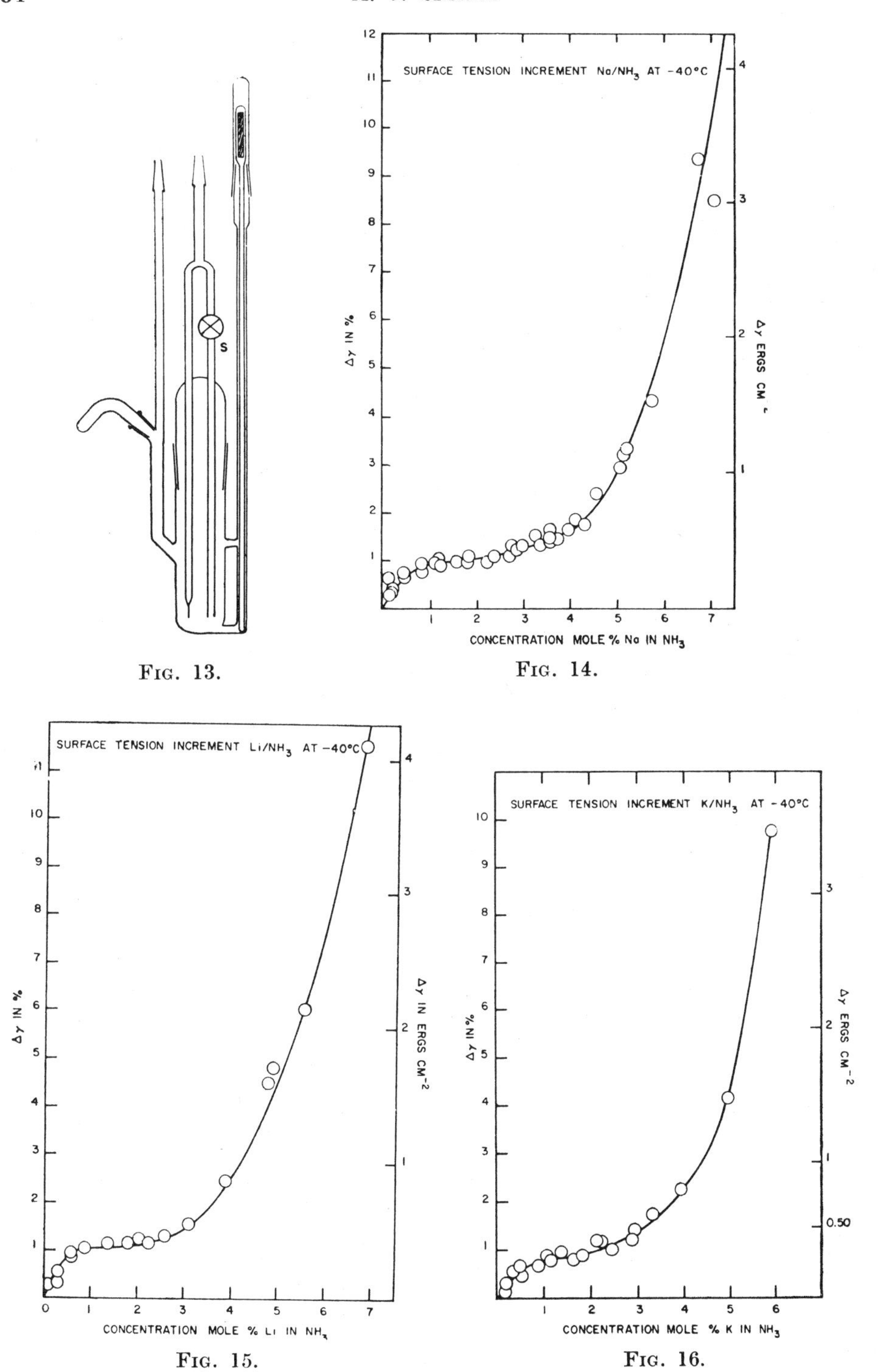

FIG. 13.

FIG. 14.

FIG. 15.

FIG. 16.

measurements was the maximum bubble pressure method. It guaranteed fresh, clean surfaces and lent itself to scrupulous cleanliness of the materials. Figure 12 shows the type of manifold required for the introduction of metal M and liquid NH_3 into the cell C and the hydrogen purification and ballast train for blowing the bubbles. Not shown but found to be absolutely neces-

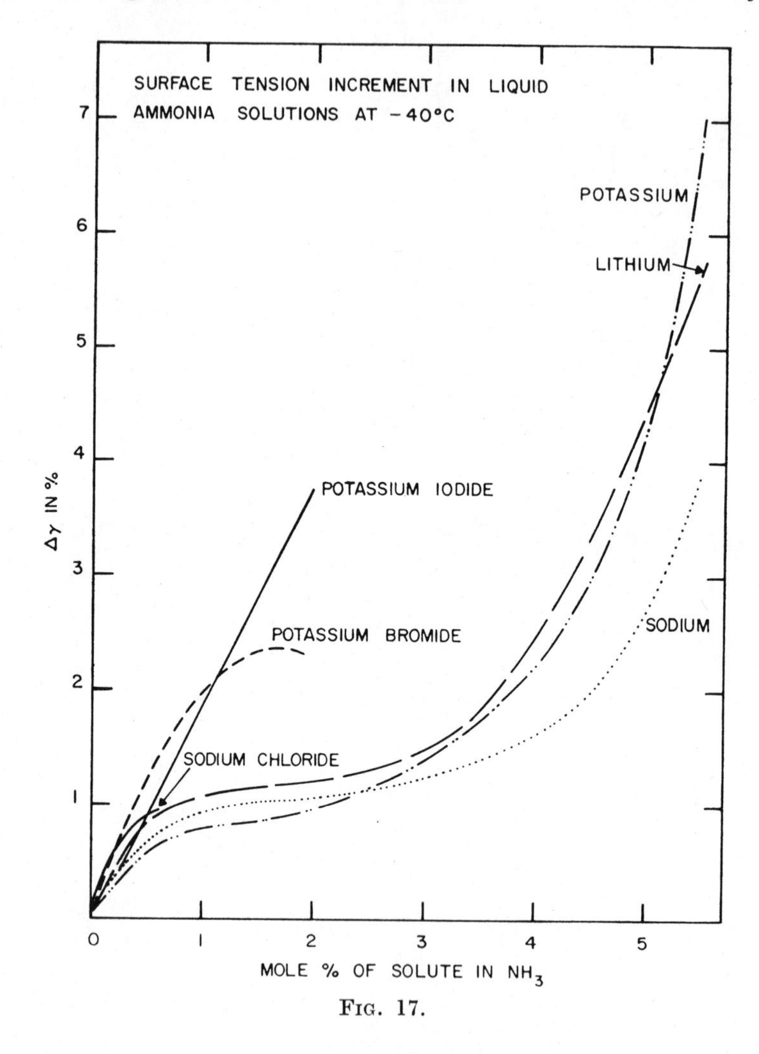

FIG. 17.

sary was a palladium type hydrogen-purifier. Figure 13 shows the cell proper with sidearm for introducing ampoules of metal M, magnetically operated stirrer, and two tubes of widely differing diameter for generating bubbles in the solution. Reproducible results are almost impossible to obtain with an all-glass cell, no matter whether it was soft glass, Pyrex, or fused silica, and it was only when we replaced the capillary tip by a noble metal (silver, gold, and platinum were used, with platinum most satisfactory because it was easiest to clean) that we were able to make reliable measurements. Otherwise, the apparent surface tension continuously increased with time, probably because of deposition of products at the capillary tip due to autocatalytic decay of the solutions.

Figures 14, 15, and 16 show, respectively, the observed increments of surface tension over that of liquid ammonia as functions of the concentration of sodium, lithium, and potassium in ammonia at — 40 °C. In Figure 17, the results are displayed again in summary and compared to those previously obtained for alkali halides [R. A. Stairs and M. J. Sienko, *J. Am. Chem. Soc.*, **78**, 920 (1956)]. From the comparison it is clearly evident that at low concentrations of alkali metal — i. e., less than 0.5 mole % — the surface-

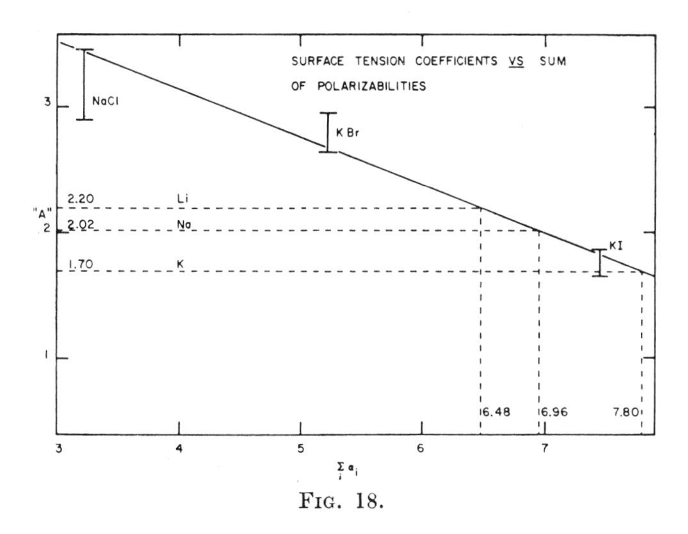

FIG. 18.

tension behavior is salt-like. In fact, if one accepts the premise suggested by Stairs and Sienko that the first virial coefficient in the power series expansion representing $\Delta\gamma$ as a function of concentration is related to the polarizability sum of the anion and cation, then, as shown in Figure 18, the previously determined straight line dependence of coefficient " A " on $\Sigma\alpha$ can be used to deduce the polarizability sum of cation and anion in dilute solutions of Li, Na, and K in NH_3. The values so obtained — 6.48, 6.96, and 7.80 Å^3 — when corrected for cation contributions of 0.03, 0.21, and 0.96 Å^3 for Li^+, Na^+, and K^+, respectively, lead to apparent ionic polarizabilities for the solvated electron in these solutions. The average value is 6.68 Å^3, which is somewhat larger than the 6.3 Å^3 characteristic of iodide ion. The implication then is that, at least in dilute solutions of metal in liquid ammonia, the surface tension behavior is such that the metals can be regarded as cation-anion pairs where the anion, the ammoniated electron, fits nicely at the lower end of the halide grouping Cl^-, Br^-, I^-, e^-.

At high concentrations of M in NH_3 there is a steep rise in surface tension. This is qualitatively what would be expected for metals, not only following the prediction of formal metal theory as expressed for example in the equation of Samoilovich [*J. Expt. and Theoret. Phys. USSR*, **16**, 135 (1946)],

$$\gamma = 0.30\rho^{7/6} - 0.76\rho^{3/2},$$

where γ is the surface tension and ρ is the charge density but also, surprisingly enough, in the phenomenological theory of H. Reiss and S. W. Mayer [*J. Chem. Phys.* **34**, 2001 (1961)]. The latter authors find that the same equation used to describe the surface tension of molten salts

$$\gamma = \left(\frac{kT}{4\pi a^2}\right)\left[\frac{12y}{1-y} + \frac{18y^2}{(1-y)^2}\right],$$

where $y = \frac{\pi a^3 n}{6}$ and a is the average hard sphere diameter of anion and cation, n being the number density of molecules in the liquid, can also be used for liquid metals. The hard-sphere radius that has to be assigned to the electron on the basis of observed surface tension data for molten sodium was 1.70 Å. In our metal-ammonia results, observed values of γ would require effective electron radii of 8.58 Å, 9.17 Å, and 9.00 Å in Na, Li, and K solutions, respectively.

The intermediate flattened portion in the $\Delta\gamma$ vs. concentration curves between the salt-like and metal extremes is reminiscent of the flattening in the van der Waals isotherms as the critical temperature is approached, and therefore they can be taken to represent incipient liquid-liquid phase separation. Presumably, in this range of concentration the chemical potentials of the individual components M and NH_3 are not changing much with variation in concentration. However, at higher temperatures the entropy difference at the various concentrations becomes more important so that the curves would be expected to be less flat as we get farther from the critical temperature. Figure 19 shows what happens to the surface tension of Na-NH_3 solutions as the temperature is raised from —40 °C (solid curve) to —35 °C (discrete points). There is the expected steepening of the van der Waals isotherm as we depart more from the critical temperature of —41.6 °C. It was not feasible to investigate the solutions at temperatures higher than —35 °C since the ammonia vapor pressure rapidly increases so as to interfere with the maxi-

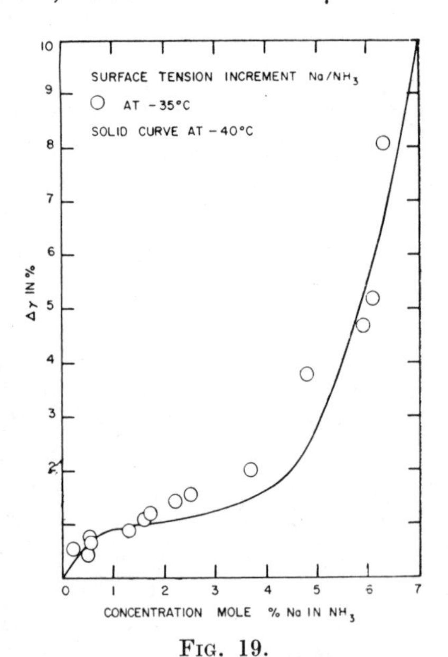

FIG. 19.

mum bubble formation. For detailed theories of molecular interactions in metal-NH_3 solutions it would be desirable to know precisely the rate of flattening as the critical temperature is approached. We heartily recommend that these experiments be done — *in someone else's laboratory.*

Ithaca, New-York, 5 June 1963.

A Model Involving Two Diamagnetic Species

Although Bingel[1] was the first to postulate an M^- species in metal–ammonia solutions, Arnold and Patterson were the first to show (in the following paper) that the available conductivity and magnetic data could be better accounted for by including an M^- species along with the M_2, M, M^+, and e^- species. However, in a later paper on this subject, Demortier and Lepoutre (see p. 372) come to a different conclusion.

---

[1]W. Bingel, *Ann. Phys.*, **12**, 57 (1953).

Reprinted from THE JOURNAL OF CHEMICAL PHYSICS, Vol. 41, No. 10, 3089–3097, 15 November 1964
Printed in U. S. A.

44

Electronic Processes in Solutions of Alkali Metals in Liquid Ammonia. I. A Model Including Two Diamagnetic Species*

EMIL ARNOLD† AND ANDREW PATTERSON, JR.
Sterling Chemistry Laboratory, Yale University, New Haven, Connecticut

(Received 8 April 1964)

The various models which have been proposed for metal–ammonia solutions are reviewed in the light of the available experimental data. It is shown that, in order to account quantitatively for the electrical and magnetic properties of these solutions, it is necessary to assume two different species containing paired electrons. A new model postulated for metal–ammonia solutions consists of the following five species in equilibrium: metal ions, polarons or e centers, M centers, M_2 centers, and M' centers. An M' center is assumed to consist of an electron trapped in the field of an M center. The equilibrium concentrations of the species which are assumed to be present in metal–ammonia solutions are calculated for different metal concentrations. The calculated values agree well with experimental transference numbers and magnetic susceptibilities. The expressions for the equilibrium constants are related to the binding energies of electrons trapped by the various centers, and the relative values of the trapping energies are estimated from the magnetic and transference data. It is concluded that the main contribution to the binding in all centers arises from the polarization of the dielectric. The estimate of the concentration at which a narrow conduction band may begin to form agrees with the experimental temperature coefficient of conductivity which has a sharp maximum at $0.8M$.

INTRODUCTION

ALTHOUGH the solutions of alkali metals in liquid ammonia have been investigated for the past hundred years, 1964 marking the centenary of Weyl's paper,[1] there is as yet no agreement on their exact structure. The fundamental properties of these solutions are well known.[2,3] It is generally agreed that in dilute solutions the metal is completely dissociated into solvated ions and solvated electrons. We refer to the latter as e centers in this paper. The configuration of e centers is believed to be stabilized by the polarization of the dielectric, and the mobility of such self-trapped electrons has been successfully accounted for in terms of ionic transport models.[4,5] As concentration increases, the accompanying decrease in the equivalent conductance indicates that ion pairing or assembly into higher aggregates takes place. The observed final transition to the metallike state has been attributed to the overlap of wavefunctions of the various centers, thus forming a conduction band. As concentration increases from the dilute region, the observed decrease in molar paramagnetic susceptibility is generally believed to be caused by the association of solvated electrons into species containing electron pairs. Several models have been devised embodying specific entities and equilibria to take into account these experimental

* Contribution No. 1755 from the Sterling Chemistry Laboratory, Yale University. This paper is taken in part from a dissertation submitted by Emil Arnold to the Graduate School, Yale University, in partial fulfillment for the requirements of the degree of Doctor of Philosophy, June 1963.

† Present Address: Philips Laboratories, Irvington-on-Hudson, New York.

[1] W. Weyl, Pogg. Ann. **121**, 601–612 (1864).

[2] E. C. Evers, J. Chem. Educ. **38**, 590 (1961).

[3] W. L. Jolly, Progr. Inorg. Chem. **1**, 235 (1959).

[4] E. C. Evers and P. W. Frank, Jr., J. Chem. Phys. **30**, 1, 61 (1959).

[5] D. S. Berns, G. Lepoutre, E. A. Bockelman, and A. Patterson, Jr., J. Chem. Phys. **35**, 5, 1820 (1961).

observations. Those of Kaplan and Kittel,[6] Becker, Lindquist, and Alder,[7] Deigen,[8–11] and Gold, Jolly, and Pitzer[12] may particularly be mentioned.

The model of Kaplan and Kittel (KK) proposes that the dissolved metal is dissociated into metal ions and electrons, the latter residing in the cavities of the liquid (e centers). At higher concentrations these electrons are assumed to pair up with opposing spins (e_2 centers), thus causing the decrease of the paramagnetic susceptibility of the system.

The Becker, Lindquist, and Alder (BLA) model pictures the solute as four species in equilibrium. According to this model the solvated ions and electrons associate to form a neutral species, called the monomer, in which the electron circulates among the ammonia molecules oriented around the ion. Monomers can dimerize reversibly to form diamagnetic hydrogenlike molecules.

A very detailed model was proposed by Deigen. The model is based on the work of Pekar,[13] who treated trapped electrons in polar crystals by considering the coupling of the electronic motion to the vibrational modes of the lattice. In Deigen's model, the electrons trapped by polarizing the surrounding medium are called polarons. The polarons can interact with metal ions to form electrically neutral F centers. At higher concentrations the F centers are assumed to associate into diamagnetic couples of F_2 centers. The wavefunctions of localized electrons are calculated by a variational method, and the concentrations of the various centers are obtained by minimizing the free energy of the system.

A model recently proposed by Gold, Jolly, and Pitzer (GJP) is a modification of the BLA model and considers the monomers and dimers as ion pairs or higher aggregates, held together by Coulomb forces. The structure of the solvated electron is assumed to remain unchanged in all species present and throughout the range of concentrations, up to the metallic range.

Although all these models seem to account for the essential features of the behavior of metal–ammonia solutions, a more careful scrutiny shows serious discrepancies between the theories and experiment. Values for the equilibrium constants obtained from electrical conductivity,[4] ionic transference numbers,[14] and from magnetic susceptibility[2,7] are quoted in the literature. However, the equilibrium constants calculated from magnetic data to fit a given model do not agree with the constants obtained from the electrochemical data. It can be demonstrated that the discrepancy between the electrical and magnetic data is quite independent of any assumptions regarding the equilibrium constants themselves or the use of the mass action law, and that it arises merely because of the particular species chosen in the proposed models. Since this discrepancy suggests a flaw common to all models, we shall discuss it in some detail before proceeding to a model which overcomes these difficulties.

FAILURE OF CURRENT MODELS

To illustrate the discrepancy we calculate the concentrations of unpaired electrons and metal ions from the available experimental data and compare the results with the predictions based on the different models. The concentration of centers with unpaired electrons can be calculated from the paramagnetic susceptibility of the solution. For localized centers the concentration N_p of the paramagnetic species is related to the paramagnetic susceptibility χ_p through the following equation:

$$\chi_p = N_p \beta^2 / kT, \qquad (1)$$

where β is the Bohr magneton. N_p can be calculated from the data of Hutchison and Pastor,[15] who measured the susceptibility in both sodium and potassium solutions using the electron spin resonance technique.

The concentration of metal ions can be estimated from the equivalent conductance L of the ions in solution. In very dilute solution the concentration of metal ions N_0 can be obtained from the equation

$$N_0 = CL/(\mu_+ + \mu_-)\mathbf{F}, \qquad (2)$$

where C is the stoichiometric concentration of the metal, L the equivalent conductance, μ_+ and μ_- the mobilities of the positive and negative species and $\mathbf{F}$ is the Faraday constant. At concentrations above $0.01M$ the metal ion concentration can be calculated from transference data. The moving boundary method used by Dye *et al.*[14] should be expected to give the true mobility of electrons, regardless of the mechanism of their transport. In terms of T_+, the cation transference number, we have

$$N_0 = CLT_+ / \mathbf{F}\mu_+ \qquad (3)$$

In order to calculate N_0 from Eqs. (2) and (3) it is necessary to know the ionic mobilities as a function of ion concentration. These were calculated from the Onsager–Kim theory[16] which gives the mobility of an

[6] J. Kaplan and C. Kittel, J. Chem. Phys. **21**, 9, 1429 (1953).
[7] E. Becker, R. H. Lindquist, and B. J. Alder, J. Chem. Phys. **25**, 5, 971 (1956).
[8] M. F. Deigen, Zh. Eksperim. i Teor. Fiz. **26**, 293 (1954).
[9] M. F. Deigen, Zh. Eksperim. i Teor. Fiz. **26**, 300 (1954).
[10] M. F. Deigen, Tr. Inst. Fiz. Akad. Nauk Ukr. SSR **5**, 105, 119 (1954).
[11] M. F. Deigen and Y. A. Tsvirko, Ukr. Fiz. Zh. **1**, 3, 245 (1956).
[12] M. Gold, W. L. Jolly, and K. S. Pitzer, J. Am. Chem. Soc. **84**, 2264 (1962).
[13] S. I. Pekar, *Untersuchungen uber die Elektronentheorie der Kristalle* (Akademie Verlag, Berlin, 1954).
[14] J. L. Dye, R. F. Sankuer, and G. E. Smith, J. Am. Chem. Soc. **82**, 4797 (1960).
[15] C. A. Hutchison, Jr., and R. C. Pastor, J. Chem. Phys. **21**, 11, 1959 (1953).
[16] L. Onsager and S. K. Kim, J. Phys. Chem. **61**, 215 (1957).

ion in an arbitrary mixture of ions. The results of the calculation for the sodium ion at $-33.5°C$ show that even for concentrations of ions as high as $1M$ the theory predicts only a few percent change in cation equivalent conductance from the limiting value L_+^0. The values of N_0 calculated from Eq. (3) are, therefore, practically independent of the conductance function used to calculate μ_+; in fact, the results would be almost unchanged if we were to replace $\mathbf{F}\mu_+$ by L_+^0. The same argument applies to Eq. (2) at very low concentrations. Of course, at concentrations approaching the minimum in the equivalent conductance curve, the use of Eq. (2) to calculate ionic concentrations is no longer safe because the behavior of μ_- probably changes drastically in this region. Equation (3), however, should give correct values of cation concentrations throughout the concentration range, the only assumption being that the positive species in solution is an ion.

The values of N_p and N_0 calculated from Eqs. (1)–(3) are shown in Fig. 1. N_p and N_0 appear to be equal below about $0.01M$, but diverge steadily at higher concentrations. At $0.1M$ the concentration of ions is three times greater than the concentration of unpaired electrons. We shall now compare these results with the predictions of the various models proposed for these solutions.

We first consider the models which assume pairing of the electrically neutral paramagnetic species (such as monomers, F centers or ion pairs) into diamagnetic aggregates (dimers, F_2 centers, or ionic quadrupoles). We shall refer to the former as the M center and to the latter as the M_2 center. Each model assumes that

$$N_p = N_e + N_1 \qquad (4)$$

and

$$N_e = N_0, \qquad (5)$$

where N_e is the number of electrons detached from the metal and N_1 the number of M centers. Equations (4) and (5) lead to the requirement that

$$N_p \geq N_0, \qquad (6)$$

which is just the opposite from the experimental results shown in Fig. 1. Also shown in Fig. 1 are values of N_p and N_0 based on calculations of Deigen[11] and those of Evers.[4] Deigen's results were derived from magnetic data, while Evers' were based on the BLA model and equilibrium constants calculated from conductance data of Kraus.[17,18] As expected, the calculated and experimental results do not agree, except in quite dilute solutions. The curve based on Evers' calculations gives fair agreement with experimental N_0, but fails to reproduce the magnetic data. Deigen's results predict N_p correctly, but not N_0. Most important, however, is the fact that Eq. (6) which is required by both models is not satisfied in reality.

It is easy to see that a model which assumes e_2 centers as the diamagnetic species is also in conflict with the experimental evidence. According to this model the following relations must hold:

$$N_p = N_e + N_1, \qquad (7)$$

$$C = N_0 + N_1, \qquad (8)$$

where we have allowed for the presence of M centers. The model thus requires that

$$N_0 + N_p = C + N_e$$

or

$$N_0 + N_p \geq C. \qquad (9)$$

It can be seen from Fig. 1 that this requirement is not satisfied at concentrations higher than $0.01M$. Other models involving only one type of diamagnetic species such as negative metal ions lead to similar contradictions with experiment.

It seems quite unlikely that either the magnetic or transference data could be in error by such a large factor. The arguments we have used involve no assumptions regarding equilibrium constants or the use of the mass action law. They merely assume the existence in solution of metal ions and other distinct species, one of which has paired electrons. The discrepancy could not be due to the wrong choice of the cation mobility in Eq. (3) because, in order to obtain agreement with magnetic data, it would be necessary to

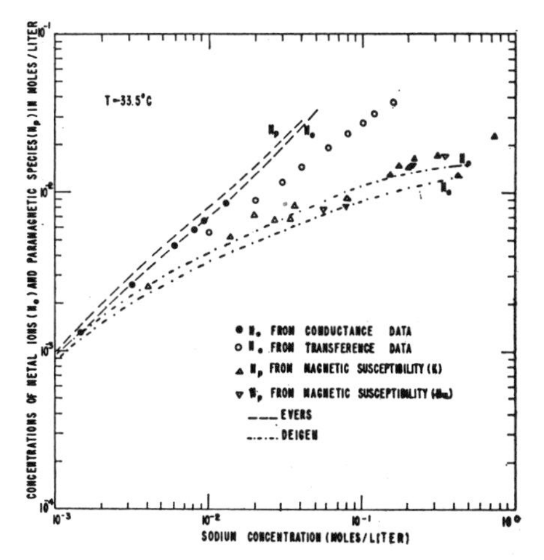

FIG. 1. Concentrations of metal ions (N_0) and paramagnetic species (N_p) in sodium/liquid ammonia at $-33.5°C$. Filled circles, N_0 calculated from conductance data of Kraus; open circles, N_0 calculated from transference data of Dye, Sankuer and Smith;[14] triangles, N_p calculated from paramagnetic susceptibility data of Hutchison and Pastor.[15] Broken lines represent theoretical calculations of Evers[4] (- - - -) and Deigen[11] (-•-•-•).

[17] C. A. Kraus, J. Am. Chem. Soc. 43, 749 (1921).
[18] C. A. Kraus and W. W. Lucasse, J. Am. Chem. Soc. 43, 2529 (1921).

make the unreasonable assumption that the mobility of cations increases with increasing concentration.

Next, we shall consider the validity of Eq. (1) which was used to calculate N_p. It can be shown[19,20] that Eq. (1) can be applied to localized electrons and also to electrons in a conduction band, provided the distribution in the latter case is nondegenerate. For degenerate electrons in a conduction band the susceptibility would be smaller. However, at concentrations close to $1M$ the paramagnetic susceptibility of metal–ammonia solutions varies approximately exponentially with reciprocal temperature.[15] This behavior is suggestive of some activation process and is quite unlike that of a degenerate electron gas.

Actually, we can show that even if a conduction band did exist, it could not possibly account for the observed discrepancy. KK have shown that the smallest carrier mobility one could reasonably expect to find in a conduction band is approximately 2 $cm^2/V\cdot sec$. Now in $0.1M$ solutions, where the discrepancy between the electrical and magnetic properties is already considerable, the conductivity is about 0.05 $(\Omega\cdot cm)^{-1}$. If we were to assume that the entire conductivity is due to carriers in a conduction band, then the concentration of carriers would be $n=1000\ \sigma/\mathbf{F}\mu$, where σ is the conductivity and μ is the carrier mobility; therefore, $n\leq 2.5\times 10^{-4}$ moles/liter. But from the electrical neutrality condition, the total number of electrons should be equal to the number of metal ions, which in a $0.1M$ solution is $0.03M$. Hence the majority of electrons cannot be in a conduction band.

Thus, the electrical properties and the temperature variation of the paramagnetic susceptibility both indicate that, in dilute and moderately concentrated solutions, we are dealing with localized electron centers, rather than with electrons in a conduction band. The discrepancy between the observed values of N_0 and N_p, therefore, cannot be attributed to the degeneracy of an electron gas.

MODEL FOR METAL–AMMONIA SOLUTIONS

It is possible to remove the discrepancy between the electrical and magnetic data by assuming the existence of two different diamagnetic species. Such an assumption makes it possible to remove the restrictions (6) and (9) which contradict the experimental results and thus offers a possibility for reasonable explanation of the data. It is shown later that both electrical and magnetic properties of sodium–ammonia solutions can indeed be explained if such an assumption is made. We chose to term the additional species an M' center and imagine it as consisting of an electron trapped in the field of an M center, but the prime requirement of the center is that it be diamagnetic and negatively charged.

[19] E. Mooser, Phys. Rev. **100**, 1589 (1955).
[20] E. Sonder and D. K. Stevens, Phys. Rev. **110**, 5, 1027 (1958).

We thus have a model based on the following five species in equilibrium: metal ions, e centers, M centers, M_2 centers, and the additional M' centers. The first four have already been postulated in the models of BLA and Deigen. We introduce here the additional concept of M' centers to reconcile the electrical and magnetic data. At this point we make no assumptions regarding the exact structure of the various centers and confine our discussion to the numerical calculation of the equilibrium constants for the various processes from the electrical and magnetic data.

In terms of the model chosen, two types of equilibrium processes can be distinguished: the association of M centers into M_2 centers, and the interaction of electrons with the various traps such as metal ions, M centers and polarization centers in the dielectric. The association of two M centers may be regarded as a chemical reaction to which the mass-action law may be applied:

$$M_2 \rightleftarrows 2M \qquad N_1^2/N_D = K_D, \tag{10}$$

where N_1 and N_D are the concentrations of M and M_2 centers and K_D is the equilibrium constant.

Applying the mass-action law to the dissociation of M centers and M' centers, we can write

$$M \rightleftarrows M^+ + e \qquad N_0 N_e f_\pm^2/N_1 = K_1, \tag{11}$$

$$M' \rightleftarrows M + e \qquad N_1 N_e/N_2 = K_2, \tag{12}$$

where N_0, N_2 and N_e are the concentrations of metal ions, M' centers, and e centers, respectively (the numerical subscripts refer to the number of electrons attached to the metal ion), and $f_\pm$ is the mean ionic activity coefficient.

The equilibrium constants K_1 and K_2, and the dimerization constant K_D, can be calculated from known concentrations of local centers; these in turn are estimated from the electrical and magnetic properties. Thus, K_1 was determined at $-33.5°C$ from conductance data at very low concentrations, where the concentration of M_2 and M' centers is negligibly small; K_2 and K_D were then calculated from transference data and from paramagnetic susceptibility. The method used in the calculation is outlined in Appendix A. For sodium solutions at $-33.5°C$ the following values were obtained: $K_1=9.9\times 10^{-3}$, $K_2=9.7\times 10^{-4}$, $K_D=1.9\times 10^{-4}$.

The method of calculating the concentrations of local centers from the known equilibrium constants is outlined in Appendix B. The calculated concentrations are shown in Table I, and in Figs. 2 and 3. Figure 4 shows the comparison between computed and experimental values of N_0 and N_p at $-33.5°C$. The agreement is good, even at the highest concentrations at which the experimental data are available, and the figure may be contrasted with Fig. 1.

TABLE I. Computed concentrations of local centers at −33.5°C.

$K_1=9.88\times10^{-3}$, $K_2=9.68\times10^{-4}$, $K_D=1.85\times10^{-4}$

(concentrations in moles/liter)

C	N_0	N_e	N_1	N_2	N_D	N_p	$f_\pm$
1×10^{-4}	9.910×10^{-5}	9.901×10^{-5}	8.033×10^{-7}	8.217×10^{-8}	3.488×10^{-9}	9.982×10^{-5}	0.8994
2×10^{-4}	1.964×10^{-4}	1.958×10^{-4}	2.897×10^{-6}	5.862×10^{-7}	4.538×10^{-8}	1.987×10^{-4}	0.8626
4×10^{-4}	3.852×10^{-4}	3.813×10^{-4}	9.886×10^{-6}	3.894×10^{-6}	5.283×10^{-7}	3.911×10^{-4}	0.8156
1×10^{-3}	8.994×10^{-4}	8.614×10^{-4}	4.274×10^{-5}	3.804×10^{-5}	9.876×10^{-6}	9.041×10^{-4}	0.7383
2×10^{-3}	1.609×10^{-3}	1.449×10^{-3}	1.070×10^{-4}	1.601×10^{-4}	6.185×10^{-5}	1.556×10^{-3}	0.6732
4×10^{-3}	2.721×10^{-3}	2.210×10^{-3}	2.241×10^{-4}	5.116×10^{-4}	2.714×10^{-4}	2.434×10^{-3}	0.6067
1×10^{-2}	5.203×10^{-3}	3.460×10^{-3}	4.874×10^{-4}	1.742×10^{-3}	1.284×10^{-3}	3.948×10^{-3}	0.5172
2×10^{-2}	8.485×10^{-3}	4.651×10^{-3}	7.979×10^{-4}	3.834×10^{-3}	3.442×10^{-3}	5.449×10^{-3}	0.4469
4×10^{-2}	1.413×10^{-2}	6.189×10^{-3}	1.242×10^{-3}	7.942×10^{-3}	8.341×10^{-3}	7.431×10^{-3}	0.3746
1×10^{-1}	2.925×10^{-2}	9.172×10^{-3}	2.119×10^{-3}	2.008×10^{-2}	2.427×10^{-2}	1.129×10^{-2}	0.2793
2×10^{-1}	5.318×10^{-2}	1.269×10^{-2}	3.090×10^{-3}	4.050×10^{-2}	5.162×10^{-2}	1.578×10^{-2}	0.2127
3×10^{-1}	7.677×10^{-2}	1.551×10^{-2}	3.825×10^{-3}	6.127×10^{-2}	7.907×10^{-2}	1.933×10^{-2}	0.1782
4×10^{-1}	1.002×10^{-1}	1.794×10^{-2}	4.439×10^{-3}	8.229×10^{-2}	1.065×10^{-1}	2.238×10^{-2}	0.1562
5×10^{-1}	1.236×10^{-1}	2.012×10^{-2}	4.979×10^{-3}	1.035×10^{-1}	1.340×10^{-1}	2.510×10^{-2}	0.1407
6×10^{-1}	1.468×10^{-1}	2.209×10^{-2}	5.466×10^{-3}	1.247×10^{-1}	1.615×10^{-1}	2.755×10^{-2}	0.1291
8×10^{-1}	1.928×10^{-1}	2.556×10^{-2}	6.333×10^{-3}	1.672×10^{-1}	2.168×10^{-1}	3.190×10^{-2}	0.1127
1×10^{0}	2.382×10^{-1}	2.858×10^{-2}	7.101×10^{-3}	2.096×10^{-1}	2.725×10^{-1}	3.568×10^{-2}	0.1015
1.2	2.829×10^{-1}	3.124×10^{-2}	7.800×10^{-3}	2.517×10^{-1}	3.288×10^{-1}	3.904×10^{-2}	0.0934

TRANSITION TO METALLIC STRUCTURE

The arguments presented above apply only as long as we are dealing with local centers. As the metal concentration is increased, a point is eventually reached when the electronic wavefunctions on the various centers begin to overlap and merge into a band. The concept of local centers then breaks down, and a description in terms of a degenerate electron gas becomes appropriate. It is desirable to estimate the concentration at which the band formation should begin. A calculation of this sort was made by Baltensperger[21] for impurity centers in semiconductors. A hydrogenlike potential was assumed for the separate centers located on a regular lattice, and the energy levels for this system were obtained as a function of the separation between the centers. According to this calculation the individual levels begin to merge into a narrow band when the concentration of impurities N_i is such that

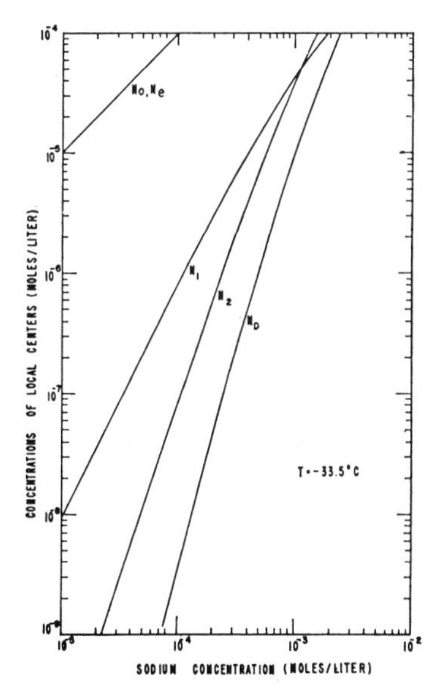

FIG. 2. Calculated concentrations of local centers in very dilute solutions. N_0 is the concentration of sodium ions; N_e, the concentration of e centers; N_1, the concentration of M centers; N_2, the concentration of M' centers; and N_D, the concentration of M_2 centers.

$$r_i \approx 4a^*,$$

where

$$(4/3)\pi r_i^3 = 1/N_i \tag{13}$$

and a^* is the Bohr radius of the electron bound to the center. The width of the band increases rapidly with further increase in concentration.

If the model can be applied approximately to metal–ammonia solutions, we can set

$$N_i \approx N_e + N_1 + N_2 + N_D. \tag{14}$$

Now at high concentrations the M_2 centers predominate. If we assume that the Bohr radius of an electron in an M_2 center is approximately the same as in an M center, then $a^* \approx 2.5$ Å.[22] A band should form, therefore, when $N_i \approx 0.5M$, or when the total metal concentration is about $1M$. This concentration may be taken as the approximate limit to the applicability of the local-center model. In the case of a random arrangement of local centers, we might expect the broad-

[21] W. Baltensperger, Phil. Mag. **44**, 1355 (1953).
[22] J. Jortner, J. Chem. Phys. **34**, 678 (1961).

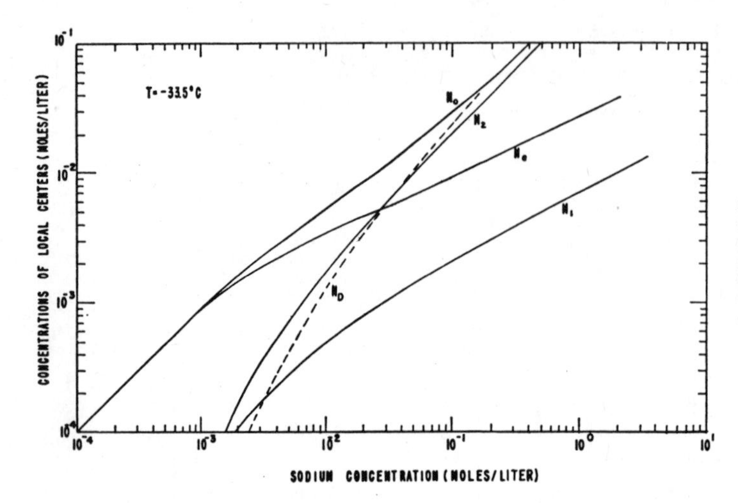

FIG. 3. Calculated concentrations of local centers at low and intermediate concentrations. N_0 is the concentration of sodium ions; N_e, the concentration of e centers; N_1, the concentration of M centers; N_2, the concentration of M' centers; and N_D the concentration of M_2 centers.

ening of energy levels to begin at somewhat lower concentrations.

The above estimate agrees qualitatively with the observed concentration dependence of the temperature coefficient of conductivity, which has a sharp maximum at 0.8M. The transition to the metallic conduction and the behavior of the conductivity in this region will be discussed in more detail in another paper.[23]

BINDING ENERGIES AND TEMPERATURE DEPENDENCE OF EQUILIBRIUM CONSTANTS

Thus far, no assumptions have been made regarding the detailed structure of the various centers. It was merely necessary to assume that the interaction between the two electrons in M' centers and M_2 centers was sufficiently strong for these species to be diamagnetic.

In order to obtain a clearer picture of the trapping processes in these solutions it seems desirable to express the equilibrium constants K_1 and K_2 in terms of the various trapping energies, which can then be determined from experimental data. We accomplish this by considering the local centers as being analogous to defects in solids in which electrons can be captured. The view taken here is similar to that employed by Landsberg[24] for multiple impurity levels in semiconductors. Here we have three distinct kinds of trapping sites: the metal ions and M centers produced by the dissociation of M_2 centers, and the centers of dielectric polarization or e center sites. We shall refer to the latter as "holes". For the purpose of the present discussion we shall assume a definite equilibrium concentration N_h of such holes, distributed on N_{h0} possible sites, each hole being capable of capturing at most one electron. We shall neglect the effect of trapped electrons upon the hole–liquid equilibrium and assume that $N_e \ll N_h \ll N_{h0}$. The number of holes is then given by

$$N_h = N_{h0} \exp(-U/kT) \tag{15}$$

where U is the increase in the potential energy of a site associated with the formation of a hole.

The equilibrium constants K_1 and K_2 can be expressed approximately in terms of the ground-state energies E_1, E_2, and E_h of electrons in M centers, M'

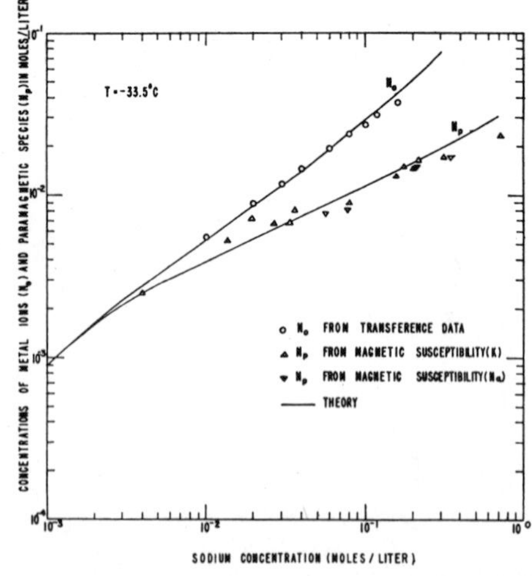

FIG. 4. Concentrations of metal ions (N_0) and paramagnetic species (N_p) at −33.5°C. Circles, N_0 obtained from transference data of Dye *et al.*;[14] triangles N_p obtained from paramagnetic susceptibility data of Hutchison and Pastor.[15] Solid lines, theory.

[23] E. Arnold and A. Patterson, J. Chem. Phys. **41**, 3098 (1964), Paper II.

[24] P. T. Landsberg, *Semiconductors and Phosphors*, edited by M. Schon and H. Welker (Interscience Publishers, Inc., New York, 1958), p. 45.

centers, and e centers, respectively, and the corresponding degeneracies g_1, g_2, and g_h:

$$K_1=\frac{N_0 N_e f_\pm^2}{N_1}=\frac{g_h}{g_1}N_{h0}\exp\left[\frac{E_1-E_h-U}{kT}\right], \quad (16)$$

$$K_2=\frac{N_1 N_e}{N_2}=\frac{g_h g_1}{g_2}N_{h0}\exp\left[\frac{E_2-E_1-E_h-U}{kT}\right]. \quad (17)$$

If we assume that the first electron can have either direction of spin, and that the second electron (in the case of the M' center) has to pair it, then $g_2=1$, and $g_1=g_h=2$.

The value of N_{h0} and the relative positions of the various energy levels in Eqs. (16) and (17) were estimated from the dependence of K_1 and K_2 on temperature.[25] It must be emphasized that the calculations are only approximate, because transference data are available at one temperature only, and the available magnetic data are quite scattered. However, from the temperature dependence of the paramagnetic susceptibility we have estimated that $N_{h0}=10$ moles/liter, which corresponds exactly to one hole site for every four molecules of ammonia.[25] (It is interesting to note here that Bennett *et al.*[26] recently concluded from the ESR spectra of sodium solutions in ice that the electron interacts equally with four protons while the interaction with the alkali metal nucleus is negligible.)

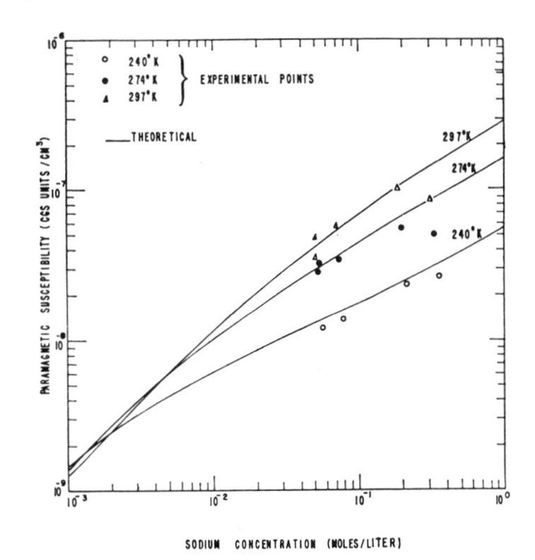

Fig. 6. Experimental and theoretical values of paramagnetic susceptibility in sodium/liquid-ammonia solutions at three temperatures. The experimental points are taken from Ref. 15.

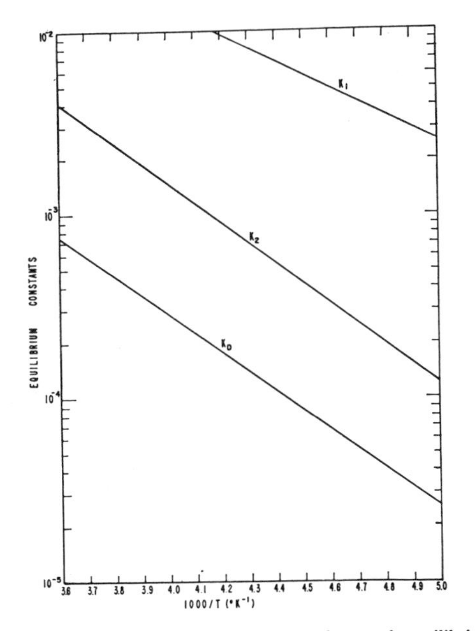

Fig. 5. Calculated temperature dependence of equilibrium constants, K_1, K_2 and K_D, for sodium/liquid ammonia solutions.

The resulting expressions for the three equilibrium constants are as follows (with energies expressed in electron volts):

$$K_1=10\exp(-0.14/kT), \quad (18)$$

$$K_2=40\exp(-0.22/kT), \quad (19)$$

$$K_D=4.1\exp(-0.21/kT). \quad (20)$$

Figure 5 shows the calculated temperature variation of the equilibrium constants for sodium solutions. The calculated and experimental values of the paramagnetic susceptibility at three different temperatures are shown in Fig. 6; the fit of the theoretical curves to the experimental data may serve as a rough indication of the reliability of the calculated equilibrium constants.

DISCUSSION

The good agreement between the present theory and experimental magnetic and transference data lends support to the model chosen for metal–ammonia solutions. By assuming the existence of two different diamagnetic species it has been possible to eliminate the inconsistencies which appeared to exist between the results obtained from electrical and magnetic measurements. The improved agreement between the electrochemical and magnetic results can be seen by comparing Figs. 1 and 4. Although the good agreement lends support to the model chosen, there is no assurance that, for example, an additional diamagnetic species is not involved; indeed, as soon as one has postulated M_2 and M' the possibility of e_2 becomes more likely. All that one can say with the data available is that with the addition of

[25] E. Arnold, Ph.D. dissertation, Yale University, 1963.
[26] J. E. Bennett, B. Mile, and A. Thomas, Nature **201**, 919 (1964).

a second diamagnetic species we have the simplest model which will allow adequate correlation of experiment and theory.

It is interesting to examine the relationship between the various binding energies in somewhat more detail. A comparison of the relative magnitudes of the exponential terms in Eqs. (18) and (19) with the equilibrium expressions (16) and (17) shows that U is the predominant term in the expressions for the equilibrium constants, and that the binding energies of electrons in e centers, M centers and M' centers probably do not differ by much more than kT. It would therefore appear that the main contributions to the equilibrium expressions (16) and (17) arise from N_{h0} and U, i.e., from the formation of the holes. Thus, our interpretation of the electrical and magnetic data leads to the conclusion that it is the polarization of the medium which gives these solutions their main character, and the other interactions which characterize the various centers are considerably weaker.

We would expect, therefore, the excited states for electrons in all species to be nearly identical and characteristic of the medium itself. This might explain why the shape of the optical absorption curve in these solutions is nearly independent of the particular alkali metal and its concentration, and that Beer's law is generally obeyed. It seems preferable, however, to treat the various local centers as distinct species, similar to impurity centers in crystals, rather than to picture them as "ion pairs" or similar aggregates.

It should be emphasized that the above conclusions follow directly if one assumes the existence of the five species postulated earlier in this paper. The calculated results have been obtained from the electrical and magnetic data alone, without resort to speculations about the exact nature of the electron's interaction with the solvent or with the metal.

It would be highly desirable to obtain additional magnetic data and transference data at several temperatures in order to arrive at a more reliable estimate of the temperature dependence of the equilibrium constants for metal-ammonia systems.

ACKNOWLEDGMENT

The support of this work by the National Science Foundation is gratefully acknowledged.

APPENDIX A. CALCULATION OF EQUILIBRIUM CONSTANTS

According to our model, and using symbols defined in the text, we have the following relations:

$$K_1 = N_0 N_e f_\pm^2 / N_1, \qquad (A1)$$

$$K_2 = N_1 N_e / N_2, \qquad (A2)$$

$$K_D = N_1^2 / N_D, \qquad (A3)$$

$$N_0 = N_e + N_2, \qquad (A4)$$

$$C = N_0 + N_1 + N_2 + 2N_D, \qquad (A5)$$

$$N_p = N_1 + N_e. \qquad (A6)$$

K_1 can be evaluated from the conductivity in very dilute solutions. Below about $2\times10^{-4}M$ the equilibria (A2) and (A3) contribute very little, and only (A1) need be considered. A trial value of $K_1 = 7.2\times10^{-3}$ was assumed, according to Evers' calculations,[4] which were based on equilibria (A1) and (A3) only. K_2 and K_D were then computed from the experimental values of N_0 and N_p shown in Fig. 1, as follows:

$$N_1 = N_p (K_1 / N_0 f_\pm^2 + 1)^{-1}, \qquad (A7)$$

$$N_e = N_p - N_1, \qquad (A8)$$

$$N_2 = N_0 - N_e, \qquad (A9)$$

$$N_D = (C - N_0 - N_1 - N_2)/2, \qquad (A10)$$

$$K_2 = N_1 N_e / N_2, \qquad (A11)$$

$$K_D = N_1^2 / N_D. \qquad (A12)$$

The activity coefficients were calculated from the Debye-Hückel theory:

$$f_\pm = \exp[e^2\kappa / 2D_s kT(1+\kappa a)] \qquad (A13)$$

where

$$\kappa^2 = 8\pi N_0 e^2 \times 6.02\times10^{20} / D_s kT \qquad (A14)$$

A value of 4 Å was arbitrarily chosen for a. It was not considered worthwhile to use a as an adjustable parameter because, in the region where it might affect the results, the validity of the Debye-Hückel theory becomes questionable. The dielectric constant was obtained from the formula[27]:

$$D_s = 66.2 \exp(-T/217). \qquad (A15)$$

An improved value for K_1 was obtained from the conductivity data in very dilute solutions. From known values of the equilibrium constants it is possible to compute the concentrations of the various species (as outlined in Appendix B) and, with the aid of the Onsager-Kim theory, the equivalent conductance. This was done, using the following data:

$$T = -33.5°\text{C}, \quad L_+^0 = 137, \quad L_-^0 = 885,$$

$$\eta = 0.00252 \text{ P}, \quad D_s = 22.$$

The result is shown as the broken line in Fig. 7. The experimental points in the figure represent three series of experiments reported by Kraus.[17] The scatter of points at the lowest concentrations appears to be caused by the decomposition of the solution, which produces the negligibly conducting sodium amide. It can be seen that the successive runs resulted in progressively higher conductivity readings which, according to Kraus, was due to the improved experimental technique in the later runs. Accordingly, K_1 was determined by fitting the conductance data in Series XIX of Ref. 17. The solid curve in Fig. 7 was obtained using $K_1 = 9.9\times10^{-3}$, with previously calculated K_2 and K_D.

[27] W. L. Jolly, Chem. Rev. **50**, 351 (1952).

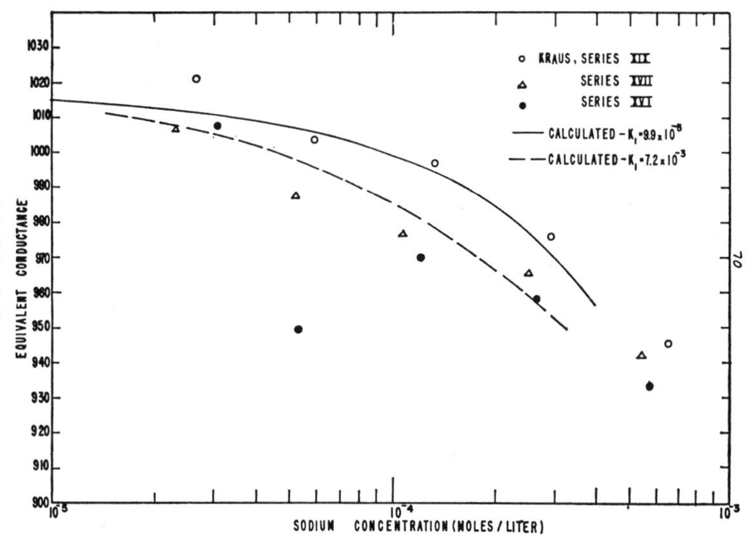

FIG. 7. Equivalent conductance in very dilute solutions of sodium in liquid ammonia. Experimental points represent data of Kraus,[17] corrected for the density of liquid ammonia. Solid line, calculated, using Onsager-Kim theory and $K_1=0.0099$. Broken line, similar calculation with $K_1=0.0072$.

The constants K_2 and K_D were recomputed from Eqs. (A7) through (A12) using the new value of K_1. The results for the various concentrations are shown in the table below:

C (mole/liter)	$K_2 \times 10^4$	$K_D \times 10^4$
0.01	9.2	2.3
0.02	9.5	2.2
0.04	9.6	2.0
0.10	10.1	1.8
0.16	10.0	1.4.

Although the calculated values appear to vary systematically, the apparent trend is difficult to interpret in view of the inaccuracies which arise from subtractions involving several experimental quantities in Eqs. (A8)-(A10). Average values were therefore arbitrarily assigned to the two equilibrium constants. These are (at $-33.5°C$): $K_2=9.7\times10^{-4}$, $K_D=1.9\times10^{-4}$.

APPENDIX B. CALCULATION OF THE EQUILIBRIUM CONCENTRATIONS OF LOCAL CENTERS

A direct calculation of the concentrations of the various local centers for an arbitrary metal concentration would be very difficult. It is much easier to start with a value of N_0 and from it to calculate the concentrations of the other centers and hence C. The concentrations of the different centers corresponding to the given N_0 are calculated from the following equations:

$$N_e=[(1+2bN_0)^{\frac{1}{2}}-1]b^{-1},$$

where

$$b=2f_\pm^2 N_0/K_1K_2, \quad \text{(B1)}$$

$$N_1=N_0N_e f_\pm^2/K_1, \quad \text{(B2)}$$

$$N_2=N_1N_e/K_2, \quad \text{(B3)}$$

$$N_D=N_1^2/K_D, \quad \text{(B4)}$$

$$C=N_0+N_1+N_2+2N_D. \quad \text{(B5)}$$

In order to obtain concentrations of local centers for any desired value of C an iteration process has been devised and programmed for the IBM 1620 computer. Details are given by Arnold.[25]

Studies of Spin Densities in Metal–Ammonia Solutions

In the following two papers, O'Reilly describes his experimental and theoretical studies of Knight shifts in metal–ammonia solutions. He extended the measurements of Acrivos and Pitzer to several alkali metals and more dilute solutions, and he showed that the data could be explained by assuming M^- to be the only spin-paired species.

Reprinted from THE JOURNAL OF CHEMICAL PHYSICS, Vol. 41, No. 12, 3729–3735, 15 December 1964
Printed in U. S. A.

45

Knight Shifts and Relaxation Times of Alkali-Metal and Nitrogen Nuclei in Metal–Ammonia Solutions*

D. E. O'REILLY
Argonne National Laboratory, Argonne, Illinois

(Received 17 July 1964)

Knight shifts of ^{7}Li, ^{23}Na, ^{87}Rb, ^{133}Cs, and ^{14}N in alkali-metal–ammonia solutions have been measured at 300°K and in the concentration range 0.03 to 1 mole liter^{-1}. ^{23}Na NMR has been observed at Na concentrations down to 0.003 mole liter^{-1} at 300°K and in the range 0.03 to 4 mole liter^{-1} at 274° and 240°K. ^{14}N shifts are reported for concentrations from 0.008 to 4 mole liter^{-1} at 300° and 240°K. Linewidths of ^{87}Rb and ^{133}Cs have been obtained at 300°K in the concentration range 0.03 to 0.8 mole liter^{-1}. At 0.35 mole liter^{-1} and 300°K the shifts of Li, Na, Rb, and Cs are 9, 72, 450, and 900 ppm, respectively, to lower magnetic field. Shifts of ^{14}N are nearly independent of metal at 300°K.

The concentration and temperature dependence of ^{23}Na Knight shift data at concentrations up to 0.4 mole liter^{-1} may be quantitatively interpreted by the following reactions:

$$M \rightleftarrows M^{+}+e^{-}, \tag{1}$$

$$M+e^{-} \rightleftarrows M^{-}, \tag{2}$$

where M represents the "monomer" species and (2) is the spin-pairing reaction. The ^{14}N spin density at unpaired electron is essentially independent of alkali metal and concentration up to 0.6 mole liter^{-1} and equal to $0.88 \pm 0.11\ a_0^{-3}$ at 300°K. The electron spin density at ^{23}Na is calculated to be $\geq 8 \times 10^{-3}\ a_0^{-3}$ approximately independent of temperature and concentration.

The lifetimes of the Rb and Cs monomers at 300°K and 0.06 mole liter^{-1} are ≤ 3 and ≤ 4 $\mu\mu$sec, respectively, as determined from NMR linewidth measurements. The monomer lifetime decreases sharply with increase in metal concentration.

I. INTRODUCTION

ALKALI-metal–ammonia solutions exhibit properties which are anomalous in contrast to those of solutions of inorganic salts or molecular solids in polar solvents. Dilute solutions are blue in color, paramagnetic, and have unusual colligative, transport and electrical properties. More concentrated solutions are bronze in color and have electrical conductivities characteristic of metals. The properties of the solutions are summarized in recent review articles by Symons,[1] Jolly,[2] and Das.[3]

Kaplan and Kittel,[4] elaborating upon the views of Kraus,[5] Freed, and Sugarman,[6] and Ogg,[7] proposed a model for the solutions in which the valence electrons of the alkali metal are completely dissociated and trapped in vacancies or "cavities" in the solvent. The cavity model can be used to account quantitatively for the density,[8] static and rf magnetic susceptibility,[9,10] and paramagnetic relaxation times [11a,11b,12] of dilute potassium–ammonia solutions, although the qualitative LCAO wavefunction originally proposed must be revised. Detailed quantum-mechanical calculations of the unpaired electronic wavefunction for the cavity model has been made by Jortner[13] based on a continuum approximation for the solvent. The results of these calculations are in reasonable agreement with optical and thermodynamic data of dilute solutions.

The cavity model, however, cannot be used to ac-

* Based on work performed under the auspices of the U. S. Atomic Energy Commission.

[1] M. C. R. Symons, Quart. Rev. (London) **13**, 99 (1959).
[2] W. L. Jolly, Progr. Inorg. Chem. **1**, 235 (1960).
[3] T. P. Das, Advan. Chem. Phys. **4**, 303 (1961).
[4] J. Kaplan and C. Kittel, J. Chem. Phys. **21**, 1429 (1953).
[5] C. A. Kraus, J. Am. Chem. Soc. **30**, 1197 (1908); see also J. Chem. Educ. **30**, 86 (1953), for references to earlier work.
[6] S. Freed and N. Sugarman, J. Chem. Phys. **11**, 354 (1943).
[7] R. A. Ogg, Jr., J. Chem. Phys. **14**, 295, 114 (1946).
[8] C. A. Hutchison Jr. and D. E. O'Reilly, J. Chem. Phys. **34**, 163 (1961).
[9] C. A. Hutchison Jr. and R. C. Pastor, J. Chem. Phys. **21**, 1959 (1953).
[10] D. E. O'Reilly, Ph.D. thesis, University of Chicago, Chicago, Illinois, 1955, Microfilm Order No. T-2964.
[11] (a) D. E. O'Reilly, J. Chem. Phys. **35**, 1856 (1961). (b) D. E. O'Reilly, Phys. Rev. Letters **11**, 545 (1963).
[12] V. L. Pollak, J. Chem. Phys. **34**, 864 (1961).
[13] J. Jortner, J. Chem. Phys. **30**, 839 (1959).

count for the presence of an appreciable Knight shift for the nuclear magnetic resonance of ^{23}Na in dilute as well as concentrated sodium–ammonia solutions reported by McConnell and Holm.[14] The ^{23}Na shift is qualitatively consistent with a model proposed for the solutions by Becker, Lindquist, and Alder[15] which postulates that electrons are trapped on metal ions as well as in cavities. Definite species, namely the "monomer" M consisting of metal ion plus electron and the "dimer" M_2 consisting of two (paired) electrons and two metal ions, were postulated to exist. This model has been used to interpret the rf susceptibility data of Hutchison and Pastor[9,15] and the electrical conductivity data of Kraus.[5,16] The sets of equilibrium constants derived from these two independent sources are not in good agreement. Quantum-mechanical calculation of the unpaired electron wavefunction for detailed monomer models has been made by Blumberg and Das.[17] Agreement of these calculations with the then available ^{23}Na, ^{14}N, and ^{1}H Knight shift data and derived electron spin densities was not completely satisfactory.

The present measurements and model calculations described in the following paper were undertaken in an attempt to clarify the nature of the interaction between the unpaired electrons and metal ions. Knight-shift measurements of Li, Na, K, Rb, and Cs solutions and also linewidth measurements of ^{87}Rb and ^{133}Cs resonances of the corresponding metal solutions are reported in Sec. IV; in Sec. VA the metal shifts are interpreted in terms of association and spin-pairing reactions and lower limits on the electron spin density at monomer metal nuclei are derived; in Sec. VB ^{14}N shifts are discussed; in Sec. VC "monomer" lifetimes are calculated from relaxation time data and related to the equilibrium constants calculated in Sec. VA; in Sec. VI the static and dynamic aspects of the electron–metal-ion interaction inferred from the present data are summarized.

II. EXPERIMENTAL

Solutions of sodium, potassium, rubidium and cesium in liquid ammonia were prepared by distillation of the metals in high vacuum. Dried ammonia was then condensed on the distilled metal and the resulting solution was allowed to run down into a capsule with a sharp constriction which was then sealed off. Solutions of lithium in liquid ammonia were prepared by condensation of ammonia on chunks of lithium situated on a fritted glass disk. The lithium solution was allowed to run through the disk into a capsule which was sealed at a constriction. After measurement, solutions were analyzed by exploding the samples in a stainless steel bomb filled with distilled water. The alkaline solution was titrated with excess hydrochloric acid, digested on a steam bath to remove carbonate, and then back-titrated with sodium hydroxide. The glass of the exploded sample was recovered and weighed; from the weight of the original sample and the weight of glass the weight of the solution was obtained.

Nuclear resonances were observed with a Varian Associates V–4200B wideline NMR spectrometer at magnetic fields in the vicinity of 10 kG. A helipot used to scan the field was calibrated with a Hewlett–Packard 524 C/D frequency counter. For intense resonances of ^{14}N or alkali metal nuclei of concentrated metal solutions shifts were measured by direct visual observation on an oscilloscope while for weaker signals of alkali metal nuclei in more dilute solutions the field was slowly scanned and resonances displayed on a graphic recorder. The Knight shift k is defined as $(H_r - H_s)/H_r$ where H_r is the magnetic field for which resonance occurs in the reference solution and H_s the field for which resonance occurs in the metal–ammonia solution at fixed frequency. For lithium and sodium, solutions of the corresponding nitrates in liquid ammonia were employed as reference solutions while for rubidium and cesium, solutions of the corresponding iodides in liquid ammonia were used for reference. Solutions of sodium rubidium and cesium salts in water were found to be shifted by 9, 96, and 161 parts per million (ppm), respectively, to higher magnetic field relative to the corresponding salts in liquid ammonia. Experimental error is about ± 4 ppm for lithium and sodium, about ± 10 ppm for rubidium and cesium and about ± 3 ppm for nitrogen shifts.

Linewidth ΔH_{ms} were measured between points of maximum slope of the derivative of the absorption mode signal. This method could not be applied to ^{7}Li or ^{23}Na resonances due to combined distortion effects associated with the audio modulation, the period of which was shorter than or comparable to the relaxation time, and the inhomogeneity of the applied static magnetic field. A Varian V–4547 continuous gas flow, variable temperature EPR accessory which was modified for use with the NMR apparatus was used for measurements at 273° and 240°K. The temperature of a sample was maintained within ± 1°C during the course of a measurement. Sample tube outside diameters were 7 mm for measurements performed at low temperature; 15 mm and 18 mm diam tubes were used in measurements at room temperature.

Linewidths and Knight shifts were found to be independent of resonance frequency over the range of frequencies available. Linewidths of ^{87}RbI and ^{133}CsI in liquid NH_3 were much smaller than the widths of the corresponding nuclei in a metal–ammonia solution, particularly for ^{133}Cs.

ESR measurements of the spin susceptibility of Li, Na, K, Rb, and Cs solutions were performed at 7

[14] H. M. McConnell and C. H. Holm, J. Chem. Phys. **26**, 1517 (1957).
[15] E. Becker, R. H. Lundquist, and B. J. Alder, J. Chem. Phys. **25**, 971 (1956).
[16] E. C. Evers and P. W. Frank, J. Chem. Phys. **30**, 61 (1959).
[17] W. E. Blumberg and T. P. Das, J. Chem. Phys. **30**, 251 (1959).

Mc/sec and 300°K. Helmholtz coils were used to supply a static magnetic field homogeneous to ±1 mG over the sample volume. The Varian wideline spectrometer was used to record the absorption derivative at a low rf level to avoid saturation. With the aid of the Kramers–Kronig relations the relative spin susceptibilities were calculated.[9] The ^{14}N nuclear resonance of each sample was also scanned and used as intensity reference for determination of absolute spin susceptibilities.

III. THEORY

A. Knight Shifts

The Knight shift of a nuclear species in a metal–ammonia solution will be considered to arise from the isotropic hyperfine interaction between unpaired electrons and the nuclei of interest. Anisotropic (dipolar or pseudodipolar) interactions need not be taken into account here since they will average to zero in a time short compared to the reciprocal of the shift expressed in radians per second and hence will not contribute to the observed shift. The Knight shift for a nuclear species N may be used to calculate the total average *electron* spin density at the nucleus N, $\langle |\psi(N)|^2\rangle_{Av}$, by means of the following relationship[14,18]:

$$k(N)=\tfrac{8}{3}\pi(\chi_v^{sp}/N_e)\langle|\psi(N)|^2\rangle_{Av}, \quad (1)$$

where $k(N)$ is the Knight shift, χ_v^{sp} the electron spin susceptibility and N_e is the number of unpaired electrons per unit volume. Equation (1) is generally valid, independent of the concentration of unpaired electron spins, and its derivation does not require that the electrons be treated in an independent particle approximation.[18]

It is also useful to consider the *nuclear* spin density at an unpaired electron $P(N)$. By its definition $P(N)$ is related to $|\psi(N)|^2$ as follows

$$N_eP(N)=N_n\langle|\psi(N)|^2\rangle_{Av}, \quad (2)$$

where N_n is the number of nuclei per cubic centimeter. For ^{14}N, combining Eqs. (1) and (2), one finds

$$Rk(^{14}N)=\tfrac{8}{3}\pi(\chi_m^{sp}/N_{Av})P(^{14}N) \quad (3)$$

where χ_m^{sp} is the molar spin susceptibility, N_{Av} is the Avogadro number, and R is the ammonia-to-metal mole ratio. In a similar fashion one obtains

$$k(M)=\tfrac{8}{3}\pi(\chi_m^{sp}/N_{Av})P(M) \quad (4)$$

for a metal nucleus M.

Let us assume that the unpaired electron exists both as a "cavity" species which is not undergoing hyperfine interaction with metal nuclei and as a "monomer" species in which the unpaired electron undergoes hyperfine interaction with a single metal nucleus. Denoting the electron density at the monomer metal nucleus by $|\psi(M)|^2$, it follows from Eq. (2) that

$$|\psi(M)|^2=(N_e/N_M)P(M), \quad (5)$$

where N_M is the number of "monomers" per unit volume. Equations (1)–(5) are used in Sec. V to derive electron and nuclear spin densities from the experimental Knight shifts.

B. Relaxation Times

The hyperfine interaction between unpaired electrons and nuclei can contribute significantly to the spin–lattice relaxation time of the metal nucleus. If the hyperfine interaction is strong compared to other relaxation-producing interactions (quadrupolar, dipole–dipole, etc.) the nuclear relaxation time will then be determined by the hyperfine interaction. Let us consider a single nuclear spin **I** interacting with electron spins **S** via the isotropic hyperfine interaction of the following form:

$$\mathcal{H}(t)=\sum_j A_j(t)\mathbf{S}_j\cdot\mathbf{I}. \quad (6)$$

The longitudinal (T_1) and transverse (T_2) nuclear magnetic relaxation times resulting from this interaction may be calculated by the method of Kubo and Tomita[19,11a] using the density matrix technique.[18] One obtains, for T_1 and T_2,

$$1/T_1=\sigma^2(\tau_0+\tau_1), \quad (7)$$

$$1/T_2=2\sigma^2\tau_1, \quad (8)$$

where

$$\sigma^2=[S(S+1)/3\hbar^2]\sum_j\langle|A_j(t)|^2\rangle_{Av},$$

$$\tau_0=a(0),$$

$$\tau_1=a(\omega_s-\omega_I),$$

$$a(\omega)=\mathrm{Re}\int_0^\infty \exp(i\omega\tau)f(\tau)d\tau,$$

and

$$f(\tau)=\langle\sum_j A_j(t+\tau)A_j(t)\rangle_{Av}/\langle\sum_j|A_j(t)|^2\rangle_{Av},$$

where the angle brackets denote time average. ω_s and ω_I are the electron and nuclear resonance frequencies (in radians per second). In the case that $f(\tau)$ is represented by the exponential relaxation function $\exp(-|\tau|/\tau_c)$, where τ_c is the correlation time, Eqs. (7) and (8) become

$$1/T_1=\sigma^2\tau_c\{1+1/[1+(\omega_s-\omega_I)^2\tau_c^2]\}, \quad (9)$$

$$1/T_2=2\sigma^2\tau_c/[1+(\omega_s-\omega_I)^2\tau_c^2]. \quad (10)$$

A_j is directly proportional to the square of the wavefunction at the nucleus N:

$$A_j=-\tfrac{16}{3}\pi(\beta\mu_N/I)|\psi_j(N)|^2,$$

where β is the Bohr magneton and μ_N is the magnetic moment of nucleus N. For metal nuclei in metal–

[18] A. Abragam, *The Principles of Nuclear Magnetism* (Oxford University Press, New York, 1961), p. 199.

[19] R. Kubo and K. Tomita, J. Phys. Soc. Japan **9**, 888 (1954).

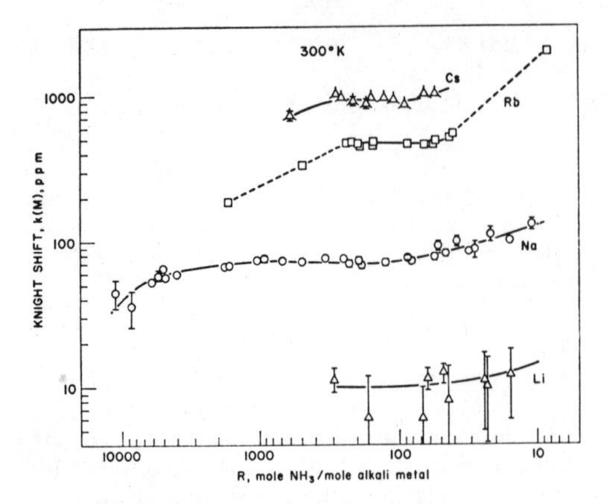

FIG. 1. Knight shift of Li, Na, Rb, and Cs–NH_3 solutions versus the mole ratio of NH_3 to alkali metal, 300°K.

ammonia solutions let us assume that a fraction x_0 of metal nuclei are in the "monomer" form and further that $(\omega_s-\omega_I)\tau_c \ll 1$. Then one finds that

$$\frac{1}{T_1}=\frac{1}{T_2}=\frac{2S(S+1)}{3\hbar^2}\left(\frac{16\pi}{3}\frac{\beta\mu_N}{I}\right)^2\langle|\psi(\mathrm{M})|^2\rangle_{\mathrm{Av}}|\psi(\mathrm{M})|^2\tau_c, \tag{11}$$

where $\langle|\psi(\mathrm{M})|^2\rangle_{\mathrm{Av}}$ is the time-average electron density at a metal nucleus and $|\psi(\mathrm{M})|^2$ is the electron density at the metal nucleus in the monomer. The relation

$$\langle|\psi(\mathrm{M})|^2\rangle_{\mathrm{Av}}=x_0|\psi(\mathrm{M})|^2 \tag{12}$$

has been used in deriving Eq. (11).

The predominant relaxation mechanism of alkali metal nuclei in solutions of salts in liquid ammonia is via the quadrupolar interaction although the dipolar interaction may contribute significantly for Li and Cs. Since these linewidths are much sharper than those in the corresponding metal–ammonia solutions the details of the line-broadening mechanism need not be considered here.

If the monomer is regarded as a unit possessing a high degree of symmetry such as spherical or cubic as discussed in the following paper then nuclear–electron interactions of the form $\mathbf{I}\cdot\mathbf{T}^{(2)}\cdot\mathbf{S}$, where $\mathbf{T}^{(2)}$ is a symmetric second-rank tensor of zero trace, are zero. Fluctuations will produce nonzero components of $\mathbf{T}^{(2)}$ which will contribute to the relaxation rate. This contribution is estimated to be small compared to that of the isotropic part of the hyperfine interaction.

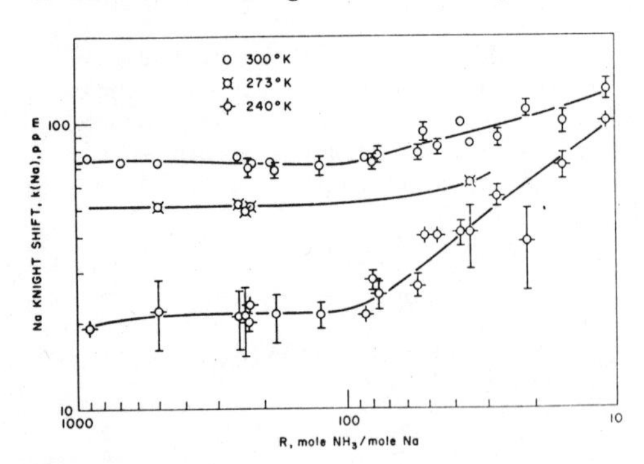

FIG. 2. Temperature dependence of the Knight shift of Na–NH_3 solutions, 240°–300°K.

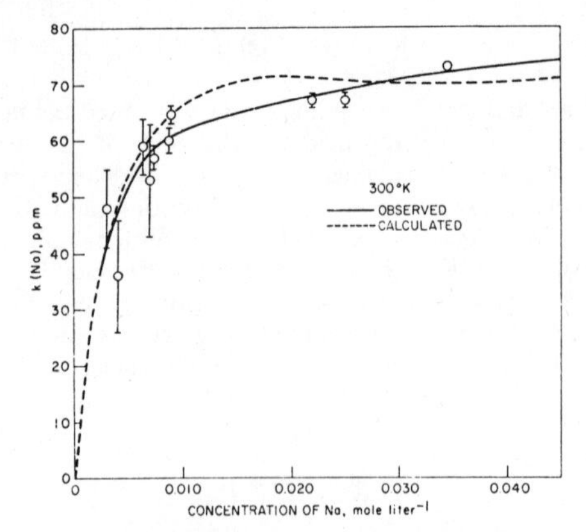

FIG. 3. Concentration dependence of the ^{23}Na Knight shift k(Na) in dilute solutions at 300°K. The dashed line represents k(Na) calculated from equilibrium constants determined from data for more concentrated solutions.

IV. RESULTS

Measured shifts of ^{7}Li, ^{23}Na, ^{87}Rb, and ^{133}Cs at 300°K are shown in Fig. 1, wherein the Knight shift of the metal nucleus k(M) is plotted versus R, the NH_3-to-metal mole ratio; both variables are on a logarithmic scale. The quantity k(M) varies by two orders of magnitude between lithium and cesium. The shifts for lithium, sodium, rubidium, and cesium at $R=100$ ($\sim$0.3 mole liter^{-1}) are 9, 72, 450, and 900 ppm,

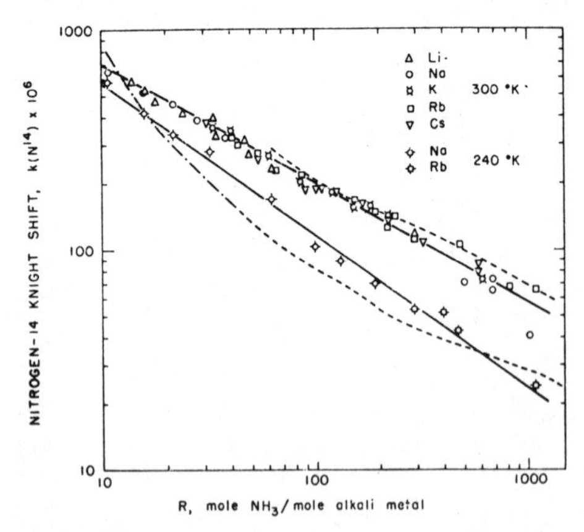

FIG. 4. Knight shift of ^{14}N in metal–ammonia solutions at 240° and 300°K. The dashed curves – – – are $\chi_m^{sp}/R\times10^8$ for K–NH_3. The extrapolated portion —•—• at 240°K is based on an estimate of χ_m^{sp}/R for concentrated solutions.

respectively. The effect of temperature on the ^{23}Na shifts of sodium solutions is given in Fig. 2 in which $k(\mathrm{Na})$ at 240° and 300°K is plotted versus R. In dilute solutions $k(\mathrm{Na})$ decreases from about 70 ppm at 300°K to about 20 ppm at 240°K. The results at 240°K for $R>100$ are somewhat larger than those reported by Acrivos and Pitzer[20] at 243°K. A decrease in the Knight shift is observed in very dilute Na solutions as shown by the data of Fig. 3. Also given in Fig. 3 is the dependence of $k(\mathrm{Na})$ on concentration calculated from equilibrium constant determined from spin susceptibility data and $k(\mathrm{Na})$ at higher concentrations as described in the following section. Values of $k(^{14}\mathrm{N})$ for various metals at 300° and 240°K versus R are presented in Fig. 4. The quantity $\chi_m^{sp}/R\times10^8$ for K solutions is given by the dashed lines of Fig. 4

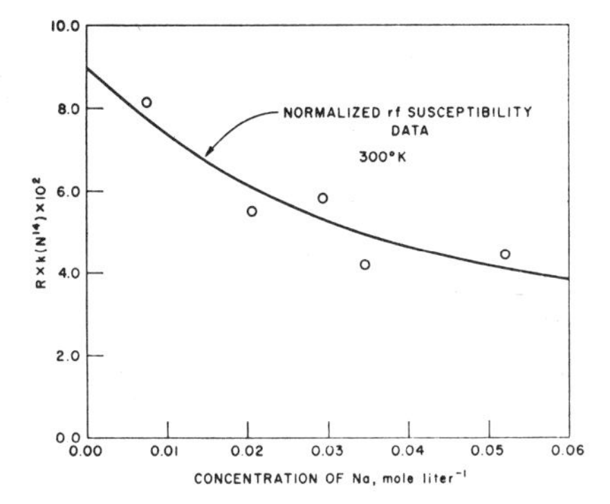

FIG. 5. Values of $R\times k(^{14}\mathrm{N})$ for dilute $Na\text{-}NH_3$ solutions at 300°K. The solid line is χ_m^{sp} for $K\text{-}NH_3$ solutions normalized to yield the best fit to the Knight shift data.

at 300°K and 240°K. The quantity $Rk(^{14}\mathrm{N})$ is plotted in Fig. 5 for dilute $Na\text{-}NH_3$ solutions versus concentration of Na in mole liter^{-1}; the quantity χ_m^{sp} for $K\text{-}NH_3$ solutions, normalized to give the best fit to the experimental points is also plotted. Linewidths between points of maximum slope ΔH_{ms} for ^{87}Rb and ^{133}Cs resonances at 300°K are shown in Fig. 6.

The spin susceptibilities of Li, Na, and K solutions were found to be the equal at the equal concentration of metal and in agreement with results reported previously[9] for K within ±10%. Spin susceptibilities of several different concentrations of Rb and Cs solutions were considerably higher than the corresponding K solution.

V. DISCUSSION

A. Metal Knight Shifts

The average electron density $\langle|\psi(\mathrm{M})|^2\rangle_{\mathrm{Av}}$ at a metal nucleus M may be calculated using the experimental Knight shifts and Eq. (1). Values so calculated are listed in Table I along with experimental values of the

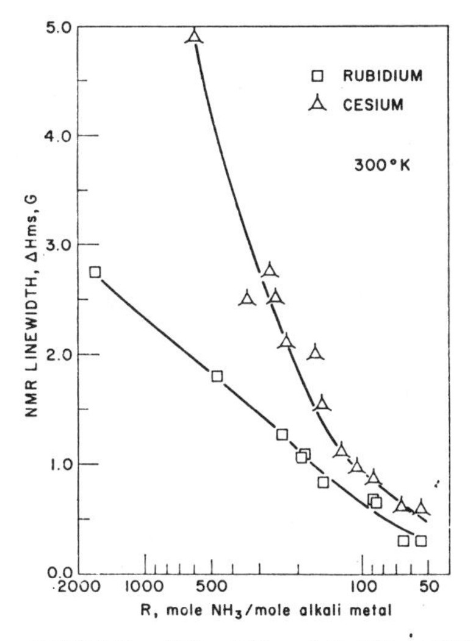

FIG. 6. NMR linewidths of Rb and $Cs\text{-}NH_3$ at 300°K.

electron density at the nucleus in the corresponding free gaseous metal atom $|\psi_0(\mathrm{M})|^2$. For Na solutions $\langle|\psi(\mathrm{Na})|^2\rangle_{\mathrm{Av}}$ is essentially independent of concentration from about 0.03 to 0.40 mole liter^{-1}. The observed concentration and temperature dependence of $k(\mathrm{Na})$ may be interpreted by assuming the existence of monomer and cavity species for the unpaired electrons as postulated by Becker, Lindquist, and Alder.[15] In addition, a pairing reaction must be assumed to completely determine the concentration of the monomer and cavity species. The data were fitted with the following reactions:

$$\underset{x_0}{\mathrm{M}}\rightleftarrows\underset{x_1'}{\mathrm{M}^+}+\underset{x_1}{e^-}, \tag{13}$$

$$\underset{x_0}{\mathrm{M}}+\underset{x_1}{e^-}\rightleftarrows\underset{x_2}{\mathrm{M}^-}, \tag{14}$$

where the mole fraction of M, e^-, etc., are given by x_0, x_1, etc., and

$$K_1=x_1x_1'y_\pm^2M/x_0, \tag{15}$$

$$K_2=x_2/x_1x_0M, \tag{16}$$

where M is the molar concentration of metal and $y_\pm$ is the mean ionic activity coefficient. The activity co-

TABLE I. Unpaired electron densities at metal nuclei.[a]

Metal	Free atom[b] $|\psi_0(\mathrm{M})|^2$	Solution average[c] $|\psi(\mathrm{M})|^2\times10^3$	Monomer $|\psi(\mathrm{M})|^2\times10^3$ (lower limit)
Li	0.231	0.077	1
Na	0.750	0.61	8
Rb	2.34	3.8	50
Cs	3.88	7.7	100

[a] In units of a_0^{-3}.
[b] Calculated from data given by W. D. Knight, Solid State Phys. 2, 93 (1956).
[c] $R=100$, 300°K.

[20] J. V. Acrivos and K. S. Pitzer, J. Phys. Chem. 66, 1693 (1962).

Table II. Equilibrium constants and associated heats and entropies of reaction.

T°K	$K_1 \times 10^2$ (mole liter^{-1})	ΔH_1 (kcal mole^{-1})	ΔS_1 (kcal mole$^{-1}\cdot$K^{-1})	$K_2 \times 10^{-2}$ (mole^{-1} liter)	ΔH_2 (kcal mole^{-1})	ΔS_2 (kcal mole$^{-1}\cdot$K^{-1})
240	3.45±0.68	−1.3±0.9	−12 ±3	37.8±6.3	−7.3±0.4	−14±1
300	2.04±0.33			1.77±0.07		

efficient of the monomer is assumed to be unity. The species M^- has two electrons which are trapped to form a singlet ground state. Mean ionic activity coefficients were calculated from the modified Debye–Hückel expression for which it is necessary to choose a mean value for the sum of the effective ionic diameters of a colliding pair of ions. A value for this quantity of 6.6 A was chosen somewhat arbitrarily for use in the calculations and was not varied to obtain the best fit to the data. Values of K_1 and K_2 were calculated at 300° and 240°K for various values of the constant of proportionality between the sodium Knight shift and the mole fraction x_0 of the monomer species. From Eqs. (4) and (5) one obtains

$$k(\text{Na}) = \tfrac{8}{3}\pi(\chi_{mo}^{sp}/N_{Av}) \,|\psi(\text{Na})|^2 x_0,$$

where χ_{mo}^{sp} is the electron spin susceptibility of a mole of unpaired electrons and $|\psi(\text{Na})|^2$ is the electron density at the electron density at the sodium nucleus in the monomer. For values of $|\psi(\text{Na})|^2 \geq 8\times10^{-3}a_0^{-3}$ the Knight shift and rf susceptibility data could be fit nearly equally well with constant values of K_1 and K_2 but for values of $|\psi(\text{Na})|^2$ about 30% less than $8\times10^{-3}a_0^{-3}$ much poorer agreement resulted. Values of equilibrium constants and associated enthalpies are listed in Table II for $|\psi(\text{Na})|^2 = 8\times10^{-3}a_0^{-3}$.

The data are not well represented by the original pairing reaction proposed by Becker, Lindquist, and Alder,[15] i.e., $M+M \rightleftarrows M_2$ for any choice of the constant of proportionality between $k(\text{Na})$ and x_0. K_2 calculated for this reaction varies by a factor of five at 300°K, and by a factor of 12 at 240°K, over the concentration range of $R=1000$ to $R=100$. The pairing reaction of Kaplan and Kittel[4] can represent the data but has not been used extensively in the present work because of the much stronger dependence of the activity coefficients on concentration and hence greater uncertainty in the derived equilibrium constants.

The behavior of the Knight shift at 300°K for concentrations of Na less than $0.036M$ ($R=1000$) as given by the equilibrium constants of Table II is shown in Fig. 3 along with the available experimental data. The agreement is considered to be satisfactory considering the lack of precise knowledge of the activity coefficients. The values of K_1 and K_2 of Ref. 15 are of the same order of magnitude as those given in Table II since they are also derived from rf susceptibility data.

The electron spin densities $|\psi(\text{M})|^2$ given in Table I are lower limits, although electron resonance linewidths of Rb and Cs solutions[21] indicate these values to be correct within a factor of two at most.

The sharp rise of $k(\text{Na})$ at 240°K with increase in concentration above about 0.4 mole liter^{-1} may be interpreted with a constant $|\psi(\text{M})|^2$ as due to an increase of $f=N_M/N_e$ as well as an increase in χ_m^{sp}. For $R=7$ the estimated spin susceptibility ($\sim130\times10^{-6}$) is nearly equal to that of a free electron gas of the same total electron concentration (70×10^{-6}). At 0.40 mole liter^{-1} $f\approx0.3$, $\chi_m^{sp}=82\times10^{-6}$ and the value of $k(\text{Na})$ at $R=10$ can be accounted for with $f\approx1$ assuming $|\psi(\text{M})|^2$ is the same as found in dilute solutions. The much more gradual increase in $k(\text{Na})$ with concentration at 300°K can be interpreted as due to an increase in f as at 240°K but a slight decrease in χ_m^{sp} from 190×10^{-6} at 0.4 mole liter^{-1} to the presumably temperature-independent value at $R=10$ ($\sim130\times10^{-6}$), the susceptibility of a degenerate free electron gas being independent of temperature.

B. Nitrogen Knight Shift

Values of the nitrogen-14 nuclear density at an unpaired electron $P(\text{N})$, determined from the ^{14}N Knight shift data of Na and K solutions are listed in Table III. The results for Na and K solutions at 300°K are in good agreement and, as indicated by the dashed curve of Fig. 4, $P(\text{N})$ appears to be independent of concentration and metal from infinite dilution to the vicinity of one mole liter^{-1}. At 240°K, $P(\text{N})$ has increased somewhat as shown in Table III and indicated in Fig. 4. Once again, $P(\text{N})$ appears to be roughly independent of concentration up to about 1 mole liter^{-1}. The remainder of the dashed curve of Fig. 4 from $R=50$ to $R=10$ is the estimated behavior of χ_m^{sp}/R from the static susceptibility data of Huster.[22] It was

Table III. Nitrogen-14 nuclear density at an unpaired electron.[a]

T	$P(^{14}\text{N})$, a_0^{-3} Na–NH_3	K–NH_3
300	0.77±0.07[a]	0.99±0.07[b]
240	1.42±0.09[c]	···

[a] Determined from extrapolated intercept at $M=0$, Fig. 5.
[b] Determined from KS data from $M=0.06$ to 0.58 mole liter^{-1}.
[c] Determined from KS data from $M=0.21$ to 1.19 mole liter^{-1}.

[21] D. E. O'Reilly (unpublished research), Argonne National Laboratory, 1964.
[22] E. Huster, Ann. Physik 33, 477 (1938).

assumed that the molar diamagnetic susceptibility of a $R=7$ solution is the same as that of a solution with $R=40$. In this way χ_m^{sp} was estimated for the $R=7$ solution. As shown in Fig. 4, the estimated value of χ_m^{sp}/R indicates that $P(\mathrm{N})$ does not vary by more than a factor of two from one molar to very concentrated metal-ammonia solutions.

C. Metal Nuclear Linewidths

An upper limit for the lifetime τ_M of the monomer can be calculated from $\langle |\psi(\mathrm{M})|^2\rangle_{\mathrm{Av}}$ and the lower limit on $|\psi(\mathrm{M})|^2$ derived in Sec. VA using Eq. (11). Values of τ_M derived in this way for Rb and Cs solutions are listed in Table IV for $R=600$ ($M=0.058$) and $R=600$ ($M=0.35$) as well as for Na at $R=600$ from the estimated linewidth. Using the spin-lattice relaxation time of pure ammonia[11b] and the quadrupolar coupling constant[23] of ^{14}N in gaseous NH_3 one calculates by the method of Ref. 24 the correlation time for Y_2^0 to be 0.52 $\mu\mu$sec and that for Y_1^0 to be 1.6 $\mu\mu$sec. The upper limit to the lifetime of the monomer in a 0.06 molar Na-NH_3 solution at 300°K is about 5 $\mu\mu$sec which is comparable to the reorientation time for the ammonia molecule.

To proceed further we may estimate the equilibrium constant K_1 from the rates k_1 and k_{-1} of the reaction

$$\mathrm{M} \underset{k_{-1}}{\overset{k_1}{\rightleftarrows}} \mathrm{M}^+ + e^-,$$

where $k_1=\tau_M^{-1}$ and k_{-1} may be estimated for dilute solutions by the following expression:

$$k_{-1}=\frac{4\pi DL}{\exp(L/R_c)-1}\left(\frac{N_{\mathrm{Av}}}{1000}\right), \quad (17)$$

where D is the sum of the diffusion coefficients of M^+ and e^-, R_c the sum of collision radii and $L=-e^2/\mathfrak{D}kT$. At 300°K, $|L|=34$Å and for reasonable values of R_c (5–10A), k_{-1} as given by Eq. (17) is essentially independent of R_c. For dilute solutions, one estimates from the equivalent conductivity data of Kraus[5] that at 300°K, $D\approx 5\times 10^{-4}$ and thus $k_{-1}\approx 1.3\times 10^{12}$ mole^{-1} liter sec^{-1}. Hence, $K_1\approx 0.1$ mole liter^{-1} which is within an order of magnitude agreement with the value given in Table II (0.020). The value of K_1 computed in this way relative to that of Table II is independent of the true value of $|\psi(\mathrm{M})|^2$. The difference may be due to approximate values of activity coefficients used in evaluation of K_1 as discussed in Sec. VA, partly due to the uncertainty in the value of τ_M for Na or possibly an inadequacy of Eq. (17) at the concentration (0.06M) and temperature which it was applied.

Table IV. Nuclear relaxation time and upper limit of lifetime of monomer species in Na, Rb, and Cs-NH_3 solutions, 300°K.

Metal	$(T_1)_M$, msec		τ_M, $\mu\mu$sec	
	$R=600$	$R=100$	$R=600$	$R=100$
Na	~1.6	...	~4.6	...
Rb	0.068	0.20	2.9	0.63
Cs	0.066	0.33	3.7	0.58

[23] C. T. O'Konski, in *Determination of Organic Structures by Physical Methods*, edited by F. C. Nachod and W. D. Phillips (Academic Press Inc., New York, 1962), Vol. 2, Chap. 11.

[24] D. E. O'Reilly and G. E. Schacher, J. Chem. Phys. **39**, 1768 (1963).

Relatively little information is available concerning the paired electron species. If one considers a metal nucleus to be associated with the electron pair as in Reaction (14) then, since only a single metal NMR line is observed, exchange reactions such as

$$\mathrm{M}^- + \mathrm{M}^{+*} \rightleftarrows \mathrm{M}^{-*} + \mathrm{M}^+ \quad (18)$$

must occur. The lifetime of a metal nucleus in any one species (i.e., M, M^+, or M^-) must be short compared to the reciprocal of the separation between the monomer and metal ion resonances $\Delta\omega=k(\mathrm{M})\omega/x_0$. For Cs solutions at $R=100$ and $\omega=2\pi\times 8\times 10^6$ cps, $\Delta\omega=7\times 10^5$ and hence the lifetime is shorter than 1 μsec. From electron spin resonance studies[10–12] it is known that the line broadening is predominantly due to hyperfine interaction with ^{14}N nuclei in dilute solutions and the lifetime of an unpaired spin τ_u with respect to spin pairing must be long compared to the inverse of the electron resonance width, i.e., 3 μsec at 300°K. Likewise, the lifetime of the paired species τ_p must be long compared to $K_2x_0M\tau_u\approx 4\tau_u$ at $R=100$. Exchange reactions such as (18) can therefore occur without electron pair dissociation.

VI. SUMMARY

Nitrogen-14 Knight shifts indicate the unpaired electron wavefunction to be nearly independent of (1) metal, (2) temperature, and (3) concentration up to rather concentrated solutions (1 mole liter^{-1}). The metal nuclear Knight shifts may be interpreted as due to a monomer species which has a spin density at the metal nucleus that is approximately independent of concentration and temperature. The monomer is, however, a weakly bound and short-lived species. This conclusion is consistent with the fact that most of the physical properties of alkali metal-ammonia solutions do not very greatly from one metal to another. Two paired electrons are energetically more stable than two unpaired electrons by 0.15 eV per electron. Metal nuclear exchange occurs rapidly between the various species compared to the reciprocal of the lifetime of the paired species.

ACKNOWLEDGMENTS

The assistance of Kirby Smith in some of the sample preparations and measurements is gratefully acknowledged. Thanks is extended to Patrick Peterson of the Chemistry Division at Argonne who performed the chemical analyses of the solutions used in this work.

Reprinted from THE JOURNAL OF CHEMICAL PHYSICS, Vol. 41, No. 12, 3736–3742, 15 December 1964
Printed in U. S. A.

Spin Densities in Alkali-Metal–Ammonia Solutions*

46

D. E. O'REILLY
Argonne National Laboratory, Argonne, Illinois

(Received 17 July 1964)

Wavefunctions for the alkali metal monomers of Li, Na, Rb, and Cs in liquid ammonia are calculated using a multipole expansion potential. The wavefunctions are made orthogonal to molecular orbital wavefunctions for the ammonia molecules surrounding the metal ion and spin densities at the metal, nitrogen, and hydrogen nuclei are calculated. A model is proposed for the cavity species which is exactly soluble within the same approximation. Spin densities are evaluated at nitrogen and hydrogen nuclei following orthogonalization to wavefunctions of ammonia molecules on the periphery of the cavity. Orthogonalization produces a large enhancement of spin density at nitrogen and a corresponding decrease in spin density occurs at hydrogen due to a node in the wavefunction produced by orthogonalization. The calculated spin density at the metal nucleus of the monomer and at nitrogen of the cavity species are in good agreement with experimental values. An explanation for the negative spin density at the protons is proposed.

I. INTRODUCTION

NUCLEAR magnetic resonance Knight shifts combined with radio-frequency susceptibility data of alkali-metal–ammonia solutions permit the determination of electron spin densities[1] at metal nuclei as well as nitrogen-14 and proton spin densities[1,2] at an unpaired electron. There are two models currently under consideration for the paramagnetic centers in dilute metal–ammonia solutions. In the monomer model one considers the electron to be trapped in the potential field of a metal ion. This model is similar to the interstitial cation model for the *F* center in alkali halide crystals. In the vacancy or cavity model one considers the electron to be trapped at a vacancy cluster or microscopic bubble in the liquid. This model is analogous to the vacancy model for the *F* center. Spin-density data provide a means of distinguishing between these species as well as criteria for assessing the validity of calculated wavefunctions for the unpaired electron.

Trial wavefunctions for the cavity and monomer species have been calculated by Jortner[3,4] using potentials derived from a continuum approximation for the liquid ammonia solvent. This approach is most useful for energetic considerations; it yields a value of spin density at the metal nucleus but not at nitrogen or hydrogen nuclei because of the nature of the approximation. Blumberg and Das[5] have calculated wavefunctions for the sodium monomer using an approach suggested by the treatment of the *F* center by Gourary and Adrian.[6] Calculated spin densities at the Na, N, and H nuclei were in order-of-magnitude agreement with the then available experimental data. However, equilibrium constants for cavity–monomer–dimer equilibria[7] were used and, in addition, it was assumed that the cavity species does not contribute to the nitrogen-14 or proton Knight shifts. It is necessary to revise these two assumptions in view of the experimental results of the preceding paper.

It is the purpose of this paper to construct wavefunctions for the monomer and cavity species and calculate spin densities for comparison with the recently obtained experimental values. A method similar to the multipole expansion scheme of Blumberg and Das[5] is used to construct potentials and the resulting wavefunctions are orthogonalized to molecular wavefunctions of ammonia molecules in the first coordination shell (Sec. II). For the cavity species a potential well parameter is selected which yields the experimental energy difference between the ground and first excited state and the heat of solution. The wavefunction is orthogonalized to molecular wavefunctions of ammonia molecules on the periphery of the cavity (Sec. III). Spin densities at metal, nitrogen, and hydrogen nuclei are evaluated for each species and compared with experimental data (Sec. IV).

II. MONOMER

Metal ions dissolved in polar solvents produce a considerable amount of local dielectric polarization of the solvent. In consideration of the interaction of an unpaired electron with such a metal ion the polarization of the medium must be taken into account. Approximate calculations indicate that about 80% of the total interaction energy of an alkali metal ion with liquid ammonia can be attributed to ion–multipole (and some covalent) forces between the ion and ammonia molecules in the first coordination shell around the ion. To conveniently solve the problem of an odd electron bound to such an ion in the self-consistent field (SCF) approximation three major additional approximations will be made. In the first of

---

* Based on work performed under the auspices of the U. S. Atomic Energy Commission.

[1] D. E. O'Reilly, J. Chem. Phys. **41**, 3729 (1964).
[2] T. R. Hughes, Jr., J. Chem. Phys. **38**, 202 (1963).
[3] J. Jortner, J. Chem. Phys. **30**, 839 (1959).
[4] J. Jortner, J. Chem. Phys. **34**, 678 (1961).
[5] W. E. Blumberg and T. P. Das, J. Chem. Phys. **30**, 251 (1959).
[6] B. S. Gourary and F. J. Adrian, Phys. Rev. **105**, 1180 (1957).
[7] E. Becker, R. H. Lundquist, and B. J. Alder, J. Chem. Phys. **25**, 971 (1956).

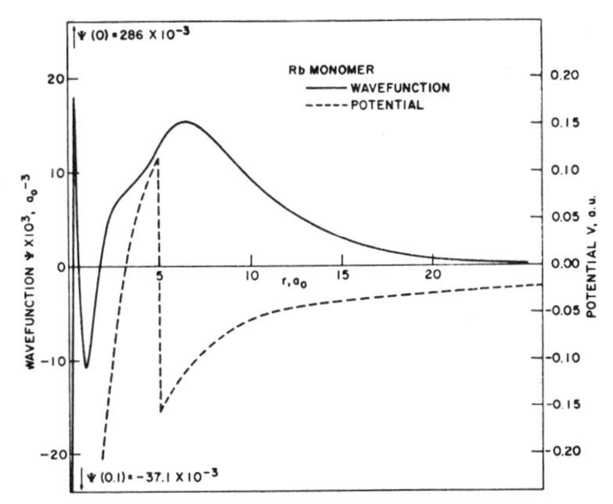

FIG. 1. Wavefunction ψ and potential V for the rubidium monomer. ψ has four nodes inside the ionic radius $r_{Rb}=2.80$ a.u. The discontinuity in V occurs at the position of the point multipoles.

B. Numerical Integration

The one-electron Schrödinger equation for Potential (1) is, in atomic units,

$$\{-\tfrac{1}{2}\nabla^2+V\}\psi=E\psi. \tag{3}$$

Since the potential is spherically symmetric ψ is a simultaneous eigenfunction of $\mathbf{L}^2$, L_z and energy and therefore Eq. (3) is separable.[13] Hence $\psi(\mathbf{r})=R(r)/(4\pi)^{\frac{1}{2}}$ for the s ground state. Placing $P(r)=rR(r)$ Eq. (3) reduces to an ordinary second-order differential equation:

$$P''+2(E-V)P=0. \tag{4}$$

This equation was numerically integrated for selected values of E using a method described by Hartree.[14] An arbitrary value of P was chosen for a large value of r (25 a.u.), and an inward integration started based on the asymptotic solution of (4), which vanishes as r approaches infinity. At $r=r_M$, P, P', and P'' were chosen to be the same as for the SCF ns wavefunctions in the ^2S ground state of the free metal atoms.[15] The inward and outward integrations were terminated at $r=a$. Logarithmic derivatives $(P'/P)_{\text{in}}$ and $(P'/P)_{\text{out}}$ were plotted versus E, and the eigenvalue E of Eq. (4) determined by the point at which the two curves crossed. The eigenfunctions determined in this way were then made continuous at $r=a$ and normalized to unity. The resulting wavefunction for the Rb monomer ($n=5$) is shown in Fig. 1. As in the free atom the wavefunction has $n-1$ nodes but differs markedly from the free atom wavefunction in that over 90% of the electronic charge density is located outside of the ionic radius. Values of the energy, the square of the renormalization constant N^{-1} for the free atom wavefunction $(r<r_M)$, and the square root of expectation value of r^2 are given in Table I for Li, Na, Rb, and Cs monomer. For values of r less than r_M,

$$\psi=N^{-\frac{1}{2}}\psi_0,$$

where ψ_0 is the ns unpaired electron SCF wavefunction of the free gaseous metal atom. Expectation values of $\langle r^2\rangle$ were calculated from the analytical representations of the wavefunctions by Slater functions discussed in Sec. II.C.

In order to be certain that an energy eigenvalue calculated by the numerical integration procedure is the lowest, i.e., corresponds to the ground state, variational calculations were performed. A potential similar to (1) and a series of trial functions of the form $r^ne^{-\alpha r}$ ($n=0$, 1, or 2) and $(1+\alpha r)e^{-\alpha r}$ were used with α as a variation parameter. These calculations indicated that the states calculated by the numerical integrations are the states of lowest energy.

C. Orthogonalization Procedure

The monomer actually consists of N_t[=63(Li), 71(Na), 99(Rb), or 117(Cs)] electrons rather than only one electron as assumed in the preceding calculation. Denoting the ion core spin-orbitals of the metal ion by ϕ_i, the molecular spin-orbitals of the αth ammonia molecule by ψ_j^α, and that of the unpaired electron calculated in Sec. II.B by ψ, the total N_t-electron SCF wavefunction Ψ is a single determinantal wavefunction

$$\Psi=\mathcal{A}\{(\prod_{i=1}^{N_c}\phi_i)(\prod_{\alpha=1}^{6}\prod_{j=1}^{10}\psi_j^\alpha)\psi\}, \tag{5}$$

where

$$\mathcal{A}=\frac{1}{N_t!}\sum_P(-1)^PP$$

is the operator which yields a wavefunction antisymmetric with respect to interchange of any two electrons and N_c is the total number of metal ion core electrons. For the present it will be assumed that (1) the functions ϕ_i and ψ_j^α are the same as those of the free gaseous ion or molecule, respectively, and (2) no

TABLE I. Energy, renormalization constant, and root mean square radius of alkali metal monomers.

Metal	E (eV)	N^{-1}	$\langle r^2\rangle^{\frac{1}{2}}$ (Å)
Li	1.68	0.015	5.2
Na	1.40	0.015	4.9
Rb	1.36	0.052	5.3
Cs	1.32	0.084	5.3

[13] E. U. Condon and G. H. Shortley, *The Theory of Atomic Spectra* (Cambridge University Press, London, 1951), p. 112.

[14] D. R. Hartree, *The Calculation of Atomic Structures* (John Wiley & Sons, Inc., New York, 1957), Secs. 4.4, 5.2, and 5.3.

[15] Alkali metal atomic SCF wavefunctions used in the present calculations are to be found in the following references:— Li: L. M. Sachs, thesis, Illinois Institute of Technology, 1961; Na: D. R. Hartree and W. Hartree, Proc. Roy. Soc. (London) **A193**, 299 (1948); Rb: J. Callaway and D. F. Morgan, Phys. Rev. **112**, 334 (1958); Cs: R. Sternheimer, Phys. Rev. **127**, 1220 (1962).

TABLE II. Calculated spin densities at monomer nuclei [unit: $(10a_0)^{-3}$].

Monomer	Metal $\|\psi(M)\|^2$	Number of coordination shells	Metal orthogonalized ψ $\|\psi_0(M)\|^2$	Nitrogen-14 $P(N)$	Nitrogen-14 orthogonalized ψ $P_0(N)$	Proton orthogonalized ψ $P_0(H)$
Li	1.9	1	0.36	0.97	20.8	0.26
		2	0.64	2.00	1300	14.3
Na	5.0	1	3.7	0.96	197	1.33
		2	9.4	2.15	2050	20.4
Rb	93	1	109	0.95	211	1.84
		2	256	2.07	1850	19.1
Cs	179	1	206	0.91	194	1.57
		2	480	2.05	1770	18.3

appreciable ion–molecule or molecule–molecule wavefunctions overlaps occur. Hence in the set $\phi_1, \cdots, \phi_N, \psi_1^1, \cdots, \psi_{10}^6$ each spin–orbital is normalized and orthogonal to every other spin–orbital. Assumption (2) is not actually used in calculations described below.

The isotropic nuclear-spin electron-spin interaction energy, which determines the Knight shift, as discussed in the preceding paper, is in first order

$$\langle\Psi\,|\,\mathcal{H}_N\,|\,\Psi\rangle = -(16\pi/3)\beta\mu_N\langle\Psi\,|\,\sum_j\delta(\mathbf{r}_N-\mathbf{r}_j)\mathbf{I}_N\cdot\mathbf{S}_j\,|\,\Psi\rangle, \quad (6)$$

where $\mathcal{H}_N$ is the Fermi contact Hamiltonian for nucleus N interacting with all the electrons j of the monomer, β is the Bohr magneton, μ_N is the magnetic moment of nucleus N, $\mathbf{S}_j$ is the spin of electron j and $\mathbf{I}_N$ is the spin of nucleus N. The matrix element (4) may be calculated directly if the orbital ψ is orthogonal to all the rest. Eq. (6) then becomes

$$-\langle\Psi\,|\,\mathcal{H}_N\,|\,\Psi\rangle = (16\pi/3)\beta\mu_N\,|\,\psi_0(N)\,|^2\,\mathbf{I}_N\cdot\mathbf{S}, \quad (7)$$

where $\psi_0(\mathrm{N})$ is the value of the orthogonalized unpaired electron orbital ψ at nucleus N. Equation (7) follows since $\mathcal{H}_N$ is a sum of one-electron operators and the set $\phi_1, \cdots, \psi_{10}^6$ constitutes a "closed shell" with zero total spin angular momentum. $|\psi_0(\mathrm{N})|^2$ is the unpaired electron spin density at nucleus N.

The orthogonalized orbital ψ_0 may be constructed from ψ by the Schmidt procedure:

$$\psi_0 = \frac{\psi - \sum_{\alpha=1}^{6}\sum_{j=1}^{5}\langle\psi\,|\,\psi_j^\alpha\rangle\psi_j^\alpha - \sum_{i=1}^{\frac{1}{2}N_c}\langle\psi\,|\,\phi_i\rangle\phi_i}{\left(1-\sum_{\alpha=1}^{6}\sum_{j=1}^{5}\langle\psi\,|\,\psi_j^\alpha\rangle^2 - \sum_{i=1}^{\frac{1}{2}N_c}\langle\psi\,|\,\phi_i\rangle^2\right)^{\frac{1}{2}}}, \quad (8)$$

where the j sum extends only over those spin–orbitals ψ_j^α which have the same spin orientation as the unpaired electron. The molecular orbitals for the ammonia molecule that are available[11,16] are expressed as a linear combination of only a few atomic orbitals $\chi_k^{N\alpha}$ (Slater functions $\times$ spherical harmonic) centered at the Nth nucleus in the αth molecule:

$$\psi_j^\alpha = \sum_N\sum_k a_{jk}^{N\alpha}\chi_k^{N\alpha}. \quad (9)$$

Overlap integrals may be conveniently calculated by expanding ψ in a finite series of Slater functions $\chi_n = r_n\exp(-\alpha_n r)$

$$\psi = \sum_n c_n\chi_n, \quad (10)$$

where χ_n is centered at the metal nucleus. The integrals $\langle\chi_n\,|\,\chi_k^{N\alpha}\rangle$ are readily calculated using elliptical coordinates or tables of integrals[17] for selected function pairs. The expansion (10) generally contained three Slater functions to represent ψ in the region $r > r_M$ and $n-1$ functions for the oscillations inside the ion core. Outside ion core Eq. (10) represented the calculated wavefunction with an accuracy of 5% or better.

For lithium and sodium, the slight nonorthogonality between the molecular orbitals ψ_j^α and the ion core orbitals ϕ_i(1s and 2s) was taken into account and an expression similar to Eq. (8) which included effects of nonorthogonality of the closed shell orbitals was used in the calculation of spin densities. For rubidium and cesium this effect of nonorthogonality is relatively small and was neglected. The molecular orbitals for ammonia given by Kaplan[16] were used for the ψ_j^α. Somewhat similar orbitals given in Ref. 11 were not used since the 2s atomic function of nitrogen employed there consisted only of a single Slater orbital with zero density at the nucleus.

The final results of the calculations are given in Table II. Values of the total spin density at the M and N nuclei are listed for both ψ and ψ_0 to show the effect of orthogonalization; the values of ψ for H are

[16] H. Kaplan, J. Chem. Phys. **26**, 1704 (1957).
[17] R. S. Mulliken, C. A. Rieke, D. Orloff, and H. Orloff, J. Chem. Phys. **17**, 1248 (1949).

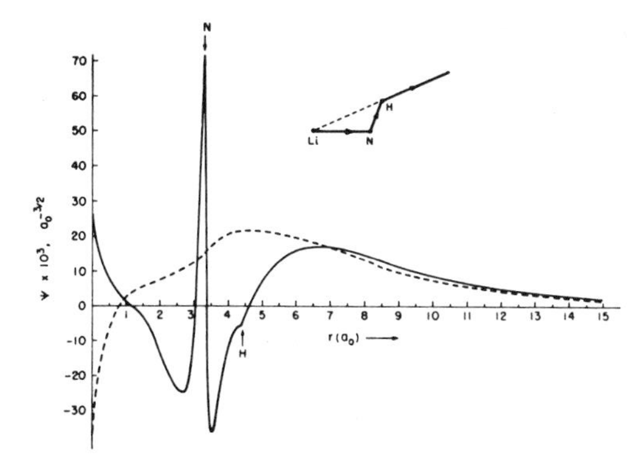

FIG. 2. One-electron and orthogonalized ψ for the lithium monomer. Cusps in ψ_0 occur at the position of N and H due to mixing of the nitrogen $2s$ and hydrogen $1s$ functions, respectively. A node of ψ_0 occurs near H, resulting in a greatly diminished spin density at H. (– – –) one-electron wavefunction ψ; —— orthogonalized ψ.

only slightly larger than those of N. In Table II $P(\text{N})=6\ |\psi(\text{N})|^2$ and $P(\text{H})=18\ |\psi(\text{H})|^2$ are the nuclear densities at the unpaired electron. For lithium the number of first coordination shell molecules is four and the preceding formulas are modified accordingly.

The effect of orthogonalization on the wavefunction of the lithium monomer may be seen in Fig. 2 where the dependence of ψ_0 is shown along a path which starts at the origin, passes through N and H and proceeds radially outward from H. The most pronounced effect of orthogonalization is the appearance of three additional nodes in the wavefunction, resulting in an over-all oscillatory appearance of the wavefunction. Near the origin, ψ_0 has opposite sign to ψ due primarily to the contribution of the $1s$ orbital of Li^+. In the vicinity of N, a sharp peak in ψ occurs due to both the $1s$ and $2s$ orbitals of N. This results in an *amplification* of the spin density at N by a factor of about 20 for the Li monomer and a factor of over 100 for the other alkali monomers. In the vicinity of H a node in ψ_0 occurs, resulting in an *attenuation* of the spin density at H by an order of magnitude.

The function ψ was also orthogonalized to ammonia molecules in the second coordination shell surrounding the ion. The nitrogen and hydrogen nuclei of these molecules were assumed to be at an average distance equal to $b+\frac{1}{2}V_m^{\frac{1}{3}}$, where V_m is the molecular volume. The number of such molecules n_2 was calculated as follows:

$$n_2=[4\pi(b+\tfrac{1}{2}V_m^{\frac{1}{3}})^2/V_m^{\frac{2}{3}}]. \tag{11}$$

The results of these calculations are also listed in Table II. Inclusion of the second coordination shell molecules alters the values of $P_0(\text{N})$ and $P_0(\text{H})$ by greater than an order of magnitude.

III. CAVITY SPECIES

In the following the cavity species are regarded in a "quasicrystalline" approximation for the liquid as an unpaired electron trapped at a vacancy cluster. From data for density[18] and change in volume on solution[19] the effective void volume of this species is known. On the average, the cavity is spherical in shape and has a number n of ammonia molecules on the periphery which are oriented towards the center of the cavity. Adopting the approximations discussed in Sec. II with n point dipoles and quadrupoles located at $r=a$, only the $l=0$ component of the potential V do not average to zero. Hence,

$$\begin{aligned} -V&=(n/a^2)[\mu-(Q_{zz}/a)]=-V_0, && r\leqslant a,\\ V&=0, && r\geqslant a. \end{aligned} \tag{12}$$

As with (1), (9) is not a SCF potential and is modified by the presence of the unpaired electron, but similar to (1), (9) provides a good approximation to the true SCF potential. This point is discussed further in Appendix A.

The potential (9) is a spherical square well for which analytical solutions to Schrödinger's equation are known.[20] There is one bound $l=0$ state if $\pi^2/8a^2<|V_0|<9\pi^2/8a^2$, two bound $l=0$ states if $9\pi^2/8a^2<|V_0|<25\pi^2/8a^2$, one bound $l=1$ state if $\pi^2/2a^2<|V_0|<2\pi^2/a^2$, etc. In Fig. 3 the ground state energy E_{1s} and energy difference between the ground and first excited p state ΔE_{1s-2p} is given as a function of V_0 for $a=9.0\ a_0$.

A value for the quantity V_0 may be estimated as follows. Let us consider the solvent to be a continuous medium. The "radius" of the cavity is experimentally determined to be 3.0 A, essentially independent of temperature. The distance a one would then estimate to be $3.0+\frac{1}{2}V_m^{\frac{1}{3}}=4.75A\ (9.0a_0)$ where V_m is the molecular volume. The number n, in the same approximation, is $4\pi a^2/A_m$, where $A_m=V_m^{\frac{2}{3}}$ is the mean area occupied by an ammonia molecule on the periphery. Further let us approximate the local electric field E_{loc} acting on an ammonia molecule due to the unpaired electron as that of a unit negative charge at the center of the cavity. Then

$$-V_0\approx(4\pi/A_m)\langle\cos\theta\rangle_{\text{Av}}(\mu-Q/a), \tag{13}$$

where

$$\langle\cos\theta\rangle_{\text{Av}}=1-kT/\mu E_{\text{loc}} \quad \text{for} \quad \mu E_{\text{loc}}/kT\gg1.$$

Equation (10) is independent of n and only weakly dependent on a. Using values of the electric moments and polarizability given in Sec. II, Eq. (10) yields—$V_0=0.17$ a.u. (4.7 eV).

[18] C. A. Hutchison Jr. and D. E. O'Reilly, J. Chem. Phys. **34**, 163 (1961).
[19] T. P. Das, Advan. Chem. Phys. **4**, 303 (1961).
[20] L. I. Schiff, *Quantum Mechanics* (McGraw-Hill Book Company, Inc., New York, 1949), p. 76.

The experimental value of ΔE_{1s-2p} is 0.8 eV assuming that the observed absorption peak in the infrared is due to this transition. The energy E_{1s} may be estimated from the heat of solution ΔH_{soln} which is 1.7 ± 0.7 eV.[3] ΔH_{soln} and E_{1s} are related by $\Delta H_{soln}=-E_{1s}-\Pi$, where Π is the energy required to create the potential well in which the electron is bound. This energy is primarily the dipole polarization energy of the solvent and may be estimated to be the negative of the Born polarization energy of the cavity, $(e^2/2a)(1-1/D)=1.4$ eV. Thus $E_{1s}=-3.1\pm0.7$ eV. From Fig. 3 at $|V_0|=4.9$ eV one obtains $E_{1s}=-3.75$ eV and $\Delta E_{1s-2p}=1.20$ eV. The dashed curves in Fig. 3 correspond to a radius 15% larger ($a=5.5$ Å) from which one obtains $E_{1s}=-3.95$ eV and $\Delta E_{1s-2p}=0.90$ eV. These values are in good agreement with the experimental values and even

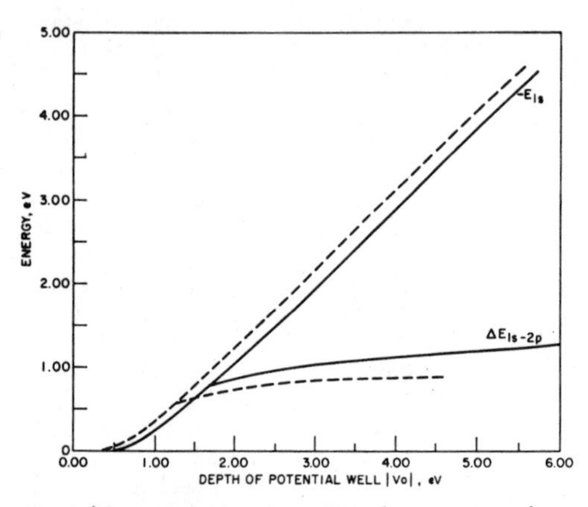

FIG. 3. The ground state energy E_{1s} and energy separation to lowest p excited state for a spherical square well of radius $a=9.0$ a.u. versus the depth of the well $|V_0|$. The dashed curves correspond to a well of radius 10.3 a.u.

better agreement is obtained for V_0 somewhat smaller. $|V_0|=3.1$ eV is used in the following calculations of spin densities.

Spin densities were calculated by orthogonalizing the ammonia molecular orbitals and the spherical square well ground state eigenfunction

$$\psi=(A\sin\alpha r)/r, \qquad r<a,$$
$$\psi=Be^{-\beta r}/r, \qquad r>a, \qquad (14)$$

where $\alpha=[2(|V_0-E_{1s}|)]^{\frac{1}{2}}$ and $\beta=[2|E_{1s}|]^{\frac{1}{2}}$. Overlap integrals were computed by representing (11) by a Slater function of the form $e^{-\gamma r}$ with γ chosen to yield the lowest energy by variation. The error in overlap integrals introduced by this procedure is less than 10%. The results are listed in Table III where values of the nuclear densities at an unpaired electron $P(N)=n|\psi(N)|^2$, $P(H)=3n|\psi(H)|^2$ are given for $|V_0|=3.1$ eV and two values of a, 4.75 Å and 5.46 Å. Ortho-

TABLE III. Values of spin density for the cavity species (atomic units).

$\lvert V_0\rvert$	a	P(N)	P_0(N)	P(H)	P_0(H)
0.115	8.97	0.0013	0.84	0.0064	0.00062
0.115	10.31	0.0010	0.49	0.0051	0.00011

gonalization increases P(N) by nearly three orders of magnitude and decreases P(H) by one order of magnitude. ψ_0 has a node in the vicinity of H and a sharp peak at N similar to that shown in Fig. 2.

IV. DISCUSSION

The calculated spin densities at the metal nuclei are compared with values calculated from experimental Knight shift data[1] in Table IV. The agreement is generally good and lends support to the model proposed for the monomer species.[5,7] The experimental nitrogen nuclear density at an unpaired electron is equal to $0.88\pm0.11\ a_0^{-3}$ independent of alkali metal at 300°K. This value compares favorably with the value calculated for the cavity species with $a=8.97$ a.u. The values of P_0(N) calculated for the monomer are not independent of metal and are as much as a factor of two larger than the experimental value. This is consistent with the monomer being a relatively unimportant species in dilute solutions from a thermodynamic point of view.

The binding energy calculated for the sodium monomer is much greater than the value obtained by Blumberg and Das.[5] The difference may be due to a different expression for the multipole expansion potential in their paper although our result is also larger than their value for a distributed charge potential. Nuclear spin densities listed in Table II for the sodium monomer (one coordination shell) are considerably smaller than those those given by Blumberg and Das for the multipole expansion potential.

The experimental proton spin density at an unpaired electron is $-(1.7\pm0.1)\times10^{-2}a_0^{-3}$ from the experimental Knight shift data of Hughes[2] on Na–NH_3 solutions. Values of P(H) calculated for both the monomer and cavity models are *positive* and hence not in agreement with the experimental result. The explanation of the *negative* proton spin density appears to be similar to that proposed for the negative electron

TABLE IV. Comparison of observed and calculated spin densities.

Metal	$\lvert\psi_0(M)\rvert^2\times10^3$, a_0^{-3} Obs	Calc
Li	1	0.64
Na	8	9.4
Rb	50	256
Cs	100	480

spin density at protons in π-electron free radicals.[21] In an unrestricted Hartree–Fock scheme, a molecular orbital ψ_i^α of the ammonia molecule will be the same for both spin orientations (i.e., $\psi_i^{\alpha+}=\psi_i^{\alpha-}$) in the absence of the unpaired electron. If the unpaired electron orbital ψ_0^+ is included then $\psi_i^{\alpha+}\neq\psi_i^{\alpha-}$ because of the negative exchange potential energy between electrons of like spin. For the cavity or monomer $|\psi_i^{\alpha+}|^2$ will be greater than $|\psi_i^{\alpha-}|^2$ near the nitrogen nucleus since $|\psi_0^+|^2$ is relatively large there. However, near the nodes of ψ^+ one expects $|\psi_i^{\alpha-}|^2>|\psi_i^{\alpha+}|^2$ and hence a negative contribution to the spin density at hydrogen. These contributions will not greatly alter the value of the large positive spin density at nitrogen but can alter the sign of the relatively small spin density at hydrogen.

ACKNOWLEDGMENTS

Stimulating discussions with Professor J. Jortner and Dr. J. Robinson contributed to the present work. The help of K. McCormick and K. Smith, who assisted in the numerical integrations, is greatly appreciated.

APPENDIX

In the following an SCF potential for an electron interacting with a positive ion (Sec. II) or a vacancy cluster (Sec. III) will be derived. Let us consider a positive charge located at the center of spherical cavity in a continuous dielectric medium. In the SCF approximation, the polarization of the dielectric medium produced by an electron interacting with the positive ion will be determined by the electronic charge density $\rho(r)=e|\psi(\mathbf{r})|^2$. By the application of Gauss' theorem the electric field strength is readily verified to be as follows

$$\mathbf{E}^-=e\{1-f(r)\}/r^2,\qquad r<a,$$
$$\mathbf{E}^+=e\{1-f(r)\}/\kappa r^2,\qquad r>a,\qquad \text{(A1)}$$

where

$$f(r)=\frac{4\pi}{e}\int_0^r \rho(t)t^2dt$$

is the fraction of the total electronic charge that is contained in a sphere of radius r. An SCF potential for the electron may be constructed by calculation of the potential energy due to the charge of the positive ion plus the static polarization of the medium produced by the positive ion and the charge distribution $\rho(r)$. The polarization $\mathbf{P}(r)$ is simply

$$4\pi\mathbf{P}(r)=\mathbf{D}-\mathbf{E}=\{e[1-f(r)]/r^3\}(1-\kappa^{-1})\mathbf{r}.\qquad \text{(A2)}$$

The SCF potential V_{SCF} is determined by

$$V(r)_{\mathrm{SCF}}=\int\frac{[e\delta(\mathbf{r}-\mathbf{r}')-\nabla\mathbf{P})}{r'}dv-\int\frac{\mathbf{P}\cdot\mathbf{r}}{r'r}dA,\qquad \text{(A3)}$$

where the second integral extends over the surface of the interface at $r=a$ and $\mathbf{r}-\mathbf{r}'$ is the radius vector to the volume element dv. Evaluation of these integrals yields

$$V^-_{\mathrm{SCF}}=\frac{e}{r}-\frac{e[1-f(a)]}{a}(1-\kappa^{-1})-e(1-\kappa^{-1})\times\int_a^\infty\frac{\partial[1-f(t)]}{\partial t}\frac{dt}{t}$$

$$V^+_{\mathrm{SCF}}=\frac{e}{r}-\frac{e[1-f(a)]}{r}(1-\kappa^{-1})-e(1-\kappa^{-1})\left(r^{-1}\int_a^r\left\{\frac{\partial}{\partial t}[1-f(t)]\right\}dt+\int_r^\infty\left\{\frac{\partial}{\partial t}[1-f(t)]\right\}\frac{dt}{t}\right).\qquad \text{(A4)}$$

Similar expressions for the SCF potential are obtained for the vacancy cluster discussed in Sec. III except that the first term is absent and $1-f(r)$ is replaced by $-f(r)$.

For the monomer, the potential (A4) was used with $f=0$ everywhere and $\kappa=\kappa_\infty$, the high-frequency dielectric constant. To obtain a true SCF solution, a second solution to $\mathcal{H}\psi=E\psi$ should be constructed using $f(r)$ calculated from the first solution and this procedure repeated until self-consistency is obtained. This procedure has not been carried out, but it is estimated that no large change in the calculated spin densities would occur. For the cavity, $f(a)\approx 0.9$ and the SCF correction should be very small in this case.

[21] H. M. McConnell, J. Chem. Phys. **24**, 764 (1956); H. M. McConnell and D. B. Chestnut, J. Chem. Phys. **28**, 107 (1958).

Phase and Conductivity Studies of the Lithium–Ammonia System

For many years it had been suspected that lithium forms a tetraammine analogous to the hexaammines of the alkaline earth metals, but the evidence was always equivocal. The electrical conductivity data presented in the following paper by Morgan, Schroeder, and Thompson constitute some of the first fairly direct evidence for a lithium–ammonia compound.

THE JOURNAL OF CHEMICAL PHYSICS VOLUME 43, NUMBER 12 15 DECEMBER 1965

Phase Changes and Electrical Conductivity of Concentrated Lithium–Ammonia Solutions and the Solid Eutectic*

J. A. MORGAN, R. L. SCHROEDER, AND J. C. THOMPSON

The University of Texas, Austin, Texas

47

(Received 29 July 1965)

The electrical conductivities of concentrated solutions of lithium in ammonia have been measured over the temperature range 65° to 300°K by both conventional dc techniques and electrodeless techniques. In the liquid phase, conductivities ranged from 400 to 15 000 $\Omega^{-1}\cdot cm^{-1}$ at 240°K; the temperature coefficient of conductivity varied from +1.0% deg^{-1} to zero. Phase changes were indicated by discontinuities in either the conductivity or the temperature derivative of conductivity. No sign of a maximum in the melting-point-vs-concentration curve near the tetra-amine lithium [$Li(NH_3)_4$] concentration was found. The eutectic was found to be 89.6°±0.3°K. The conductivity of the solid was 6.7 times as large as that of the liquid at the melting point. In addition to phase changes in the liquid, two solid-phase changes were observed; one at 82°K and the other at 69°K. Conductivities in the solid were decreasing functions of the temperature in each solid phase, yet were higher in the high-temperature solid phases. The highest conductivity observed was 90 000 $\Omega^{-1}\cdot cm^{-1}$ (at 83°K). The solid does not behave as do the eutectics of sodium-ammonia or potassium–ammonia solutions. A comparison with other metal–ammonia solutions is made.

INTRODUCTION

ONE of the most striking features of dilute metal–ammonia solutions is the lack of variation in most of the properties of the solutions with solute.[1] This behavior is presumably the result of the dominant role of the solvated electron.[2] On the other hand, in concentrated solutions the valence electron is free and there are few solvent molecules to screen the metal ion. As a result, variations from one solute to another appear in such properties as the solubility, the eutectic composition, and the conductivity, as well as the phase boundaries.[1] Since Kraus' early work[3] and the excitement of the later 1940's little has been reported on concentrated metal–ammonia solutions.[4] Indeed, even the phase boundaries have not been accurately measured for most of these solutes. We report here a portion of the results of a series of measurements on concentrated metal–ammonia solutions.

These solutions are of interest as examples of liquid metals. They offer a wide range of electron concentrations and ion densities in addition to considerable experimental convenience relative to the pure alkali metals in that they are molten in a somewhat more convenient temperature and pressure range. There is further interest generated by the fact that these systems exhibit a Mott transition between a metallic and a nonmetallic state.[5] Kyser and Thompson[6] have observed such a transition in both lithium– and sodium–ammonia solutions. There is no reason to doubt that such a transition also occurs for the other alkali-metal solutes. Furthermore, the Mott transition as observed in metal–ammonia solutions is not complicated by influence from the band structure of the host material as with semiconductors. As the ions are mobile, there is no problem of inhomogeneities in the material.

For these reasons we have attempted measurements on the electron transport properties of concentrated metal–ammonia solutions. Thermal conductivities[7] and Hall-effect data[6] have already been reported. Since lithium is thought not to precipitate from the ammonia solutions upon freezing,[4] and since Sienko[8] has suggested that there should be a maximum in the melting curve at the $Li(NH_3)_4$ freezing point, we have chosen to examine in some detail the melting points near the eutectic composition and to extend the measurements into the solid phase at the eutectic concentration. The freezing point of this eutectic, is the lowest known for a metallic material.[9]

EXPERIMENTAL

It is well known that a reaction occurs between the solvent and solute. This reaction removes metal from solution and produces hydrogen. Rigorous cleaning reduces the reaction rate and is a necessary preliminary of any successful experiment. Metal electrodes are also

* Supported by the Robert A. Welch Foundation and the U.S. Office of Naval Research.

[1] T. P. Das, Advan. Chem. Phys. **4**, 303 (1962).
[2] J. Jortner, S. A. Rice, and E. G. Wilson, in *Metal-Ammonia Solutions*, edited by G. Lepoutre and M. J. Sienko (W. A. Benjamin, Inc., New York, 1963).
[3] C. A. Kraus and W. W. Lucasse, J. Am. Chem. Soc. **43**, 754, 2533 (1921); C. A. Kraus and W. W. Lucasse, J. Am. Chem. Soc. **44**, 1941, 1949 (1922); and others. See any review about metal-ammonia solutions such as Ref. 1 above.
[4] A. J. Birch and D. K. C. MacDonald, Trans. Faraday Soc. **44**, 735 (1948); J. W. Hodgins, Can. J. Research **27B**, 861 (1949).
[5] N. F. Mott and W. D. Twose, Advan. Phys. **10**, 107 (1961).
[6] D. S. Kyser and J. C. Thompson, J. Chem. Phys. **42**, 3910 (1965).
[7] P. G. Varlashkin and J. C. Thompson, J. Chem. Phys. **38**, 1974 (1963).
[8] M. J. Sienko, in *Metal-Ammonia Solutions*, edited by G. Leportre and M. J. Sienko (W. A. Benjamin, Inc., New York, 1963).
[9] H. Jaffe, Z. Physik **93**, 741 (1935).

to be avoided as they are not so easily cleaned as glass, and, even when clean, they serve as catalysts for the decomposition reaction. Nevertheless, conventional probe measurements of voltage and current offer advantages over electrodeless techniques. We therefore began with four-probe measurements and continued them until confidence in other techniques was attained. Both techniques are discussed, as are the cleaning procedures employed.

Cleaning

Although the conductivities were measured by two independent means, the cleaning procedure for each method was roughly the same. The first part of the cleaning procedure employs a group of reagents for cleaning the surface of the glass sample cells. The second part of the cleaning procedure consists of a wash of the interior of the sample cells with dry ammonia to remove or replace water absorbed on the surface. These two steps virtually eliminate decomposition in the electrodeless measurements as is indicated below. The cleaning somewhat decreases the decomposition in the four-probe measurements, but the tungsten electrodes themselves are a considerable source of decomposition. Their effect is further detailed below. The electrodes were cleaned by electrolytic action in KOH as well as by the glass-cleaning procedure. The glass-cleaning procedure, suggested to us by Naiditch,[10] consists of a soak for several minutes in a solution consisting of 5% HF, 33% HNO_3, 2% Alconox, and 60% H_2O (all by volume). The soak is followed by a rinse with aqua regia and six or more rinses with distilled water. Finally the interior of the sample cell is washed with ammonia which has been dried over sodium.[11]

We have evidence both from internal consistency and from data taken on cesium-ammonia solutions that, in the absence of electrodes, the above cleaning procedure is sufficient to keep decomposition at a very low rate. In the electrodeless measurements, lithium-ammonia solutions were kept for many days over which period the conductivity showed no appreciable change. As the conductivity shows an extremely strong dependence on concentration, the constancy indicates that little or no decomposition took place. Measurements have also been made on cesium-ammonia solutions utilizing the electrodeless techniques and the same cleaning procedure.[12] A chemical analysis performed after these measurements, indicated that over a period of five weeks less than 0.03% per day of the metal was removed from the solution by decomposition. We made one measurement which employed both the probe and electrodeless techniques simultaneously on the same lithium solution. From this experiment we concluded that the presence of the electrodes caused a fast initial decomposition followed by a very slow steady decomposition. The net decomposition varied from run to run and was in all cases such as to lower the calculated concentration by less than 1.0 mole percentage point, 0.6 mole % on the average. The solutions used for studies in the neighborhood of the eutectic were kept at low temperatures and showed no signs of decomposition. The probe data have been corrected to take such decomposition into account.

Sample Preparation

Lithium metal for the solutions was cut from stock supplied by A. D. Mackay, Inc. This metal, quoted at 99.8% pure, was cut and trimmed under mineral oil, washed in petroleum ether, weighed, and introduced into the sample cell. Care was taken throughout the preparation of a metal sample to prevent contact with the atmosphere.

The ammonia for the solutions was first dried by distillation onto sodium metal. The desired amount of ammonia for a given solution was measured by either a liquid-volume measurement or a gas-volume, temperature, and pressure measurement. The gas-phase measurements are both more convenient and more accurate; and in addition they allow solutions to be diluted in very small steps. Dilutions were only used in the electrodeless system. In either case, after the ammonia was distilled onto the lithium-metal sample, the solution was made homogeneous by mixing or stirring. The four-probe measurements employed a sample cell with a U-tube configuration and mixing was accomplished by introducing Grade-A helium gas up to atmospheric pressure and bubbling the gas through the solution by means of a small rubber-bulb hand pump. The electrodeless technique measurements employed a sample cell design, which may be described as a cylinder connected to a ball. The solution was mixed by pouring the solution back and forth from the cylinder to the ball.

Temperature Control

Temperature control for the four-probe measurements was accomplished in one of two ways, depending on the temperature range. For data taken in the range 140° to 240°K a 4-liter Freon bath in contact with the sample cell was used as a thermostat. The bath was warmed by electric heaters and cooled by contact with liquid nitrogen. For data taken in the range 65° to 160°K a liquid-nitrogen cryostat was used. Temperatures above 77°K were obtained by isolating the system from the nitrogen bath and heating electrically. Temperatures below 77°K were obtained by pumping on the liquid-nitrogen bath. Temperature control for the electrodeless measurements was obtained by using a Freon bath in the same manner as for the high-temperature data with the four-probe technique.

---

[10] S. Naiditch (private communication).
[11] J. J. Lagowski suggested this procedure to us.
[12] R. L. Schroeder and J. C. Thompson (to be published).

Measurement of the Conductivity

For the four-probe dc technique the sample cell was calibrated against the known value of the conductivity of mercury[13] and the sample resistance was measured in the standard manner.

The electrodeless technique is a standard ac method in which a cylindrical sample is inserted into a solenoid carrying an ac current.[14] The sample becomes a core for the solenoid and the eddy currents induced in this core give rise to an apparent increase in the resistance and decrease in the inductance of the solenoid. At the frequencies employed (5–20 kc/sec), the relative resistance change on insertion of the core was about 10 times the relative inductance change and thus the resistance measurements alone were used. McLachlan[15] gives the resistance change, $R=R_{\rm in}-R_{\rm out}$, as

$$\frac{cR}{\omega L_0}=\frac{2}{x}\frac{\mathrm{ber}(x)\mathrm{ber}'(x)+\mathrm{bei}(x)\mathrm{bei}'(x)}{\mathrm{ber}^2(x)+\mathrm{bei}^2(x)},$$

where $x^2=\omega\rho\sigma a^2$, ω is the angular frequency of the current, ρ is the permeability of the core (assumed to be free-space value), σ is the conductivity of the core, a is the radius of the core, c is the ratio of the coil cross section to core cross section, and L_0 is the inductance of the coil when empty. The ber(x) and bei(x) are the Bessel real and imaginary functions and the primes denote derivatives with respect to the argument. Since the equation cannot be easily inverted, a computer was used to generate several curves of conductivity vs cR/L_0 for various values of the parameters ω and a. The solenoid resistance was measured with a Maxwell bridge for sample in and sample out, the change in resistance R was calculated, and the conductivity was then determined from the graph of conductivity vs cR/L_0. The parameter c/L_0 was obtained from a mercury calibration,[13] and a was determined by direct measurement.

RESULTS

The various curves in Fig. 1 represent the conductivity of the lithium–ammonia solutions at the listed concentrations. Note that no low-temperature data were taken with the electrodeless technique. The concentration is quoted as the mole fraction of lithium; this number (×100) will be quoted as atomic % or mole %. The slopes vary more or less smoothly from zero for the 20.0% solution to $0.044\times10^3\Omega^{-1}\cdot\mathrm{cm}^{-1}\cdot\mathrm{deg}^{-1}$ at 6.2%. In all cases the conductivity appears to be a *linear* function of the temperature. As the slope is small, the conductivity also appears to be linear on the logarithmic graph. The discontinuities in the slopes of some of the curves define points on the phase diagram. For the data presented on this graph the phase-diagram points represent the temperature and concentration at which some ammonia is first frozen out of the solution. Cooling below these points produces a more concentrated solution in addition to crystals of frozen ammonia. Figure 2 presents a continuation of the conductivity-vs-temperature curves for several concentrations exceeding the solubility limit. The curvature exhibited by these data implies that in contrast to the previous data shown in Fig. 5, the saturation line slopes towards lower metal concentrations at temperatures above 240°K. Figure 3 is a plot of α, the temperature coefficient of conductivity vs concentration. We have used

$$\alpha=(100/\sigma_2)[(\sigma_1-\sigma_2)30]\%\ \mathrm{deg}^{-1}$$

as an ordinate in order to facilitate comparison with Kraus' data.[3] The conductivities at −33.5° and −63.5°C are denoted, respectively, by σ_1 and σ_2. It may be noted that

$$-\alpha=(100/R_1)[(R_1-R_2)/30]\%\ \mathrm{deg}^{-1},$$

which is approximately what Kraus reported. The resistivities at −33.5° and −63.5°C are denoted, respectively, by R_1 and R_2. The scatter in the data is a good measure of the precision of these numbers. This scatter is due to the small size of the difference, $\sigma_1-\sigma_2$, which is of the order of the precision of the conductivity measurement, especially for the high-concentration data. Figure 4 is again conductivity vs temperature; however, these two curves are typical data for concentrated solutions at low temperatures and were taken with the four-probe technique. Data were also taken at 20.0%, 20.5%, and 22.6% concen-

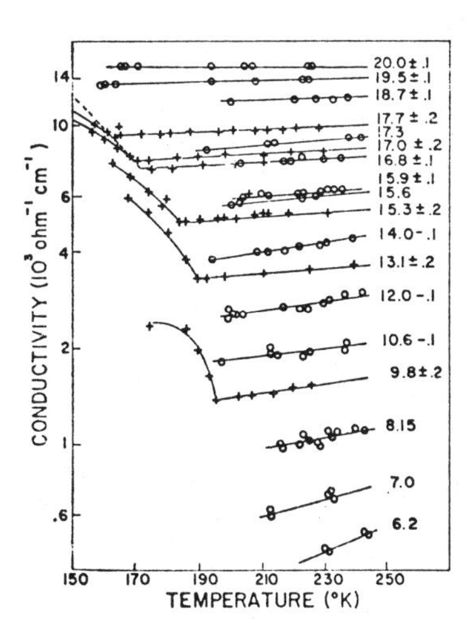

FIG. 1. Conductivity of lithium–ammonia solutions vs temperature. The circles are the electrodeless data; the crosses are the probe data. The concentrations in mole percent lithium are listed in the column to the right. This graph obscures the actual linear relation between conductivity and temperature.

[13] N. E. Cusack, Rept. Progr. Phys. **26**, 361 (1963).
[14] J. E. Zimmerman, Rev. Sci. Instr. **32**, 402 (1961).
[15] McLachlan, *Bessel Functions for Engineers* (Clarendon Press, Oxford, England, 1934).

trations in the same temperature range. These data virtually duplicate the 21.0% data insofar as phase boundaries are concerned. For the 21.0% data the curve slopes rather gently down from the right, extending beyond the break point at 89.6°K, Point A. This behavior is also typical of all the data and represents the fact that the solutions exhibit supercooling at the freezing point. The supercooling did not persist on further cooling, rather the potential across the sample dropped rapidly indicating a decrease in the resistance. When the resistance fell, warming up to 89.6°K immediately occurred. This spontaneous warming is taken as evidence of a latent heat of fusion. The conductivity rises rapidly during the transition and very little data could be taken along the near-vertical line. Below 87°K, data were again taken in detail and supercooling was again observed as indicated by the overshoot near 80°K, Point B, this time in the solid material.

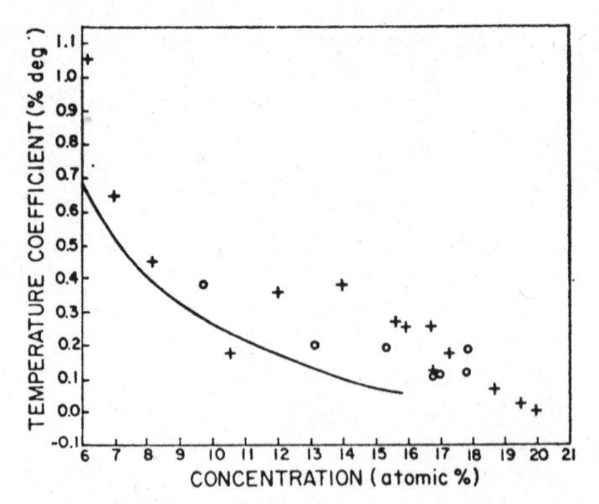

FIG. 3. Temperature coefficient of conductivity vs lithium concentration. The circles are probe measurements; the crosses are electrodeless measurements. The solid line is data for Na quoted in Ref. 1.

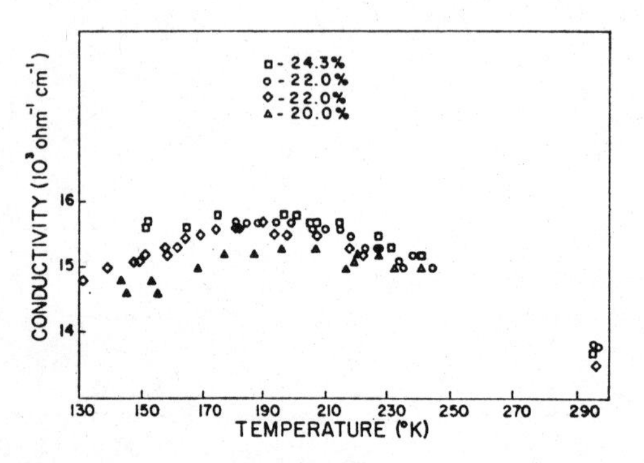

FIG. 2. Conductivity of lithium-ammonia solutions vs temperature for several concentrations exceeding the solubility limit. All of these data are from electrodeless measurements.

That the sample is solid below 89.6°K is evident from the failure of attempts to stir the solution. Measurements for the 21.0% solution end at about 80°K, as shown in Fig. 4; however, the data for the 20.0% and 20.5% solutions extend to about 78°K. In the interval 78° to 80° these curves almost exactly duplicate the shape of the 19.5% curve as shown on Fig. 4, although they are some 10% higher in conductivity. The 19.5% solution shows the typical linear behavior above approximately 142°K. At this point, ammonia begins to freeze out of the solution and the solution in effect "slides down the phase boundary," becoming more concentrated and eventually freezing as a mixture composed of the eutectic and frozen ammonia. The lower conductivity exhibited at the eutectic freezing point and below for the 19.5% solution is due to alteration of the geometric form factor by the frozen pure ammonia. We have nevertheless presented the data as conductivities to facilitate comparison in the liquid state. The first solid-solid phase transition for the 19.5% solution agrees as expected with the other data.

For the 19.5% solution the experimental apparatus was arranged to extend the range of measurements down to 65°K and yet one more solid-solid phase transition was observed. We place this transition at 69°K which corroborates fairly well the transition temperature of 67°K as reported by McDonald.[16] Figure 5 is the phase diagram for the lithium-ammonia solutions from 0% to 23%. Note that the measurements of Birch and MacDonald[4] can be brought into agreement with ours by assuming their concentration to be in error by a consistent 3 at. %. We have chosen the data of Johnson and Piskur[17] over that of Ruff and Geisel[18]

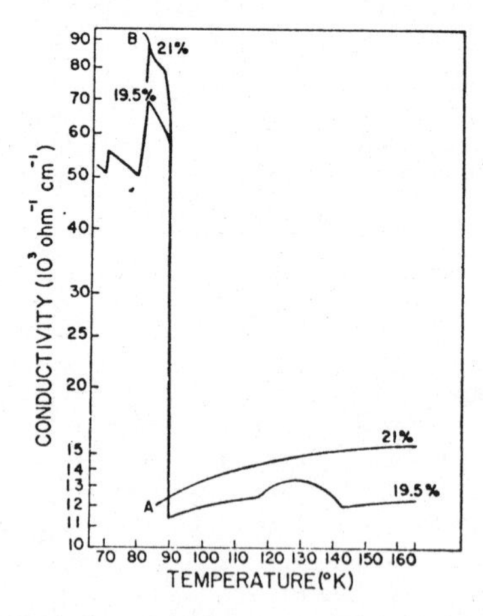

FIG. 4. Typical conductivity-vs-temperature curves for concentrations near the eutectic composition and at low temperatures. The curves are identified by their lithium concentrations in mole percent.

[16] W. J. McDonald, D. E. Bowen, and J. C. Thompson, Bull. Am. Phys. Soc. 9, 735 (1964).
[17] W. C. Johnson and M. M. Piskur, J. Phys. Chem. 37, 93 (1933).
[18] O. Ruff and E. Geisel, Ber. 39, 828 (1906).

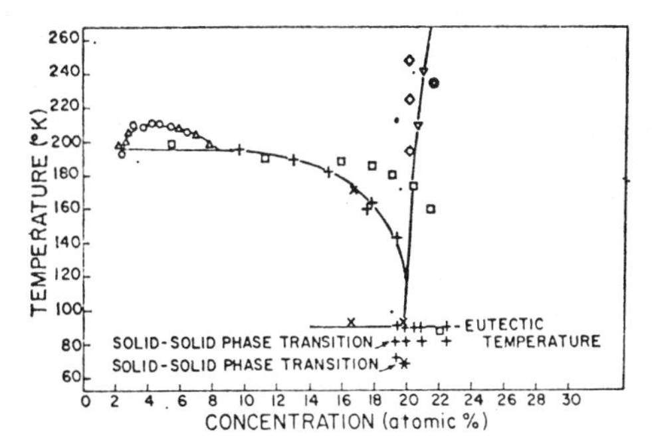

FIG. 5. Phase diagram for lithium-ammonia solutions. The triangles (base down) are data from P. D. Schettler and A. J. Patterson, J. Phys. Chem. **68**, 2865, 2870 (1964). The circles are data taken from a report of otherwise unpublished data of Ogg and Loeffler quoted in Ref. 8, the points represented by x's are from Ref. 9, the boxes present data taken from Ref. 4, the diamonds locate data taken from Ref. 18, the triangles (base up) are data taken from Ref. 17, the asterisk represents data taken from Ref. 16, and the double circle presents data from C. A. Kraus and W. C. Johnson, J. Am. Chem. Soc. **47**, 731 (1925). The crosses are our own probe measurements.

for the saturation line mostly on the basis of age, the Johnson and Piskur data being some 30 years younger. The freezing points and the first solid-solid phase-transition points which we report are plotted at their nominal concentration. We found it unnecessary to make any correction in the concentration for the probe measurements for the data taken in the low-temperature region. The curve for each of the other concentrated solutions was extrapolated to 240.2°K, and comparison was made with the conductivity as measured by the electrodeless technique.

Figure 6 is a graph of conductivity vs concentration for lithium-, sodium-, and potassium-ammonia solutions at 240°K. Note that the conductivities at all concentrations are in the order Li: Na: K. The measurements of cesium-ammonia solutions,[12] show the conductivity of Cs to be below that of K.

DISCUSSION

The conductivities reported here clearly place the concentrated lithium-ammonia solutions in the metallic range. The linear dependence of conductivity on temperature further excludes the possibility of a semiconducting material for which an exponential temperature dependence would be expected. Moreover, the Hall-effect data of Kyser and Thompson[6] indicate free-electron behavior above 4 at. %. Thus the positive temperature coefficient of conductivity in the liquid-metal-ammonia solutions, which some recent contributors have taken as an indication of thermally activated carriers, must instead be associated with structure in the solutions.[19] It should be noted that liquid zinc also shows a positive temperature coefficient of conductivity near its melting point.[13] One may further bolster the contention that these solutions are of a metallic character by adducing the *negative* temperature coefficient of conductivity in the solid which is just that expected for a metal. As shown in Fig. 6, the solution conductivity depends on a high power of the mole fraction x (approximately x^3), but decreases to $\sigma \propto x$ above 16 mole %. It is impossible to explain this strong variation of the conductivity with concentration by any simple model in the absence of thermally excited carriers.[6]

We confirm the eutectic point as reported by Jaffe,[9] Birch, and MacDonald,[4] and others; however, we do not confirm the location of the line at which ammonia freezes out of more dilute solutions nor the eutectic composition reported by Birch and MacDonald. With the amount of decomposition known in our four-probe experiments and unknown in the work of Birch and MacDonald, we feel that the present report is to be taken as the more reliable. The conductivity ratio between solid and liquid is over twice as high as any reported for a pure metal.[13]

Relative to the suggestion by Sienko[8] that a maximum in the melting curve should be expected at the $Li(NH_3)_4$ concentration, we find no evidence in the structure of the conductivity-vs-temperature curves that would suggest more than one minimum at the eutectic freezing point. Our data scatter by less than 0.5 K°. Indeed, all of our data show that a solution when cooled will "slide down the phase boundary," freezing out ammonia and becoming more concentrated in a smooth fashion until the $Li(NH_3)_4$ composition is reached. Such a solution then freezes in a fashion to be expected at a eutectic. Further, in examinations of solutions with concentrations in excess of the eutectic composition, no change in the freezing point of the solutions was observed. For solutions with nominal concentrations

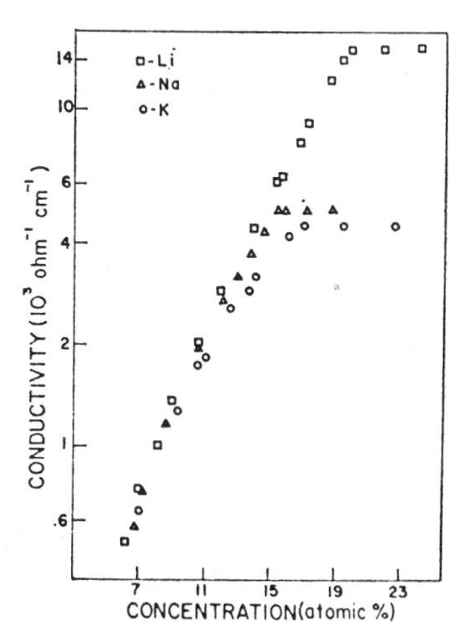

FIG. 6. Conductivity of metal-ammonia solutions vs metal concentration. The lithium data are our electrodeless data. The sodium and potassium data are from Ref. 3.

[19] J. M. Ziman, Phil. Mag. **6**, 1013 (1961).

above 20%, data were taken at high enough temperatures to have crossed the metal saturation line should solubility in excess of 20% occur at low temperatures. As no variation in the freezing point of these solutions above 20% was found, and as no structure indicating a variation in the location of the saturation line at low temperatures and high concentrations was observed in the conductivity-vs-temperature curves in this region, we conclude that solutions with concentrations slightly above or below 20% have higher freezing points. That is, solutions slightly less concentrated freeze out solid NH_3 as they cool, finally reaching the 20% concentration at 89.6°K. Solutions with a nominally higher metal content actually have an essentially constant metal concentration of 20% at temperatures below 0°C, and also freeze at 89.6°K.

The nature of the solid material is nevertheless unclear. The material frozen at the Li–NH_3 eutectic differs from that frozen at the sodium or potassium eutectic in at least three ways. First, the solid retains the bronze color and metallic luster of the liquid. Solid sodium–ammonia eutectic is gray-white in color and of low reflectivity. Second, the conductivity of Li–NH_3 increases on freezing. Frozen sodium– or potassium–ammonia has a conductivity lower than the liquid.[4] Third, the vapor pressure of the lithium–ammonia solutions is lower than that of sodium– or potassium–ammonia solutions.[9] These observations are in better accord with compound formation than with the existence of a normal lithium–ammonia eutectic or solid solution. The compound need not be $Li(NH_3)_4$; if the 20%–89°K point is a true eutectic, then compounds richer in metal than $Li(NH_3)_4$ are indicated. On the other hand, a "suppressed maximum" in the melting curve (peritectic)[20] may occur below 89°K at 20%, whereupon $Li(NH_3)_4$ would exist though not in equilibrium with the liquid. Sienko based his suggestion of a hump in the melting point curve at $Li(NH_3)_4$ on the conventional behavior of binary systems. The steepness of the melting curves on either side of the eutectic is unconventional, so also may be the character of the solid formed at the eutectic.

SUMMARY

The properties of the solid are not those of a mixture or solid solution of lithium and ammonia. They are, rather, consistent with the existence of a compound of lithium and ammonia with a mole ratio greater than or equal to 1:4. Information on the phase diagram at higher temperatures and pressures and higher metal concentrations will be required to determine the composition of this compound.

We report two distinct transitions in the solid phase, one at 82°K which is first order, and the other near 69°K. The presence of a latent heat for the 82°K transition indicates that it is first order. The experimental arrangement precluded the observation of a latent heat for the 69°K transition. Such transitions are unlikely for a eutectic mixture. It is not possible for the discontinuity in conductivity, by which the second solid–solid phase transition was located, to be explained on the basis of the frozen solution contracting or breaking away from the electrodes, as McDonald's[16] report of this transition is based on an electrodeless measurement of the conductivity. In the absence of x-ray studies of the various phases reported here, no attempt at predicting the crystal structure is made. Somewhat similar transitions are observed in gallium,[21] barium,[22] and the tungsten bronzes,[23] though the first two require high pressures to reveal the other phases and the last shows transitions near 600°K.

Comparison of the solutions of different alkali metals shows that up to about 16% the conductivities decrease only slightly as the metal is changed from Li to Na to K to Cs. In the pure-liquid metals the order of the conductivities is Na, K, Li, Cs; that is, Li is anomalous. We infer that in the solutions the relevant property of the metal ion is its size and not its potential. The high solubility and consequent high conductivity of Li is thus attributed to its small size. Further, we can infer that the coordinated ammonias around the ion tend to smooth the potential of the ion and thus the electron sees a potential approximately independent of the metal. At 240°K the temperature coefficient of conductivity is positive and slightly higher in Li solutions than in the other solutions and tends to zero at saturation (20%). Na and K solutions also have positive temperature coefficients which tend toward zero as their respective saturation points are approached (about 16%). Naiditch[24] has shown that the coefficient changes sign at about 350°K in unsaturated sodium–ammonia solutions, and similar behavior probably occurs in the lithium–ammonia solutions. This is reminiscent of zinc, though metal–ammonia solutions are the only "monovalent metals" with this behavior.[19]

[20] J. E. Ricci, *The Phase Rule* (D. Van Nostrand Company, Inc., New York, 1951), p. 75.

[21] A. Jayaraman, W. Klement, Jr., R. C. Newton, and G. S. Kennedy, J. Phys. Chem. Solids **24**, 7 (1963).

[22] D. E. Bowen and B. C. Deaton, Appl. Phys. Letters **4**, 97 (1964).

[23] H. R. Shanks, P. H. Sidles, and G. C. Danielson, Advan. Chem. **39**, 237 (1963).

[24] S. Naiditch, in *Metal Ammonia Solutions*, edited by G. Lepoutre and M. J. Sienko (W. A. Benjamin, Inc., New York, 1963).

An Optical and ESR Study
of Europium in Ammonia

Solutions of europium in ammonia are similar in most respects to those of alkaline earth metals, but are different in that the cation, Eu^{2+}, is paramagnetic. In the following paper, Catterall and Symons describe their observation of separate esr spectra for Eu^{2+} and e^{-}_{am}, leading them to the conclusion that these species exist as pairs of solvated ions rather than as centrosymmetric units. Thompson, Hazen, and Waugh,[1] in an essentially independent study of the same phenomena, obtained similar results but interpreted them somewhat differently.

---

[1]D. S. Thompson, E. E. Hazen, and J. S. Waugh, *J. Chem. Phys.*, **44**, 2954 (1966).

689. *Unstable Intermediates. Part XXXI.* Solvated Electrons: Solutions of Europium in Ammonia*

48

By R. Catterall and M. C. R. Symons

The blue solution that results when europium metal dissolves in ammonia has been studied by optical and electron spin resonance spectroscopy. It is found that dilute solutions can best be described in terms of europium(II) cations and solvated electrons. That separate spin resonance spectra can be observed for these two species leads to the conclusion that they are well separated in the form of loose ion-pairs rather than as centrosymmetric units similar to the monomer described by Becker, Lindquist, and Alder.[1]

Europium metal has been reported to dissolve in liquid ammonia to form a blue solution,[2] and one aim of this research has been to determine whether this solution is comparable to those of the alkali and alkaline-earth metals.[3] It has been suggested [2] that the solubility of europium is related to its relatively low density and high volatility, and to the high stability of the europium(II) ion ($4f^7$ configuration). This solution thus appears to be analogous to those of the alkaline-earth metals, but is unique in that the cation itself is paramagnetic.

A second aim has been to probe the nature of the paramagnetic, non-conducting species which has been postulated [4] to account for the magnetic [5] and conductometric [4] properties of metal solutions. Two models have been suggested for this unit, a loose ion-pair,[6] and a monomer [1] having a relatively high spin-density on the cation. Although recent nuclear magnetic resonance studies of ammonia solutions [7] and electron spin resonance studies of amine solutions [8] appear to favour a centrosymmetric monomer, this conclusion is not compelling.[9] Addition of excess of cations (as alkali halides) to the solutions in an attempt to constrain the equilibria was complicated by a strong electron–anion interaction.[10] In europium solutions, the paramagnetic cation can act as a probe for differentiating between the two models.[11]

Interactions between paramagnetic ions of opposite charge in aqueous solutions at room temperature have been studied by Pearson and Buch,[12] who used paramagnetic metal salts of diamagnetic anions and potassium nitrosyldisulphonate, $2K^+[ON(SO_3)_2]^{2-}$. The three narrow lines ($\Delta H_{ms} = 0{\cdot}35$ gauss, $|A|_N = 13$ gauss) obtained from the dilute nitrosyldisulphonate solutions are broadened by increasing concentration of paramagnetic cations [such as the hexa-aquochromium(III) ion] at an initial rate of about 1000 gauss mole^{-1}. This contrasts markedly with the initial self-broadening of 10 gauss mole^{-1} associated with increase of anion concentration in the absence of a paramagnetic cation. Lanthanide ions in general were much less efficient in broadening the anion resonance, with the exception of the hexa-aquogadolinium(III) ion [$4f^7$, isoelectronic with europium(II)], which showed a strong broadening effect identical with that of the transition-metal ions.

The ground state of the europium atom ($4f^75s^25p^66s^2$) and the bivalent ion ($4f^75s^25p^6$) are both spherically symmetrical ($^8S_{7/2}$). Hyperfine structure of the electron spin resonance absorption arising from interaction with the nuclear magnetic dipole was first observed by Bleaney and Low,[13] but fluid solutions have not been studied by this technique. Magnetic parameters for the ion in various host crystals are collected in Table 2. Individual hyperfine components in crystals are sufficiently narrow to allow resolution of two sextets from the two naturally occurring europium isotopes (Table 2).

The two broad absorption bands of the europium(II) ion in the near ultraviolet, which are attributed [14,15] to the $4f^7 \longrightarrow 4f^65d^1$, or possibly [14,16] to the $4f^7 \longrightarrow 4f^66s$ transitions, show a marked dependence on the nature of the host lattice.[14–17] The lower-energy band is characterised by a tail that extends into the visible and sometimes shows underlying

* Part XXX, M. J. Blandamer, L. Shields, and M. C. R. Symons, preceding Paper.

structure. In addition, sharp bands in the 31,000 cm.$^{-1}$ region have sometimes been recorded (Table 1).

Solutions of the europium(III) ion in water, methanol, and ethanol show characteristic visible absorptions [18] at 17,240, 19,050, and 21,510 cm.$^{-1}$.

Experimental

Materials.—Europium metal of 99·9% purity and reagent grade liquid ammonia were employed. The purification of the ammonia and the treatment of glassware have been described elsewhere.[10] Since europium metal cannot be distilled in Pyrex vessels, clean pieces of metal were cut and transferred under dry, oxygen-free petroleum ether, which was finally removed under high vacuum.

Preparation of Solutions.—Solutions of europium metal in liquid ammonia were prepared by conventional high-vacuum techniques and a modification of the method described for alkali-metal solutions.[10] The solutions strongly resemble those of calcium in their affinity for ammonia, and in their tendency to phase-separation at high temperatures. 0·1M-Solutions were prepared by mixing known amounts of ammonia and metal; more dilute solutions were prepared by dilution of 0·1M-solutions, or by decomposition of metal solutions to the amide. Samples for optical and electron spin resonance spectroscopy were sealed in 4 mm. diameter Pyrex tubing and stored at 77°K.

Solutions of europium(II) iodide were prepared under conditions of high vacuum by decomposing solutions of the metal in liquid ammonia with an approximately equimolar quntity of ammonium iodide. A bulky, yellow, crystalline solid was precipitated, leaving a pale yellow solution. The iodide was observed to be considerably more soluble at higher temperatures, and samples for spectroscopy were prepared by decanting the saturated solution at −78°. More dilute samples were prepared from these solutions by the addition of an approximately five-fold excess of ammonia. Samples were sealed in 4 mm. diameter tubing before being warmed to room temperature. It has been observed that europium(II) salts decompose slowly in aqueous solution,[14] but no such behaviour was observed in liquid ammonia, and samples were stored at room temperature.

Electron Spin Resonance Apparatus.—The 3 cm. apparatus used has been described previously.[10,19] Measurements at −23 and −78° were made by surrounding the microwave cavity with a bath of solid–liquid carbon tetrachloride or of solid carbon dioxide ethanol.

Optical Absorption Measurements.—Optical spectra from the near ultraviolet to the near infrared were measured at room temperature with a Unicam S.P. 700 double-beam, recording spectrophotometer. The position of the infrared maximum was located by reference to the ammonia absorptions near 5000 cm.$^{-1}$. Samples were contained in cylindrical Pyrex cells of approximately 4 mm. internal diameter, the reference beam passing through a similar cell (made from the same length of tubing) containing liquid ammonia.

Results

Europium(II) *Iodide Solutions.—Optical absorption spectra.* The solutions showed a weak absorption band at 27,400 cm.$^{-1}$ with a broad, structureless, low-energy tail extending into the visible. This band is identified with the lower-energy transition reported previously for the europium(II) ion (Table 1) and has an extinction coefficient of approximately 4: the work of Low [15] indicates a value of about 10. A sharp band at 31,400 cm.$^{-1}$, similar to bands observed in aqueous solutions [14] and in calcium fluoride [15] (Table 1), is attributed to a weak, forbidden transition within the f^7 configuration of europium(II). The optical cut-off at 34,000 cm.$^{-1}$ is presumably due to the intense iodide charge-transfer-to-solvent absorption [20] (37,000 cm.$^{-1}$ at room temperature in liquid ammonia), the higher-energy europium(II) absorption, and solvent absorption.

No visible absorption bands characteristic of the europium(III) ion were observed.

Electron spin resonance spectra. The parameters (Table 2) derived from the resolved hyperfine components of the spectra at room temperature and −78° (Figures 1*a* and *b*) are very similar to those found for the europium(II) ion in other media (Table 2). These spectra are identical with those obtained by Lorentzian reconstruction (Figure 1*c*) from the observed hyperfine splitting constants and *g*-factor (Table 2) and line-widths of 15

TABLE 1

Absorption bands of the europium(II) ion in various media

	$\nu_{max.}$ (cm.$^{-1}$)				
	Strong, broad bands				
Medium	I	II	Weak, narrow bands		Ref.
CaF_2	29,000	45,050	24,200 35,400	32,300 37,000	15
CaF_2	27,150	45,050			16
SrF_2	27,870	43,480			16
BaF_2	28,470	42,700			16
NaCl	29,410	41,530			*a*
KCl	29,730	41,370			*a*
KCl	29,150	40,130			17
KBr	29,730	38,140			*a*
KBr	29,000	40,210			17
$SrCl_2$	30,570	—			*b*
H_2O	31,200	40,320	30,960 31,450	31,200 31,700	14
EuI_2–NH_3	27,400	*c*	31,400		*d*
Eu metal–NH_3 [e]	27,400				*d*

[a] J. Kirs and L. A. Niilisk, *Trudy Inst. Fiz. i Astron. Akad. Nauk Est., S.S.R.*, 1962, **18**, 36. [b] S. Freed and S. Katcoff, *Physica*, 1948, **14**, 17. [c] Masked by absorption of iodide ions and solvent. [d] This work. [e] An additional band, at 5600 cm.$^{-1}$, arises from a transition of the solvated electron.

and 25 gauss, respectively. The widths of the hyperfine components in solution are appreciably greater than those observed for the ion in strontium sulphide [13] (4 gauss) and in barium fluoride [21] (6 gauss).

Europium Metal Solutions.—Decomposition products. The fading of the blue colour of dilute solutions is accompanied by the precipitation of an orange, flocculent solid, presumably europium(II) amide. No trace of the optical absorptions of amide or of

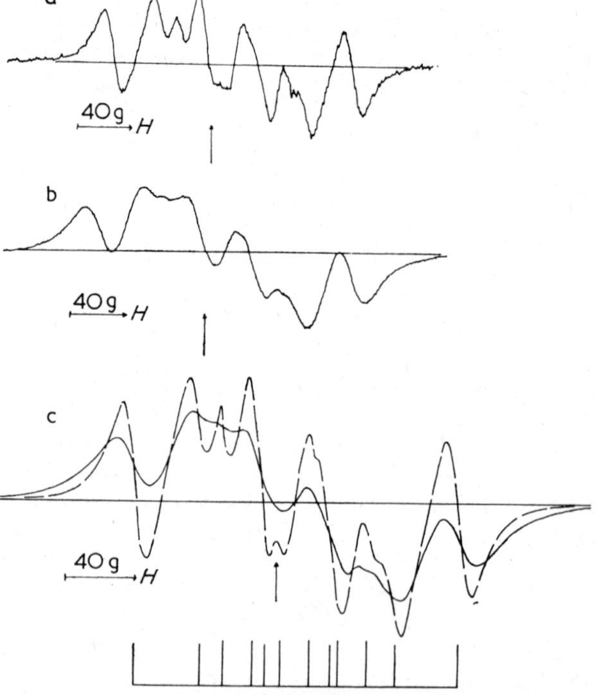

FIGURE 1. Electron spin resonance spectra of dilute solutions of europium(II) iodide in liquid ammonia: *a*, room temperature; *b*, −78°; *c*, obtained by Lorentzian reconstruction from the observed hyperfine splitting constants and *g*-factor (Table 1) and line widths of 15 and 25 gauss; $g = 2{\cdot}000$ is indicated by the vertical arrow.

europium(II) ions could be detected in the fully decomposed solutions. Also, no electron spin resonance absorption due to the europium(II) ion could be found. We conclude that the solubility of the decomposition products is negligible for the purpose of this work.

TABLE 2

Electron spin resonance characteristics of the europium(II) ion in various media

Medium	Temp. (°K)	g-Factor	^{151}Eu|A| [a] (10^{-4} cm.$^{-1}$)	^{153}Eu|A| [b] (10^{-4} cm.$^{-1}$)	Ref.
SrS	R.T.[c]	1·9912(10) [d]	30·82(20)	13·79(20)	13
BaF_2	77	1·993(1)	33·5(6)	14·9(6)	21
CaF_2	13	1·9926(3)	34·3263(4)	15·2350(8)	e
EuI_2–NH_3	R.T.	1·9921(4)	32·90(10)	14·60(10) [f]	g
EuI_2–NH_3	195	1·993(1)	33·0(3)	—	g
Eu metal–NH_3	R.T.	1·992(1)	33·1(3)	—	g
Eu metal–NH_3	195	1·9922(4)	32·96(10)	—	g

[a] ^{151}Eu; natural abundance 47·77%, nuclear spin × ($h/2\pi$) = 5/2, nuclear moment × ($eh/4\pi$) = 3·4. [b] ^{153}Eu; natural abundance 52·23%, nuclear spin × ($h/2\pi$) = 5/2, nuclear moment × ($eh/4\pi$) = 1·5. [c] Room temperature, 21° ± 2°. [d] Figures in parentheses denote the original authors' estimate of the uncertainty in the last figure. [e] J. M. Baker, J. P. Hurrell, and F. I. B. Williams, "Paramagnetic Resonance," ed. W. Low, Academic Press, New York, 1963, **1**, 202; J. M. Baker and F. I. B. Williams, *Proc. Roy. Soc.*, 1962, *A*, **267**, 283. [f] ^{153}Eu|A| calculated from the measured ^{151}Eu|A| and the hyperfine splitting ratio determined by electron nuclear double resonance; *e* above. [g] This work.

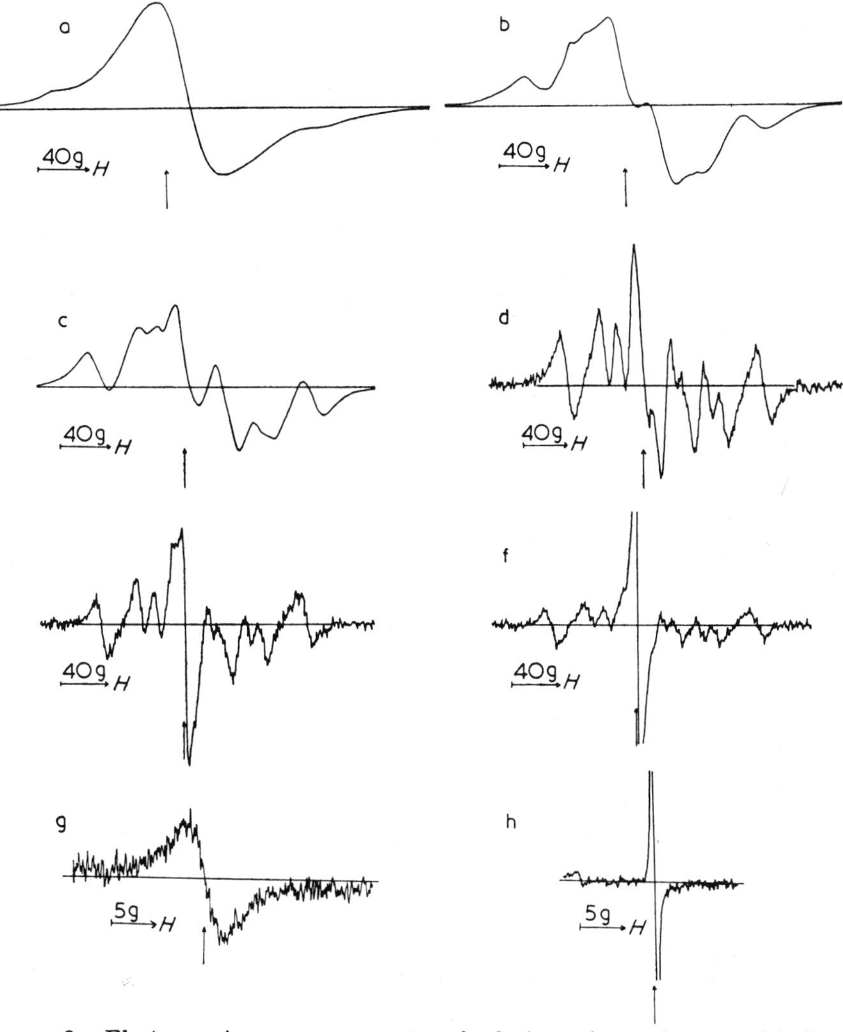

FIGURE 2. Electron spin resonance spectra of solutions of europium metal in liquid ammonia at room temperature: *a*, 0·1M; *b*, 0·001M; *c* to *f*, progressive dilution; *g*, approximately 5 × 10^{-5}M; *h*, approximately 10^{-7}M; $g = 2·0004$ is indicated by the vertical arrow.

Optical absorption spectra. Solutions of europium metal had a very strong absorption in the near infrared at all concentrations up to 0·1M (the most concentrated solution examined). This absorption had a maximum at 5600 cm.$^{-1}$ and a broad high-energy tail extending into the visible and giving rise to the blue colour. The band was completely absent in the decomposed solutions. A second optical band at 27,400 cm.$^{-1}$, detected only in the more concentrated solutions, was identical with the lower-energy transition of the europium(II) ion (Table 1).

Electron spin resonance spectra. The spectra were strongly dependent on both concentration and temperature. The effect of changing metal concentration is illustrated in Figure 2 for solutions below 0·1M at room temperature. The resonance signal obtained from a 0·3M-solution consisted of a symmetrical, Lorentzian line having $\Delta H_{ms} =$ 28·5 gauss and $g = 1{\cdot}9925 \pm 0{\cdot}0004$. Solutions 0·1M in metal had asymmetric spectra ($\Delta H_{ms} = 42{\cdot}6$ gauss, $g = 1{\cdot}9857$) (Figure 2*a*) and showed traces of the outermost ^{151}Eu hyperfine components. Below this concentration, the asymmetry was lost (Figures 2*b* and *c*), the spectra being similar to those obtained from europium(II) iodide solutions, except for an enhancement of one of the inner lines (Figure 1). On further dilution, this feature was better resolved (Figures 2*d* and *e*) and finally the europium(II) spectrum became too weak to detect, leaving a single, symmetrical line (Figure 2*g*) having ΔH_{ms} ~4 gauss, $g = 2{\cdot}0004 \pm 0{\cdot}0002$, which narrowed on further dilution to about 0·3 gauss, without any change in the *g*-factor, before also becoming undetectable.

This line strongly resembles that observed for solutions of alkali metals in ammonia. Hutchison and Pastor [5] found widths of about 0·025 gauss and a *g*-factor of $2{\cdot}0012 \pm 0{\cdot}0002$ for potassium solutions, but we have recently [22] compared electron spin resonance spectra for all the alkali metals at 0·1m and find the *g*-factors somewhat lower, and markedly greater line-widths for cæsium solutions (0·2 and 0·4 gauss at 0·1 and 0·2m, respectively).

Metal:	Li	Na	K	Rb	Cs	Eu	Ca, Ba, Sr [a]
g-Factor [b] (room temp.)...	2·0006	2·0004	2·0006	2·0004	2·0002	2·0004	2·001
,, (−78°)	2·0008	2·0008	2·0009	2·0005	2·0005	—	—

[a] K. D. Voss, Ph.D. Thesis, Michigan State University, 1962 (estimated error ±0·001). [b] Estimated error ±0·0002, with reproducibility ±0·00005.

The effect of lowering the temperature for a 0·001M-solution may be seen by comparing Figures 2*b* and 3. The 0·1M-solutions showed a very similar improvement in resolution,

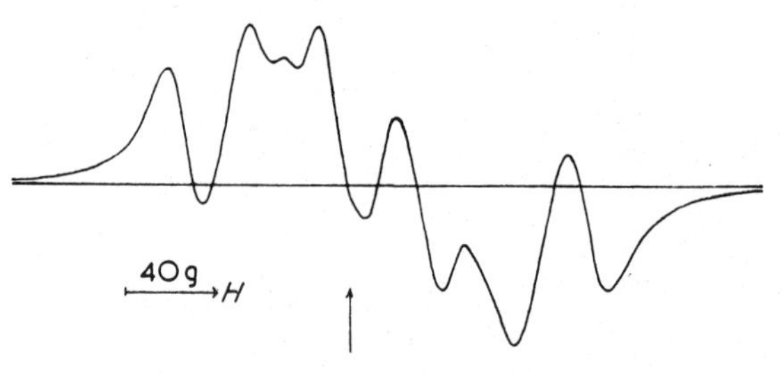

FIGURE 3. Electron spin resonance spectrum of a 0·001M-solution of europium metal in liquid ammonia at −78°. $g = 2{\cdot}000$ is indicated by the vertical arrow.

and spectra at −23° had intermediate resolution. This improvement in resolution with decreasing temperature contrasts with the decreasing resolution in europium(II) iodide solutions (Figure 1).

Correlation of optical and electron spin resonance spectra. The decay of europium metal solutions was followed in the extremely dilute range by concurrent optical and electron spin resonance spectroscopy. The relationship between electron spin resonance line-widths and optical density at 5600 cm.$^{-1}$ was almost linear. If an extinction coefficient of 10^5 is assumed,[9] the optical data can be used as a measure of concentration and hence the equation

$$\Delta H_{ms} = 0{\cdot}30 + (7{\cdot}8 \times 10^4)\text{M}$$

can be derived, M being the concentration of the solution between 5×10^{-5} and 10^{-7}M.

Discussion

The Nature of Europium Metal Solutions.—The present results are in good accord with the concept that the metal in dilute solutions of europium in ammonia is fully converted into europium(II) cations and solvated electrons. Thus, the hyperfine coupling constant, the g-factor, and the optical spectra indicate extensive formation of europium(II). Europium atoms are known to have a smaller hyperfine coupling constant,[23] whilst europium(I) and europium(III) would not be expected to give rise to detectable electron spin resonance spectra in fluid solutions. Absence of the latter ion is confirmed by the absence of its optical absorption bands.

In addition, electron spin resonance results are characteristic of solvated electrons in ammonia, except that the resonance spectra are considerably broader than usual for all but the most dilute solutions. The optical spectra in the near-infrared region also show no differences and, in particular, the unusual band-shape thought to be characteristic of solvated electrons [24] is a prominent feature of the europium solutions. The following discussion is accordingly based on the concept that the metal ionises completely to europium(II) and solvated electrons. We stress that, although the solutions slowly decomposed, this did not give rise to an accumulation of europium(II) ions, since the product of decomposition, presumably the amide, was quantitatively precipitated.

Cation–Electron Interactions.—The properties of alkali-metal solutions in ammonia are generally discussed in terms of the equilibria

$$M^+ + e^-_{solv.} \rightleftharpoons M \qquad (1)$$

$$M \rightleftharpoons \tfrac{1}{2}M_2 \qquad (2)$$

$$e^-_{solv.} \rightleftharpoons \tfrac{1}{2}(e_2{}^{2-})_{solv.} \qquad (3)$$

$$M + e^-_{solv.} \rightleftharpoons M^- \qquad (4)$$

where M^+ is an alkali-metal cation, M is either an " ion-pair " ($M^+_{solv.}$ - - $e^-_{solv.}$) [6] or the expanded metal monomer of Becker *et al.*[1] M_2 is either a cluster of four ions,[9] the electrons in separate cavities being paired, or an expanded metal dimer.[1] The species $(e_2{}^{2-})_{solv.}$ comprises two electrons paired in a single solvent cavity,[24,25] and M^- the two-electron, spin-paired analogue [26] of the ion-pair ($e^-_{solv.}$ - - - $M^+_{solv.}$ - - - $e^-_{solv.}$), or of the monomer. These species can be suitably modified to accommodate a bivalent cation such as calcium(II) or europium(II), M then being written as $M(II)^+$ and M^- as M(II).

Currently, equilibria (1) and (2) are generally invoked to explain magnetic and conductometric results, though it has been suggested [26] that the additional equilibrium (4) is necessary to reconcile the equilibrium constants derived from the two approaches. Also, O'Reilly [27] has used a scheme involving equilibria (1) and (3) or (4) to accommodate his recent nuclear magnetic resonance results.

Just as the observation of hyperfine coupling to a single cation nucleus in various alkali-metal–amine solutions [8] seems to show that monomer formation is the dominating mode for cation–electron interaction in those solvents, so the detection of separate electron spin resonance spectra for europium(II) ions and solvated electrons in europium solutions under conditions of extensive ion-pairing shows, in contrast, that in ammonia, cation–electron interaction involves loose ion-pairing. The marked broadening, as the concentration is increased, of the singlet assigned to solvated electrons confirms that such weak interaction is occurring.

We conclude that centro-symmetric monomer units are not important constituents of these solutions, and hence that they are even less likely to be significant in solutions of univalent metals.

Electron–Electron Interactions.—The present results also give some information about electron pairing in solutions of metals in ammonia. Thus we find that the unit $Eu^{2+}_{solv.}$ - - - $e^-_{solv.}$ does not give rise to a singlet or a triplet state, but remains effectively as two doublet-state species. If the electron-paired species were either ($Eu^{2+}_{solv.}$ - - $e^-_{solv.})_2$

or $e^-_{solv.}$ - - $Eu^{2+}_{solv.}$ - - $e^-_{solv.}$, it would be difficult to understand why the two $e^-_{solv.}$ units should " combine " to give a deep singlet state (deep because the triplet level of the unit is not thermally populated), whilst neither electron interacts in this way with the neighbouring europium(II) cations. If this conclusion is correct, we are forced to consider the $(e_2^{2-})_{solv.}$ species as being of importance. Other factors in favour of this unit, which will, of course, generally be loosely associated with cations, will be presented later.[28]

Line-broadening Factors.—The mechanism by which the electron spin resonance lines are mutually broadened presumably arises from magnetic dipole forces since, for the loose ion-pair, spin-exchange effects will be unimportant.

The marked broadening of the electron spin resonance line assigned to the solvated electrons is similar to that found for the anion $[ON(SO_3)_2]^{2-}$ in the presence of various paramagnetic cations,[12] but the rate of broadening with increase in concentration is far greater (80,000 gauss mole^{-1} compared with 1010 gauss mole^{-1}). This is presumably due both to the greater mobility of solvated electrons and to their far greater delocalisation relative to the unpaired electron in $[ON(SO_3)_2]^{2-}$

By arbitrarily assuming a line-width of about 5 gauss for the solvated electron in the ion-pair, the relative concentrations of free and ion-paired solvated electrons in europium solutions can be estimated very roughly and compared with those in alkali metal solutions.

Europium solutions

Concentration (M)	10^{-6}	10^{-5}	3×10^{-5}	5×10^{-5}
$\frac{100[M]}{[e^-] + [M]}$	4	11	16	42

Alkali-metal solutions

Concentration (M)	5×10^{-4}	5×10^{-3}	$1{\cdot}25 \times 10^{-2}$	0·4
$\frac{100[M]}{[e^-] + [M]}$ (i) from conductance data [a]	4	12	13	30
(ii) from magnetic data [b] ...	1·5	9	13·5	30

[a] D. S. Berns, G. Lepoutre, E. A. Bockelman, and A. Patterson, jun., *J. Chem. Phys.*, 1961, **35**, 1820. [b] Ref. 1.

Concentrated Solutions.—The onset of marked broadening and asymmetry of the electron spin resonance band as the concentration of metal is increased is taken as an indication of the formation of solvent-deficient ion-clusters that are of sufficient size to have the character of colloidal metal particles. This occurs for europium solutions at a far lower concentration than is found for solutions of alkali metals (Figure 2), but the behaviour seems to be comparable with that of calcium solutions.[29]

It is significant that this onset moves to higher concentrations of metal as the temperature is lowered. Since at lower temperatures electron-pairing is favoured, this result is somewhat surprising. It is presumably a result of the decrease in the dielectric constant of the solvent on warming, with a consequent increase in ion-clustering.

We thank the D.S.I.R. for financial support.

DEPARTMENT OF CHEMISTRY, THE UNIVERSITY, LEICESTER. [*Received, December 29th*, 1964.]

[1] E. Becker, R. H. Lindquist, and B. J. Alder, *J. Chem. Phys.*, 1956, **25**, 971.
[2] J. C. Warf and W. L. Korst, *J. Phys. Chem.*, 1956, **60**, 1590.
[3] M. C. R. Symons, *Quart. Rev.*, 1959, **13**, 99.
[4] C. A. Kraus, *J. Amer. Chem. Soc.*, 1908, **30**, 1323; 1921, **43**, 749.
[5] C. A. Hutchison, jun., and R. C. Pastor, *J. Chem. Phys.*, 1953, **21**, 1959.
[6] C. A. Kraus, *J. Chem. Educ.*, 1953, **30**, 83.
[7] H. M. McConnell and C. H. Holm, *J. Chem. Phys.*, 1957, **26**, 1517; J. V. Acrivos and K. S. Pitzer, *J. Phys. Chem.*, 1962, **66**, 1693.
[8] K. D. Vos and J. L. Dye, *J. Chem. Phys.*, 1963, **38**, 2033; K. Bar-Eli and T. R. Tuttle, jun., *ibid.*, 1964, **40**, 2508.
[9] M. Gold, W. L. Jolly, and K. S. Pitzer, *J. Amer. Chem. Soc.*, 1962, **84**, 2264; M. Gold and W. L. Jolly, *Inorg. Chem.*, 1962, **1**, 818; C. Hallada and W. L. Jolly, *ibid.*, 1963, **2**, 1076.
[10] R. Catterall and M. C. R. Symons, *J.*, 1964, 4342.

[11] R. Catterall and M. C. R. Symons, *J. Chem. Phys.*, 1965, **42**, 1466.
[12] R. G. Pearson and T. Buch, *J. Chem. Phys.*, 1962, **36**, 1277.
[13] B. Bleaney and W. Low, *Proc. Phys. Soc.*, 1955, *A*, **68**, 55.
[14] F. D. S. Butement, *Trans. Faraday Soc.*, 1948, **44**, 617.
[15] W. Low, *Nuovo cim.*, 1960, **17**, 607.
[16] A. A. Kaplyanskii and P. P. Feofilov, *Optics and Spectroscopy*, 1963, **13**, 129.
[17] R. Reisfeld and A. Glesner, *J. Opt. Soc. Amer.*, 1964, **54**, 331.
[18] E. V. Sayre, D. G. Miller, and S. Freed, *J. Chem. Phys.*, 1957, **26**, 109.
[19] J. A. Brivati, N. Keen, and M. C. R. Symons, *J.*, 1962, 237.
[20] T. R. Griffiths and M. C. R. Symons, *Trans. Faraday Soc.*, 1960, **56**, 1125.
[21] V. M. Vinokurov, M. M. Zaripov, V. G. Stepanov, G. E. Chirkin, and L. Ya Shekun, *Fiz. Tverd. Tela.*, 1963, **5**, 1936 (translation in *Soviet Physics-Solid State*, 1964, **5**, 1415).
[22] R. Catterall and M. C. R. Symons, unpublished results.
[23] P. G. H. Sandars and G. K. Woodgate, *Proc. Roy. Soc.*, 1960, *A*, **257**, 269.
[24] M. J. Blandamer, R. Catterall, L. Shields, and M. C. R. Symons, *J.*, 1964, 4357.
[25] S. Freed and N. Sugarman, *J. Chem. Phys.*, 1943, **11**, 354; R. A. Ogg, jun., *ibid.*, 1946, **14**, 114.
[26] E. Arnold and A. Patterson, jun., "Metal–Ammonia Solutions," eds. G. Lepoutre and M. J. Sienko, Benjamin, New York, 1964, p. 285.
[27] D. E. O'Reilly, *J. Chem. Phys.*, 1964, **41**, 3729.
[28] R. Catterall and M. C. R. Symons, unpublished results.
[29] R. A. Levy, *Phys. Rev.*, 1956, **102**, 31; D. Cutler and J. G. Powles, *Proc. Phys. Soc.*, 1962, **80**, 130.

PRINTED IN GREAT BRITAIN BY RICHARD CLAY (THE CHAUCER PRESS) LTD.,
BUNGAY, SUFFOLK.

The Conductivity of Lithium–Ammonia Solutions

In the following paper, Evers and Longo report their conductance data for dilute lithium solutions at −71°. They showed, by using a calculational method of Evers and Frank,[1] that the data are quantitatively in accord with the Shedlovsky equation for conductance and the equilibria proposed by Becker, Lindquist, and Alder (p. 227).

[1] E. C. Evers and P. H. Frank, Jr., *J. Chem. Phys.*, **30**, 61 (1959).

49

The Conductance of Dilute Solutions of Lithium in Liquid Ammonia at -71°[1,2]

by E. Charles Evers and Frederick R. Longo

The John Harrison Laboratory of Chemistry, University of Pennsylvania, Philadelphia, Pennsylvania
(Received July 26, 1965)

The conductance of solutions of lithium in liquid ammonia at −71° has been measured from 0.000309 to 0.14 *N*. The results are compared with the data for sodium in liquid ammonia at −34° and for lithium in methylamine at −78°. A mechanism for conduction based on the inversional motion of the ammonia molecule is proposed for dilute metal solutions.

The properties of metal–amine solutions have been investigated extensively for many years. Unfortunately, because these systems are metastable and the reaction between metal and ammonia is subject to catalysis, it has been difficult to obtain precise data in very dilute solutions; however, such data are essential since these solutions represent a region in which the systems exhibit quasi-electrolytic properties that gradually transform into those representing a metallic system as the metal concentration is increased. If we may assume that the metal solutions at low concentrations do conform to laws which govern the behavior of normal electrolytes, then, by measurement of various physical properties, it should be possible to establish some mode of the electrical interactions which occur among the species. One method of approaching this problem is to study the conductance of these systems as a function of solute concentration, for, in the case of electrolytes, we have a fairly comprehensive theory of the interaction of ions subject to various parameters such as temperature and dielectric constant.

Precise conductance data have been reported for very dilute sodium in ammonia solutions by Kraus[3] and lithium in methylamine solutions by Berns, Evers, and Frank.[4] These data have been analyzed[4,5] using a

(1) Taken in part from a thesis by F. R. Longo, presented in partial fulfillment of the requirements of the Ph.D. degree, Dec 1962.

(2) Sponsored in the main by the Office of Ordnance Research, U. S. Army, and partially by the Advanced Research Project Agency, Contract SD-69.

(3) C. A. Kraus, *J. Am. Chem. Soc.*, **43**, 749 (1921).

(4) D. S. Berns, E. C. Evers, and P. H. Frank, Jr., *ibid.*, **82**, 310 (1960).

conductance function based upon a modified form of the Shedlovsky[6] equation for conductance and a mass action model proposed by Becker, Lindquist, and Alder.[7] An equation relating the conductance to the two equilibrium constants for the reactions proposed by Becker, *et al.*, is

$$\frac{1}{\Lambda S(Z)} = \frac{1}{\Lambda_0} + \frac{S(Z)Nf^2\Lambda}{\Lambda_0^2 k_1}\left[1 + \frac{2k_2^2\Lambda^2 S(Z)^2 N^2 f^2}{k_1 \Lambda_0^2}\right]$$

where Λ is the equivalent conductance, Λ_0 is the equivalent conductance at infinite dilution, $S(Z)$ includes mobility corrections as defined by Shedlovsky, f is the mean ionic activity coefficient for the ionized metal, N is the normality, k_1 is the equilibrium constant for the ionization, $M = M^+ + e^-$, and k_2 is the equilibrium constant for the dimerization, $M = 0.5\ M_2$.

More recently, Dewald and Dye[8] have reported a conductance study of dilute solutions of alkali metals in ethylenediamine at room temperature. They noted that solutions in this solvent are less stable than those in ammonia or methylamine. The data were treated by the Shedlovsky analysis, and values for Λ_0 and the ionization constant were obtained for solutions of cesium, potassium, and rubidium. These values were compared with those calculated by Evers, *et al.*, for Na in NH_3 and Li in $MeNH_2$. Dewald and Dye[9] augmented their conductance studies with a spectral investigation and, on the basis of this work, have concluded that amine solutions differ essentially from ammonia solutions.[10] They postulate the existence of "gaslike covalent dimers and electrons trapped by one-electron bonded M_2^+ ions," in addition to the species which are generally accepted as being present in ammonia solutions.

In order to obtain further, precise data for dilute solutions it was decided to measure the conductance of solutions of lithium in liquid ammonia. At −34° decomposition was so rapid in dilute solutions that it was not possible to obtain reproducible results; hence, the measurements were finally made at −71°. The data obtained are compared with the data for sodium in liquid ammonia at −34° and for lithium in methylamine at −78°.

Experimental Section

At −71.0° the rate of decomposition was negligible provided that extreme care was taken to guarantee the cleanliness of the equipment, but even at this low temperature it was noted that the shiny platinum electrodes became blackened during some experiments. The black material could not be removed by strong mineral acids or by alcoholic KOH. However, it was found that the electrodes could be cleaned by elec trolysis in concentrated HCl solutions.

The conductance cells used in this investigation were large enough to accommodate 1 l. of solution. The bright platinum ball electrodes, previously described by Hnizda and Kraus,[11] were small in area, and the cell constants varied from 1 to 2 cm^{-1}. The cell was filled with a 50 vol % mixture of concentrated sulfuric acid and fuming red nitric acid and steamed out before using. The cell constants are determined by established procedures.[12]

A thermocouple well was sealed through the body of the cell in such a way that the temperature could be measured close to the electrodes. A 23-mm Pyrex tube rose vertically from the top of the cell and communicated with a "doser" stopcock assembly which was used to provide the cell with weighed samples of lithium. The cell also communicated with the vacuum system through this tube. The entire cell, including electrodes, was sealed through a desiccator-like lid. The lid was then cemented with black Apiezon wax to a large jar which served as a bath for temperature control. Details concerning construction may be found in the thesis by Longo.[1]

Measurements were made at −71.00 ± 0.03°. This temperature was maintained by controlling the vapor pressure of CHF_2Cl (Freon 22), which was distilled into the bath around the conductance cell. The bath was stirred by permitting a stream of Freon vapor to pass through the liquid at all times. The vapor pressure of this liquid was controlled at approximately 145 ± 0.05 mm with the use of a Micro-Set Manostat (Catalog No. 63273, Precision Scientific Co.). A copper–constantan thermocouple was used for temperature measurements in conjunction with a Leeds and Northrup portable precision potentiometer, Model No. 8662.

Lithium was obtained from the Lithium Corp. of America, (analysis: Li, 99.95; Na, 0.005; K, 0.01; Cu, 0.002; N, 0.01; Fe, 0.01; Si, 0.002; Cl, 0.001%). Metal samples were cut under argon-saturated mineral oil in a heavy petri dish. The cut samples were taken from the mineral oil with forceps and placed in a

(5) E. C. Evers and P. H. Frank, Jr., *J. Chem. Phys.*, **30**, 61 (1959).
(6) T. Shedlovsky, *J. Franklin Inst.*, **225**, 739 (1938).
(7) E. Becker, R. H. Lindquist, and B. J. Alder, *J. Chem. Phys.*, **25**, 971 (1956).
(8) R. R. Dewald and J. L. Dye, *J. Phys. Chem.*, **68**, 128 (1964).
(9) R. R. Dewald and J. L. Dye, *ibid.*, **68**, 121 (1964).
(10) R. R. Dewald and J. L. Dye, *ibid.*, **68**, 135 (1964).
(11) V. F. Hnizda and C. A. Kraus, *J. Am. Chem. Soc.*, **71**, 1956 (1949).
(12) E. C. Evers and A. G. Knox, Jr., *ibid.*, **73**, 1739 (1951).

bottle, where the mineral oil was washed away with dry toluene by a procedure described by Longo.[1]

After being washed, the lithium samples were transferred under argon into weighing bottles which were designed to fit on a vacuum system. The toluene was pumped off, the samples were weighed under vacuum, and transferred into the four doser stopcocks under an argon atmosphere. For solutions of higher concentration, a large metal sample was sealed in a side arm above the cell. It was pushed into the cell with a glass-covered magnet.

Ammonia was obtained from the National Ammonia Co. (The company furnished the following analysis: H_2O, 50 to 100 ppm; hydrocarbon oils, 3 ppm; nonvolatiles, 1 ppm.) It was purified in a manner which was previously found successful for monomethylamine.[4]

The bath was filled with Freon 22 by distillation from a commercial cylinder. Ammonia was condensed in the cell from a weighed storage can and was stirred by an externally driven glass-covered magnet. The vapor pressure of the Freon was adjusted, and the ammonia came to temperature equilibrium within 1 hr after the distillations.

The lithium samples were dropped individually from the four doser stopcocks. The resistances of the resulting solutions were measured at 2000 cps with a Leeds and Northrup Jones bridge using earphones as a null detector.

The concentration of metal is reported in g-atoms of metal per liter of solvent. For the calculation of solvent volume, the density data of Cragoe and Harper[13] were used.

In order to apply the method of Evers and Frank[5] it was necessary to know the dielectric constant and viscosity coefficient of the solvent. The dielectric constant was obtained by a graphical interpolation of data in the literature covering the temperature range from -77.70 to $35°$.[14] At $-71.0°$, the dielectric constant, D, is 25.1. Using the method of Nissan,[15] we calculated the viscosity at $-71°$ by extrapolation of the data of Fredenhagan[16]; at $-71.0°$, $\eta = 0.00500$ poise.

Discussion

Figure 1 is a plot of Λ *vs.* $\sqrt{N}$ for the conductance of Li in liquid ammonia at $-71°$. At the lowest concentration measured, 0.000309 N, the value of Λ is 445 Kohlrausch units. The equivalent conductance decreases sharply with increasing metal concentration until it reaches a minimum value of about 220 at approximately 0.025 N. From this point it rises to a value of 304 at 0.14 N, the concentration at which a second liquid phase was observed to separate.[17] A plot of the conductance data for this system at concentrations below the minimum is formally similar to that obtained for other metal–amine systems which have been studied and for normal electrolytes. The method of Evers and Frank[5] was applied to the conductance data. A fit of the data to their function requires that $\Lambda_0 = 558.70 \pm 15.90$, $k_1 = 1.28 \pm 0.16 \times 10^{-3}$, and $k_2 = 4.33 \pm 1.29$.

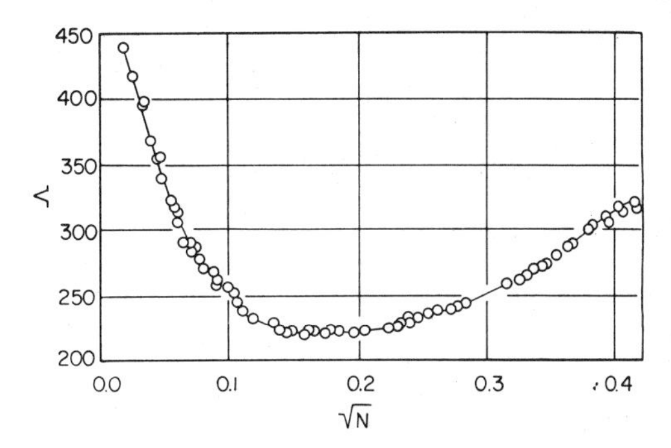

Figure 1. Λ *vs.* $\sqrt{N}$ for Li in liquid NH_3 at $-71°$.

A comparison of the experimental data with calculations from the above function is shown graphically in curve 2, Figure 2, where Λ/Λ_0 is plotted *vs.* the square root of the concentration. The solid line was calculated from the function, and the open circles are experimental data. The agreement is fairly good up to 0.02 N, the highest concentration shown. The significance of the other plots in Figure 2 will be discussed below.

Also given in Figure 2 are plots for sodium in liquid ammonia at $-34°$, curve 1, and for lithium in liquid methylamine at $-78°$, curve 3.

Table I summarizes some of the important results derived from conductance measurements in various systems, which have been treated according to the method of Evers and Frank. In this table are presented the equivalent conductance at infinite dilution,

(13) C. S. Cragoe and R. Harper, National Bureau of Standards, Scientific Papers, No. 420, U. S. Government Printing Office, Washington, D. C., 1921, p 313.

(14) National Bureau of Standards Circular 514, U. S. Government Printing Office, Washington, D. C.

(15) A. H. Nissan, *Phil. Mag.*, **32**, 441 (1941).

(16) K. Fredenhagan, *Z. Anorg. Allgem. Chem.*, **186**, 1 (1930).

(17) P. D. Schettler, Jr., and A. Patterson, Jr., *J. Phys. Chem.*, **68**, 2865 (1964), have recently reported results of a phase study of the Li-NH_3 system. They found that at $-70°$, the concentration at which a second liquid phase appears is 1.6 M. Furthermore, Sienko, "Metal-Ammonia Solutions," G. Lepoutre and M. J. Sienko, Ed., W. A. Benjamin, Inc., New York, N. Y., 1964, has reported that separation occurs at about 2 M at this temperature.

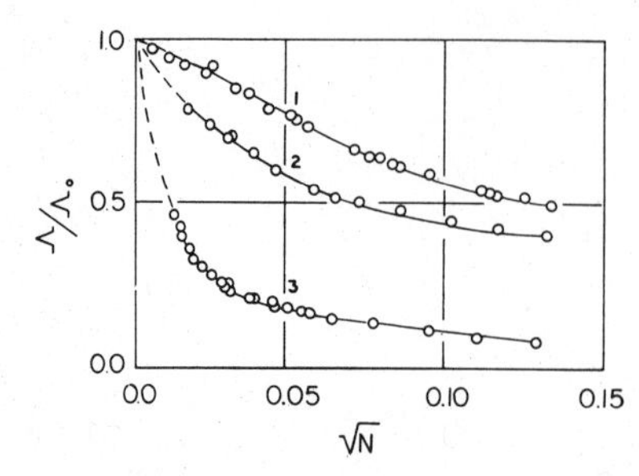

Figure 2. Curve 1: Na in NH_3 at −34°. Curve 2: Li in NH_3 at −71°. Curve 3: Li in CH_3NH_2 at −78°.

Λ_0, the equilibrium constants for the ionization and dimerization processes, k_1 and k_2, and the Walden product $\Lambda_0\eta_0$. Also listed are properties of the solvents, the viscosity coefficient η, and the dielectric constant D.

Table I: Constants Derived from Conductance Data

	System		
	Na–NH_3, −34°	Li–$MeNH_2$, −78°	Li–NH_3, −71°
Λ_0	1022.0	228.3	558.7
Concn. at minimum	0.04	0.13	0.025
Λ minimum	540	17	220
10^3k_1	7.23	0.0479	1.28
k_2	27.0	5.42	4.33
Solvent $\eta \times 10^3$	2.54	9.12	5.00
Solvent D	22	17.12	25.1
$\Lambda_0\eta_0$	2.6	2.1	2.8

The conductance of sodium in ammonia at −34° is approximately twice the conductance of lithium in ammonia at −71° at corresponding metal concentrations, a result not evident from the plot because the values of Λ have been divided by the respective Λ_0 values. The slope of the curve in the dilute region is steeper at −71°. Another important aspect is the position of the minimum. For sodium in ammonia at −34°, it occurs at 0.04 N and for Li in ammonia at −71° it occurs at 0.025 N. In an attempt to interpret these observations, we have assumed that dilute lithium and sodium solutions possess identical electrical properties at −34° since data which are available[3] for lithium indicate that the conductance values are approximately equal; also the equivalent conductance curves are approximately parallel. These data suggest that at −34° the values of k_1 for sodium and lithium solutions are approximately equal.

The fact that the conductance is higher at the higher temperature indicates that the system is predominantly nonmetallic. This conforms to the idea of quasi-electrolytic behavior where conductance is enhanced by an increase in temperature. This is also supported by the fact that the data seem to be in fair agreement with Walden's rule; *i.e.*, $\Lambda_0\eta_0$ is fairly constant for these systems, suggesting that the conductance mechanism is viscosity dependent at infinite dilution.

The steeper slope for the curve for lithium in ammonia at −71° (Figure 2) indicates that the efficiency of conduction is decreasing more rapidly with concentration at this temperature. This may be interpreted as being due to different causes depending upon the model considered. Using the model of Becker, *et al.*, and discounting the concentration of dimers in this dilute region, one may contend that the equilibrium $M = M^+ + e^-$ determines the state of the system. According to Becker, *et al.*,[7] this ionization is endothermic by about 0.1 ev, and therefore the equilibrium constant, k_1 = (slope $\Lambda_0^3)^{-1}$, should be lower at −71° than at −34°. It is not possible to compare the constants k_1 (−34°) and k_1 (−71°) because data for lithium in ammonia at −34° are not available at sufficiently low concentrations. However, since there is not too great a difference between the conductance of sodium and lithium solutions at −34°, one might use Evers' value for sodium–ammonia at −34°.[5] He found k_1 (−34°) = 7.23×10^{-3}, and from this investigation k_1 (−71°) = 1.28×10^{-3}. The values of k_1 can be used to obtain the energy of ionization. The result is 0.2 ev and is in fair agreement with the value of Becker, *et al.*

The behavior of the three systems is in qualitative agreement with predictions based on electrolyte theory for dilute solutions. Table I and Figure 2 show that parameters such as temperature, viscosity, and dielectric constant affect the conductance in a manner which we would predict to be normal for electrolytes of classes such as weak acids in water, *e.g.*, HIO_3. However, we feel that the minimum must be attributed to causes other than those ascribed to electrolytic solutions. We agree with others that this may be attributed to the onset of metallic character which depends upon a relatively small separation of metal ions and a fairly regular, periodic-potential field. The minimum value of Λ found in this investigation occurs at 0.025 N which corresponds to an average distance between metal atoms of approximately 37 A. This distance is too great to allow ordinary metallic conduction, but the separation of a golden phase at low concentration (0.14 N) suggests that "cybotactic metallic groups" are present before phase separation is observed.

If these groups begin to form well below 0.025 N, the minimum and the eventual increase in Λ are easily understood since the electrical conductance of such groups would be very high.

Conductance minima occur, respectively, at concentrations of 0.025, 0.04, and 0.13 N in Li–NH_3 (−71°), Na–NH_3 (−34°), and Li–$MeNH_2$ (−78°). This is exactly what might be expected if one considers the effect of temperature and the nature of the solvent molecules on the formation of a metallic structure. A lower temperature would favor order, but the unsymmetrical $MeNH_2$ molecules would oppose this. Hence, metallic structure is more easily achieved in NH_3 than in $MeNH_2$ and more readily at −71° than at −34°.

The values of k_2 calculated from the Evers and Frank[5] conductance function must be suspected since the dimerization equilibrium becomes important only above concentrations where the calculated activity coefficients and the mobility corrections apply. However, if the relative values of k_2 are meaningful, they indicate that lithium solutions in methylamine at −78° and in ammonia at −71° have a smaller tendency to dimerize than the sodium–ammonia solutions at −34°. This is not in disagreement with paramagnetic susceptibility data[18,19] which indicate that the atomic susceptibility of lithium solutions is very high.

A Model for the Dilute Region

The nuclear magnetic resonance (nmr) investigations of McConnell and Holm[20] on solutions of sodium in ammonia at room temperature indicate a substantial Knight shift for the nitrogen nuclei and a definite, but much smaller, shift for sodium nuclei, whereas proton shifts are very small. Their data also indicate that in dilute solution the contact interaction between nitrogen nuclei and electrons is independent of metal concentration, but the interaction between sodium nuclei and electrons increases with increasing concentration.

The above considerations, together with others which will be discussed below, have suggested to us a conductance mechanism based on a cavity model in which the electron is in a molecular orbital on the nitrogen nuclei adjacent to the cavity. Our mechanism is based on an accepted molecular motion of the ammonia molecule, namely, inversion. When an ammonia molecule at the cavity wall undergoes inversion with the nitrogen atom moving toward the center of a trap, two events occur simultaneously: (1) the trap tends to collapse, and (2) an adjacent one tends to form. If the bulk solvent molecules near the cavity are favorably oriented at the time of inversion the electron can be transported quickly and trapped in the new cavity adjacent to the collapsing one. Since several solvent molecules are involved in the cavity unit, the probability of this type of transport may be large.

Some support for the inversion mechanism comes from the fact that the inversional frequency (4.5×10^{10} sec^{-1}) is about equal to perturbation frequency (3.5×10^{10} sec^{-1}), as calculated by Kaplan and Kittel[21] and by Pollak[22] for rotation and diffusion of ammonia molecules at the cavity wall. Hence, if these workers had used inversional motion as the perturbation of the electron–nitrogen interaction, they would have obtained equally good agreement with experiment.

As the metal concentration increases from infinite dilution, it becomes probable that some of the ammonia molecules which normally are involved in the cavity unit are also participating in metal ion solvation. Metal ion solvation should impede the inversional motion of the ammonia molecule, and the equivalent conductance would be reduced. The aggregate of the cavity and metal ion is the formal equivalent of the monomer of Becker, Lindquist, and Alder[7] and is very similar to the monomer proposed by Gold, Jolly, and Pitzer.[23] In dilute solution the monomer is considered to be participating in two equilibria, a dissociation and a dimerization

$$M = M^+ + e^-$$

$$M = 0.5M_2$$

In the dimer some of the molecules which solvate a particular metal ion are present at two cavity walls.

It is obvious from the above equilibria that the success of the method of Evers and Frank[5] in analyzing the conductance data can be explained equally well by this cavity model or by the model proposed by Becker, Lindquist, and Alder[7] since the equilibria proposed are formally identical. Our model also conforms to the nmr implications which Pitzer[24] has discussed: the model must permit rapid exchange of

(18) A. Charru, *Compt. Rend.*, **247**, 195 (1958).

(19) R. A. Levy, *Phys. Rev.*, **102**, 31 (1957).

(20) H. McConnell and C. Holm, *J. Chem. Phys.*, **26**, 1517 (1957).

(21) J. Kaplan and C. Kittel, *ibid.*, **21**, 1856 (1953).

(22) V. L. Pollak, *ibid.*, **34**, 864 (1961).

(23) M. Gold, W. L. Jolly, and K. S. Pitzer, *J. Am. Chem. Soc.*, **84**, 2264 (1962).

(24) K. S. Pitzer, "Metal-Ammonia Solutions," G. Lepoutre and M. J. Sienko, Ed., W. A. Benjamin, Inc., New York, N. Y., 1964, p 193.

electrons between monomer and e^- species, and the monomer must be of such a nature that it will bring the Na^{23} nucleus in contact with electrons without making an appreciable change in N^{14}–electron contact.

The Miscibility Gap in Sodium–Potassium–Ammonia Solutions

The following study of the liquid–liquid phase separation in solutions containing both sodium and potassium gives evidence of interactions between the different metals and points out the need for further studies of this type.

[Reprinted from the Journal of Physical Chemistry, 71, 3540 (1967).]

Liquid–Liquid Phase Separation in Alkali Metal–Ammonia Solutions. IV. Sodium and Potassium

50

by Patricia White Doumaux and Andrew Patterson, Jr.

Sterling Chemistry Laboratory, Yale University, New Haven, Connecticut 06520

Accepted and Transmitted by The Faraday Society (February 20, 1967)

The phenomenon of liquid–liquid phase separation in ammonia solutions containing both sodium and potassium has been examined over a range of concentrations of both metals at two temperatures, −56.39 and −75.00°. Separate analyses were made for both metals with a precision of about 1%. Perceptible, but not large, $metal_1$–$metal_2$ interactions occur which cannot, from these experiments alone, be assigned with certainty solely to the metal atoms or to the positive ions.

In earlier papers[1,2] we have examined the effect on phase separation in sodium–ammonia solutions when metal salts are added. The results indicated that the effect of the salt depended upon its anionic constituent, and it seemed most likely (from other experimental measurements) that interactions between the electrons in the solution and the anion of a solvated ion-paired salt molecule were responsible for the observed broadening of the miscibility gap and raising of the consolute temperature. In spite of this at least self-consistent interpretation of our data, the kinds of models for behavior of alkali metal–ammonia solutions which one invokes are not all in agreement and are often dependent upon the kind of experiment employed and the system studied. Thus, conductivity measurements on alkali metal salt–liquid ammonia systems by Hnizda and Kraus[3] indicated a much smaller interaction of the halide ion with the solvent than is the case with the cation. On the other hand, Catterall and Symons[4] have found the effect of the cation on esr *g* shifts in solutions containing both metal and salt to be greatly subsidiary to that of the anion of the salt used, as was also our finding in our phase separation measurements.[1,2] While these results are not necessarily in conflict—in one case salt only is present, while in the other both salt and metal are involved—and while it is likely that ion solvation was the important factor in Kraus' work but a quite different interaction responsible for the results in that of Catterall and Symons, we felt it desirable to investigate the importance of possible $metal_1$–$metal_2$ or metal ion_1–metal ion_2 interactions on phase separation. We have done this through a study of phase separation in solutions containing both sodium and potassium. It is of interest that Professor Kraus, in his prefatory note to the "Colloque Weyl" presentations,[5] suggested just such an investigation. We plan to extend these measurements to other alkali metal systems.

Experimental Section

The procedures used were essentially those earlier described.[1,2] Total alkalinity was determined on the mixed metal hydroxide residues by titration with nitric acid. Potassium was determined in this solution remaining from the titration by a gravimetric procedure developed by Kohler,[6] employing sodium tetraphenylborate. Measurements were made at −56.39 and −75.00°. One determination was made at −32.90° to afford a check on the precision of the analyses; as might have been expected, phase separation was not observed.

(1) P. D. Schettler and A. Patterson, *J. Phys. Chem.*, **68**, 2870 (1964).

(2) Part III: P. W. Doumaux and A. Patterson, *ibid.*, **71**, 3535 (1967).

(3) V. F. Hnizda and C. A. Kraus, *J. Am. Chem. Soc.*, **71**, 1565 (1949).

(4) R. Catterall and M. C. R. Symons, *J. Chem. Soc.*, 4342 (1964).

(5) C. A. Kraus, Foreword, "Solutions Métal-Ammoniac—Propriétés Physicochemiques, Colloque Weyl," G. Lepoutre and M. J. Sienko, Ed., W. A. Benjamin, Inc., New York, N. Y., 1964.

(6) M. Kohler, *Z. Anal. Chem.*, **138**, 9 (1953).

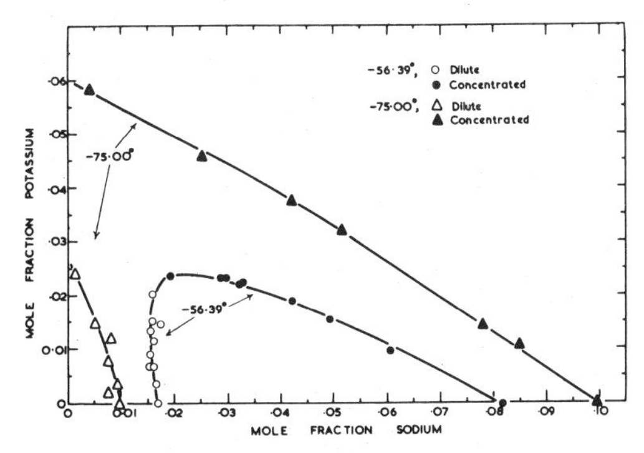

Figure 1. Plot of mole fraction of potassium *vs.* mole fraction of sodium for data on phase separation of solutions containing sodium, potassium, and liquid ammonia. Measurements were made at the two temperatures indicated in the legend.

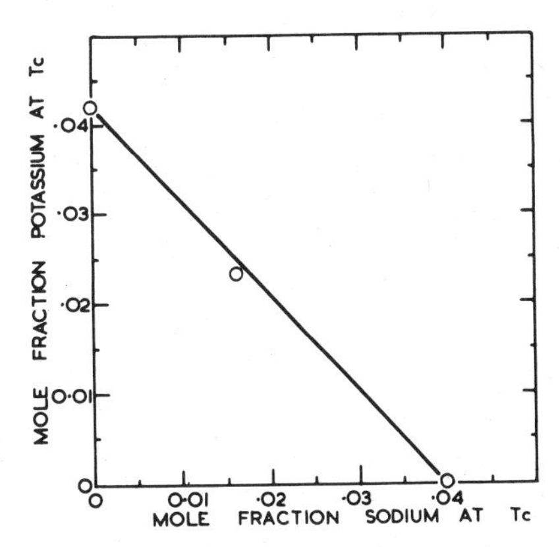

Figure 2. Plot of mole fraction of potassium *vs.* mole fraction of sodium at the consolute temperature, with the extrapolated value for a solution containing both sodium and potassium, obtained from the data of Figure 1 at −56.39°, at center. This is a one-dimensional projection of the mole fraction of potassium–mole fraction of sodium–consolute temperature surface.

Table I: Liquid–Liquid Phase Separation Data on Sodium–Potassium–Ammonia Solutions

Dilute phase		Concd phase	
N_{Na}[a]	N_K	N_{Na}	N_K
	Temperature, −56.39°		
0.0153	0.00677	...	...
0.0155	0.00919	0.0421	0.0188
0.0162	0.0118	0.039[b]	0.025[b]
0.0156	0.0135	0.0328	0.0224
0.0162	0.00694	0.0493	0.0154
0.0175	0.0147	0.0322	0.0223
0.0166	0.00392	0.0603	0.00985
...	...	0.0299	0.0233
0.0159	0.0151	0.0289	0.0234
0.0159	0.0202	0.0194	0.0236
	Temperature, −75.00°		
0.00812	0.0120	0.0420	0.0376
0.00768	0.00824	0.0515	0.0319
0.0077[b]	0.0020[b]	0.0849	0.0109
0.00507	0.0148	0.0252	0.0459
0.00954	0.00342	0.0777	0.0144
0.00128	0.0242	0.00394	0.0584
0.00986	None[c]	0.0997	None[c]

[a] All concentrations are in mole fractions. [b] Data given in two significant figures were estimated due to small loss of ammonia before analysis. [c] The notation "none" indicates a determination on sodium alone.

Results

The data are shown in Table I and in Figure 1 in the form of plots of mole fraction sodium in dilute and concentrated phases at the two temperatures studied. Because the sodium phase separation data available[7] did not extend as low as −75.00°, a determination was made on sodium alone at that temperature. Earlier, Schettler[7] had extrapolated data on solutions containing sodium iodide to zero concentration sodium iodide; his extrapolated values and the present results check closely, the preferred experimental values being N_{Na}(dilute) = 0.00986, N_{Na}(concd) = 0.0997. In Figure 2 of ref 7 the data are plotted as mole ratios, while mole fractions are used here. Data for potassium alone were read from a large-scale plot of data of ref 7 since a number of points had been determined near −75.00°. In the determination at −32.90° on separated samples of the same solution, the results were: "dilute phase" sample, N_K = 0.00631, N_{Na} = 0.0474; "concentrated phase" sample, N_K = 0.00625, N_{Na} = 0.0470.

Discussion

The reproducibility of the data is of the same order as that in the study of bromide and azide where multiple determinations are made on the same sample,[2] about 1%; the data of Schettler on sodium–sodium iodide are appreciably better, but this is a much more favorable system, from the point of view of analysis, on which to work.

In Figure 1, the −75° data pass smoothly and with but little curvature from the potassium to the sodium axis; the curvature of the data at −56.39° is more pro-

(7) P. D. Schettler and A. Patterson, *J. Phys. Chem.*, **68**, 2865 (1964).

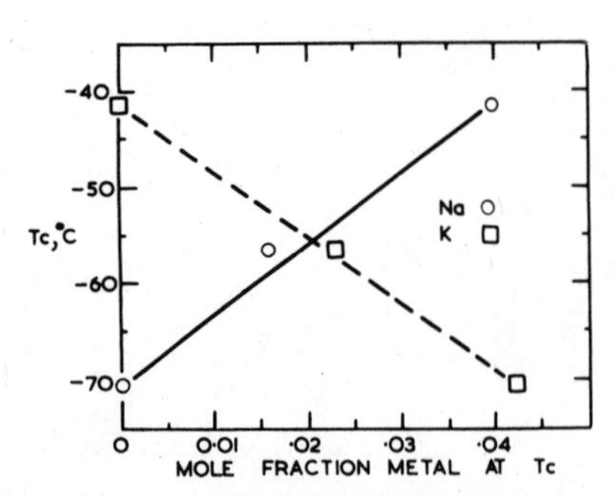

Figure 3. Plot of consolute temperature *vs.* mole fraction of metal at the consolute temperature for sodium and potassium in mixed solutions as indicated in the legend. As in Figure 2, this is a one-dimensional projection of the three-variable surface.

nounced, but this in part is presumably due to the fact that this temperature is above the consolute temperature for potassium, in which connection the data on sodium–sodium iodide[2] below and above the consolute temperature of sodium should be recalled. At $-56.39°$, by extrapolating the midpoints of the tie lines, we obtain a consolute point for this temperature of N_K = 0.0232 and N_{Na} = 0.0159. Using these values and those for sodium and potassium two-component systems, $-41.55°$, N_{Na} = 0.0398, $-70.05°$, N_K = 0.0421, one can construct a two-dimensional projection of the temperature, N_{Na}, N_K surface in two graphs, Figures 2 and 3. These show small but significant deviations from a straight line connecting the extreme two-component points: about 1.4° in the temperature *vs.* mole fraction metal, and 8% in the mole fraction metal at consolute temperature plots. Such deviations from linearity could be due to differences between interactions of the two metals with the solvent, to specific interactions between the two ions themselves, or to a combination of both. The present data are not sufficient to decide this point. The deviations, though small, are significant, for they exceed by some tenfold the precision of the measurements. Kraus' suggestion[5] that mixed-metal studies be extended to include the alkaline earth metals is then of much interest since the charge on the metal ion becomes one of the variables available for examination. We propose to pursue this point.

Acknowledgment. This work was supported by the National Science Foundation.

The Problem of the Sharp Minima in the Volumes of Metal–Ammonia Solutions

Evers and his co-workers, in studies of the apparent molar volume of metals in liquid ammonia as a function of concentration, found sharp minima at concentrations around 0.02 *M*. This amazing behavior was completely unexpected and defied explanation. The following paper is the last of several by Gunn in which he reports the inability to confirm the sharp minima reported by Evers *et al.*

Reprinted from THE JOURNAL OF CHEMICAL PHYSICS, Vol. 47, No. 3, 1174–1178, 1 August 1967
Printed in U. S. A.

Absence of a Volume Change upon Dilution of Dilute Sodium–Ammonia Solutions at −45° and Potassium–Ammonia Solutions at −34°*

51

STUART R. GUNN
Lawrence Radiation Laboratory, University of California, Livermore, California

(Received 10 April 1967)

Dilatometric measurements of the volume change upon mixing metal–ammonia solutions at concentrations from 0.01 to 0.06*N* with an approximately equal quantity of pure ammonia have been performed. The observed changes for both sodium solutions at −45° and potassium solutions at −34° are essentially zero throughout the concentration range, and provide no support for existence of sharp minima in the apparent molar volumes at ~0.02*N* which have been reported by other workers.

Conflicting observations upon the behavior of the apparent molar volume (ϕ_v, milliliters per gram-atom) of dilute solutions of alkali metals in liquid ammonia as a function of concentration have been reported in recent years. Evers and his students[1–6] have found large, sharp minima, centered at 0.022*N* for sodium at −45°C, at 0.0075*N* for potassium at −45°, and at 0.020*N* for potassium at −34°. These data are reproduced in Figs. 1, 2, and 3, the data points and curves being taken from the theses.[1–3] Gunn and Green,[7–9] on the other hand, found ϕ_v to be essentially constant through these concentration regions, for lithium and sodium at 0° and for sodium at −45°. The latter set of points is included in Fig. 1.

Both Evers *et al.* and Gunn and Green used dilatometric techniques, although of a somewhat different nature. The dilatometers of Evers *et al.* consisted of two bulbs, the upper of which was somewhat larger, connected by a capillary tube. The metal was distilled into the upper bulb, and then enough ammonia was distilled in to fill the lower bulb and part of the capillary. The lower bulb and capillary were immersed in the thermostat, and the level of the liquid column in the capillary was measured. The entire dilatometer was then immersed in another bath at −70° and repeatedly inverted to wash the metal into the lower bulb. The dilatometer was then returned to the thermostat and the liquid level again read after thermal equilibration. Possible difficulties in this method include hysteresis of the Pyrex dilatometer and incomplete drainage of solution back to the lower bulb. None of the papers[1–6] report "blank" measurements, wherein the dilatometer would be filled with ammonia, but no metal, and manipulated as in metal–ammonia measurements to check the return of the liquid level to the original position.

Gunn and Green used a dilatometer in which was placed a fragile glass bulb containing the metal. An iron armature enclosed in glass was manipulated by a ring magnet surrounding the dilatometer body to break the bulb and stir the solution; the volume change determined from the change in the level of the meniscus, plus the previously determined internal volume of the sample bulb, represented the apparent volume of the metal sample in solution. This method involved a much smaller mechanical and thermal perturbation of the dilatometer than that of Evers *et al.* It did, however, leave fragments of broken glass in the dilatometer which might catalyze ammonolysis of the dissolved metal and perhaps trap bubbles of hydrogen.

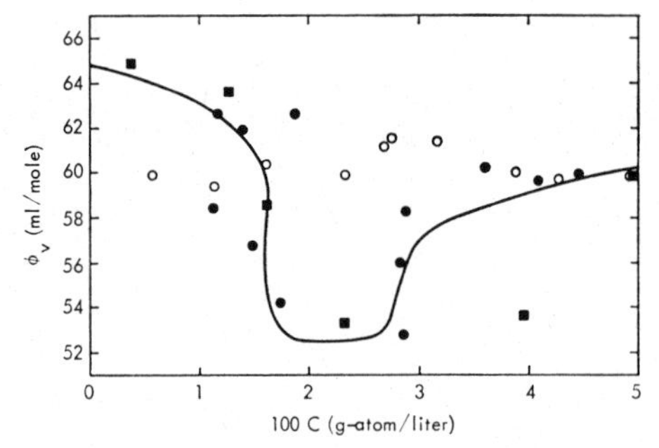

FIG. 1. Apparent molar volume of sodium at −45°. ■, Filbert[1]; ●, Orgell[2]; and ○, Gunn and Green.[7]

There is no obvious error in either method which would give an artificial minimum or fail to detect a real one. Whichever is the case, one method or the other would seem to include two errors of opposite sign, one a more sensitive function of concentration than the other.

It may be noted that there is considerable scatter in the data of Evers *et al.* shown in Fig. 1 and, particularly, in Fig. 2. The first publications of Evers and Filbert[1,4] gave a curve in which the minimum was defined by the two points at 0.0232 and 0.0396*N*. In latter publications[2,5] the second of these points was neglected, and the curve shown in Fig. 1 was drawn

* This work was performed under the auspices of the U.S. Atomic Energy Commission.
[1] A. M. Filbert, thesis, University of Pennsylvania, 1962.
[2] C. W. Orgell, thesis, University of Pennsylvania, 1962.
[3] W. H. Brendley, Jr., thesis, University of Pennsylvania, 1965.
[4] E. C. Evers and A. M. Filbert, J. Am. Chem. Soc. **83**, 3337 (1961).
[5] C. W. Orgell, A. M. Filbert, and E. C. Evers, *Solutions Metal–Ammoniac: Properties Physicochemiques, Colloque Weyl, Lille, 1963*, G. Lepoutre and M. J. Sienko, Eds. (W. A. Benjamin, Inc., New York, 1964), p. 67.
[6] W. H. Brendley, Jr., and E. C. Evers, *The Solvated Electron* (The American Chemical Society, Washington, D.C., 1965), p. 111.
[7] S. R. Gunn and L. G. Green, J. Chem. Phys. **36**, 363 (1962).
[8] S. R. Gunn and L. G. Green, J. Am. Chem. Soc. **85**, 358 (1963).
[9] S. R. Gunn, Ref. 5, p. 76.

with the minimum at a lower concentration. The curve given in Fig. 2 depends heavily on the single point at 0.0079*N*. Furthermore, the experimental errors would be expected to be greater in the more dilute region where the presumed minimum occurs in Fig. 2. The curve of Fig. 3, however, is quite well defined, and the minimum is about twice as deep as for the other two systems.

The existence or nonexistence of this minimum is a matter of some consequence in evaluation of theories of the nature of metal–ammonia solutions. However, in the intervening years, there seem to have been no relevant studies published by any other laboratories, with perhaps one exception. Schindewolf, Böddeker, and Vogelsgesang[10] measured the conductivity of sodium–ammonia solutions at −35° as a function of concentration and pressure up to 1500 atm. From the fact that the pressure coefficient of conductivity, which is negative, showed a broad minimum around 0.1*N*, they concluded that pressurization favored nonconducting species, whose volume must hence be smaller, in accord with the interpretation of Evers. In view of the complex nature of the solutions and other effects of pressure upon the solvent properties, such as viscosity, this explanation would, however, seem to be somewhat less then compelling. Furthermore, the minimum in the pressure coefficient of conductivity is not nearly as sharp as the volume minima found by Evers, and is at a concentration greater than that of the volume minimum at −45° by a factor greater than would be anticipated for the 10° difference in temperature on the basis of the potassium volume measurements at −45° and −34°.

In the present investigation a dilution-dilatometric method was used to investigate the concentration dependence of the apparent molar volume of sodium at −45° and potassium at −34°. The method involves simply an observation of the volume change upon mixing a metal–ammonia solution with an approximately equal volume of pure ammonia. While this does not give absolute values of ϕ_v, it does give improved reliability in detecting changes in ϕ_v. Thus, according to the curve of Fig. 3, a measurement at an initial concentration of 0.04*N* should give a volume decrease of 21 ml/g-atom; and starting at 0.02*N* should give an increase of 13 ml/g-atom. If, on the other hand, ϕ_v is constant, both dilutions would give zero volume change. This latter behavior is what has in fact been observed in the present work.

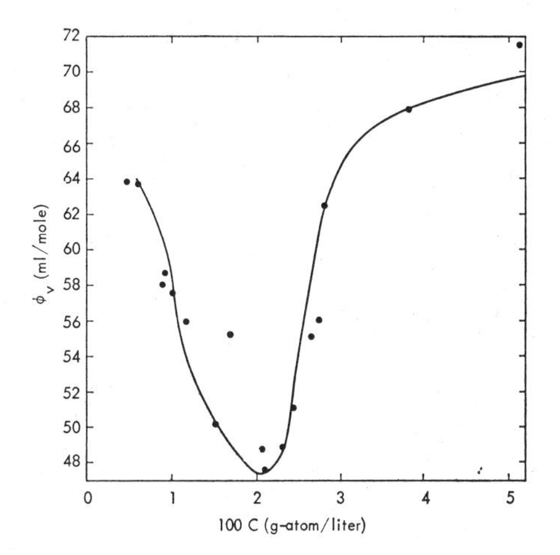

FIG. 3. Apparent molar volume of potassium at −34°.

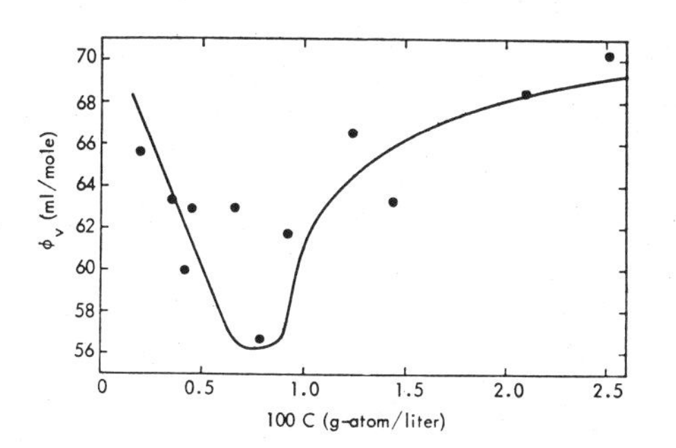

FIG. 2. Apparent molar volume of potassium at −45°.

[10] V. Schindewolf, K. W. Böddeker, and R. Vogelsgesang, Ber. Bunsenges. Physik. Chem. **70**, 1161 (1966), English translation available from Lawrence Radiation Laboratory.

EXPERIMENTAL

Although the experimental method is quite simple, it is described in considerable detail to permit evaluation of the reliability of the results and to permit other investigators to reproduce the conditions closely, should they wish to do so.

The Pyrex dilatometer is illustrated in Fig. 4. Three of these were used, but the dimensions of all were essentially identical. The measuring tube A was a thick-walled capillary, 2 mm i.d. The upper and lower compartments B and C were connected by a 29/42 standard-taper ground joint. This was closed by a solid glass plug D which was made from a 29/42 bottle stopper with about half of its length at the lower end sawed off, leaving a section about 2 cm long. Twelve longitudinal grooves, about 1 mm wide and 0.3 mm deep, were cut on the surface of this with a carborundum wheel. These grooves served to pass gas and liquid past the plug with adequate rapidity during evacuation and filling of the dilatometer, but prevented any mixing of the solutions in the two compartments once thermal equilibrium was established. A glass-encased iron armature E, having a volume of 16 to 20 ml, served to stir the solution and to knock out the plug D at the desired time. The net volume of the upper compartment, after sidearm F was sealed off at G, was 55 to 60 ml.

In later work, a plug only 0.5 cm long and having 24 grooves was used. The three dilatometers used with this plug are designated 1B, 2B, and 3B; with the 2-cm plugs, 1A, 2A, and 3A.

A procedure for tapping Plug D into place with a reproducible force was used. The armature was inserted

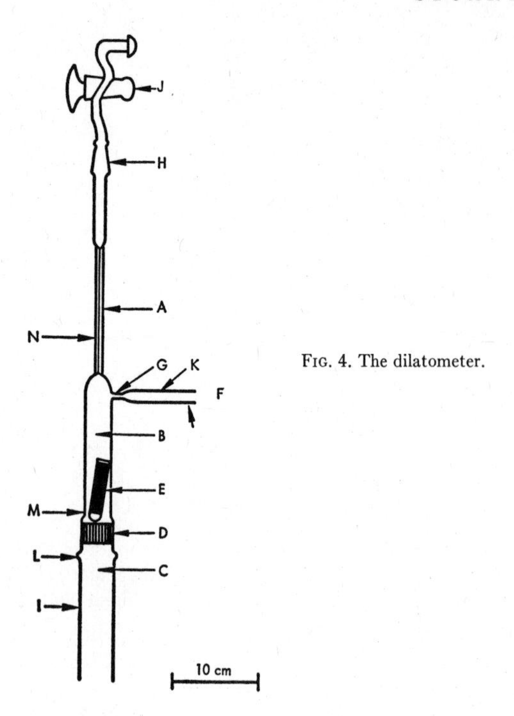

FIG. 4. The dilatometer.

into the upper compartment and the plug was moved into position, an index mark among the grooves being used to give the same angular orientation for each run. Then the dilatometer was inverted and clamped in a vertical position with taper H resting on the bench top. A test tube loaded with lead shot, weighing 120 g, was suspended by a wire in C, with its lower end 1.0 cm above the plug, and was dropped by cutting the wire.

The lower part of the dilatometer was sealed off with a flat bottom at I, giving a lower compartment of volume approximately equal to, or slightly greater than, the upper. Stopcock J was attached with Apiezon-W wax on joint H, and sidearm F was temporarily closed with a pinched-off rubber tube. The dilatometer was attached to the vacuum line through an articulated linkage and evacuated.

A lump of sodium or potassium was cut under petroleum ether, dried in a stream of helium, and weighed in a helium-filled tube. The dilatometer was filled with helium and the metal was quickly placed in F, which was sealed off at K. The dilatometer was then evacuated overnight.

The thermostat used was a Dewar, 6 in. i.d., having a lucite plug in the bottom to make the liquid depth 18 in. Two vertical unsilvered strips 1 in. wide, 180° apart, were provided to permit observation of the dilatometer. A perforated Lucite tube adjacent to one of these held the dilatometer in position; a fluorescent lamp behind the other provided back illumination. The Lucite tube also served as a guide for a surrounding ring magnet, which was manipulated by strings from above the bath. The level of the column of liquid was observed, with reference to a centimeter scale attached to the capillary, by means of a cathetometer telescope. The position could be readily measured to an accuracy of 0.01 cm; since the solution in the capillary was a dark blue, both before and after the compartments were mixed, there was no problem with a change of appearance of the meniscus.

The thermostat fluid was *n*-heptane. Cooling was provided through a copper rod, soldered to a liquid-nitrogen tank, which dipped into the thermostat.[11] A Hallikainen thermoregulator was used, and the temperature was measured with a Hewlett–Packard quartz thermometer.

After the dilatometer had been evacuated overnight, the metal was distilled into the upper chamber and the sidearm was sealed off at G. A flask of sodium–ammonia solution was put on the vacuum line and pumped for repeated intervals; the ammonia was then distilled into another flask on the vacuum line, stirred, and again repeatedly pumped. Finally, it was distilled into the dilatometer at a pressure of about 2 atm, the dilatometer being first immersed in the heptane up to Level L and then gradually to Level M to fill the lower compartment, and then immersed to N to fill the upper compartment and dissolve the metal.

A small amount of the metal—no more than a few tenths of 1%, enough to give a faint blue color—was transferred to the lower compartment by warming the bath 1 or 2 deg above the operating temperature and then letting it cool back to the operating temperature; contraction of the liquid in the lower compartment sucked a little solution down from the upper compartment. Although there was no provision for stirring the lower compartment, this color soon became uniformly distributed. The purpose of this was to ensure the absence of any impurities which might react with the metal when the contents of the two compartments were mixed.

When the liquid extended a few centimeters up into the capillary at the operating temperature, stopcock J was closed, the ammonia supply was shut off, the vacuum line was evacuated and then pressurized to 2.0 atm with helium, and stopcock J was reopened. The purpose of thus pressurizing the system was to inhibit the formation of bubbles by any slight ammonolysis which might occur; in principle, bubbles would not form until the concentration of hydrogen dissolved in the ammonia was greater than that required to give a partial pressure greater than the difference between the total pressure—2.0 atm—and the vapor pressure of ammonia at the operating temperature.

The dilatometer was kept immersed with the ammonia level in the capillary a few centimeters lower than the heptane level in the bath. The ammonia level was measured intermittently, generally for 30 to 90 min, with occasional stirring; then the plug was knocked out and the observations were continued for a similar

[11] S. R. Gunn, Rev. Sci. Instr. **33,** 880 (1962).

period. The bath temperature was recorded continually and corrections to the measurements were applied for any small temperature fluctuations or drifts. At the end of this period, in several runs with each dilatometer, the temperature was raised $\sim\frac{1}{2}$ deg and the new level was measured after equilibrium was attained; and finally, in all runs, the helium pressure was reduced from 2.0 to 1.2 (at $-34°$) or 1.0 atm (at $-45°$), and the rise in liquid level again measured.

Upon completion of the measurement, the dilatometer was removed from the bath and chilled briefly in liquid nitrogen to bring the liquid level below the lower end of the capillary. Joint H was then warmed, Stopcock J removed, and the joint cleaned of wax. The dilatometer was inclined with Joint H somewhat lower, and the solution was forced out into a flask by heating the body of the dilatometer slightly. The dilatometer was washed twice with water, which was sucked in and expelled by cooling and heating. The ammonia was evaporated and the residue combined with the washings.

The solution was evaporated and ignited in a platinum dish to expel ammonia and carbon dioxide. The alkali hydroxide was titrated and the metal was also determined gravimetrically (as sodium magnesium uranyl acetate or potassium tetraphenylborate); the potassium solutions were also checked for sodium by flame photometry, about 0.5% by weight being found.

The net internal volume of the dilatometer—upper and lower chambers combined—and then of the upper chamber only were determined by filling with water and weighing.

The measurement of the change of liquid level with temperature—5 to 6 cm/deg—was used directly to correct the height readings for small changes in the thermostat temperature. It was also used, in conjunction with the total volume of the dilatometer and literature values for the coefficient of expansion of ammonia, to calculate the cross-sectional area of the capillary. The values thus obtained for Dilatometers 1, 2, and 3 were, respectively, 3.96, 3.76, and 4.00 mm^2, the individual values for each lying in a range of $\sim 2\%$. No great accuracy was needed in this figure because of the very small magnitude of the observed volume changes upon mixing.

Heat-of-solution data for sodium and potassium are not available at the present temperatures. Heats of solution of sodium at various concentrations have, however, been measured at 25°, 5°, and $-15°$.[9] Dilution is endothermic, and the heat of dilution increases at lower temperatures. The data at $-15°$ indicate that dilution from 0.02 to 0.01N under adiabatic conditions would give a temperature decrease of about 0.01°. Since the thermal time constant of the dilatometer, even without stirring, with respect to the bath is only a little over 1 min, this temperature change would soon disappear. The dilatometer was in practice stirred for about 1 min after the plug was knocked out, and no significant changes in the liquid level were seen thereafter.

The depressurization measurements were performed to check for the presence of bubbles in the solution at the end of the measurements. Any bubbles present would expand immediately upon decrease of the helium pressure to give a greater-than-normal rise in the level of the liquid. For all of the runs, including blank runs with pure ammonia, except as noted below in the Results section, dv/vdp was in the range 80 to 105×10^{-6} atm^{-1}, corresponding to a liquid-level rise of about 0.25 to 0.30 cm. This showed that the measurements had not been significantly perturbed by bubble formation and, further, that the coefficient of compressibility of the solutions did not vary significantly with concentration nor from the value for pure ammonia. Thus, the measurements of $\Delta\phi_v$ were unaffected by having been performed at a total pressure of 2 atm rather than at the saturation vapor pressure.

RESULTS

Results of the measurements are given in Table I. All of the runs performed are included, except those wherein there were gross failures, such as breakage of the armature. The amounts of metal listed are the values determined from the analyses of the solutions; in general, these are $\sim 5\%$ to 10% lower than the amount determined by weighing, which is reasonable in view of the oxidation of the metal occurring during cutting, drying, weighing, and transfer. The difference tended to be lower and more reproducible for potassium compared with sodium. The analytical samples for Runs 16, 19, and 20 were lost; the amount of metal is assumed to be 5% less than that weighed. The concentrations are calculated from the amounts of metal and the measured volumes of the chambers; the effects due to the small amount of metal initially in the lower chamber, and the probable incomplete mixing of solution below the plug when it falls to the bottom of the lower chamber, are neglected. Δv is the observed volume change at the time the plug was knocked out, obtained by extrapolating the liquid-level observations before and after this time; the limits of error are estimates based on the fluctuations of these observations. The observed change in apparent molar volume $\Delta\phi_v$(obs) is compared with values of $\Delta\phi_v$(calc) obtained from the curves of Figs. 1 and 3.

Several "blank" runs were performed at $-45°$; the dilatometer was loaded with ammonia, without any metal present, and the volume change upon knocking out the plug was observed. The volume changes observed were as follows: Dilatometer No. 1A, -1.6 ± 0.8; 2A, -4.5 ± 1; 3A, 0 ± 0.8; 1B, 0 ± 0.8; 2B, 0 ± 0.8; and 3B, 0 ± 0.8 μl. The cause of the effect with the A series is not known with certainty. If the standard taper joining upper and lower chambers were distorted when the plug was forced into it, one would expect to observe a rise rather than a fall of the liquid level when the plug was knocked out. Frequently, during filling of the dilatometer, a small bubble remained below the plug as the liquid level rose above it. The size of this bubble gradually decreased, and it had always disappeared by

TABLE I. Volume change upon dilution of metal–ammonia solutions.

Run	Dilatometer No.	Metal (mg-atom)	C_1 (g-atom/liter)	C_2 (g-atom/liter)	Δv (μl)	$\Delta\phi_{v(obs)}$ (ml/g-atom)	$\Delta\phi_{v(calc)}$ (ml/g-atom)
			Sodium at −45°				
1	1A	3.13	0.0567	0.0242	−6.8±1.2	−2.2±0.4	−8
2	2A	2.57	0.0475	0.0218	−4.8±1.2	−1.9±0.5	−8
3	3A	2.32	0.0385	0.0203	−4.0±1.6	−1.7±0.7	−6
4	3A	2.28	0.0385	0.0193	0.0±2.0	0.0±0.9	−6
5	1A	1.69	0.0307	0.0147	−1.2±0.8	−0.7±0.5	+4
6	3A	1.42	0.0238	0.0106	+1.2±0.8	+0.8±0.6	+10
7	1A	1.28	0.0229	0.0098	−3.6±2.0	+2.8±1.6	+11
8	3B	3.49	0.0614	0.0290	0.0±1.2	0.0±0.3	−5
9	1B	2.71	0.0486	0.0221	0.0±0.8	0.0±0.3	−8
10	3B	2.24	0.0394	0.0202	+1.2±0.8	−0.5±0.4	−7
11	1B	1.57	0.0277	0.0139	+1.2±0.8	+0.8±0.5	+8
12	2B	1.21	0.0218	0.0107	+0.8±0.8	+0.7±0.7	+10
13	3B	1.23	0.0217	0.0107	+0.8±0.8	+0.7±0.7	+10
14	1B	0.75	0.0133	0.0061	+0.8±0.8	+1.1±1.1	+2
15	3B	0.65	0.0115	0.0056	0.0±1.2	0.0±1.9	+2
			Potassium at −34°				
16[a]	2A	2.46	0.0316	0.0186	−4.8±3.6	−2.0±1.5	−18
17	3A	1.78	0.0298	0.0155	−5.2±2.0	−2.9±1.1	−15
18	2A	1.18	0.0214	0.0096	−2.9±1.8	−2.5±1.5	+12
19	3A	1.19	0.0198	0.0096	0.0±1.2	0.0±1.0	+12
20	2B	3.29	0.0659	0.0302	0.0±3.7	0.0±1.1	−6
21	1B	2.43	0.0437	0.0201	+0.8±0.8	+0.3±0.3	−21
22	3B	2.47	0.0434	0.0205	+1.2±1.2	+0.5±0.5	−21
23	3B	1.75	0.0310	0.0148	+0.4±0.8	+0.2±0.5	−15
24	1B	1.16	0.0207	0.0095	+1.2±0.8	+1.0±0.7	+12
25	2B	1.09	0.0198	0.0088	+1.2±0.8	+1.1±0.7	+13
26	2B	0.71	0.0126	0.0062	+0.7±1.1	+1.0±1.5	+11

[a] In this run, the metal was initially washed into the lower chamber.

the time measurements were started. Presumably the bubble was hydrogen which was not completely removed from the ammonia before the dilatometer was filled, and presumably it disappeared as it gradually dissolved in the liquid ammonia in the lower chamber. However, if any of the hydrogen were trapped between the ground surfaces of the plug and joint, it would diffuse only slowly into the ammonia and might dissolve only after the plug was knocked out. It would be expected that use of the 0.5-cm plug would reduce this effect, relative to use of the 2-cm plug. As noted previously, the volume changes dv/vdp upon depressurization of the dilatometer at the end of the runs were in the range 80 to 105×10^{-6} atm^{-1}, with a few exceptions for potassium runs. Run 23 showed a slightly larger change; Runs 17 (wherein the period from filling to knocking out the plug was extended to over 3 h), 18, and 20 showed much larger increases, presumably indicating that bubbles formed when the pressure was reduced.

To decrease the already low probability that some unknown trace impurity in the ammonia might be affecting the measurements, two different stocks of ammonia were used. A recently received small cylinder of Matheson 99.99% material was used for Runs 10, 12, 21, and the blank run with dilatometer 3B; another similar cylinder was used for Run 25. The remaining runs were done with ammonia from a 150-lb tank which was also used some six years earlier for the prior studies.[7,8] Melting-curve analysis of a sodium-dried sample from this tank has previously been reported[12]; an upper limit of 10^{-5} mole fraction impurity was set.

The random, negative volume changes evidenced by the blank runs for the A series have not been subtracted from the data in Table I; if they were, the volume changes would average about zero at all concentrations. The B series are considered substantially more reliable. The data set upper limits upon the volume changes which are more than an order of magnitude smaller than the values of Evers *et al.* shown in Figs. 1 and 3, and are in agreement with the earlier data of Gunn and Green shown in Fig. 1.

ACKNOWLEDGMENT

I thank Norman Smith for performance of the analyses.

[12] S. R. Gunn, Anal. Chem. **34**, 1292 (1962).

Thermal Properties of $Li(NH_3)_4$

In the following note, Mammano and Coulter present thermal data which give further support to the existence of a compound $Li(NH_3)_4$. They also postulate the existence of a quadruple point at 88.8°K in the Li–NH_3 phase diagram—a point at which $Li_{(s)}$, $NH_{3(s)}$, $Li(NH_3)_{4(s)}$, and saturated solution coexist. However the possibility of the quadruple point seems to be ruled out by recent work which shows that the 88.8°K transition temperature varies with the applied pressure.[1]

---

[1]S. Zolotov and M. J. Sienko, private communication, 1972.

Reprinted from THE JOURNAL OF CHEMICAL PHYSICS, Vol. 47, No. 4, 1564–1565, 15 August 1967
Printed in U. S. A.

52

Thermal Properties of Solid Lithium Tetrammine

N. MAMMANO* AND L. V. COULTER
Department of Chemistry, Boston University, Boston, Massachusetts
(Received 20 February 1967)

A recent study[1] of the electrical conductivity in concentrated lithium–ammonia solutions appears to have established the existence of a compound, probably "$Li(NH_3)_4$," in the solid below the "eutectic" temperature of the solution, 89.6°K. The bronze color, high electrical conductivity, and appearance of transitions at 82° and 69°K, in this solid, are evidence for the presence of this "tetrammine."

However, the phase equilibria of the system remain ambiguous. If 89.6°K is a simple binary eutectic temperature as suggested in the recently proposed phase diagram,[1] then it represents the low-temperature intersection of the solubility curves for pure solid lithium and pure solid ammonia in equilibrium with the saturated solution. Consequently, the solid freezing isothermally at the 89.6°K eutectic should be a two-phase eutectic mixture of solid lithium and solid ammonia crystals, and not "$Li(NH_3)_4$."

We wish to report here some preliminary thermal data for the lithium–ammonia system, as well as some considerations which bear on the eutectic-vs-compound question. Adiabatic heat capacities have been measured for several compositions in the 12% to 22% range from about 60°K to 110–200°K. The following observations are relevant at this time.

(a) Two first-order transitions have been observed, one at 88.79°K and one at 82.18°K, which we associate with a "eutectic" and a solid-state transition, respectively. There is no experimental basis for differentiating between these two temperatures of the transitions indicated by the conductivity measurements.[2] Our measurements were made with a platinum resistance thermometer[3] under equilibrium conditions and are believed to be more reliable than the temperatures found in the conductivity studies. A third transition at 69°K, detected in the conductivity work,[1] was not observed in these thermal studies. No transitions were observed at temperatures above 89°K.

(b) The results on the 88.79°K transition are somewhat incomplete, but the present data indicate just a single experimentally detectable transition in the 89°K temperature range at 88.79°K. The 82.18°K transition is quite sluggish, several hours being required for thermal equilibration under adiabatic conditions. The 88.79°K transition is clearly more "rapid," thermal equilibrium being achieved in approximately 10 to 15 min.

(c) The enthalpy change associated with each transition was determined for each composition studied by diluting the solution in the calorimeter with known increments of ammonia and measuring the transition heats following each increment. We have found at each transition temperature that, for compositions with lithium in excess of $NH_3/Li=4.15$, the enthalpy change was directly proportional to the amount of NH_3 present; for compositions with NH_3 in excess of $NH_3/Li=4.15$, the enthalpy change was directly proportional to the amount of Li present. This composition dependence (with enthalpy maximization at $NH_3/Li=4.15$) rules out solid solution formation[4] and also indicates that the horizontal constant-temperature tie lines for each transition extend across the phase diagram from pure NH_3 to pure lithium. For the enthalpy per mole of $Li(NH_3)_4$ (or per mole of $Li(NH_3)_{4.15}$) transforming: $\Delta H(82.18°K)=520.3\pm0.7$ cal; $\Delta H(88.79°K)=553.2\pm 1.1$ cal.

(d) C_p for the solid below the 88.79°K eutectic is 20% to 50% higher than the sum of the heat capacities of the pure components, in agreement with the proposal that a compound exists below this transition.

These data can be explained if one postulates that the observed transition at 88.79° consists of a normal eutectic at which pure crystals of ammonia and lithium separate, followed by a peritectoid reaction of the pure solid phases to give a solid lithium ammine at an "undetermined" temperature only slightly below the eutectic temperature; or that for the prevailing pressure (sufficiently augmented by hydrogen from slight decomposition of the system) the nonvariant quadruple point[5] has been reached at 88.79°K at which four phases [Li(s), NH_3(s), solid lithium ammine, and saturated solution] are in equilibrium. This equilibrium would occur at only a single temperature and pressure and would represent on a pressure–temperature projection either: (a) the intersection of the solid lithium ammine decomposition curve and the eutectic curve for the three-phase equilibrium involving the saturated solution and two solids (incongruent-melting case) or (b) the intersection of the solid lithium ammine melting curve (solid and melt of the same composition) with the three-phase eutectic curve (congruent-melting case).

We have approached to within several tenths of a degree of the 88.79°K "eutectic" from the low-temperature side; our failure to resolve a distinct temperature for the peritectoid transition suggests that the explanation in terms of a quadruple point is plausible. The relatively rapid thermal equilibration time for the 88.79°K transition compared to the 82.18° solid–solid transformation implies that a liquid phase is present in the former. This would indicate that the 88.79° transition is not the simple solid-state peritectoid with a true eutectic at some higher temperature. Furthermore, compound formation at 82°K from solid Li and solid NH_3 (freezing at the 88.79°K eutectic) is not in accord with the bronze color, high electrical conductivity, and high heat capacity of the solid above 82°K.[1]

The fact that the constant-temperature line for the 88.79°K transition extends across the phase diagram from pure NH_3 to pure Li is consistent with the suggestion that the compound does not exist in equilibrium with the saturated solution above 88.79°K.[1] If it did so exist, then the solubility curve for such an equilibrium, compound (s) = compound (soln), would descend to 88.79°K (and to the eutectic composition) from either a maximum melting point or from an incongruent melting point (peritectic). In either case, the 88.79°K eutectic enthalpy change would vanish at the compound stoichiometry and not extend to the composition axes for the pure components. Enthalpy maximization at $NH_3/Li = 4.15$ for the 82.18°K solid–solid transformation in the compound indicates, of course, that the composition of the compound corresponds to a nonstoichiometric tetrammine. It should be noted that similar deviations from perfect stoichiometry have been reported for the alkaline-earth hexammines.[6]

* National Science Foundation Cooperative Fellow, 1962–1965. Present address: Department of Chemistry, Cornell University, Ithaca, N.Y.

[1] J. A. Morgan, R. L. Schroeder, and J. C. Thompson, J. Chem. Phys. **43**, 4494 (1965); W. J. McDonald and J. C. Thompson, Phys. Rev. **150**, 602 (1966).

[2] J. C. Thompson (private communication).

[3] L. V. Coulter, J. R. Sinclair, A. G. Cole, and G. C. Roper, J. Am. Chem. Soc. **81**, 2986 (1959).

[4] J. E. Ricci, *The Phase Rule* (D. Van Nostrand, Co., Inc., New York, 1951), pp. 191–192.

[5] Reference 4, pp. 130–131.

[6] P. R. Marshall and H. Hunt, J. Phys. Chem. **60**, 732 (1956).

The Optical Spectroscopy of Tetraalkylammonium Solutions

Any doubts about the "normalcy" of tetraalkylammonium–ammonia solutions or of the absence of "tetraalkylammonium radicals" in such solutions were dispelled by the following paper by Quinn and Lagowski.

[Reprinted from the Journal of Physical Chemistry, 72, 1374 (1968).]

Metal–Ammonia Solutions. III. Spectroscopy of Quaternary Ammonium Radicals

53

by R. K. Quinn and J. J. Lagowski

Department of Chemistry, The University of Texas, Austin, Texas 78712 *(Received October 30, 1967)*

Techniques for the electrolytic generation and the spectroscopic characterization of quaternary ammonium radicals in liquid ammonia have been developed. The spectra of seven tetraalkylammonium radicals in liquid ammonia have been examined in the range 200–2500 mμ. The effects of concentration and temperature, as well as dissolved salts, upon the spectra have been examined. These data indicate that the absorbing species is the same in each case and has the same characteristics as the absorbing species present in metal–ammonia solutions. The implications of these results with respect to the current theories of metal–ammonia solutions are discussed.

Introduction

The addition of active metals to liquid ammonia yields solutions whose physical properties are apparently independent of the identity of the metal; the properties of dilute solutions ($<5 \times 10^{-3}$ M) are attributed to the "solvated electron,"[1] a species that is still poorly defined despite the considerable data that have been gathered on these systems. Two basic models for the solvated electron have been used to interpret the properties of these solutions. In the "cavity" model,[2–5] the electron is trapped in a cavity created in the bulk solvent; the system is stabilized by the orientation of the ammonia dipoles on the periphery of the cavity. The "expanded-metal" model[6–8] is conceptually related to the "cavity" model in that the electron in question occupies an "expanded orbital" defined by the hydrogen atoms of the first solvent sphere about the cation.

The transmission spectra of several alkali metal solutions in the concentration range 10^{-1} to 10^{-3} M indicate the presence of a single intense band at approximately 1500 mμ.[9–11] The band maximum shifts to lower energies with an increase in temperature and to higher energies if sodium iodide is added to sodium–ammonia solutions. Similar results were obtained for dilute solutions of alkali and alkaline earth metals in liquid deuterioammonia.[12] According to the description of the two models for the negative species present in metal–ammonia solutions, the "cavity" model suggests that the energy for the transition should be essentially independent of the identity of the metal cation. The energy of transition based on the "expanded-metal" model should show a more marked dependence on the size of the cation. However, the variation in the size of the alkali and alkaline earth metal ions may not be sufficient to allow discrimination between the two models.

The electrolysis of quaternary ammonium salts in liquid ammonia produces blue solutions[13–15] which have chemical properties similar to those of the metal–ammonia solutions[16–18] as well as comparable oxidation potentials.[19,20] These observations suggested that a spectroscopic study of the cathodic products formed in the electrolysis of tetraalkylammonium salts in liquid ammonia might aid in understanding the nature of metal–ammonia solutions because of the possibility of varying the size of the cation associated with the electron in these solutions.

Experimental Section

Tetraalkylammonium salts, the purest available commercially, were dried *in vacuo* at 100°, recrystallized from anhydrous liquid ammonia, heated again *in vacuo* at 100°, and stored in a helium-filled drybox the at-

(1) C. A. Kraus, *J. Franklin Inst.*, **212**, 537 (1931).
(2) R. A. Ogg, *J. Am. Chem. Soc.*, **68**, 155 (1946).
(3) R. A. Ogg, *J. Chem. Phys.*, **14**, 144 (1946).
(4) R. A. Ogg, *ibid.*, **14**, 295 (1946).
(5) J. Jortner, *ibid.*, **30**, 839 (1959).
(6) E. Becker, R. H. Lindquist, and B. J. Alder, *ibid.*, **25**, 971 (1956).
(7) E. Arnold and A. Patterson, Jr., *ibid.*, **41**, 3089 (1964).
(8) H. M. McConnell and C. H. Holm, *ibid.*, **26**, 1517 (1957).
(9) M. Gold and W. L. Jolly, *Inorg. Chem.*, **1**, 818 (1962).
(10) C. J. Hallada and W. L. Jolly, *ibid.*, **2**, 1076 (1963).
(11) R. C. Douthit and J. L. Dye, *J. Am. Chem. Soc.*, **82**, 4472 (1960).
(12) D. F. Burow and J. J. Lagowski, Advances in Chemistry Series, No. 50, American Chemical Society, Washington, D. C., 1965, p 125.
(13) W. Palmaer, *Z. Elektrochem.*, **8**, 729 (1902).
(14) H. H. Schlubach, *Ber.*, **53B**, 1689 (1920).
(15) H. H. Schlubach and F. Ballauf, *ibid.*, **54B**, 2811 (1921).
(16) H. H. Schlubach and G. V. Zwehl, *ibid.*, **56B**, 1889 (1923).
(17) H. H. Schlubach and H. Miedel, *ibid.*, **65B**, 1892 (1923).
(18) S. Goldschmidt and F. Nagel, *ibid.*, **64B**, 1744 (1931).
(19) G. S. Forbes and C. E. Norton, *J. Am. Chem. Soc.*, **48**, 3233 (1926).
(20) H. A. Laitinen and C. J. Nyman, *ibid.*, **70**, 3002 (1948).

mosphere of which was equilibrated with Na–K liquid alloy.[21]

Preliminary studies of the spectra of electrolytically generated tetraalkylammonium radicals were made using a modified version of an apparatus described previously.[12] The results of these experiments indicated the desirability of modifying several of the design features of the original optical cell to increase the reliability of the data. Accordingly, a multicompartmented dewar vessel (Figure 1) was constructed. The winch mechanism used to introduce solids into the cell has been described[12] previously. The outer vacuum jacket was continually evacuated with an oil diffusion pump backed by a mechanical pump; under these conditions condensation did not form on the optical windows. The electrolytic–optical cell was attached to the vacuum jacket by means of an O-ring joint which permitted ready access to the tri-compartmented Pyrex electrolysis cell carrying a fused quartz optical cell; this feature was indispensible in the early experiments. The electrode compartments (A and B) were separated by medium porosity frits from an intermediate compartment (C) which was used as a salt bridge. The optical cell was attached to the cathode compartment with graded seals.

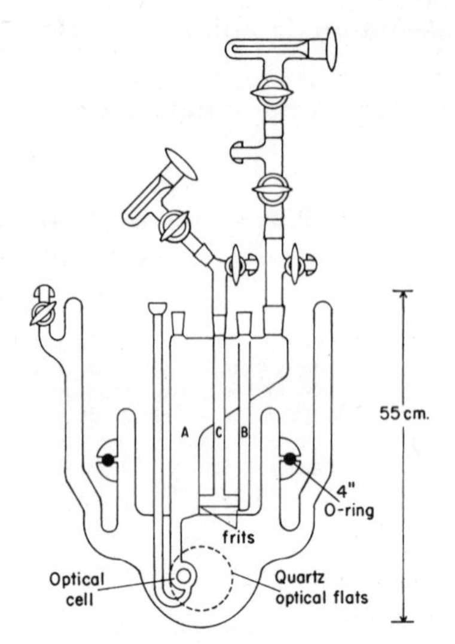

Figure 1. A schematic diagram of the electrolytic–optical dewar system. An "unfolded" cross section of the cell is shown for ease of description.

The large extinction coefficient ($\epsilon \simeq 4 \times 10^4$) of the broad absorption band characteristic of metal–ammonia solutions[9–12] and the very intense solvent bands for liquid ammonia in the region of interest[12] required that the optical path length be $\leq$0.5 mm for best results; the path length of the cell (Figure 2) used in these experiments was 0.421 mm. The platinum gauze cathode was positioned to permit the electrolysis products to diffuse rapidly into the light path.

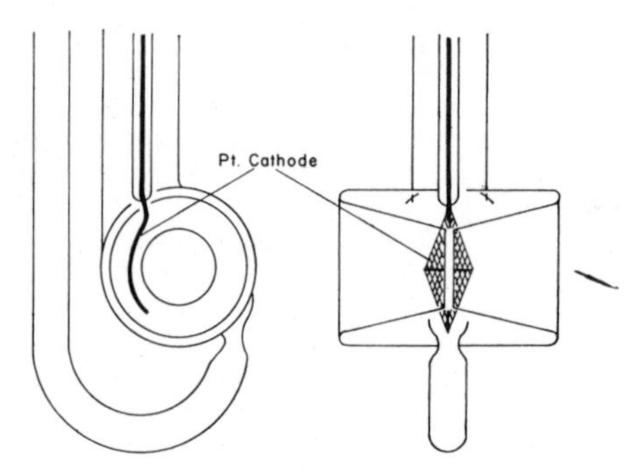

Figure 2. Quartz optical cell with platinum cathode in position.

The entire dewar system was permanently installed in the cell compartment of a Beckman DK-2 ratio-recording spectrophotometer. The optical quartz windows were secured to ground-glass flanges on the dewar vacuum jacket with picein sealing compound and transmitted both the sample and reference beams of the spectrometer. The outer dewar jacket, O-ring, and inner electrolytic–optical cell were oriented to allow the sample beam to pass through the inner optical cell while the reference beam passed through the evacuated portion of the dewar.

The reaction vessel was cleaned by a special rinsing procedure[12] which involved keeping a sodium–ammonia aging solution in the vessel for several hours as the key step. Qualitatively, ammonia solutions of tetraalkylammonium radicals appear to be less stable than those of the alkali metals. However, there was no appreciable decomposition in the time required to determine the spectrum of a given system. If the absorbance at a given wavelength decreased by less than 0.025 unit in the time required to record the spectrum from 200 to 2500 mμ, the spectrum was considered acceptable since the region of particular interest (800–2000 mμ) required only a fraction of the total scan time. The concentration of the absorbing species was determined coulometrically using a graphical integration method accurate to within $\pm$0.005 coulomb.[22]

Results and Discussion

Spectral measurements on the rate of diffusion of the electrolytically generated species indicated that diffusion had virtually ceased about 10–12 min after the electrolysis was terminated. After this time slow de-

(21) D. F. Burow, Ph.D. Dissertation, The University of Texas, 1966.

(22) R. K. Quinn, Ph.D. Dissertation, The University of Texas, 1967.

composition of the radical was observed as reflected by a decrease of <0.025 absorbance unit/hr at the band maximum. Metal–ammonia solutions decompose to form metal amide and hydrogen in the presence of platinum,[23] and by analogy, a similar reaction would be expected for ammonium radicals.

$$NR_4\cdot + NH_3 \longrightarrow NR_4^+ + NH_2^- + {}^1/_2H_2 \quad (1)$$

However, the decomposition of the cathodically generated species in the presence of platinum is not as important as the procedure used to clean the cell; a much higher rate of decomposition is observed if the apparatus is only rinsed with anhydrous ammonia. Quaternary ammonium ions are unstable in the presence of amide ions at the boiling point of liquid ammonia.[24] If the decomposition of the cathodically generated species corresponds to the process shown in eq 1, the decrease in absorbance is equivalent to the formation of less than 10^{-5} mole/l. of NH_2^- per hour, an amount which could not be detected spectrophotometrically[25] with the apparatus used in this investigation. Decomposition of the quaternary ammonium ions by NH_2^- arising from eq 1 should be significant under these conditions.

Electrolysis of concentrated ammonium chloride solutions maintained at −78 to −80° for several hours produced large quantities of hydrogen and nitrogen as verified by mass spectroscopy. Previous attempts to produce hydrogen-substituted ammonium radicals were also unsuccessful.[14,15,26,27] Quaternary ammonium salts containing an aromatic group did not yield blue solutions after electrolysis for several hours, in agreement with the results of other investigators.[14,15]

The spectra of the blue solutions formed in the electrolysis of seven tetraalkylammonium salts were determined at 200–2500 mμ. A single broad band with a maximum at 1440 ± 4 mμ at −70° was observed in all cases (Figure 3). The band parameters were the same within experimental error for all the radicals studied in the concentration range 2–200 × 10^{-5} *M* (Table I). Although the bands appear symmetrical with respect to wavelength, they are asymmetrical on an energy scale. Calculation of the extinction coefficients of the radicals based on the number of coulombs of electricity passed through the cell gave an average extinction coefficient of 4 ± 1 × 10^4 l. mole^{-1} cm^{-1}. Using the approximation method suggested by Dunn,[26] a value of 0.65 ± 0.06 was obtained for the oscillator strength of the species giving rise to the transition. An increase in wavelength of the absorbance maximum with increasing concentration (1–100 × 10^{-3} *M*) for solutions of alkali metals in ammonia has been reported;[9,11] however, this dependence was not observed for more dilute (5–50 × 10^{-5} *M*) metal–deuterioammonia solutions.[12] The spectra of quaternary ammonium radicals throughout the concentration range (2–200 × 10^{-5} *M*) studied show a random scattering of λ_{max} that is within the error

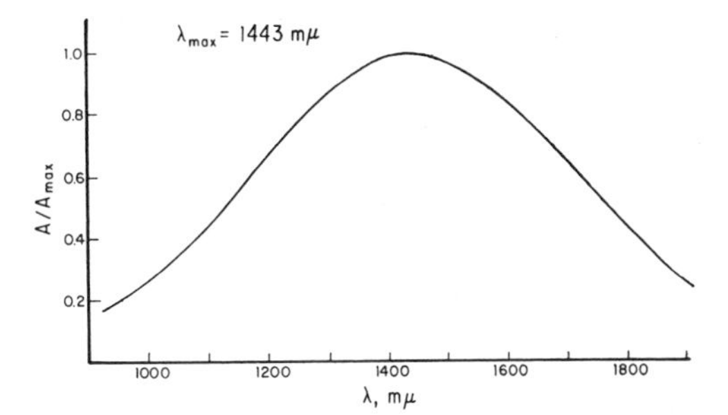

Figure 3. The absorption spectrum (normalized) of 3.48 × 10^{-4} *M* tetraethylammonium radical at −70°.

Table I: Position of Absorption Maxima of Quaternary Ammonium Radicals in Ammonia at −70°

$NR_4\cdot$	λ_{max}, mμ	$\bar{\nu}_{max}$, cm^{-1}	$W_{1/2}$, mμ
$N(Me)_4$	1445	6920	653
$N(Et)_3Me$	1438	6954	627
$N(Et)_4$	1443	6930	644
$N(Et)_3Pr^n$	1437	6959	660
$N(Pr^n)_4$	1434	6974	629
$N(Bu^n)_4$	1434	6974	660
$N(Me)_3$dodecyl	1446	6916	655
Average	1440 ± 4	6944 ± 21	647 ± 12

of the experiments, indicating that there is no dependence of the band position on the concentration.

Previous investigations of metal–ammonia solutions indicate a perturbation of the spectrum in the presence of alkali metal halides.[9–12,27] Since all the quaternary ammonium radicals were generated electrolytically from solutions of their respective salts, it should be possible to observe these effects in the experiments reported here. However, the position and the shape of the band for all the tetraalkylammonium radicals studied were unaltered as the concentration of tetraalkylammonium salt was varied in the range 1–1000 × 10^{-4} *M*.

The effect of temperature on the absorption band was studied using a solution of 2.85 × 10^{-4} *M* tetraethylammonium radical, generated in the presence of 0.03 *M* $(C_2H_5)_4NBr$, in the temperature range −33 to −73°. The band maximum moves to longer wavelengths as the temperature increases. The linear relationship

(23) G. W. Watt, G. D. Barnett, and L. Vaska, *Ind. Eng. Chem.*, **46**, 1022 (1954).

(24) W. L. Jolly, *J. Am. Chem. Soc.*, **77**, 4958 (1955).

(25) R. E. Cuthrell and J. J. Lagowski, *J. Phys. Chem.*, **71**, 1298 (1967).

(26) T. M. Dunn, "Modern Coordination Chemistry," J. Lewis and R. G. Wilkins, Ed., Interscience Publishers Inc., New York, N. Y., 1960, p 276.

(27) H. C. Clark, A. Horsfield, and M. C. R. Symons, *J. Chem Soc.*, 2478 (1959).

between temperature and position of the band maximum (Figure 4) gives a temperature coefficient of 2.7 ± 0.4 mμ deg^{-1} (−12.7 ± 2.0 cm^{-1} deg^{-1}). Temperature coefficients of −9.1,[28] −9.7,[11] −12.7,[9] and −12.0[22] cm^{-1} deg^{-1} have been reported for solutions of alkali metals in liquid ammonia. Analysis of the spectrum of the tetraethylammonium radical at several temperatures indicates that the band width at half-height (647 ± 12 mμ) varies in a random manner with temperature. Thus, the shape of the band for the electrolytically generated species appears to be independent of the temperature. Electrolytically generated alkali metal solutions in the same concentration range show a single band with a maximum centered between 1430 and 1454 mμ, with an average width at half-height of 639 mμ, and an average extinction coefficient of 4.8 × 10^4 l. $mole^{-1}$ cm^{-1}.[22] Thus it appears that the species giving rise to the spectral transition in solutions of tetraalkylammonium radicals is the same as that in metal–ammonia solutions, *i.e.*, the "solvated electron."

The most outstanding feature of the spectra of liquid ammonia solutions of tetraalkylammonium radicals is the constancy of the position of the absorption band at 1440 mμ throughout the range of cation size studied (Figure 5). An estimate of the volumes for the symmetrical and nearly symmetrical tetraalkylammonium ions was obtained by assuming that the alkyl branches were in the least sterically hindered conformation; this procedure gives a result which represents the minimum volume that the tetraalkylammonium cations could occupy. Spectral data for dilute solutions of the alkali and alkaline earth metals are also included in Figure 5: the volumes of the corresponding cations have been calculated from Pauling's ionic radii.[28] Errors in the calculation of the volumes introduced by these approximations are relatively unimportant in view of the fact that no significant variation in the position of the absorption band was observed from the smallest alkali metal ion to the largest tetraalkylammonium ion. The slope of the line shown in Figure 5, which was obtained from a least-squares analysis, is indistinguishable from zero within experimental error. Of special interest are the results obtained with solutions containing dodecyltrimethylammonium cations; the band maximum for this species was observed at 1446 mμ. Elworthy[29] has described the shape of the solvated trimethyloctylammonium ion in aqueous solution as a prolate ellipsoid with a major axis of 9 Å and a minor axis of 3.6 Å, yielding an approximate volume of 500 $Å^3$. In the case of dodecyltrimethylammonium cation, the major axis, at least, would have to be longer than that for trimethyloctylammonium ion, leading to a larger volume for the former species. Thus, one of the tetraalkylammonium cations used in this investigation must be unsymmetrical with respect to the center of positive charge.

Basically, two models for the monomer species have

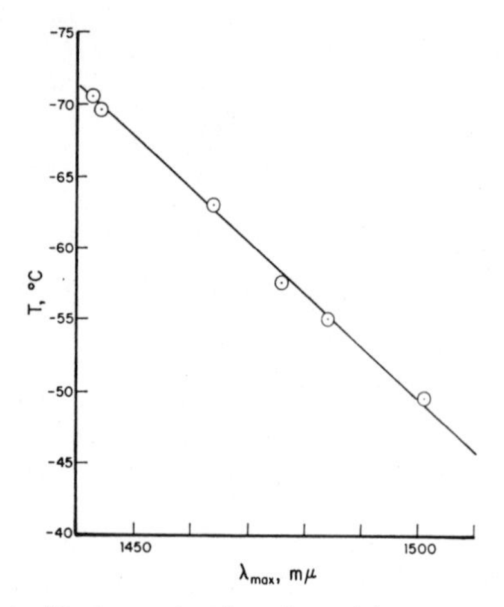

Figure 4. The temperature dependence of the position of the band maximum for 2.85 × 10^{-4} *M* $(C_2H_5)_4N\cdot$ in liquid ammonia.

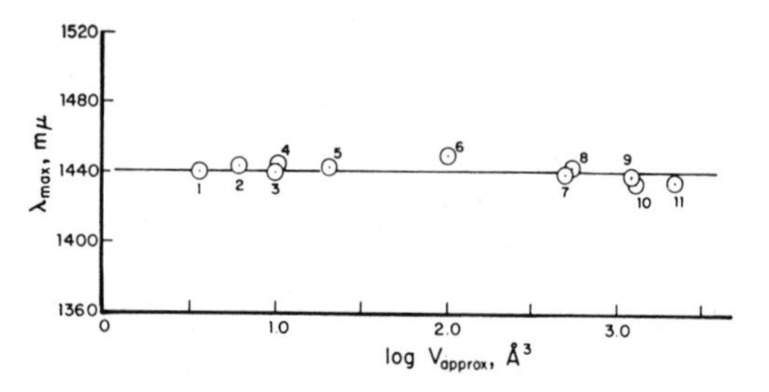

Figure 5. The position of the absorption maximum in solutions containing the solvated electron associated with cations of various sizes: 1, Na^+; 2, Sr^{2+}; 3, K^+; 4, Ba^{2+}; 5, Cs^+; 6, $N(CH_3)_4^+$; 7, NEt_3Me^+; 8, NEt_4^+; 9, NEt_3Pr^+; 10, NPr_4^+; 11, NBu_4^+.

been suggested: (a) an electrostatic association of the solvated electron and a solvated cation,[30] and (b) the "expanded-metal" model.[6-8] It has been suggested that the monomer in the "expanded-metal" model is in equilibrium with solvated electrons ($M = M^+ + e^-$, $K = 0.03$ at −33°).[6] Although in metal–ammonia solutions the concentration of the monomer is negligibly small in dilute solutions (*i.e.*, a total metal concentration of 1 × 10^{-3} *M*), in the presence of a large excess of cation the concentration of monomer should increase. Under the conditions of our experiments (*e.g.*, 0.1 *M* MX and 1 × 10^{-3} *M* solvated electrons) and assuming that the value of the constant reported[6] for the equilib-

(28) L. Pauling, "The Nature of the Chemical Bond," 3rd ed, Cornell University Press, Ithaca, N. Y., 1960, p 514.

(29) P. H. Elworthy, *J. Chem. Soc.*, 388 (1963).

(30) M. Gold, W. L. Jolly, and K. S. Pitzer, *J. Am. Chem. Soc.*, **84**, 2264 (1962).

rium between monomers and solvated electrons for the alkali metals is essentially the same for tetraalkylammonium cations, some solutions should have contained at least 60% of the electrons as monomers. Moreover, the monomers, if they are formed containing tetraalkylammonium cations, should contain cations with widely differing geometrical requirements which should be reflected in the spectra of these solutions.

Making the reasonable assumption that our solutions contain a significant fraction of electrons as NR_4 monomers, the spectral data suggest that in the relatively dilute concentration range ($2–200 \times 10^{-5}$ M) the solvated electron is not associated with the solvated cationic species in solution in the manner required by the "expanded-metal" model.[6–8] Rather the results seem to favor the "cavity model"[2–5] in which the trapped electron is electrostatically associated with the solvated cations present,[30] much as is any ammonia solution containing charged species.[31]

Acknowledgment. We gratefully acknowledge the financial support of the Robert A. Welch Foundation and the National Science Foundation.

(31) V. F. Hnizda and C. A. Kraus, *J. Am. Chem. Soc.*, **71**, 1565 (1949).

Sodium Ions from Pyrex Glass

The discovery that sodium can be introduced into metal–amine solutions by exchange from Pyrex glass, reported in the following communication by Hurley, Tuttle, and Golden, was extremely important in eliminating much of the confusion from previous studies of metal–amine solutions and indicated the precautions necessary in such studies.

Reprinted from THE JOURNAL OF CHEMICAL PHYSICS, Vol. 48, No. 6, 2818–2819, 15 March 1968
Printed in U. S. A.

54

Origin of the 660-mμ Band in the Spectra of Alkali-Metal–Amine Solutions

IAN HURLEY, T. R. TUTTLE, JR., AND SIDNEY GOLDEN
Department of Chemistry, Brandeis University, Waltham, Massachusetts
(Received 9 October 1967)

Investigations of the optical spectra of alkali metals in amines[1–8] have established the existence of three bands. We here present experimental evidence in support of our contention that the 660-mμ band is to be associated with a presence of *sodium* in alkali-metal–amine solutions which heretofore have been examined regardless of the actual metal used in their preparation.

In *Pyrex apparatus* we have carried out experiments on the potassium–ethylamine system with the following results:

(1*p*) In agreement with results of others[2,3] the optical spectra exhibited two principal absorbances at 660 and 870 mμ.

(2*p*) Addition of KI, sublimed *in situ*, to a solution exhibiting a dominant 870-mμ band yielded spectrum exhibiting essentially 660-mμ band.[9] Subsequently, we found that the sublimation leads to more than half sodium salt in the sublimate. The supposed addition of KI was undoubtedly an addition of a mixture containing sodium.

(3*p*) A solution prepared by reacting ethylamine with a 1:2 Na–K exhibited essentially 660-mμ band. Flame photometric analyses of the residue from these solutions gave an Na/K mole ratio of 0.24.

In *quartz apparatus*, by contrast, the following results were obtained with the potassium–ethylamine system:

(1*q*) The optical spectrum consisted of a dominant 870-mμ band; *no significant 660-mμ band was observed.*

(2*q*) Added KI, sublimed *in situ*, produced no essential change in the spectrum.

(3*q*) A solution of NaI which was allowed to react with *K*-metal exhibited a spectrum dominated by 660-mμ band.

In addition, experiments were performed on the potassium–ethylamine system in apparatuses of the following constructions: (A) Pyrex except for optical cell of quartz, (B) Pyrex except for quartz sidearm for metal distillation and quartz optical cell, and (C) quartz. Results of these experiments are summarized in Table I. The Na/K mole ratios in Table I show that appreciable sodium is picked up in distilling potassium in Pyrex, compare entries (Aa) and (Ab) with (Ca) and (Cb) or (Ba) and (Bb), and also that in Pyrex apparatus additional sodium is picked up by the potassium solution presumably by its reaction with Pyrex, compare (Bb) and (Bc) with (Cb) and (Cc). Comparison between Columns c and d of Table I reveals a correlation between the Na/K mole ratio found in the residues from solutions and their optical spectra. [See also (3*p*)]. The larger the proportion of sodium, the larger the proportion of 660-mμ band compared to 870-mμ band. In summary, we detect no 660-mμ band unless appreciable sodium is present, the 660-mμ band may be produced by addition of a source of sodium, and the 660-mμ band is apparent when appreciable sodium is present. In fact, the more sodium, the more 660-mμ band.

In quartz apparatus, ethylenediamine solutions of sodium metal, of sodium metal plus KI, and of potas-

TABLE I. Results from flame photometry[g] and spectrophotometry.

(h)	a	b	c	d
A	(0.000154)[f]	(0.0054)[e]	0.053	1.0±0.01
B	(0.000154)[f]	0.00049	0.0150	0.30±0.03
C	(0.000154)[f]	0.00122	0.00134	0.24±0.03

[a] Na/K mole ratio in metal before distillation.
[b] Same after distillation.
[c] Same in residue from all solutions decomposed in apparatus.
[d] Ratio of optical density of typical metal solution at 650 mμ to that of 900 mμ; for these solutions absorbance at 280 mμ ranged from 0.04 to 0.40.
[e] Refers to a separate experiment in which K was distilled in Pyrex in the usual manner.
[f] Refers to K used in Footnote e before distillation.
[g] Mole ratios estimated to be reliable to within about 5%; great care was taken to assess possible sources of sodium and avoid introduction of sodium during analytical procedure.
[h] Apparatuses were thoroughly washed with distilled water and reagent-grade acetone only.

sium metal plus NaI all display essentially the same spectra consisting of an intense band at 660 mμ and a weak, broad one at 1300 mμ. A potassium ethylenediamine solution in quartz exhibits *no detectable 660-*mμ *band*, but exhibits both 870-mμ and 1300-mμ bands. A previously reported spectrum obtained in Pyrex apparatus exhibits intense 660-mμ absorbance in addition to those at 870 and 1300 mμ. The solutions compared both have absorbance divided by path length of about 3 at 1300 mμ. Recalling the results of our experiments on ethylamine solutions we suggest that the 660-mμ band here is to be attributed to sodium inadvertently introduced by distillation of the potassium in Pyrex and by reaction of the potassium–ethylenediamine solution with the Pyrex vessel.

When the source of the 660-mμ band is designated as a species of sodium the available data may be simply summarized in terms of an oxidation–reduction equilibrium

$$Na^{+}+X_{M}\rightleftarrows M^{+}+X_{Na}, \quad (1)$$

where M represents an alkali metal, X_M reducing species which occur in a solution of metal, M and M^+ oxidized forms of metal. In terms of Eq. (1), the appearance of the 660-mμ band in potassium solutions, even when the proportion of dissolved sodium is rather small, i.e., 5% (see Table I) indicates that the equilibrium constant for Reaction (1), K_1, is large. Most of the dissolved potassium is apparently present as ethylamide. The fact that KI in a sodium solution does not appreciably affect its spectrum also indicates that K_1 is large when M is K. The appearance of the 660-mμ band in the spectra of amine solutions of Li, Rb, and Cs[1,2,8] suggests that K_1 is probably large in these systems also.

We are indebted to Professor H. Linschitz for helpful comments. This work has been supported in part by a grant from the Petroleum Research Fund of the American Chemical Society.

[1] R. R. Dewald and J. L. Dye, J. Phys. Chem. **68**, 121 (1964).
[2] M. Ottolenghi, K. Bar-Eli, H. Linschitz, and T. R. Tuttle, Jr., J. Chem. Phys. **40**, 3729 (1964).
[3] M. Ottolenghi, K. Bar-Eli, and H. Linschitz, J. Chem. Phys. **43**, 206 (1965).
[4] L. R. Dalton, J. D. Rynbrant, E. M. Hanson, and J. L. Dye, J. Chem. Phys. **44**, 3969 (1966).
[5] R. Catterall and M. C. R. Symons, J. Chem. Soc. **1965**, 6656.
[6] R. Catterall, M. C. R. Symons, and J. W. Tipping, J. Chem. Soc. **1966**, 1529.
[7] H. Blades and J. W. Hodgins, Can. J. Chem. **33**, 411 (1955); G. Fowles, W. McGregor, and M. Symons, J. Chem. Soc. **1957**, 3329.
[8] S. Windwer and B. R. Sundheim, J. Phys. Chem. **66**, 1254 (1962).
[9] I. Hurley and T. R. Tuttle, Jr., unpublished work, in part reported at the Symposium on Electron Spin Resonance at Michigan State University. See comments by T. R. Tuttle, Jr., J. Phys. Chem. **71**, 191 (1967).

An X-Ray Diffraction Study of $Li(NH_3)_4$

In the following paper, Mammano and Sienko present X-ray diffraction data which leave little doubt about the existence of the compound tetraamminelithium(0).

55

Low-Temperature X-Ray Study of the Compound Tetraamminelithium(0)[1]

Nicholas Mammano and M. J. Sienko

Contribution from the Baker Laboratory of Chemistry, Cornell University, Ithaca, New York 14850. *Received June 6, 1968*

Abstract: The compound $Li(NH_3)_4{}^0$ has been prepared by cooling solutions of lithium in liquid ammonia. X-Ray powder studies at 77°K indicate that $Li(NH_3)_4$ exists in two phases: a cubic form with $a_0 = 9.55$ Å, stable between 82 and 89°K, and a hexagonal form having $a = 7.0$ Å and $c = 11.1$ Å, stable below 82°K. At 77°K the measured density is 0.57 g/cc, corresponding to two molecules of $Li(NH_3)_4$ per hexagonal unit cell. Reflections of the type $00l$ (l odd) were not observed, suggesting that the space group is $P6_3$ or $P6_3mc$ with four ammonia molecules tetrahedrally disposed about each lithium atom. The c/a ratio together with reasonable assumptions for bond lengths suggests the $Li(NH_3)_4$ tetrahedra are in hexagonal close packing. Arguments are given for believing that the molecule $Li(NH_3)_4$ is unstable to dissociation in the gas phase but that the solid compound is stable because of a large electron delocalization energy. The energy change for the reaction $Li(s) + 4NH_3(s) \rightarrow Li(NH_3)_4(s)$ is estimated to be -20 kcal/mole.

The phase diagram of the lithium–ammonia system shows a deep eutectic at about 20 mole % lithium.[2] However, unlike the sodium and potassium systems, where the solid formed is a gray-white mixture of metal and solid ammonia, the lithium system gives a golden, conducting solid more like the compounds $M(NH_3)_6$ of the alkaline earth metals.[3] Although the characterization of the hexaammine alkaline earth metals is complicated by large deviations from stoichiometry arising from ammonia vacancies or interstitial ammonia, compound formation is rather clearly supported by maxima in the temperature–composition curves. No such maximum-melting composition exists in the lithium–ammonia case, so this primary criterion for compound formation is lacking. Still there is considerable indirect evidence that a compound (usually taken to be $Li(NH_3)_4$, although deviations from perfect stoichiometry analogous to alkaline earth hexaammines have also been observed[4]) actually exists in the lithium–ammonia system. Such evidence includes, for example, the large negative heat of solution for lithium in ammonia at intermediate concentrations,[5] the low vapor pressure of the saturated solution,[6] the persistence of the bronze color of the liquid into the solid state, and the high conductivity of the solid.[7] More direct evidence comes from breaks observed in the conductivity–temperature curve of the solid, suggesting solid–solid phase transitions,[8] and from the fact that the heat capacity of the solid is 20–50% higher than the sum of the heat capacities of the pure components.[4]

(1) This research was sponsored by the National Science Foundation through Grant No. GP-6246 and was supported in part by AFOSR and ARPA.

(2) See, for example, the review article by M. J. Sienko in "Metal Ammonia Solutions," G. Lepoutre and M. J. Sienko, Ed., W. A. Benjamin, Inc., New York, N. Y., 1964, p 25.

(3) H. J. Holland and F. W. Cagle, Jr., presented at the 145th National Meeting of the American Chemical Society, New York, N. Y., Sept 1963.

(4) N. Mammano and L. V. Coulter, *J. Chem. Phys.*, **47**, 1564 (1967).

(5) L. V. Coulter and L. Monchick, *J. Am. Chem. Soc.*, **73**, 5687 (1951).

(6) P. R. Marshall and H. Hunt, *J. Phys. Chem.*, **60**, 732 (1956).

(7) H. Jaffe, *Z. Physik*, **93**, 741 (1935).

(8) J. A. Morgan, R. L. Schroeder, and J. C. Thompson, *J. Chem. Phys.*, **43**, 4494 (1965).

The present investigation was undertaken in an attempt to find X-ray evidence supporting the existence of the compound $Li(NH_3)_4$ in the solid state.

Experimental Section

Preparation of Material. The ammonia was obtained from Matheson and was their highest grade anhydrous 99.99%. The lithium, obtained from Lithium Corporation of America, was 99.9%. The lithium was cleaned mechanically under argon, weighed in a sealed tube, and transferred under argon into one arm of a modified Faraday tube. The tube was evacuated, and ammonia, dried over sodium, was condensed on the lithium in sufficient amount to give a 20 mole % lithium solution. The arm of the Faraday tube was then cooled in liquid nitrogen and evacuated to remove any traces of decomposition hydrogen. The solution was then allowed to warm up; the Faraday tube was turned and a drop of solution allowed to run over into the other arm which was drawn into a capillary. When an appropriate bead had collected, the capillary tip was quenched in liquid nitrogen and sealed off for transfer to the X-ray camera. The inner diameter of the capillary was about 1 mm.

X-Ray Data. A conventional Debye camera of 57-mm diameter was modified to allow for spray cooling of the target sample. The metal–ammonia capillary prepared as above was rotated and sprayed with liquid nitrogen during the 4–6-hr exposures. Radiation was Cu Kα. Film shrinkage was corrected for by tungsten calibration. The X-ray pictures showed no evidence of any decomposition products such as amide, imide, oxide, or hydroxide. However, many of the X-ray patterns obtained were of poor quality.

Density Determination. The density of $Li(NH_3)_4$ was measured at 77°K by gas displacement. A large sample of $Li(NH_3)_4$ was prepared in a 100-ml bulb of calibrated volume, cooled slowly through the 89°K solidification point, and then cooled very slowly (*vide infra*) through the 82°K solid–solid transition. After thorough annealing of the sample at 77°K, helium was allowed to expand from a known volume into a new volume including the sample. The pressure drop was measured and used to calculate the sample volume. The measured density, 0.57 g/cc, is believed to be accurate to within 0.03 g/cc.

Results and Discussion

Table I shows the results of a typical X-ray exposure. Diffraction lines of observable intensity were exceedingly difficult to obtain, particularly for the relatively weak lines of the supercooled cubic phase. All pictures could be accounted for by assuming that two phases were present: a low-temperature hexagonal phase and a high-temperature cubic phase. The persistence of the cubic phase to lower temperature is not surprising, since the 82°K transition is a very sluggish

Table I. X-Ray Data for $Li(NH_3)_4(s)$

2θ	Intensity	d_{obsd}, A	d_{calcd}, Å[a]	(*hkl*)
13.35	vvw	6.63	6.75 c	(110)
14.55	s	6.09	6.09 h	(100)
16.70	vw	5.26	5.31 h	(101)
18.75	w	4.73	4.78 c	(200)
21.88	m	4.06	4.10 h	(102)
26.52	w	3.36	3.37 h	(111)
27.93	m	3.20	3.18 c	(300)
30.55	ms	2.93	2.93 h	(201)
32.44	vw	2.76	2.77 h, c	(004), (222)
35.14	vw	2.55	2.55 h, c	(113), (321)
38.95	mw	2.31	2.30 h	(210)

[a] (c) cubic, a_0 = 9.55 Å; (h) hexagonal, a = 7.0 Å, c = 11.1 Å.

one. The heat capacity studies of Mammano and Coulter[4] indicated that their 82.18°K transition required several hours for thermal equilibration under adiabatic conditions. Resistivity studies, carried out at Cornell in collaboration with the Cryogenic Physics group, suggest that at least a day may be required to pass through the transition on cooling. There is some evidence that the transition, which may involve a homogeneous shear as in a martensitic transformation, can be facilitated by tapping.

The principal phase present at 77°K, believed to be stable at this temperature, has hexagonal symmetry with a = 7.0 Å and c = 11.1 Å. Two molecules of $Li(NH_3)_4$ per unit cell lead to a calculated density of 0.53 g/cc, in acceptable agreement with the experimentally determined density at 77°K of 0.57 g/cc. The cubic phase, which appears to be stable between 82 and 89°K, has a unit cell with a_0 = 9.55 Å. Assignment of four molecules of $Li(NH_3)_4$ per unit cell leads to a calculated density of 0.57 g/cc.

Although a unique space group cannot be assigned to the hexagonal phase, the absence of reflections of the type (00*l*) with *l* odd together with a probable tetrahedral configuration of the NH_3 molecules around the lithium suggests P6$_3$ or P6$_3$mc. A tentative assignment of the eight ammonia molecules and two lithium atoms per unit cell is as follows: six NH_3 in the *c* positions, two NH_3 in *b* positions at *z*, and two lithium in *b* positions at *z'*. These positions correspond to two tetrahedra related to each other by a twofold screw axis through the center of the primitive cell.

A tetrahedral arrangement of NH_3 molecules about lithium has previously been suggested by Sienko[2] on the basis of a simple valence-bond model for $Li(NH_3)_4$. Promotion of the 2s electron of lithium to the 3s orbital opens up the 2s orbital for sp^3 hydridization. The lone pair of an NH_3 molecule can occupy an sp^3 hybrid of the lithium to form a σ bond between the NH_3 and the Li. The dimensions of the $Li(NH_3)_4$ molecule can be crudely estimated by taking the Li–N distance to be about the same as the 1.94 Å in Li_3N, the N–H distance to be 1.01 Å as in NH_3, and the Li–N–H bond angle to be 111° (calculated to fit with the observed H–N–H bond angle of 107° in NH_3). The Li-to-H distance so computed comes to about 2.5 Å. Using 1.2 Å for the van der Waals radius of hydrogen, the effective radius of the $Li(NH_3)_4$ molecule would be about 3.7 Å. Assuming hard-sphere contact in the solid, the closest Li-to-Li distance would be 7.4 Å, which in close packing of spheres would lead to an interplanar spacing of 6.0 Å. As seen from Table I, the most intense X-ray line observed is at 6.09 Å. It is probably more than just a coincidence that Schmidt,[9] in his study of the X-ray scattering of concentrated *liquid* solutions of lithium in ammonia, found an intense scattering peak at 0.117 radian, which, for the Mo Kα radiation used, corresponds to 6.04 Å. Schmidt's experiments were done at −75° with 20.6 atom % Li in NH_3. Thus, there is the strong possibility that some of the structure observed in the solid at 77°K persists in the liquid state at higher temperature. Morgan, Schroeder, and Thompson,[8] in fact, suggest that the positive temperature coefficient of conductivity observed in liquid metal–ammonia solutions is associated with structure in the solutions. It might be noted also that Schmidt's concentration data, when combined with a not unreasonable estimate 0.5 g/cc for the solution density, lead to a calculated layer spacing of 5.7 Å if it is assumed the liquid consists of close-packed lithium atoms coordinated by four NH_3 molecules.

An alternate view of the $Li(NH_3)_4$ molecule can be obtained from a hydrogenic 3s wave function with equivalent screening. Using Slater rules to calculate an effective nuclear charge of 1.30, we find that Ψ_{300} has maxima at 1.4 and 4.7 Å and nodes at 0.8 and 2.9 Å. Given that the bond distances suggest an Li-to-H distance of 2.5 Å, it appears that the great bulk of the 3s electron density is outside the shell of the 12 hydrogen atoms of $Li(NH_3)_4$. As a result, the molecule $Li(NH_3)_4$ can perhaps be better visualized as an $Li(NH_3)_4^+$ ion enclosed in a spherical sheath of one-electron charge density. The observed metallic properties would then be a consequence of the considerable overlap of expanded 3s orbitals.

Free-energy considerations indicate that the molecule $Li(NH_3)_4$ is probably unstable to dissociation in the gas phase, at least at 238°K. The reaction Li(g) + $4NH_3(g) \rightarrow Li(NH_3)_4(g)$ can be considered the sum of three steps: (a) ionization; (b) ammoniation; (c) electron affinity.

(a) The ionization energy $Li(g) \rightarrow Li^+(g) + e^-$ is 124 kcal/mole.

(b) The ammoniation energy $Li^+(g) + 4NH_3(g) \rightarrow Li(NH_3)_4^+(g)$ is estimated to be about −72 kcal/mole. (This value is obtained as follows: Marshall's value[10] of −9.2 kcal for $\Delta G°$ of $Li(s) \rightarrow Li^+(am) + e^-(am)$ is combined with other standard data and Jolly's value[11] of −39 kcal for $\Delta H°$ of $e^-(g) \rightarrow e^-(am)$ to get an experimental value of −123 kcal for the process $Li^+(g) \rightarrow Li^+(am)$. An estimated value of −51 kcal is obtained for $Li(NH_3)_4^+(g) \rightarrow Li(NH_3)_4^+(am)$ by applying the Born equation to a 2.5-Å ion in liquid ammonia, for which the Latimer cation-radius correction appears to be 0.6 Å. The difference between the solvation energies of $Li^+(g)$ and of $Li(NH_3)_4^+(g)$ is taken as a measure of the interaction energy between Li^+ and NH_3 in the gas phase.)

(c) The electron affinity $Li(NH_3)_4^+(g) + e^- \rightarrow Li(NH_3)_4(g)$ is estimated to be −44 kcal, *i.e.*, the difference between the 2s → 3s promotion energy[12] and the 2s → ∞ ionization energy for lithium.

(9) P. W. Schmidt, *J. Chem. Phys.*, 27, 23 (1957).
(10) P. R. Marshall in ref 2, p 107.
(11) W. L. Jolly, *Progr. Inorg. Chem.*, 1, 264 (1959).
(12) C. E. Moore, "Atomic Energy Levels as Derived from the Analyses of Optical Spectra," Circular 467, National Bureau of Standards, Washington, D. C., 1958.

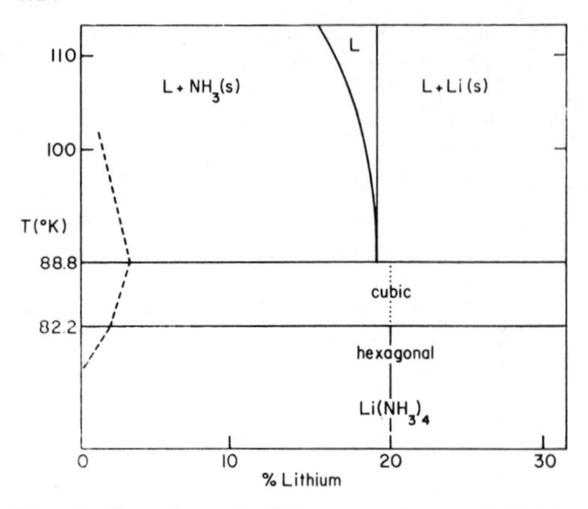

Figure 1. Phase diagram for lithium–ammonia system in the low-temperature region. The dashed lines on the left suggest possible solid–solution formation. The dotted line at 20 mole % in the cubic region may actually be a solid line if $Li(NH_3)_4$ compound formation occurs at 88.8°K. The eutectic has been placed at 19.4 mole % lithium to conform with the findings of Mammano and Coulter[4] that the heat effects peak at $Li(NH_3)_{4.15}$.

The sum of a, b, and c suggests +8 kcal for the free-energy change of the reaction $Li(g) + 4NH_3(g) \rightarrow Li(NH_3)_4(g)$ at 238°K.

In the solid state, possibly also in the liquid state, $Li(NH_3)_4{}^0$ would have an additional large stabilization energy arising from electron delocalization. This would be analogous to the metal lattice cohesive energy associated with, for example, the condensation of infinitely attenuated sodium gas to sodium metal crystal. It can be viewed as the quantum-mechanical lowering of kinetic energy that accompanies expansion of a particle-in-a-box, *viz.*, from an electron confined to a single $Li(NH_3)_4$ molecule to an electron spread over all $Li(NH_3)_4$ molecules of the entire sample. The magnitude of the delocalization energy can be estimated from a modified Hartree model[13] as equal to $[(-1.80/r_s) + (2.21/r_s{}^2)]$ rydbergs, where r_s is the radius of the Wigner–Seitz sphere in units of Bohr radii. Given that there are two molecules of $Li(NH_3)_4$ and hence two metallic electrons per unit cell, the Wigner–Seitz radius comes out to be $7.25a_0$, and the corresponding energy is −65 kcal. This energy does not include any correction for correlation energy, which, following the interpolation method of Nozières and Pine,[14] can be estimated to add an additional −28 kcal. The total stablization energy, −93 kcal, more than compensates for the vaporization energy of lithium (+37 kcal/mole) and that of ammonia (+7 kcal/mole), so, even assuming that the binding energy is unfavorable by 8 kcal in the gas phase, we find for $Li(s) + 4NH_3(s) \rightarrow Li(NH_3)_4(s)$ a favorable energy change that amounts to −20 kcal.

Because the above X-ray work was confined to 77°K, it is not possible at this time to make a definitive case for compound formation in the range 82–89°K. The results for this temperature region are, however, consistent with the high heat capacity, high electrical conductivity, and golden color of the solid, in that they uphold the view that the solid freezing at 89°K is not a simple binary eutectic of lithium and ammonia crystals. On the other hand, the formation of a solid solution in this temperature range is not precluded. Figure 1 shows the most probable phase relations, given that the dotted lines represent conjecture. Two principal possibilities present themselves.

(a) Case I. The vertical dotted line at 20 mole % lithium may in fact not exist in the cubic region. In such case the 88.8°K line would represent a eutectic where liquid solution separates into solid lithium and a solid solution of lithium and excess ammonia. (A solid solution of ammonia in excess lithium is possible but unlikely in view of structural considerations.) The 82.2°K line would then be where the solid solution and solid lithium react to form the compound $Li(NH_3)_4$ in its hexagonal form.

(b) Case II. The vertical dotted line at 20 mole % lithium may actually be a solid line in the cubic region. In such case the 88.8°K line represents an equilibrium between the liquid solution, solid lithium, solid ammonia, and the compound $Li(NH_3)_4$ in its cubic form. The 88.8°K line might actually be two closely spaced lines: the upper of which corresponds to a eutectic temperature for the equilibrium, liquid $\rightleftarrows$ $Li(s) + NH_3(s)$ and the lower, to a peritectoid temperature for the equilibrium $Li(s) + 4NH_3(s) \rightleftarrows Li(NH_3)_4(s)$. If the eutectic temperature and the peritectoid temperature coincide, there is a quadruple point with zero degrees of freedom. For case II, the 82.2°K line represents the temperature at which the cubic and hexagonal forms of $Li(NH_3)_4$ convert into each other.

The present evidence, summarized by Mammano and Coulter,[4] favors case II. X-Ray studies in the range 82–89°K would be informative, but these are difficult because of the greater problem encountered in keeping sample temperature constant at other than liquid nitrogen temperature. Further studies on this surprising material will be continued, but the case for compound formation seems to be substantially supported.

(13) See, for example, C. Kittel, "Quantum Theory of Solids," John Wiley and Sons, Inc., New York, N. Y., 1963, p 88.
(14) P. Nozières and D. Pines, *Phys. Rev.*, **111**, 442 (1958).

Reflection Spectra of Potassium–Ammonia Solutions

The enormous extinction coefficient of metal–ammonia solutions ($\epsilon_{max} \sim 4 \times 10^4$ M^{-1} cm^{-1}) prohibits the use of conventional transmission techniques for studying the absorption spectra of solutions more concentrated than 3×10^{-2} *M*. Beckman and Pitzer[1] studied relatively concentrated solutions by reflection from the surface of solutions. Koehler and Lagowski, whose work is reported in the following paper, studied the intermediate concentration region (0.04–0.4 *M*) by internal reflection using a hemicylindrical prism. The data showed that the band maximum frequency becomes essentially constant at concentrations above 10^{-2} *M* and that Beer's law is at least approximately obeyed; the logical conclusion is that the absorbing species in moderately concentrated solutions is the same as that in dilute solutions.

---

[1]T. A. Beckman and K. S. Pitzer, *J. Phys. Chem.*, **65**, 1527 (1961); also see R. B. Somoano and J. C. Thompson [*Phys. Rev. A*, **1**, 376 (1970)] for the results of an ellipsometric study of concentrated metal–ammonia solutions.

Metal-Ammonia Solutions. V. Optical Properties of Solutions in the Intermediate Concentration Region

56

by W. H. Koehler and J. J. Lagowski

Department of Chemistry, The University of Texas at Austin, Austin, Texas 78712 *(Received December 9, 1968)*

The reflection spectra of potassium-ammonia solutions in the intermediate concentration region were determined at −50° using a specially modified recording spectrophotometer. The reflection data were converted into the equivalent transmission spectra using an explicit solution of the Fresnel equations. These data indicated that the band maximum achieves an essentially constant position at metal concentrations greater than $\sim 10^{-2}$ *M*. The remaining band parameters (ϵ_{max} and $W_{h/2}$) are unchanged over a 4000-fold range of concentrations. The implications of these data on the nature of the solvated electron are discussed.

Introduction

The optical properties of dilute ($<10^{-3}$ *M*) alkali metal-liquid ammonia solutions have been extensively investigated.[1-5] The results of these investigations indicate that the spectra of these solutions are characterized by a single broad, intense ($\epsilon_{max} \sim 4 \times 10^4$ l. mol^{-1} cm^{-1}) asymmetric band with a maximum at $\sim$6800 cm^{-1}.

In dilute solutions, it is generally accepted that the optically important species is the solvated electron—an electron trapped in a cavity formed by solvent molecules.[6] Considerable controversy currently exists concerning the nature of the diamagnetic species; some investigators[7,8] prefer to attribute the decrease in paramagnetism of the metal-ammonia solutions with increasing metal concentration to the formation of dimers similar to those described by Becker, *et al.*,[9] whereas others have suggested the formation of solvated e_2^{2-} centers[10] or an association of solvated electrons resulting from an overlapping of the electronic wave functions.[11] There is general agreement that increasing the metal concentration leads to some type of association.

The optical properties of solutions exhibiting a high degree of association ($>10^{-2}$ *M*) are not well established. Jolly and Gold[2] were able to study the entire absorption band in solutions up to 0.03 *M;* the spectra reported for more concentrated solutions did not include the band maximum. The results indicate that the spectra are essentially unchanged at these concentrations, the band maximum having moved to lower energies with increased concentration. Beckman and Pitzer[12] investigated solutions of sodium in ammonia in the region of the metal-nonmetal transition using specular reflection techniques; however, no attempt was reported to calculate the optical constants of the solutions from these data.

The large extinction coefficient of these solutions ($\epsilon_{max} \simeq 4 \times 10^4$ l. mol^{-1} cm^{-1}) make conventional transmission techniques virtually useless for studying metal-ammonia solutions of concentrations greater than 3×10^{-2} *M*. The use of internal reflection techniques circumvents many of the problems associated with both transmission and specular reflection techniques. The applicability of internal reflection techniques to the study of metal-ammonia solutions has been demonstrated qualitatively in a study of the

(1) D. F. Burow and J. J. Lagowski, Advances in Chemical Series, No. 50, American Chemical Society, Washington, D. C., 1965, p 125.

(2) M. Gold and W. L. Jolly, *Inorg. Chem.*, **1**, 818 (1962).

(3) R. C. Douthit and J. L. Dye, *J. Amer. Chem. Soc.*, **82**, 4472 (1960).

(4) R. K. Quinn, Ph.D. Dissertation, The University of Texas at Austin, 1967.

(5) H. Blades and J. W. Hodgins, *Can. J. Chem.*, **33**, 411 (1955).

(6) J. Jortner, *J. Chem. Phys.*, **30**, 839 (1959).

(7) E. Arnold and A. Patterson, *ibid.*, **41**, 3089 (1964).

(8) S. Golden, C. Guttman, and T. Tuttle, *ibid.*, **44**, 3791 (1966).

(9) E. Becker, R. H. Lindquist, and B. J. Alder, *ibid.*, **25**, 971 (1956).

(10) R. Catterall and M. C. R. Symons, *J. Chem. Soc.*, 13 (1966).

(11) M. Gold, W. L. Jolly, and K. S. Pitzer, *J. Amer. Chem. Soc.*, **84**, 2264 (1962).

(12) T. A. Beckman and K. S. Pitzer, *J. Phys. Chem.*, **65**, 1527 (1961).

solvent spectrum in metal-ammonia solutions.[13] The general technique has been extensively reviewed by Harrick.[14]

Experimental Section

The spectrometer used in this investigation was a Beckman DK-1 instrument which was extensively modified to accommodate the reflection cell and dewar assembly (Figure 1).[15] The monochromator was rotated 90° about its optical axis and mounted in a housing which permitted the angle of incidence to be varied continuously in the range 45–90° (±0.1°). A Glan-Thompson polarizer (Karl Lambrecht Co., Chicago, Ill.) was placed in the light path at the exit slit so that rotation of the polarizer produced a beam with the electric vector oriented either perpendicular or parallel to the incident plane. The double-beam features of the original instrument were preserved by installing two specially designed chopper mirrors in the light path. The sample beam was focused at a point outside the surface of the prism to assure a well-defined incident angle at the reflecting face of the prism.[16]

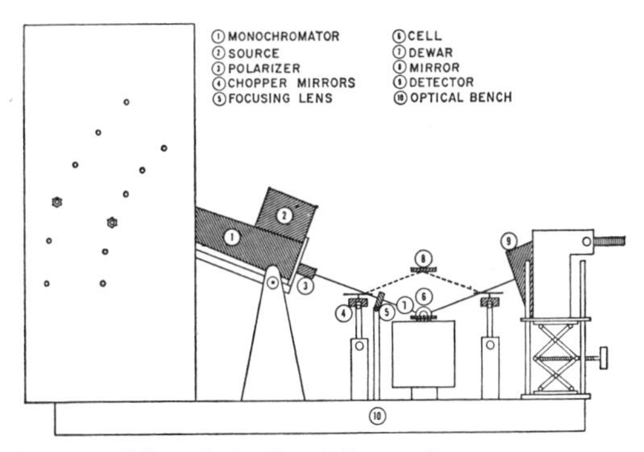

Figure 1. Schematic drawing of the recording reflection spectrophotometer.

Thus, the beam was alternately reflected from the prism and the plane mirror, which was used as a reference, into a standard DK-1 detector assembly which was also rotated 90° and mounted in another adjustable housing (Figure 1). Two auxiliary lenses (not shown in Figure 1) were used to produce identical slit images on the detector surface. The electronics and recorder supplied with the spectrophotometer as standard equipment were used without modification. After alignment and calibration, all performance features except the useable wavelength range and the baseline behavior were within the manufacture's specifications for this instrument. The hemispherical prism and auxiliary lenses restricted the wavelength range to 350–2200 mμ and created a nonequivalent optical path for the reference and sample beam which resulted in a nonlinear baseline.

The design of the reflection cell was primarily influenced by three factors: (1) the shape necessary for and physical properties of the prism, and the capability of interchanging prism materials, (2) the reactive nature of the metal–ammonia solutions, and (3) a system which would not introduce strain in the optical elements leading to a distortion of the polarization. The reflection element in the cell was a hemicylindrical prism which allowed a single reflection and was nonrefracting for angles of incidence in the range 0–90°. The prism was fabricated from fine annealed flint glass (Optical Instruments Laboratories, Houston, Texas), the optical properties of which have been discussed previously.[17] The hemicylindrical prism had a rectangular reflecting face (1.0 in. × 0.8970 in.) polished flat to a quarter of the wavelength of sodium D light and a radius of curvature of 0.4485 ± 0.0005 in.

The body of the cell incorporating two concentric cavities was constructed of 316 stainless steel (Figure 2). The inner cavity could be filled with a metal–ammonia solution introduced through the bottom of the cell. The prism was arranged over the central cavity on a slotted plate surrounded by a 1-mm retaining ring. The retaining ring assured that the solution could be

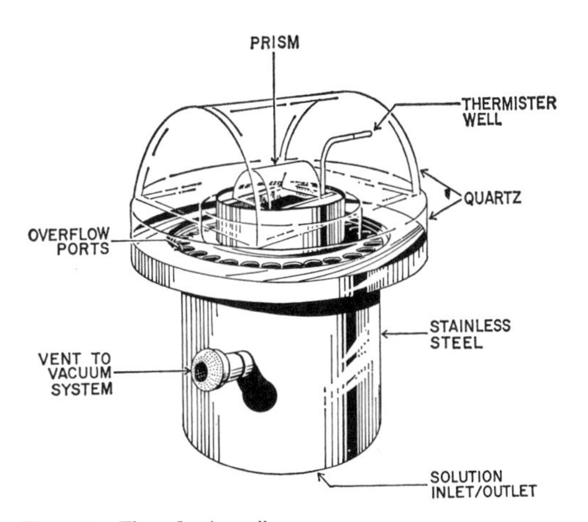

Figure 2. The reflection cell.

raised sufficiently high to make optical contact with the base of the prism before overflowing into the exterior cavity which also served as an overflow reservoir and provided a vent to the vacuum system.

The top of the cell, consisting of a plate with a hemi-

(13) D. F. Burow and J. J. Lagowski, *J. Phys. Chem.*, **72**, 169 (1968).

(14) N. J. Harrick, "Internal Reflection Spectroscopy," Interscience Publishers, New York, N. Y., 1967.

(15) For detailed drawings, specifications, and procedures, see W. H. Koehler, Ph.D. Dissertation, The University of Texas at Austin, 1969.

(16) J. Fahrenfort, *Spectrochim. Acta*, **17**, 698 (1961).

(17) R. E. Stephens and W. S. Rodney, *J. Res. Natl. Bur. Std.*, **52**, 303 (1954).

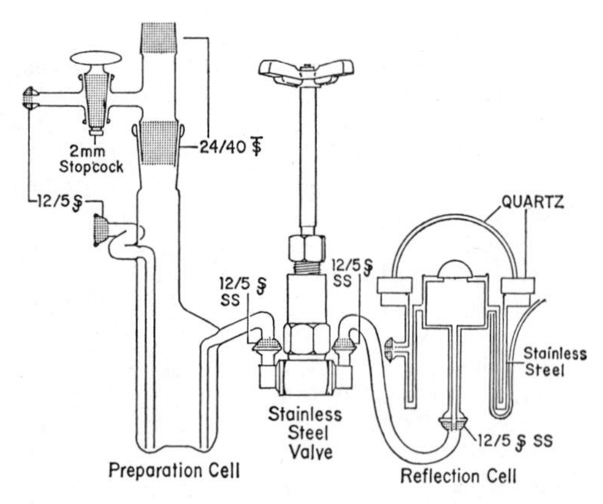

Figure 3. Reflection cell and solution preparation vessel.

cylindrical cover made from precision bore tubing, was fabricated from optical grade quartz (Ruska Instruments, Inc., Houston, Texas). A thermistor well passing through one end of the hemicylinder cover was oriented so that the temperature of the solution near the solution–prism interface could be monitored.

The solution preparation cell, valve, and reflection cell (Figure 3) were contained in a specially designed rectangular, stainless steel dewar. Incorporation of these three items into a common dewar system permitted the preparation, transfer, and spectra determination of the solutions under isothermal conditions. The temperature of the heat-transfer medium (isopropyl alcohol) in the dewar was controlled by circulating methanol, cooled in an acetone–Dry Ice bath, using the system described by Nasby.[18]

Prior to the preparation of the solutions to be studied, the solution preparation and the reflection cell were treated with a potassium–ammonia solution as described previously.[1] In a typical experiment, a quantity of anhydrous ammonia was distilled from sodium into a calibrated measuring vessel where its volume was determined at −70°. This known quantity of ammonia was then distilled into the previously dried preparation cell and a known quantity of potassium (99.95%, Alfa Inorganics) was added by means of an assembly which could accommodate large quantities of metal (Figure 4). The potassium metal was obtained in argon-filled ampoules and was used without further purification. A baseline for both the reflectivities of parallel and perpendicular radiation was determined on the prism in the empty reflection cell; the solution was then transferred into the reflection cell by applying a differential helium pressure. The parallel and perpendicular reflectivities at the solution–prism interface were recorded. No hydrogen gas was detected in the system during the time required to make the measurements (about 30 min). Typical reflection spectra for a 1.0 mol % (0.4 M) and

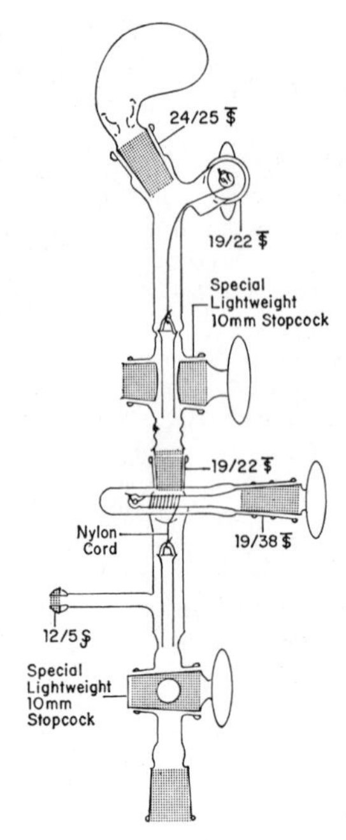

Figure 4. Winch assembly for introducing large quantities of solids into a closed system.

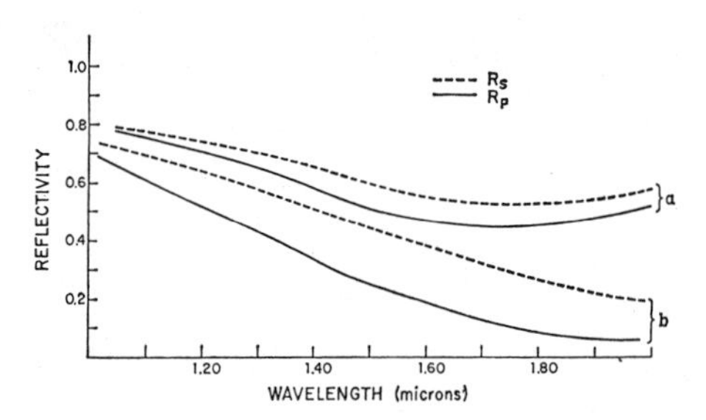

Figure 5. The reflectivities of parallel (R_P) and perpendicularly (R_S) polarized light from a 0.04 M (a) and a 0.4 M (b) potassium solution in liquid ammonia.

a 0.1 mol % (0.04 M) solution of potassium are presented in Figure 5.

Results and Discussion

The optical constants n and κ of a substance are related to the reflectivity at a known polarization at a given angle of incidence; the symbols R_s and R_p are used for the reflectivity of perpendicular and parallel polarized light, respectively. In the case of reflection from an interface formed by a dielectric and a conduc-

(18) R. Nasby, Ph.D. Dissertation, The University of Texas at Austin, 1968.

tor, the relations, expressed by the Fresnel equations, can be written as

$$R_s = \frac{(\cos\theta - a)^2 + b^2}{(\cos\theta + a)^2 + b^2} \quad (1)$$

$$R_p = \frac{[n^2(1-\kappa^2)\cos\theta - a]^2 + [2n^2\kappa\cos\theta - b]^2}{[n^2(1-\kappa^2)\cos\theta + a]^2 + [2n^2\kappa\cos\theta + b]^2} \quad (2)$$

where the quantities a and b are given by eq 3 and 4.

$$a = \left[\frac{n^2(1-\kappa^2) - \sin^2\theta + [(n^2(1-\kappa^2)\sin^2\theta)^2 + 4n^4\kappa^2]^{1/2}}{2}\right]^{1/2} \quad (3)$$

$$b = \left[\frac{-n^2(1-\kappa^2) + \sin^2\theta + [(n^2(1-\kappa^2) - \sin^2\theta)^2 + 4n^4\kappa^2]^{1/2}}{2}\right]^{1/2} \quad (4)$$

In eq 1–4, θ is the angle of incidence of the beam and the quantity n is defined as

$$n = n_2/n_1 \quad (5)$$

The quantity n_1 is the refractive index of the transmitting medium (the prism) and n_2, the real part of the complex refractive index of the absorbing medium, is given by eq 6

$$n_2 = n_2(1 - i\kappa) \quad (6)$$

where κ is the attenuation index of the absorbing medium. Thus, experimental determination of R_s and R_p at a given oblique angle of incidence should permit n and κ to be computed from eq 1–6 if an explicit solution could be found. The general form of an explicit solution was first suggested by Heilman;[19] unfortunately, certain formulas in the reported solution were incorrect.[20] Accordingly, the correctly derived equations for the explicit solution are presented here.

For convenience, the functions V, Y, W, Z, and A are defined in terms of experimental variables R_s, R_p, and θ by eq 7-11.

$$V = \frac{R_s - R_p}{R_s + R_p} \quad (7)$$

$$Y = V\frac{1 + R_s}{1 - R_s} \quad (8)$$

$$W = \tan\frac{Y - 1}{\tan^2\theta - Y} \quad (9)$$

$$Z = \left[\left(\frac{\tan^2\theta - 1}{(2\cos\theta)^{1/2}}\right)\left(\frac{V}{\tan^2\theta - Y}\right)\right]^2 \quad (10)$$

$$A = \frac{Z}{\cos\theta} \quad (11)$$

Equations 7-11 can be combined with eq 1 and 2 to yield two expressions in terms of n and κ (eq 12 and 13).

$$n = [(A - W + (2A\sin^2\theta + W^2)^{1/2}]^{1/2} \quad (12)$$

$$n\kappa = [(-A + W + (2A\sin^2\theta + W^2)^{1/2}]^{1/2} \quad (13)$$

Since V, Y, W, Z, and A are related to experimentally determinable quantities, both n and κ can be calculated.

The refractive index of the glass prism used in this investigation is given[17] by eq 14, where $q^2 = 2.5551726$, $j = 0.00825919$, $P = 0.000894995$, $h = 0.010023789$, $l = 0.0432000$, and λ is the wavelength in microns.

$$n_1^2 = q^2 - j\lambda^2 - P\lambda^4 + \frac{h}{\lambda^2} + \frac{m}{\lambda^2 - l^2} \quad (14)$$

Thus, the refractive index and attenuation index of the absorbing material can be calculated from eq 12, 13, and 14. The real and imaginary parts of the dielectric constant and the absorptivity (α) were then calculated from n and κ by using eq 15–17.

$$\epsilon_1 = n_2^2(l - \kappa^2) \quad (15)$$

$$\epsilon_2 = 2n_2^2\kappa \quad (16)$$

$$\alpha = \frac{4\pi n_2\kappa}{\lambda} \quad (17)$$

All computations were performed on a CDC 6600 digital computer.[15] The results of these calculations are presented in Figures 6–8.

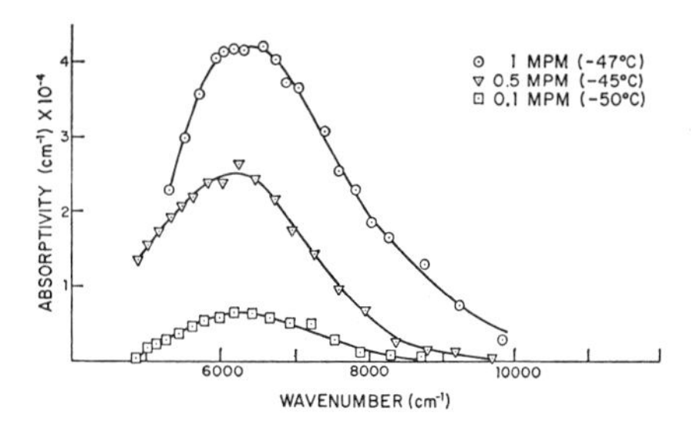

Figure 6. The absorptivity of potassium-ammonia solutions as a function of wavelength.

The absorptivity α of a sample is related to the absorbance obtained in transmission experiments through eq 18. Hence, the reflection data can be used to obtain the transmission spectra of highly absorbing substances (Figure 8).

$$\alpha(\text{cm}^{-1}) = 2.303\epsilon(\text{l. mol}^{-1}\text{ cm}^{-1})C(\text{mol l.}^{-1}) \quad (18)$$

The results of our reflection experiments on metal–ammonia solutions of intermediate concentration (Table I) indicate the presence of a single band with characteristics similar to that observed by numerous investigators[1-5] in the dilute region by using trans-

(19) G. Heilman, *Z. Naturforsch.*, **16a**, 714 (1961).

(20) J. Fahrenfort, private communication, 1965.

Table I. Band Parameters for Potassium–Ammonia Solutions

Concentration, mol % (*M*)	λ_{max}, μ (cm^{-1})	ϵ_{max}, l. mol^{-1} cm^{-1}	$W_{h/2}$, μ	Temp, °C	f^{a}
1.0 (0.4)	1.613 (6200)	$4.6 \pm 0.3 \times 10^4$	0.64	−47	0.56
0.5 (0.2)	1.623 (6160)	$5.4 \pm 1.0 \times 10^4$	0.73	−45	0.64
0.1 (0.04)	1.616 (6190)	$6.9 \pm 1.7 \times 10^4$	0.50	−50	0.62

[a] Calculated according to the method given by T. M. Dunn, "Modern Coordination Chemistry," J. Lewis and R. G. Wilkins, Ed. Interscience, New York, N. Y., 1960.

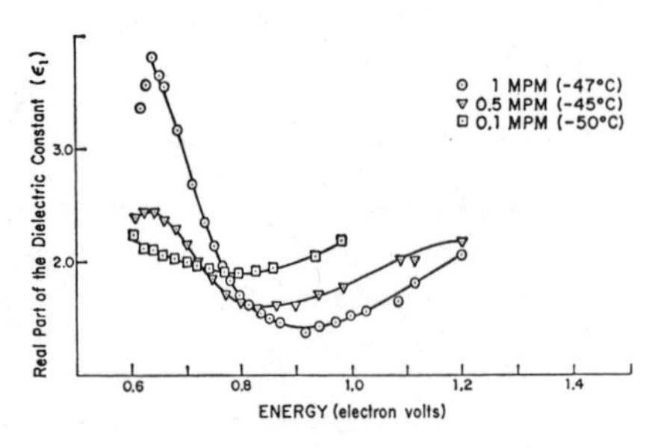

Figure 7. The real part of the dielectric constant for potassium–ammonia solutions as a function of energy.

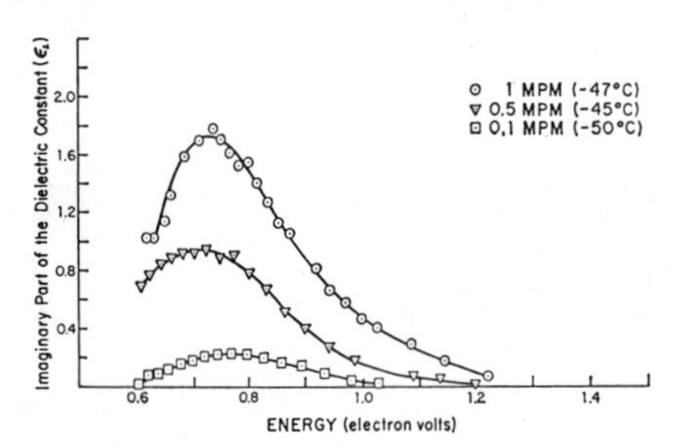

Figure 8. The imaginary part of the dielectric constant for potassium–ammonia solutions as a function of wavelength.

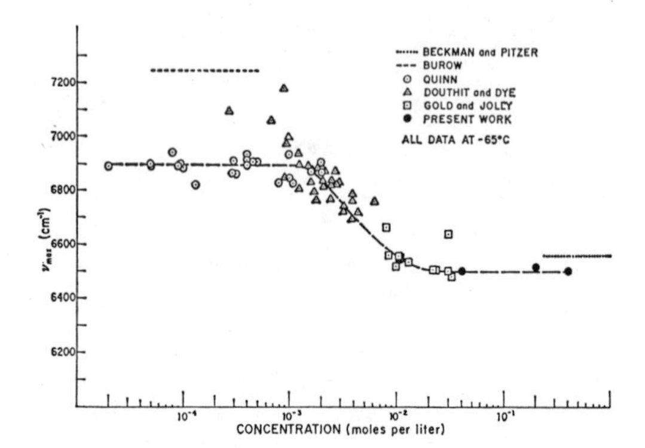

Figure 9. The position of the band maximum for potassium–ammonia solutions at different analytical concentrations of metal.

mission measurements. The band observed for the more concentrated solutions exhibits the same asymmetry on the high-energy side as that observed in the dilute region. The band maximum (6300 cm^{-1}) appears to be independent of concentration at these relatively high concentrations. Extinction coefficients at the band maximum are in good agreement with those reported previously for the dilute solutions[2,3,21] which have been reported in the range $4\text{–}5 \times 10^4$ l. mol^{-1} cm^{-1}. The band observed in the near-infrared region in this concentration range appears to be the same band observed in the more dilute region; however, the position of the band maximum has been shifted to lower energies. The dependency of the band maximum on concentration is shown in Figure 9. As is evident from Figure 9, a rather abrupt shift in the position of the maximum occurs between $2\text{–}20 \times 10^{-3}$ *M*. At concentrations on either side of this region, the position of the band appears to be invariant with concentration. This trend was observed at lower concentrations;[1,3] however, the concentration range available in the previous experiments reported was not sufficiently high to observe the fact that the position of the band maximum becomes invariant at higher concentrations. At 3×10^{-4} *M* it has been reported that only 5% of the electrons have paired spins compared to 90% spin pairing at 0.1 *M*.[22] It would appear that the shift in position of the band maximum may be related to the spin-pairing process as suggested by Catterall and Symons[10] and by Thompson and Cohen.[23]

Various attempts have been made to account for the band observed for metal–ammonia solutions at about 1.5 μ in terms of competing equilibria involving singularly and doubly occupied cavities[10] or monomeric and dimeric species of the type described by Becker, *et al.*,[9] or Golden, *et al.*[8] In an attempt to resolve the arguments concerning the origin of the band, the spectra reported by various investigators at a variety of concentrations were subjected to a mathematical analysis which involved an attempted resolution of the experi-

(21) R. K. Quinn and J. J. Lagowski, *J. Phys. Chem.*, **72**, 1374 (1968).

(22) S. Freed and N. Sugarman, *J. Chem. Phys.*, **11**, 354 (1943).

(23) J. C. Thompson and M. H. Cohen, *Advan. Phys.*, in press.

Table II: Band Resolution Results

Number of bands assumed	$C = 10^{-4}$ M[a] λ_{max}, μ	A_{max}	W, μ	$C = 2 \times 10^{-3}$ M[b] λ_{max}, μ	A_{max}	W, μ	$C = 6 \times 10^{-3}$ M[b] λ_{max}, μ	A_{max}	W, μ	$C = 4 \times 10^{-1}$ M[c] λ_{max}, μ	A_{max}	W, μ
1	1.466	0.497	0.678	1.465	0.790	0.667	1.473	1.627	0.640	1.58	1.890	0.606
2	1.282	0.150	0.462	1.404	0.384	0.605	1.413	1.125	0.639	1.44	1.366	0.476
	1.534	0.418	0.625	1.526	0.427	0.692	1.581	0.589	0.545	1.73	1.161	0.388
3	1.189	0.188	0.428	1.232	0.326	0.520	1.145	0.448	0.433	1.27	0.742	0.318
	1.461	0.224	0.364	0.492	0.558	0.458	0.396	0.779	0.359	1.50	1.251	0.299
	1.641	0.277	0.657	1.791	0.211	0.477	1.622	1.130	0.486	1.75	1.373	0.344
4	1.087	1.141	0.351	1.073	0.194	0.390	1.107	0.424	0.178	1.19	0.420	0.254
	1.290	0.213	0.305	1.303	0.392	0.323	1.296	0.465	0.140	1.39	0.977	0.290
	1.493	0.323	0.334	1.527	0.557	0.365	1.472	0.981	0.348	1.60	1.463	0.306
	1.743	0.232	0.504	1.788	0.276	0.442	1.686	0.890	0.520	1.82	0.941	0.268
Experimental data	1.470	0.504	0.690	1.450	0.787	0.700	1.470	1.615	0.650	1.60	0.835	0.640

[a] Burow, Ph.D. Dissertation, The University of Texas at Austin, 1966. [b] R. C. Douthit and J. L. Dye, *J. Amer. Chem. Soc.*, **82**, 4472 (1960). [c] This investigation.

mentally determined band envelope into two or more bands. The spectra of potassium–liquid ammonia solutions (1×10^{-4}, 2×10^{-3}, 6×10^{-3}, and 4×10^{-1} M) reported by Burow,[24] Douthit and Dye,[3] as well as those obtained in this investigation were employed as typical examples of data obtained across a wide range of concentrations. The analysis was performed by mathematically obtaining a set of bands, the sum of which gave the observed envelope to within experimental error, using the Levenberg method of damped least squares, a modification of the least-squares method. This technique has been incorporated in a computer program (RESØL) originally developed by Dr. D. D. Tunnicliff of Shell Development Co., Emeryville, Calif., and modified by P. F. Rusch; details of the program are discussed in detail by Takemoto.[25] The program resolves a band envelope into symmetric bands which can be described by Gaussian, Lorentzian, or a linear combination of these two types of distributions; there are no provisions for generating asymmetric bands. Since the 1.5-μ band observed in metal–ammonia solutions is asymmetric when plotted in terms of energy but symmetric on a wavelength plot, the symmetric envelopes obtained from the wavelength plot were subjected to analysis. The resolved symmetric bands are asymmetric when plotted against energy. Each spectrum was resolved by assuming the presence of one, two, three, and four bands. The results are presented in Table II.

Inspection of Table II reveals that the experimental band envelopes cannot be decomposed into component bands exhibiting behavior characteristic of a system at equilibrium. Although all the spectra subjected to analysis were not at the same temperature, a consideration of the temperature effects reported[13] suggest that the shift in the position of the band maxima of the resolved bands cannot be attributed solely to a temperature difference. If an equilibrium existed between two or more species that are spectroscopically distinguishable, the band parameters for the resolved bands (which presumably are characteristic of individual species) should show a predictable relationship with increasing concentration. For example, the band positions should be constant with concentration to within experimental error; none of the assumed models, *i.e.*, two, three, or four bands (species), follow this behavior. Our calculations show that the "best" fit of the experimental envelope is either one band or a very large number of bands; however, the positions of the latter are not constant with concentration.

The observation that the band width at half-height is remarkably constant over so wide a concentration range and the fact that the band is asymmetrical at the lowest concentration presently attained (in which the species present is presumably the simplest in the system) suggest the absence of a spectroscopically detectable equilibrium.

Based on the results of these analyses, there seems to be no apparent spectroscopic justification for the existence of the equilibria previously described. The consistency of the band shape and the adherence to Beer's law at the band maxima over a 4000-fold concentration range lead to the conclusion that the absorbing species present in the intermediate concentration region is basically the same species as that present in the very dilute region. There is, however, a perturbation of the species in either its ground state or excited state or both. Since the evidence for the existence of a

(24) D. F. Burow, Ph.D. Dissertation, The University of Texas at Austin, 1966.

(25) J. H. Takemoto, Ph.D. Dissertation, The University of Texas at Austin, 1968.

"solvated electron" in these systems is persuasive, these spectroscopic results suggest that such a species remains basically intact over a wide concentration range. One possible explanation for a shift of the band maximum to lower energies with increasing concentration of metal may be the association of the charged species present in solution because the solvent has an intermediate dielectric constant which promotes such processes; the concept of ion clusters has been suggested by other investigators.[2,3,11] As the concentration of the metal is increased, association occurs leading to the formation of dipoles, quadrupoles, and larger ionic clusters, the number of free solvent molecules available for solvation of additional species decreases, and the system becomes more ordered. The magnetic data which indicate that diamagnetic species are favored as the concentration of the metal increases can be incorporated into these ideas if the electron wave functions of adjacent cavities can overlap sufficiently to permit pairing. Thus, the interaction of a solvated electron with its nearest neighbors results in a perturbation of the electron wave function but does not drastically alter the integrity of this species. Whether this perturbation arises from coulombic interactions between solvated cations and solvated anions or from spin–spin interactions between adjacent cavities has not been determined. More probably, both effects are important. In any event, the perturbation apparently reaches a limiting value when the maximum number of neighbors around any one cavity has been established.

Acknowledgments. We gratefully acknowledge the financial assistance of the Robert A. Welch Foundation and the National Science Foundation. One of us (W. H. K.) acknowledges fellowships from Union Carbide Company and Socony-Mobil Oil Company.

A Reconsideration of the Becker–Lindquist–Alder Equilibria

In the following paper, Demortier and Lepoutre give the results of a least-squares fitting of Hutchison and Pastor's magnetic susceptibility data for sodium (see p. 204) to the equilibrium constants of the Becker–Lindquist–Alder reactions (see p. 227). They obtain constants which are quite different from those calculated from the same data by Becker, Lindquist and Alder and which are close to the values calculated for sodium from conductivity data. Therefore they suggest that it was unnecessary for Arnold and Patterson (p. 291) to propose another species, M^-, to account for the magnetic and conductivity data of dilute metal–ammonia solutions. In this paper, Demortier and Lepoutre also report new susceptibility data for sodium solutions that are consistent with the conductivity data. A full account of this work is given in a more recent publication.[1] In yet another recent paper,[2] Demortier and Lepoutre have shown that the Knight shift data are consistent with the Becker–Lindquist–Alder equilibria and that it is not necessary to include an equilibrium involving an M^- species, as did O'Reilly (pp. 301, 308).

---

[1] A. Demortier, M. De Backer, and G. Lepoutre, *J. Chim. Phys.*, 380 (1972).
[2] A. Demortier and G. Lepoutre, *J. Chim. Phys.*, 179 (1972).

C. R. Acad. Sc. Paris, t. 268, p. 453-456 (10 février 1969). Série C

57

CHIMIE PHYSIQUE. — *Constantes d'association dans les solutions diluées métal-ammoniac.* Note (*) de MM. **Antoine Demortier** [1] et **Gérard Lepoutre**, présentée par M. Georges Champetier.

Les équilibres d'association présentés par Becker, Lindquist et Alder pour interpréter les propriétés des solutions diluées métal-ammoniac ont été critiqués à plusieurs reprises. De nouveaux calculs sur les données expérimentales antérieures et de nouvelles mesures expérimentales nous amènent à confirmer la validité des équilibres proposés par B. L. A.

Introduction. — On cherche habituellement à rendre compte des propriétés des solutions diluées métal-ammoniac par deux équilibres d'association proposés par Becker, Lindquist et Alder (B. L. A.) [2] :

$$M \rightleftharpoons M^{+} + e^{-} \quad (K_1),$$

$$M \rightleftharpoons \frac{1}{2} M_2 \quad (K_2).$$

Les couples de constantes K_1, K_2 ont été calculés vers — 35°C pour le sodium à partir de mesures assez précises de conductance [3] et pour le potassium à partir de mesures moins précises de susceptibilité magnétique [4]. Ces deux couples de constantes ont été trouvés fort différents, alors que les susceptibilités magnétiques [4] d'une part, les conductances [5] d'autre part, étaient à peu près identiques pour le sodium et pour le potassium vers — 35°C.

Nous avons recalculé le couple de constantes pour le potassium à partir des données antérieures; nous n'avons pas retrouvé les valeurs affirmées par les auteurs précédents. Les valeurs que nous trouvons sont beaucoup plus proches de celles du sodium.

Nous avons effectué des premières mesures magnétiques sur le sodium; les susceptibilités trouvées laissent prévoir une différence significative entre les comportements magnétiques du sodium et du potassium. Les équilibres B. L. A. ne seraient donc pas sujets à critique.

1. Les équilibres B. L. A. — En solution infiniment diluée dans l'ammoniac liquide, les métaux alcalins se dissocient entièrement en ions M^+ solvatés et en électrons solvatés. Quand la concentration augmente, la conductivité équivalente diminue fortement; ceci indique une interaction entre les espèces chargées, avec formation d'une espèce neutre non conductrice. La solution, qui contient des électrons et une espèce associée

possédant un électron non apparié, est donc paramagnétique. La susceptibilité magnétique molaire atteint à dilution infinie la valeur correspondant à une mole d'électrons libres. Lorsque la concentration augmente, la susceptibilité molaire diminue; il faut donc admettre une interaction entre des espèces paramagnétiques pour produire une nouvelle espèce diamagnétique. Ces deux propriétés peuvent être interprétées qualitativement par deux équilibres présentés par B. L. A.

$$\text{(I)} \qquad M \rightleftharpoons M^+ + e^-, \qquad K_1 = \frac{(M^+)\,(e^-)\,f_\pm^2}{(M)};$$

$$\text{(II)} \qquad M \rightleftharpoons \frac{1}{2} M_2, \qquad K_2 = \frac{(M_2)^{\frac{1}{2}}}{(M)}.$$

De ce modèle, nous ne retiendrons que la partie phénoménologique, sans faire aucune hypothèse sur la nature des espèces considérées.

2. Calculs des couples de constantes. — Sur la base de ce modèle, B. L. A. ont calculé les constantes d'équilibre pour les solutions de potassium à partir des mesures de résonance paramagnétique de Hutchison et Pastor ([4]). En supposant le coefficient d'activité égal à l'unité, on déduit des formules (I) et (II) l'équation suivante :

$$\text{(III)} \qquad \frac{N_p}{(M_2)^{\frac{1}{4}}} = \frac{1}{K_2}(M_2)^{\frac{1}{4}} + \left(\frac{K_1}{K_2}\right)^{\frac{1}{2}},$$

où

$$M_2 = \frac{C - N_p}{2} \qquad \text{et} \qquad N_p = (M) + (e^-) = \chi_p \frac{kT}{N_0 \beta^2},$$

(C) étant la concentration globale, (N_p) la concentration en espèces paramagnétiques, χ_p la susceptibilité magnétique, β le magnéton de Bohr, k la constante de Boltzman, N_0 le nombre d'Avogadro). La relation (III) est linéaire, et permet d'obtenir graphiquement les constantes K_1 et K_2. Pour — 33°C, B. L. A. ont donné les valeurs suivantes : $K_1 = 0{,}03$, $K_2 = 99$.

Evers et Frank ([2]) ont supposé que les solutions diluées métal-ammoniac se comportent comme de véritables électrolytes. En utilisant les données expérimentales de Kraus ([5]) pour les solutions de sodium, les auteurs ont calculé les constantes des équilibres B. L. A. pour — 34°C :

$$K_1 = (7{,}23 \pm 0{,}50) \,.\, 10^{-3}, \qquad K_2 = 27{,}03 \pm 2{,}40.$$

A — 34°C, les solutions diluées de sodium et de potassium ont sensiblement la même conductivité. Les mesures de susceptibilité magnétique de Hutchison et Pastor pour les solutions de potassium couvrent une zone de concentration allant de 0,01 à 0,8 M. Dans le même article, les auteurs donnent la susceptibilité de quelques solutions de sodium comprises entre 0,06 et 0,3 M. A — 33°C, les solutions de sodium et de potassium auraient même susceptibilité. Les couples de constantes devraient donc être les mêmes pour les deux métaux; or, on les trouve nettement différentes. Ceci a amené plusieurs auteurs [6] à proposer d'autres équilibres que ceux de B. L. A. pour interpréter les propriétés des solutions diluées de métal dans l'ammoniac.

Les résultats expérimentaux de Hutchison sont très dispersés. Nous avons refait le calcul de B. L. A. en vue d'estimer la zone d'erreur et de voir si les constantes d'Evers tombaient dans cette zone. Les paramètres de l'équation (III) ont été calculés par la méthode des moindres carrés en minimisant le carré des erreurs suivant chacun des axes de coordonnées,

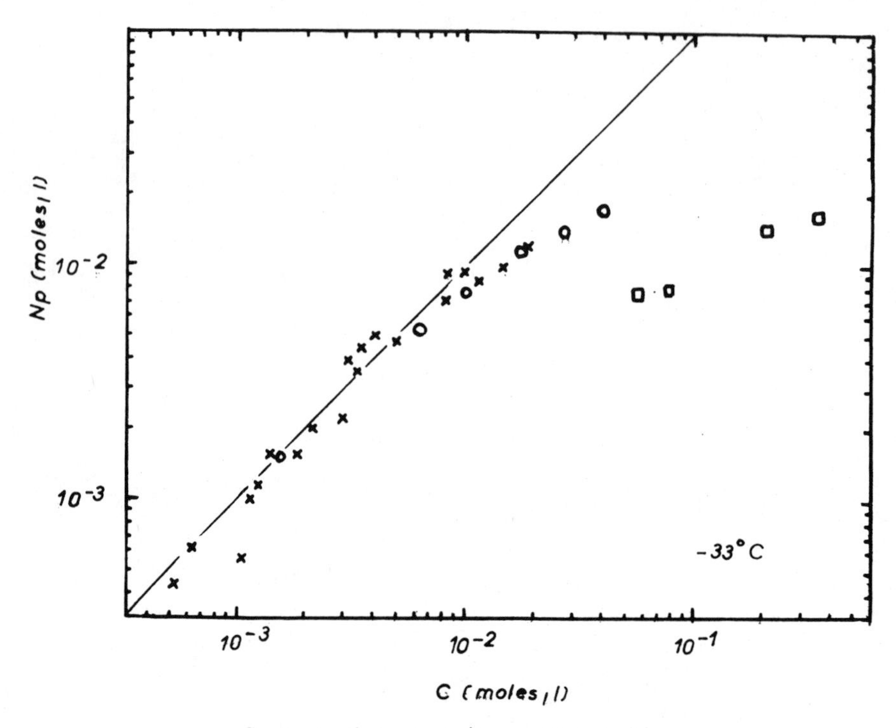

Concentration en espèces paramagnétiques en fonction de la concentration totale en sodium (échelles logarithmiques).

□ données de Hutchison; × nos mesures;
○ valeurs calculées d'après les constantes d'Evers.

car les deux variables sont entachées d'erreurs. On obtient ainsi deux couples de constantes dont on prend la valeur moyenne. La zone d'erreur ainsi obtenue est vraisemblablement majorée. Le calcul nous a donné

$$K_1 = (5,7 \pm 1,2) \,.\, 10^{-3} \qquad \text{et} \qquad K_2 = 39 \pm 3.$$

Ces chiffres sont très différents de ceux de B. L. A., et nous ne voyons pas d'où peut provenir un tel écart. Par contre, ces constantes calculées pour le potassium sont du même ordre de grandeur que les constantes du sodium, mais leur différence reste sans doute significative.

3. Susceptibilité magnétique des solutions de sodium [7]. — Nous avons effectué des mesures de susceptibilité magnétique pour des solutions

diluées de sodium. Les premiers résultats à — 33°C sont en contradiction avec les données de Hutchison. Les deux gammes de concentration ne se recouvrent pas, mais l'on voit nettement sur la figure que les résultats des deux séries sont incompatibles. Nous avons également porté sur la figure la concentration en espèces paramagnétiques calculée à partir des constantes K_1 et K_2 données par Evers pour le sodium à — 34°C. Ces valeurs recouvrent bien les résultats que nous avons obtenus. Ces premières mesures indiquent que les solutions sodium-ammoniac sont moins associées que les solutions de potassium. Dans ce cas, la constante K_2 du sodium doit être inférieure à celle du potassium. A — 34°C, les valeurs de K_2, 27 pour le sodium et 39 pour le potassium divergent dans le bon sens.

Conclusion. — Sous réserve de résultats expérimentaux plus complets, la conductivité et la susceptibilité magnétique des solutions diluées de sodium et de potassium dans l'ammoniac liquide peuvent s'interpréter par les équilibres B. L. A. en adoptant pour constantes d'équilibre, à — 34°C, les valeurs suivantes :

pour Na :

$$K_1 = (7{,}23 \pm 0{,}50) \cdot 10^{-3}, \qquad K_2 = 27{,}03 \pm 2{,}40;$$

pour K :

$$K_1 = (5{,}7 \pm 1{,}2) \cdot 10^{-3}, \qquad K_2 = 39 \pm 3.$$

(*) Séance du 23 décembre 1968.

(1) Boursier du C. E. A.

(2) E. C. Becker, R. H. Lindquist et B. J. Alder, *J. Chem. Phys.*, 25, 1956, p. 971-975.

(3) E. C. Evers et P. W. Frank, *J. Chem. Phys.*, 30, 1959, p. 61-64.

(4) C. A. Hutchison et R. C. Pastor, *J. Chem. Phys.*, 21, 1953, p. 1959-1971.

(5) C. A. Kraus, *J. Amer. Chem. Soc.*, 43, 1921, p. 749-770.

(6) E. Arnold et A. Patterson, dans *Solutions Métal-Ammoniac*, édité par G. Lepoutre et M. J. Sienko, distribué par Benjamin Inc., New York, 1964.

(7) Ce travail a été réalisé dans les laboratoires de l'Université de Leicester (professeur M. C. R. Symons).

(*Laboratoire des Métaux alcalins dans* NH_3 *liquide,*
Équipe de Recherche associée au C. N. R. S.,
Faculté libre des Sciences et Hautes Études Industrielles,
13, *rue de Toul*, 59-*Lille, Nord.*)

179131. — Imp. Gauthier-Villars. — 55, Quai des Grands-Augustins, Paris (6e).
Imprimé en France.

The Effect of Pressure on Metal–Ammonia Solutions

An extensive investigation of the effects of pressure on various properties of metal–ammonia solutions (volumes, equilibrium constants, spectra, conductivity, and miscibility gap parameters) is reviewed in the following paper by Schindewolf.

PRESSURE EFFECTS IN METAL–AMMONIA SOLUTIONS*

58

U. SCHINDEWOLF
Institut für Kernverfahrenstechnik der Universität und des Kernforschungszentrums Karlsruhe (Germany)

A study of the pressure and temperature variation of the equilibrium constant or the molar volume of a system can be used to obtain an insight into the nature of the species present. A detailed description of two experiments (compressibility and optical measurements) is presented to illustrate the techniques and operating principles employed in the high-pressure study of metal–ammonia solutions. The results of (1) measurements of the isothermal compressibility of K– and KI–NH_3 solutions, (2) an optical study of the reaction of solvated electrons with ammonia, (3) the absorption spectrum of dilute K– and KI–NH_3 solutions, (4) electrical conductivity experiments, (5) reflectivity measurements, (6) experiments on the miscibility gap, and (7) studies of the reaction of the solvated electron in water–ammonia mixtures are used to discuss the nature of metal–ammonia solutions.

THE temperature dependence of the properties of metal–ammonia solutions[1] in a relatively restricted temperature interval is well known. In contrast, little has been reported on the pressure dependence of the properties of the solutions. In principle, however, pressure dependences should give as valuable information as do temperature dependences; compare, for example, the temperature- and pressure-dependence of the equilibrium constant, K, and the rate constant, k, of chemical reactions (equations 1 to 4),

$$d(\ln K)/dT = \Delta H/RT^2 \tag{1}$$

$$d(\ln K)/dp = Q - \Delta V/RT \tag{2}$$

$$d(\ln k)/dT = \Delta H^{\neq}/RT^2 \tag{3}$$

$$d(\ln k)/dp = -\Delta V^{\neq}/RT \tag{4}$$

which yield the reaction and activation enthalpy (ΔH and $\Delta H^{\neq}$) and the reaction and activation volume (ΔV and $\Delta V^{\neq}$). The reaction and the activation volume are defined as the differences in volume between products or activated complex and the reactants. These quantities in general have the same significance as the reaction enthalpy and the activation enthalpy in chemical reactions.

The pressure dependence of the properties of the metal–ammonia solutions should be particularly pronounced in view of the large apparent volume of the solvated electron. With versatile high pressure equipment one should also be able to obtain data over a wider temperature range extending beyond the

---

* A review based on the experimental work performed in collaboration with K. W. Böddeker, H. Kohrmann, G. Lang, K. Maurer, R. Olinger and R. Vogelsgesang.

critical point of ammonia, so that the solvated electrons can be studied even in a dense gaseous phase.

During the last four years we have investigated the more important properties of some metal–ammonia solutions under pressure. The techniques of modern high pressure research start at 5000 to 10000 atm, and extend to hundreds of thousands or even millions of atmospheres. Metal–ammonia solutions possess the highest pressure coefficients observed in condensed phases; accordingly pressures of at most 2000 atm are sufficient for most experiments. The experimental apparatus to achieve these pressures costs only a few hundred dollars, but the techniques still require some skill.

The high pressure experiments reported here involve the following areas of metal–ammonia chemistry: (1) isothermal compressibility; (2) chemical equilibria; (3) absorption spectrum; (4) electrical conductivity; (5) reflection spectrum; (6) miscibility gap; (7) magnetic properties; and (8) ammonia–water mixtures. In addition we have conducted some experiments on electrolyte solutions for comparison purposes.

In the interpretations of the experimental results we were guided by the following simple model. Unpaired solvated electrons (e^-_{sol}) exist in dilute solutions ($c < 0{\cdot}001$ mole/l.). At higher concentrations the electrons form pairs e^{2-}_{2sol}.

$$2e^-_{sol} \rightleftharpoons e^{2-}_{2sol} \tag{5}$$

The diamagnetic pair e^{2-}_{2sol} might include cations to compensate for the electrostatic repulsion. With a further increase in concentration ($c > 0{\cdot}5$ mole/l.) the transition to the metallic state occurs.

$$\tfrac{1}{2}e^{2-}_{2sol} \rightleftharpoons e^-_{met} \tag{6}$$

Magnetic data give evidence for equilibrium 5. Electric conductivity and reflectivity data give evidence for the transition 6, which, according to the theory, is interrelated to the miscibility gap.

In dilute solutions the electrons are thought to be trapped in a cavity, the size of which determines the position of the light absorption maximum[2]. The molar volume of the spin-paired and metallic electrons obtained from density measurements of medium to concentrated solutions is 60 to 90 ml[1-3]. Unfortunately, the molar volume of the unpaired solvated electron is not known because density measurements on the dilute solutions have not been carried out with the required precision. According to Ogg[4] the molar volume of electrons in the highly dilute metal solutions might be as high as 800 ml. Data obtained by Evers and his group[5] indicate that the molar volume is concentration dependent, going through a pronounced minimum at about 0·01 mole/l.; Gunn's results[6] on the other hand indicate that the volume in concentrated solutions is little changed from that in dilute solution.

The results of our experiments, in terms of this model, can be summarized as follows. The metal–ammonia solutions are more compressible than pure ammonia[7], and the absorption spectrum of the dilute solutions is shifted to the blue with increasing pressure[7,8]. Both phenomena arise from a compression of the solvent structure associated with dissolved electrons. In concentrated solutions the increased compressibility may partly be caused by a shift of equilibrium 6 from the metallic to the non-metallic state since the latter requires less volume. Evidence for this shift is derived from the decrease in conductivity[9] and reflectivity[7]; in addition the displacement of the miscibility gap[10] to higher concentrations with increasing pressure also supports this suggestion. Equilibrium 5 is not affected by pressure[11], indicating equal volumes for the paramagnetic and diamagnetic species. However, other possible equilibria involving solvated electrons[12,13] are strongly affected by pressure because of the high molar volume of the solvated electrons. Using the position of the near infra-red absorption band as a criterion, the volume of the solvated electron in water–ammonia mixtures increases with increasing ammonia content[14]. The mean lifetime of the electron increases exponentially with ammonia content in these mixtures. The results suggest a correlation between the molar volume of the solvated electron and its reactivity which has been confirmed by kinetic measurements under pressure.

EXPERIMENTAL EQUIPMENT

Two typical examples will suffice to demonstrate the operating principles of high pressure equipment applied to liquid ammonia work: the high pressure apparatus for (1) determining the isothermal compressibility of metal–ammonia solutions and (2) for the radiolytic formation and optical detection of solvated electrons[14] under pressure.

(1) Compressibility measurements[7]

Figure 1 shows the apparatus employed for the compressibility measurements. All essential parts are fabricated of stainless steel and, as far as possible, are electroplated with gold to decrease the rate of decomposition of the metal solutions. The autoclave is 320 mm long and has an outer diameter of 70 mm. The autoclave contains a piston (1) which is ground and fitted with two O-rings to provide for complete sealing, in the upper portion (10 mm diameter by 130 mm long). The upper end of the piston rod (2) extends through a 3 mm hole in the autoclave lid (3). A ferritic magnet (5) is attached to the end of this rod. The upper part of the autoclave is connected to the pressure generator and to the manometer which are filled with pentane as the pressure transmitting fluid. The lower part of the autoclave (20 mm in diameter by 80 mm long) is connected to the storage autoclaves containing the solution to be investigated and pure ammonia respectively in glass cylinders.

Connections are made with steel capillary tubing with conical ends screwed into corresponding conical bores. The storage autoclaves are maintained under a low overpressure to ensure fast and complete filling of the evacuated compression autoclave. By increasing the pressure in the upper chamber of the compression autoclave the piston moves downward and compresses the solution.

The displacement of the piston is followed with a differential transformer which surrounds the capillary (4) in which the ferrite magnet moves. The differential transformer consists of three coils; an alternating current (5 kHz) flows through the central primary coil. The voltage induced in the two outer secondary coils is rectified and is fed via a differential amplifier to a null

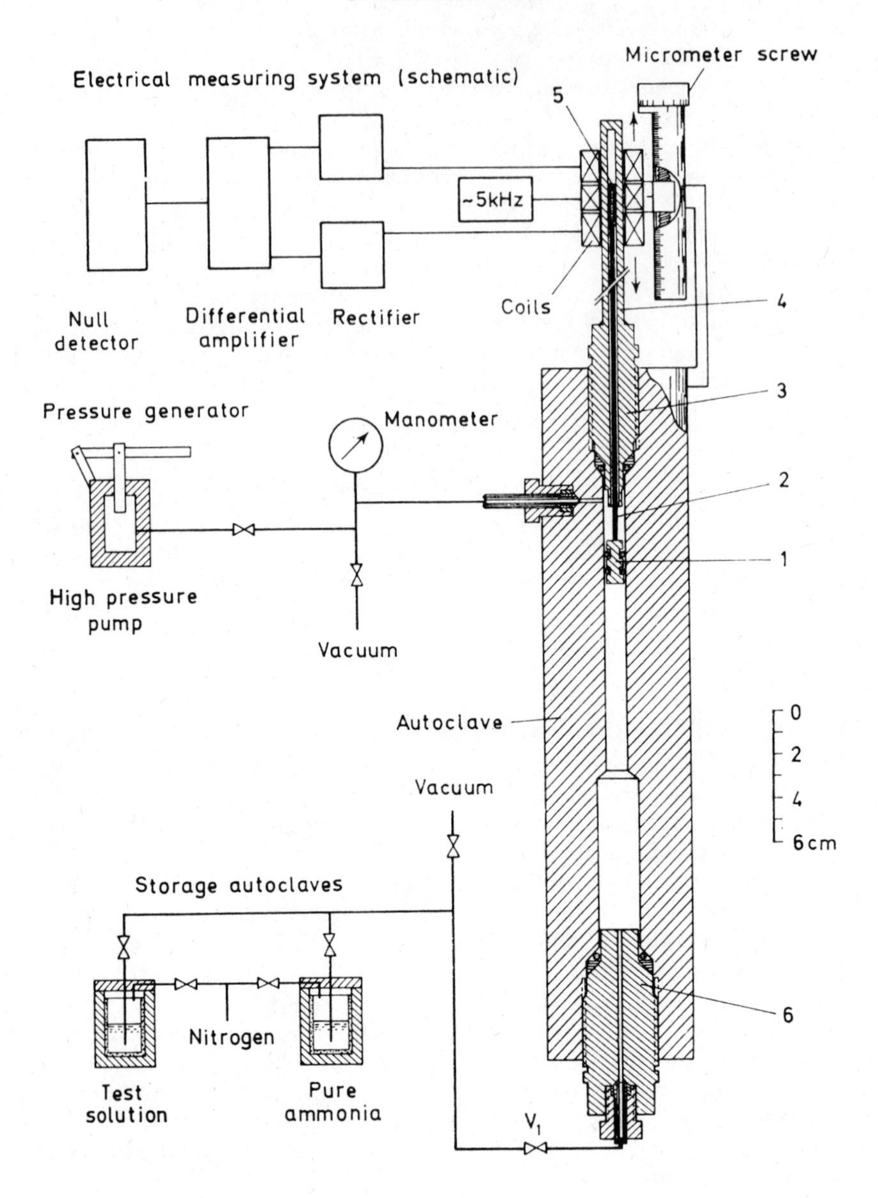

FIGURE 1. Experimental apparatus for compressibility measurements[7]

detector. As long as the ferrite magnet is exactly positioned in the centre of the differential transformer the null detector shows no deflection. A displacement of the magnet brought about by a compression of the solution results in a deflection of the null indicator which can be compensated for by a downward displacement of the transformer along a micrometer screw until a new null position is attained. The volume decrease of the solution, ΔV, obtained from the application of a pressure difference, Δp, can be obtained from the displacement of the piston. The isothermal compressibility, β, of the solution is given by equation 7, where V_0 is the inner volume of the autoclave (32·8 ml). The volume change of the autoclave due to expansion (about 0·1 per cent per 1 000 atm)

$$\beta = \Delta V / V_0 \Delta p \tag{7}$$

is negligible compared to the volume change of the solution (3 to 10 per cent per 1 000 atm). The apparatus can be used for pressures up to 4 000 atm. This equipment was calibrated with water at temperatures between 30° and 90°C, the measured compressibility being in good agreement with literature data.

(2) Optical measurements[15]

Figure 2 gives the experimental apparatus for optical studies of solvated electrons produced by pulse radiolysis under high pressure. The equipment consists of the high pressure section, the optical section, and the linear electron accelerator.

The optical section is a modified conventional spectrophotometer with lamp, lenses, monochromator and a fast responding light detector. The light detector is connected to a wideband oscilloscope which is triggered by the linear electron accelerator. The accelerator delivers up to 300 pulses per second of electrons (5 μsec, 10 MeV, 0·5 A) which induce the formation of solvated electrons in the solution contained in the high pressure cell. The high pressure optical cell is connected to the pressure generator and to the storage autoclave via capillary tubing. A compromise had to be made with respect to the wall thickness of the optical cell. The walls, on the one hand, had to be sufficiently thick to withstand a pressure of at least 1 000 atm, but they had to be transparent to the electrons from the accelerator. The range of 10 MeV electrons in steel is about 6 mm. To ensure even exposure of the solution to the electron beam, a wall thickness of only 2 mm was used. With this thickness and a twofold factor of safety the outer diameter of the cell must not exceed 14 mm. Since it is impossible to incorporate optical windows and capillaries into a steel tube of given diameter and wall thickness, only the 70 mm long centre portion (1) of the cell which is exposed to the electron beam has the critical dimensions. Two axially bored steel blocks (2) housing the window holders (3) were welded to both sides of this tube; conically formed sapphire windows (4) were ground into the holders. The window holders and washers (5) are tightened into the housings with hollow screws (6). The cell was tested at pressures up to 1 500 atm.

This apparatus is useful for photometric studies of the solutions of electrons with concentrations up to approximately 3×10^{-5} mole/l. that are easily produced by pulse radiolysis. For the study of metal–ammonia solutions with metal concentrations up to 3×10^{-2} mole/l. another similarly constructed autoclave[7] was used; however, the optical path length could be adjusted to between 20 and 0·1 mm. Other parts of the high pressure hardware used in these studies have been described elsewhere[9, 13].

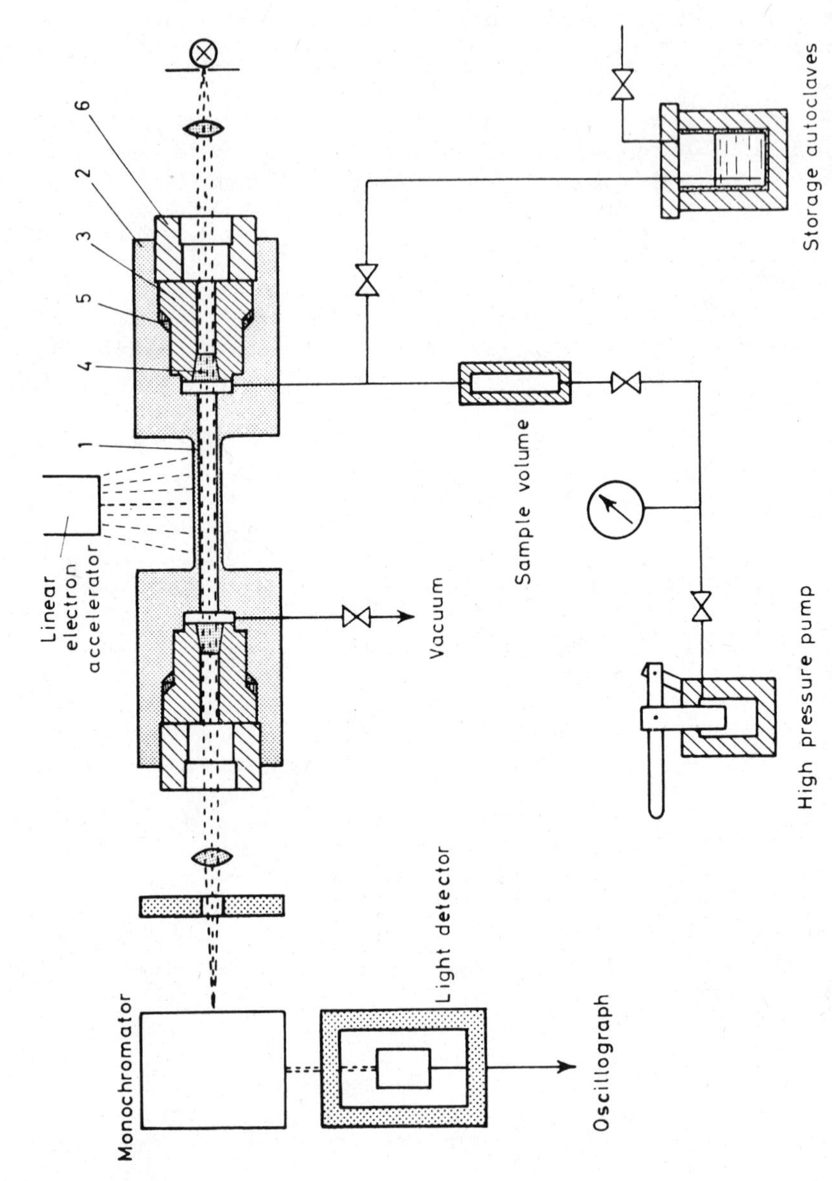

FIGURE 2. Experimental apparatus for the optical detection of solvated electrons produced by pulse radiolysis under pressure[15]

EXPERIMENTAL RESULTS AND THEIR INTERPRETATION

(1) Isothermal compressibility of potassium- and potassium iodide–ammonia solutions[7]

Figure 3 shows the compressibility of potassium–ammonia solutions and, for comparison, of potassium iodide–ammonia solutions at $-30°C$ and

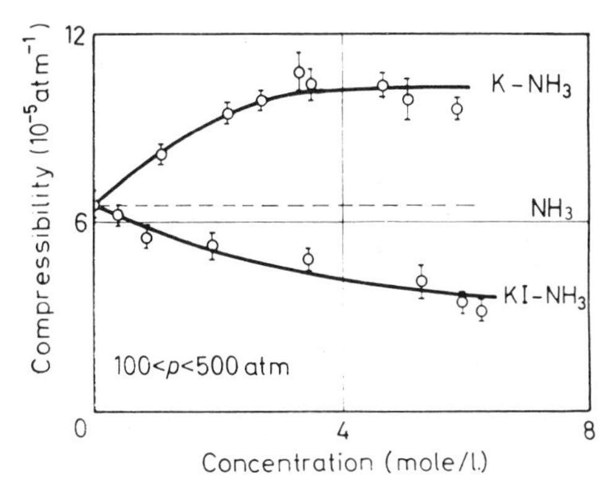

FIGURE 3. The isothermal compressibility of $K–NH_3$ and $KI–NH_3$ solutions determined at $-30°C$ between 100 and 500 atm[7]

100–500 atm. The isothermal compressibility of pure ammonia, also at $-30°C$, is indicated by the broken line. Obviously the dissolved metal and the electrolyte have an adverse effect on the compressibility. The compressibility of the iodide solution decreases with increasing concentration, which is to be expected for 'normal' electrolytes in which the reorientation of the solvent molecules by the electrostatic field of the ions (electrostriction) gives rise to a more compact, i.e. less compressible, structure of the solvent. In the metal solution this effect of the cations is overcompensated by the volume expansion of the solvated electrons present which leads to an increased compressibility with increasing concentration. The trend of the isothermal compressibility with concentration is the same as that of the adiabatic compressibility measured by Maybury and Coulter[16]. However, the isothermal data are, as expected, higher (by about 60 per cent) than the adiabatic data.

(2) Equilibrium reaction of solvated electrons with ammonia[12]

An equilibrium easily investigated using high pressure techniques is that of the reaction of solvated electrons with ammonia which leads to molecular hydrogen and amide ions

$$e^-_{sol} + NH_3 \rightleftharpoons \tfrac{1}{2}H_2 + NH^- \tag{8}$$

According to general experience, the metal–ammonia solutions tend to decay completely, equilibrium 8 being displaced far to the right. However, Jolly and Kirschke[17] showed that the concentration of solvated electrons is still measurable with an equilibrium constant of ~ 70 (mole/l.)$^{1/2}$ and a reaction enthalpy of -16 kcal/mole. The reaction is exothermic as written.

Reaction 8 was studied by observing the absorption spectrum of a 0·23 mole/l. potassium amide–ammonia solution with and without added hydrogen (solution saturated at 100 atm) in the optical high pressure cell (path length, 4 mm); these experiments were conducted at static pressures of 200 to 1 100 atm and at temperatures of 23° to 147°C. The difference between the spectra of the two solutions shows the broad absorption spectrum characteristic of solvated electrons with a maximum at 1·6 to 1·8 μ (Figure 4). The intensity of this band increases with temperature and decreases with pressure. The equilibrium concentration, C, of solvated electrons as calculated from the intensity as a function of pressure for seven different temperatures is plotted in Figure 5. In these calculations it is assumed that the extinction coefficient for solvated electrons ($\epsilon = 48000$ l./mole-cm) is independent of temperature and pressure.

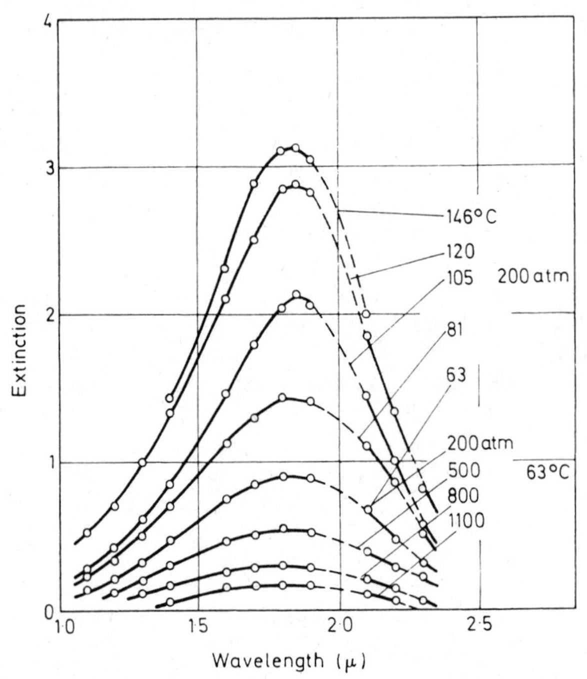

FIGURE 4. The absorption spectrum of solvated electrons in a 0·23 mole/l. KNH_2–NH_3 solution saturated under 100 atm with hydrogen at various pressures and temperatures[12]

Depending on the conditions, the equilibrium concentration of solvated electrons ranges between 5×10^{-7} and 2×10^{-4} mole/l., i.e. the concentration is so low that unpaired electrons predominate rather than spin-compensated electron pairs. From the temperature and pressure dependence of the equilibrium concentration (equations 1 and 2) a reaction enthalpy of -12 ± 1 kcal can be calculated using the data obtained between 23° and 63°C at 200 atm, which is somewhat less negative than the value reported by Jolly and Kirschke[17]. A reaction volume, ΔV, of -64 ± 3 ml is obtained from the data between 200 and 1 100 atm at 23°C. With increasing temperature and increasing pressures both ΔH and ΔV become less negative.

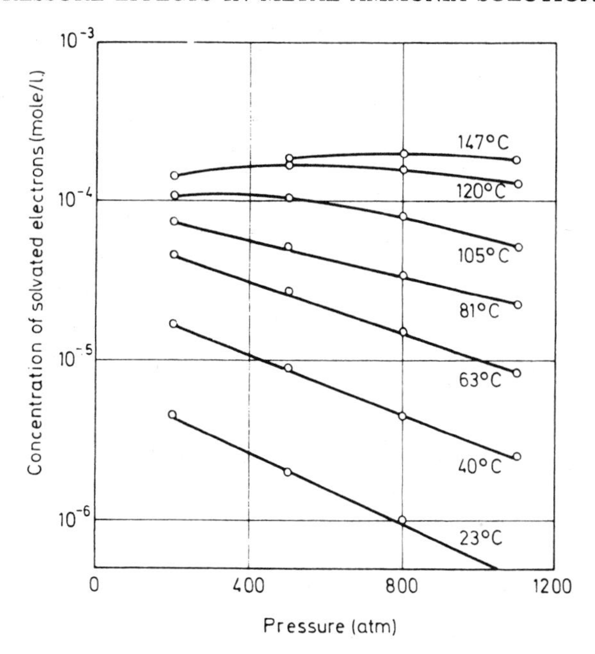

FIGURE 5. Pressure and temperature dependence of the equilibrium concentration of solvated electrons in a KNH_2–H_2–NH_3 mixture[12]. The concentrations correspond to those in Figure 4

The large reaction volume reflects the large molar volume of the solvated electron. From the reaction volume and the molar volume of the other species present the molar volume of the solvated electrons can be calculated as 98 ± 15 ml at 23°C, which, taking into consideration the thermal expansion (see below), can be reduced to about 75 ± 13 ml at -33°C. This value, within the limits of error, is the same as that obtained from density measurements on more concentrated solutions in which spin compensated electron pairs are predominant. So we conclude that solvated electrons in the spin compensated state and in the single state have about the same molar volume. A corroboration of the result using magnetic measurements will be given in a subsequent paper in this colloquium by Dr Böddeker[11].

(3) Absorption spectrum of dilute potassium- and potassium iodide-ammonia solutions[7,8]

The absorption spectrum of a dilute potassium–ammonia solution is shifted to shorter wavelengths with increasing pressure and, as has been known for some time[18], to longer wavelengths with increasing temperature. The experimental data are displayed in Figure 6 in which the position of the absorption maximum is plotted against the temperature and the pressure, respectively. According to the simple model of an electron-in-a-cavity with a square well potential the wavelength of maximum light absorption is proportional to the square of the cavity radius (equation 9).

$$\lambda_{\text{max.}} \propto R^2 \tag{9}$$

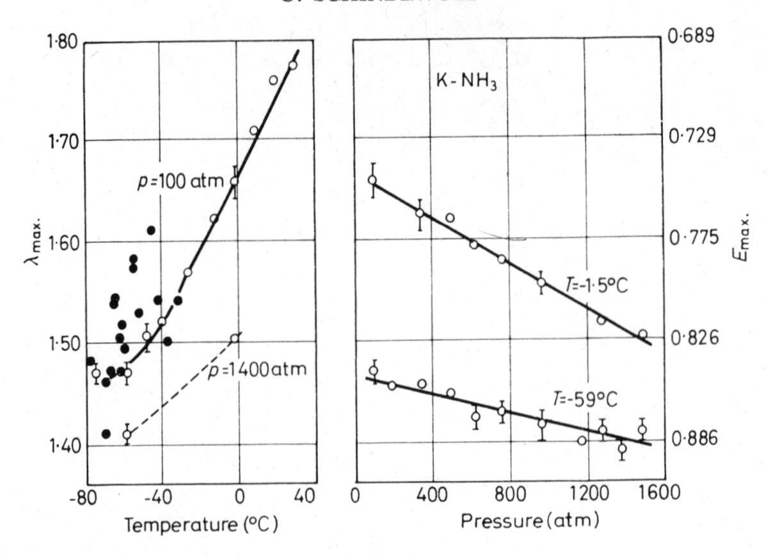

FIGURE 6. Position of the absorption maximum of solvated electrons in dilute K–NH_3 solutions as a function of pressure and temperature: o data from this laboratory[7], • data from the literature[18]

The more sophisticated models due to Jortner[2] lead to a more complex relationship but still show the same tendency. Thus the blue shift of the spectrum with increasing pressure and the red shift of the spectrum with increasing temperature can be interpreted in terms of a compression and a thermal expansion of the electron cavity.

Differentiation of equation 9 with respect to pressure and temperature, respectively, and replacement of the cavity radius R by the cube root of the cavity volume V leads to equations 10 and 11:

$$\alpha = \frac{\mathrm{d}V}{\mathrm{d}T \times V} = \frac{3}{2}\frac{\mathrm{d}\lambda}{\lambda \times \mathrm{d}T} \tag{10}$$

$$\beta = -\frac{\mathrm{d}V}{V \times \mathrm{d}p} = -\frac{3}{2} \times \frac{\mathrm{d}\lambda}{\lambda \times \mathrm{d}T_p} \tag{11}$$

where β and α are the coefficients of the isothermal compressibility and thermal expansion, respectively. Thus the shifts of the absorption band are a measure of the volume changes of solvated electrons. The numerical factor $\frac{1}{2}$ arises from the square well potential, and the factor 3 comes from the *cubic* co-efficients of compressibility and thermal expansion. Using these equations and the data in Figure 6 we obtain values for α and β of $(3{\cdot}4$ to $4{\cdot}1) \times 10^{-3}$ deg^{-1} and $(4{\cdot}4$ to $5{\cdot}9) \times 10^{-5}$ atm^{-1}, respectively, which are of the same magnitude as for pure ammonia $(1{\cdot}7 \times 10^{-3}$ deg^{-1}, $7{\cdot}5 \times 10^{-5}$ atm$^{-1})$. This coincidence of the α and β values for ammonia and for solvated electrons in ammonia, however, probably is accidental. The corresponding data for hydrated electrons and water do not agree.

The absorption spectra of the halide ions and of a number of other ions, like those of the solvated electrons, depend strongly on the solvent. The light absorption of these ions, according to the theory of Platzman and Frank[19] arises from a transition of an electron from the ion to the solvent molecules of the solvation shell (charge-transfer-to solvent: ctts). The radius of the solvation shell required determines the position of the absorption maximum (cf. equation 9). As with solvated electrons we may expect a blue shift of the

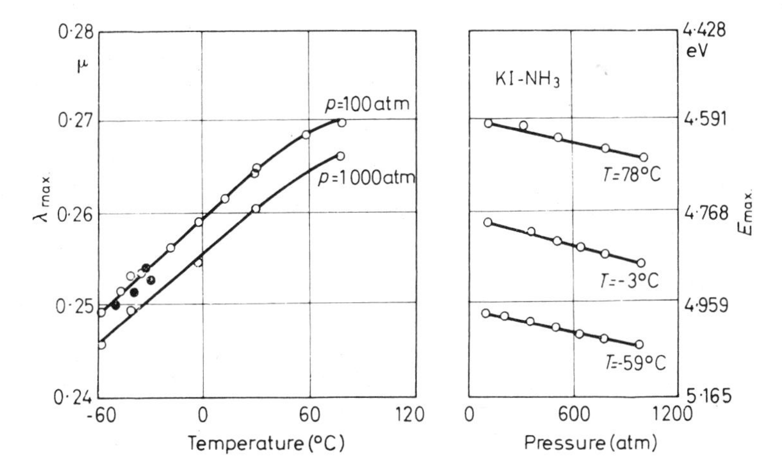

FIGURE 7. Position of the absorption maximum of iodide ions in dilute $KI–NH_3$ solutions at various pressures and temperatures: o data from this laboratory, • data from the literature[20]

spectrum with increasing pressure and a red shift with increasing temperature. The experimental results for iodide ions in Figure 7, and for amide ions (not shown) comply with these ideas and show that there is close parallelism in the optical behaviour of solvated electrons and ions possessing ctts-spectra.

(4) Electrical conductivity[9]

When pressure is applied to a solution of a strong electrolyte, the specific conductivity changes only slightly because the major effects more or less compensate each other; that is the increase in ion concentration arising from a compression of the solution is offset by a decrease of ion mobility due to the increased viscosity of the solvent. Potassium nitrate in ammonia is a typical strong electrolyte in this respect; up to 1 500 atm, only a small change in specific conductivity occurs (Figure 8)[21]. Potassium amide on the other hand shows a substantial increase in conductivity with pressure[21]. The latter behaviour is typical of weak electrolytes (the dissociation constant for $KNH_2 \rightleftharpoons K^+ + NH_2^-$ is given[22] as $K = 7{\cdot}3 \times 10^{-5}$) for which the dissociation constant increases with increasing pressure. Dissociation leads, because of electrostriction, to a volume decrease of the system ($\Delta V < 0$; cf. equation 2).

The conductivity of metal solutions shows rather complex behaviour when they are compressed. In dilute solutions ($c < 0{\cdot}5$ mole/l.) the conductivity changes only slightly with pressure (10 per cent decrease at 1 500 atm), but in the concentration range where the conductivity increases rapidly with concentration, i.e. 0·5 to 2 mole/l., the conductivity decreases markedly with

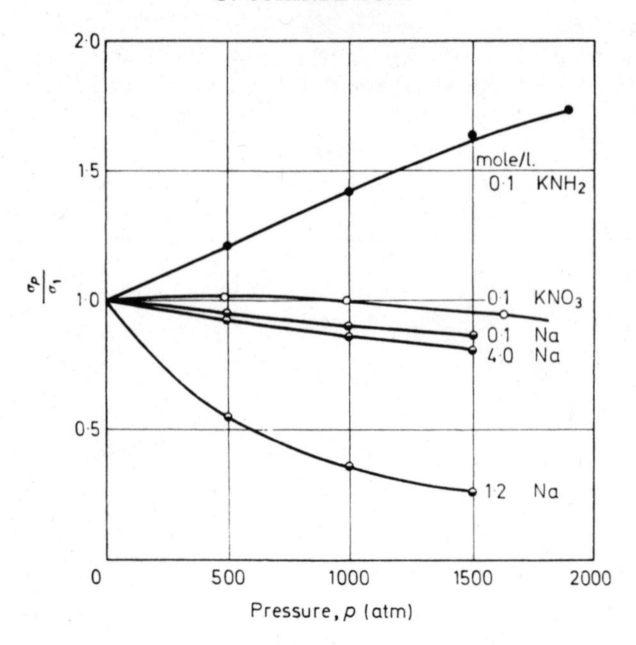

FIGURE 8. The change in specific conductivity with pressure of 0·1 mole/l. KNO_3 and KNH_2–NH_3 solutions[21] and of Na–NH_3 solutions at various concentrations[9]

increasing pressure (a factor of three at 1 500 atm). At higher metal concentrations an increase in pressure again yields only a small conductivity decrease. The data collected for sodium–ammonia solutions are displayed in Figure 9, where the pressure coefficient of the conductivity $k_P = \Delta\sigma/\sigma \times \Delta p$ is plotted against the concentration. The coefficient is negative for all concentrations and has a minimum at a concentration of about 1 mole/l. Comparison with the temperature coefficient[23] of the conductivity in Figure 9 reveals that pressure and temperature have adverse effects on the conductivity of the metal solutions. Evaluation of the data shows that the steep conductivity increase in the conductivity versus concentration diagram (or the nonmetal–metal transition described by equation 6) is shifted to smaller concentrations with increasing temperature and to higher concentrations with increasing pressure. That is, a solution of given concentration becomes more metallic with increasing temperature and less metallic with increasing pressure. Whereas the temperature dependence of the equilibrium expressed in equation 6 is known[24] and is understood on the basis of the Mott theory[25] of metallic transition, the pressure dependence of the equilibrium is surprising. One might expect that the increase of concentration as a result of the compression should make the solutions more metallic instead of less so. The results obtained can be explained on the basis of equation 6 if it is assumed that the dissolved metal in the metallic state occupies a larger volume than when it is in the nonmetallic state; and, therefore, the system, according to the principle of Brown and Le Chatelier, tries to compensate for the applied pressure by reverting to the nonmetallic state.

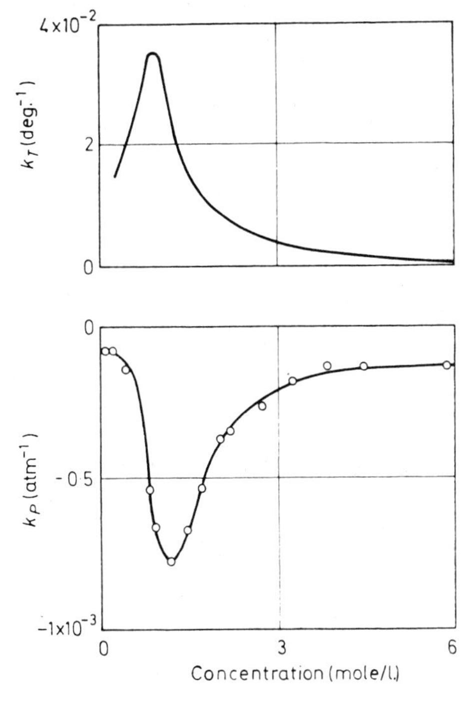

FIGURE 9. Pressure[9] and temperature[23] coefficients of the electrical conductivity of Na–NH_3 solutions, calculated from conductivity data at 1 and 500 atm (−40°C) and −34° and 50°C (1 atm)

(5) Reflectivity[7]

The bronze lustre of concentrated metal–ammonia solutions which agrees with their metallic behaviour is caused by the high reflectivity of these systems to red light. In agreement with data from the literature we find that the reflectivity (measured at an angle of nearly 90°) decreases with decreasing metal concentration[26, 27] and approaches the value for pure ammonia with increasing dilution. The concentration dependence of the reflectivity corresponds qualitatively to the theory which relates the reflectivity to the free electron concentration or to the electrical conductivity. Consequently we may expect the reflectivity to change with temperature and pressure in the same manner as does the conductivity. This expectation is borne out by the experiments which show an increase of reflectivity with temperature and a decrease with pressure. The temperature and pressure coefficients of the reflectivity are plotted versus concentration in Figure 10. The curves with maxima or minima around 1 mole/l., that is, in the concentration range of the nonmetal $\rightleftharpoons$ metal transition, parallel those of the conductivity and, like these, can be explained by the corresponding equilibrium.

(6) Miscibility gap[10]

According to the hypothesis of Mott[25] the miscibility gap of metal–ammonia solutions is closely related to the nonmetal $\rightleftharpoons$ metal transition. Since the

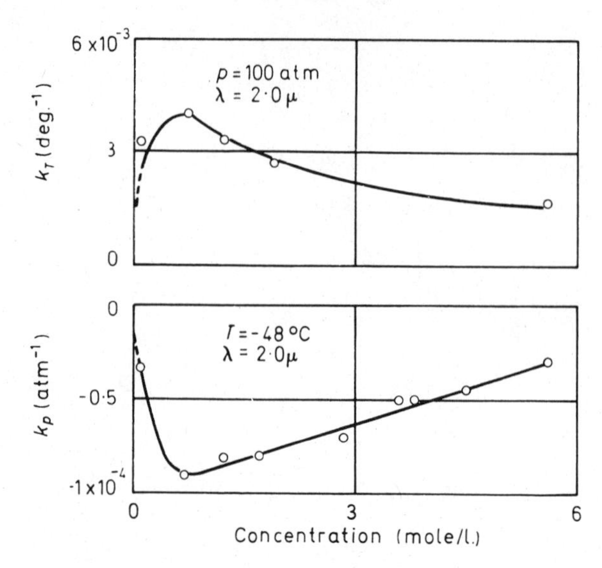

FIGURE 10. Pressure and temperature coefficients[7] of the reflectivity (wavelength 2 μ) o K–NH_3 solutions, calculated from experimental data at 100 and 1000 atm (−48°C) and at 0° and −48°C (100 atm)

position of this transition is shifted to higher concentrations by pressure the miscibility gap also should be shifted to higher concentrations with increasing pressure. This behaviour is, indeed, found in the results of experiments in which the pressure influence on the miscibility gap was investigated using conductivity measurements. Figure 11 shows the results of these experiments;

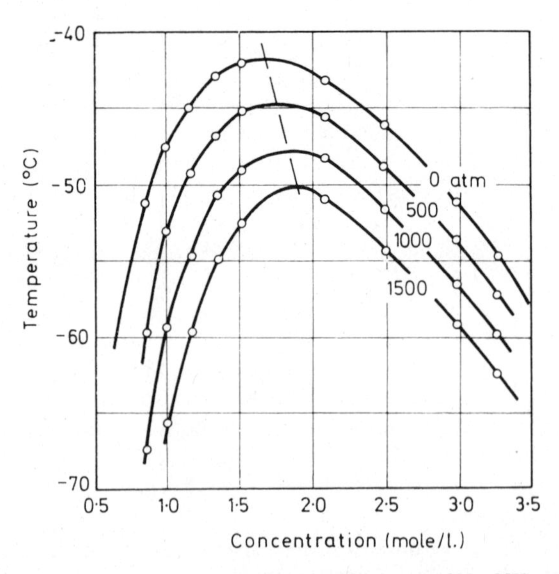

FIGURE 11. Influence of pressure on the miscibility gap of Na–NH_3 solutions[10]

the critical concentration of the miscibility gap increases with increasing pressure and the critical temperature decreases.

The change of critical temperature with pressure warrants further discussion. The thermodynamics of binary mixtures yields, for a miscibility gap[28] with upper consolate point, equation 12 as the relationship between the critical temperature, T_c, pressure, p, the molar volume, V, and concentration expressed as the mole fraction, x.

$$dT_c/dp \sim -d^2V/dx^2 \tag{12}$$

The molar volume is expressed as

$$V = x_1V_{01}+x_2V_{02}+\Delta V(x) \tag{13}$$

where V_{01} is the molar volume of the pure component 1 and $\Delta V(x)$ the volume change caused by mixing. Since dT_c/dp in our system is negative, d^2V/dx^2 must be positive. According to density measurements[29] $\Delta V(x)$ is positive at all concentrations. Therefore $\Delta V(x)/x$ or the apparent molar volume of the metal, $\Delta V(x)/x+V_0$, must increase with concentration in the concentration range of the miscibility gap. Thus the pressure shift of the critical temperature of the miscibility gap represents an independent proof of the hypothesis that the transition to the metallic state is connected with a volume increase of the system, as deduced earlier from the pressure dependences of the electrical conductivity and reflectivity.

(7) Solvated electrons in ammonia–water mixtures[14]

Because of its fast reaction with water, the solvated electron in ammonia–water mixtures cannot be studied by dissolving alkali metal. In this system solvated electrons can, however, very conveniently be produced by radiolysis. Preliminary results of some optical and kinetic studies of solvated electrons in ammonia–water mixtures will be outlined briefly.

(a) *Absorption spectrum*

The position of the electron absorption band shifts linearly with increasing ammonia content from the position in water (0·7 μ[1b,e]) to that in pure ammonia (1·7 μ). On the basis of equation 9 this means that the size of the solvent cavity associated with electrons increases with increasing ammonia content of the mixtures. The magnitude of the red and blue shifts of the spectra in the solvent mixtures which occurs with increasing temperature and pressure, respectively, lies between the corresponding shifts observed in the pure solvents.

(b) *Lifetime*

Due to the slow reaction 8 the lifetime of solvated electrons in pure ammonia is larger than 10^6 sec. In water it is less than 100 μsec[1b,e] because of the fast reaction with water, formally described by equation 14.

$$e^- + H_2O \rightleftharpoons \tfrac{1}{2}H_2 + OH^- \tag{14}$$

In ammonia–water mixtures the lifetime of solvated electrons increases essentially exponentially with increasing mole fraction of ammonia; e.g. in a

20 per cent water–80 per cent ammonia mixture we observe a lifetime of about 100 sec. Thus a fivefold dilution of the more reactive water species with relatively inert ammonia leads approximately to a millionfold decrease of the rate of reaction 14. Similar results were obtained by Dewald[30]. The rate constant of the reaction, $\varkappa$, is given by equation 15 according to transition state theory,

$$\varkappa = (kT/h) \exp(\Delta S^{\ddagger}/R) \exp(-\Delta H^{\ddagger}/RT) \quad (15)$$

where $\Delta S^{\ddagger}$ and $\Delta H^{\ddagger}$ are entropy and energy of activation, respectively. The energy of activation of reaction 14 occurring in pure water is about 4 kcal/mole[1b,e]. For mixtures with an ammonia content between 20 and 80 mol-per cent energies of activation are 1 ± 3 kcal/mole. Thus the decrease in reaction rate with increasing ammonia content must be due to a decrease in activation entropy. Since the entropy is a function of volume in the sense that the entropy of a system increases when its volume increases, the reactivity of solvated electrons with water can be related to their molar volume. As the ammonia content of the mixtures increases, the volume associated with the solvated electron increases, the volume of activation becomes more negative, the entropy of activation becomes more negative, and the reaction rate decreases. As a consequence of this relationship we expect (equation **4**) little change in the reaction rate with increasing pressure for the fast reaction of solvated electrons in water-rich mixtures (in which the electrons occupy a relatively small volume). A higher increase in the reaction rate with increasing pressure would be expected for the slower reaction of solvated electrons in the ammonia-rich mixtures, in which the electrons occupy a large volume. Indeed we find that there is no pressure influence[31] and the volume of activation is close to zero for the rate of reaction 14 in water, whereas in some preliminary experiments with ammonia–water mixtures the rate of disappearance of solvated electrons is accelerated two- to four-fold by a pressure increase of 1 000 atm yielding an activation volume of 17 to 35 ml/mole.

REFERENCES

[1] For recent reviews see:
(a) G. Lepoutre and M. J. Sienco (Eds.), *Métal–Ammonia Solutions*. W. A. Benjamin: New York (1964).
(b) R. F. Gould (Ed.), Advances in Chemistry Series No. 50, *Solvated Electron*, American Chemical Society: Washington, D.C. (1965).
(c) J. Jander in *Chemie in nichtwässrigen ionisierenden Lösungsmitteln*, Vol. I, Part 1, G. Jander, H. Spandau and C. C. Addison (Eds.), Vieweg: Braunschweig (1966).
(d) J. C. Thompson in *The Chemistry of Non-Aqueous Solvents*, Vol. II, J. J. Lagowski (Ed.), Academic Press: New York (1967).
(e) U. Schindewolf, *Angew. Chemie*, **80**, 165 (1968); *Angew. Chem.* (*Internat. Ed.*), **7**, 190 (1968).

[2] J. Jortner, *J. Chem. Phys.* **30**, 839 (1959); *Radiation Res. Suppl.* **4**, 24 (1964).
J. Jortner, S. A. Rice and E. G. Wilson, p 222 in Ref. 1a.

[3] W. N. Lipscomb, *J. Chem. Phys.* **21**, 52 (1953).

[4] R. A. Ogg, *J. Am. Chem. Soc.* **68**, 155 (1946).

[5] C. W. Orgell, A. M. Filbert and E. C. Evers, p 67 in ref. 1a; W. H. Brendley and E. C. Evers, p 111 in ref. 1b.

[6] S. R. Gunn, p 76 in ref. 1a; *J. Chem. Phys.* **47**, 1174 (1967).

[7] R. Vogelsgesang, *Dissertation*, University of Karlsruhe (1969).

[8] U. Schindewolf, *Angew. Chem.* **79**, 585 (1967); *Angew. Chem.* (*Internat. Ed.*), **6**, 575 (1967).

[9] U. Schindewolf, K. W. Böddeker and R. Vogelsgesang, *Ber. Bunsenges. phys. Chem.* **70**, 1161 (1966).
[10] U. Schindewolf, G. Lang and K. W. Böddeker, *Z. phys. Chem.*, N.F., **66**, 86 (1969).
[11] K. W. Böddeker, G. Lang and U. Schindewolf, this volume, page 219.
[12] U. Schindewolf, R. Vogelsgesang and K. W. Böddeker, *Angew. Chem.* **79**, 1064 (1967); *Angew. Chem.* (*Internat. Ed.*), **6**, 1076 (1967).
[13] K. W. Böddeker, G. Lang and U. Schindewolf, *Angew. Chem.* **81**, 118 (1969); *Angew. Chem.* (*Internat. Ed.*), **8**, 138 (1969).
[14] U. Schindewolf and R. Olinger, Bunsentagung 1969, Frankfurt.
[15] U. Schindewolf, G. Lang and H. Kohrmann, *Chem. Ing. Tech.*, **41**, 830 (1969),
[16] R. H. Maybury and L. V. Coulter, *J. Chem. Phys.* **19**, 1326 (1951).
[17] E. J. Kirschke and W. J. Jolly, *Science*, **147**, 45 (1965); *Inorg. Chem.* **6**, 855 (1967).
[18] H. Blades and J. W. Hodgins, *Canad. J. Chem.* **33**, 411 (1955); R. C. Douthit and J. L. Dye, *J. Am. Chem. Soc.* **82**, 4472 (1960); M. Gold and W. L. Jolly, *Inorg. Chem.* **1**, 818 (1962).
[19] R. Platzman and J. Frank, *Z. Physik*, **138**, 411 (1954).
[20] M. Smith and M. C. R. Symons, *Trans. Faraday Soc.* **54**, 338, 346 (1958);
J. T. Nelson, R. E. Cuthrell and J. J. Lagowski, *J. Phys. Chem.* **70**, 1492 (1966).
[21] R. Vogelsgesang, *Thesis*, University of Karlsruhe (1965).
[22] W. W. Hawes, *J. Am. Chem. Soc.* **55**, 4422 (1933).
[23] C. A. Kraus, *J. Am. Chem. Soc.* **43**, 758 (1921);
C. A. Kraus and W. W. Lucasse, *J. Am. Chem. Soc.* **44**, 1946 (1922).
[24] R. Catterall, *J. Chem. Phys.* **43**, 2262 (1965).
[25] N. F. Mott, *Phil. Mag.* **6**, 287 (1961);
N. F. Mott *et al. Physica Status Solidi* (*Berlin*), **21**, 343 (1967).
[26] T. A. Beckman and K. S. Pitzer, *J. Phys. Chem.* **65**, 1527 (1961).
[27] J. C. Thompson, unpublished results.
[28] R. Rehage, *Z. Naturforsch.* **10a**, 316 (1955);
R. Haase, *Thermodynamik der Mischphasen*, Springer: Berlin (1956); G. Schneider, *Ber. Bunsenges. phys. Chem.* **70**, 497 (1966).
[29] E. Huster, *Ann. Physik*, **33**, 477 (1938).
[30] R. Dewald and R. V. Tsina, *Chem. Commun.* 647 (1967).
[31] U. Schindewolf, H. Kohrmann and G. Lang, *Angew. Chem.* **81**, 496 (1969); *Angew. Chem.* (*Internat. Ed.*), **8**, 512 (1969).

Studies of Concentrated Metal–Ammonia Solutions

The following two papers exemplify recent efforts to understand the properties of concentrated "metallic" metal–ammonia solutions. In the first, Ashcroft and Russakoff discuss electron scattering mechanisms, and in the second Nasby and Thompson describe the Hall effect. The recent papers by Schroeder, Thompson, and Oertel[1] (on conductivity in concentrated solutions) and by Acrivos and Mott[2] (on the metal–nonmetal transition in sodium–ammonia solutions) are also recommended.

---

[1]R. L. Schroeder, J. C. Thompson, and P. L. Oertel, *Phys. Rev.,* **178**, 298 (1969).

[2]J. V. Acrivos and N. F. Mott , *Phil. Mag.,* **24**, 19 (1971).

Reprinted from

PHYSICAL REVIEW A VOLUME 1, NUMBER 1 JANUARY 1970

Incoherent Scattering and Electron Transport in Metal-Ammonia Solutions. I*

59

N. W. Ashcroft and G. Russakoff
Laboratory of Atomic and Solid State Physics, Cornell University, Ithaca, New York 14850
(Received 22 August 1969)

Electron transport is considered in the metallic range of metal-ammonia solutions; the scattering centers are (i) ions and ammonia dipoles weakly bound in solvated ion complexes, and (ii) free ammonia dipoles. They form a binary liquid mixture, but only one of the three possible partial structure factors is required. From the orientational degrees of freedom of the dipoles, it can be shown that much of the scattering is incoherent. This, combined with the observation that solvated ions are relatively weak scatterers, accounts in large measure for the high conductivities measured in metal-ammonia solutions.

I. INTRODUCTION

Above a certain concentration of the metallic component (about 8 mole % metal[1]), metal-ammonia solutions are, by all the usual definitions, good metals. One feature they exhibit is, however, quite unusual: We refer to the extraordinary dependence of their conductivities on the concentration of metallic ions. They increase sharply as a function of the concentration of metal, a behavior which at the very least is difficult to reconcile with elementary Drude theory incorporating a single relaxation time τ. For if, at a given concentration, n_e is the assumed density of free electrons, then the conductivity σ is given as usual by

$$\sigma = n_e e^2 \tau / m \; ; \tag{1}$$

and while it is, of course, clear that an increase in n_e (in proportion to metal concentration) should be followed by an increase in σ, we cannot escape the accompanying consequence that an increase in the number of ions regarded as scatterers should shorten τ. An elementary interpretation of (1) leads to a very weak dependence of σ on n_e, something like $n_e^{1/3}$.

The experiments, on the other hand, suggest a dependence more like n_e^3 (depending on the system); and to account for this curious behavior we propose that the over-all scattering of electrons proceeds by two basically independent scattering mechanisms to be described in detail in the following sections. The basic point concerns the presence of orientable dipoles as electron scatterers, both "free" and bound up in so-called solvated ions. Calculations of the conductivity for the metallic solutions clearly involve the execution of configuration averages of both density fluctuations and dipolar moment fluctuations. The latter, being to a good approximation independent of the former, reduces much of the total cross section to that expected from a set of independent or incoherent scatterers.

The formalism is set up in a pseudopotential approximation.[2] As a consequence, the usual one-electron potentials seen by conduction electrons are canceled and much reduced in the core-state regions of ions. This is also true in molecules, and we assume it to be true in the NH_3 molecules, whether bound or free. As it happens, the practical ramifications of this assumption are minimal, since for these metals the maximum momentum transfer $2k_F$ is quite small and only the screened long-range parts of the various potentials are pertinent. Further, there is experimental evidence[3] that electron scattering from ammonia is well reproduced by a point-dipole potential using (as we shall here) the Born approximation.

In the present paper, we shall outline the calculation of σ: While the bulk of the numerical work for a variety of systems will be given in a later work, we will illustrate the physical points as they arise with the system $Li(NH_3)_x$.

II. ELECTRON SCATTERING MECHANISMS

Even in the most concentrated metallic range, it appears that the metallic ions are solvated: The field of an ion polarizes the NH_3 molecules, a few of which (say, λ) subsequently bind themselves weakly to the ion. (Freely moving conduction electrons are largely excluded by orthogonalization from the interior of these complexes and therefore cannot entirely screen out the ion-ammonia bond.) The following physical picture of the solution now seems to be established: As the metal is dissolved into solution, the valence electrons dissociate, each resulting ion combining with λ ammonia molecules[4] and concomitantly decreasing the number of unbound (which we call "free") molecules.

Metal-ammonia solutions (for general metal-metallic concentrations) may be viewed as binary mixtures, the components being solvated ions and free-ammonia molecules. Schroeder and Thompson[5] have already exploited this model in a calculation of the thermodynamic functions of metal-ammonia solutions.

We take the solvated ion and the dipolar ammonia molecule to be the two basic scatterers, whose cross sections will be evaluated (in accordance with our assumptions pertaining to a pseudo Hamiltonian) in first Born approximation. For a binary system of *spherically symmetric* scatterers, the resistivity is proportional to[6]

$$\int_0^1 dy\, y^3[(1-x)V_1^2(y)S_{11}(y) + 2[x(1-x)]^{1/2}V_1(y)$$

$$\times V_2(y)S_{12}(y) + xV_2^2(y)S_{22}(y)] \,, \qquad (2)$$

where $V_1(y)$, $V_2(y)$ are the Fourier transforms[7] of the spherically symmetric electron-ion interactions for the two species, $(1-x)$ and (x) are their concentrations, and the $S_{ij}(y)$ are the partial-structure factors. We now outline the necessary modifications to (2) which result from the obvious anisotropy inherent in electron scattering from metal-ammonia systems.

To elucidate the important physical differences, let us initially consider the first term in (2). It arises in the Born approximation[8] from configuration averages over Fermi-surface scattering matrix elements of the kind

$$|\langle \vec{k}'|\textstyle\sum_{\vec{r}_1} V_1(\vec{r}-\vec{r}_1)|\vec{k}\rangle|^2 \,, \qquad (3)$$

where $\vec{r}_1$ are the instantaneous positions of the centers of the potential $V_1(\vec{r})$. Equation (3) gives

$$\langle \rho_{\vec{k}-\vec{k}'}^{(1)}\,\rho_{\vec{k}'-\vec{k}}^{(1)}\rangle\,|V_1(\vec{k}'-\vec{k})|^2 \,,$$

$$\text{where}\quad \rho_{\vec{k}-\vec{k}'}^{(1)} = \textstyle\sum_{\vec{r}_1} e^{-i(\vec{k}-\vec{k}')\cdot\vec{r}_1} \qquad (4)$$

is a Fourier component of the density fluctuation of ions of species (1), and the quantity $\langle \rho_{\vec{k}-\vec{k}'}^{(1)}\rho_{\vec{k}'-\vec{k}}^{(1)}\rangle$ leads to the partial-structure factor S_{11}.

We repeat the argument for a potential $V_1(\vec{r})$ that is no longer spherically symmetric. In fact, let us associate $V_1(\vec{r})$ with the potential of a point dipole[9] of moment $\vec{\mu}$. For a collection of these at points $\vec{r}_1$, the interaction energy with an electron at point $\vec{r}$ is

$$(-e)\textstyle\sum_{\vec{r}_1}\vec{\mu}_1\cdot(\vec{r}-\vec{r}_1)/|\vec{r}-\vec{r}_1|^3 \,;$$

its matrix elements in plane wave states are[10]

$$(-e)\langle\vec{k}'|\textstyle\sum_{\vec{r}_1}\vec{\mu}_1\cdot(\vec{r}-\vec{r}_1)/|\vec{r}-\vec{r}_1|^3|\vec{k}\rangle$$

$$= \textstyle\sum_{\vec{r}_1} e^{-i\vec{r}_1\cdot(\vec{k}'-\vec{k})}\dfrac{i4\pi e}{(\vec{k}'-\vec{k})^2}\,\vec{\mu}_1\cdot(\vec{k}'-\vec{k}) \,,$$

and its square with $\vec{K}=\vec{k}'-\vec{k}$ is

$$\left(\frac{4\pi e}{K^2}\right)^2 \sum_{\vec{r}_1,\vec{r}_1{}'} e^{i(\vec{r}_1{}'-\vec{r}_1)\cdot\vec{K}}\,(\vec{\mu}_1\cdot\vec{K})(\vec{\mu}_1{}'\cdot\vec{K}). \qquad (5)$$

It is (5) which must now be averaged over all spatial configurations (S), in any particular one of which *all* angular orientations (A) of the dipoles are assumed equally likely.[11] It follows that in

$$\left(\frac{4\pi e}{K^2}\right)^2 \Big\langle \sum_{\vec{r}_1,\vec{r}_1{}'} e^{i(\vec{r}_1{}'-\vec{r}_1)\cdot\vec{K}} \times (\vec{\mu}_1\cdot\vec{K})(\vec{\mu}_1{}'\cdot\vec{K})\Big\rangle_{S,A} \,, \qquad (6)$$

the terms with $\vec{r}_1{}'\neq\vec{r}_1$ may be written

$$\left(\frac{4\pi e}{K^2}\right)^2 \Big\langle \sum_{\vec{r}_1\neq\vec{r}_1{}'} e^{i(\vec{r}_1{}'-\vec{r}_1)\cdot\vec{K}} \times (\vec{\mu}_1\cdot\vec{K})(\vec{\mu}_1{}'\cdot\vec{K})\Big\rangle_{S,A}$$

$$= \left(\frac{4\pi e}{K^2}\right)^2 \Big\langle \sum_{\vec{r}_1\neq\vec{r}_1{}'} e^{i(\vec{r}_1{}'-\vec{r}_1)\cdot\vec{K}} \times \langle\vec{\mu}_1\cdot\vec{K}\rangle_A\langle\vec{\mu}_1{}'\cdot\vec{K}\rangle_A\Big\rangle_S$$

$$= 0,$$

since $\langle\vec{\mu}\cdot\vec{K}\rangle_A$ vanishes. With N_d the number of free dipoles, the left-hand side of (6) becomes

$$N_d\left(\frac{4\pi e}{K^2}\right)^2 \langle(\vec{\mu}_1\cdot\vec{K})^2\rangle_A \,, \qquad (7)$$

and therefore represents independent scattering.

We turn now to the equivalent of the third term in Eq. (2). In the present case, the scattering is given by

$$\langle|\langle\vec{k}'|\textstyle\sum_{\vec{r}_2} V_2(\vec{r}-\vec{r}_2)|\vec{k}\rangle|^2\rangle_{S,A} \,, \qquad (8)$$

where, in accordance with the model of the solvated atom,

$$V_2(\vec{r}) = V_i(\vec{r}) + \sum_{l=1}^{\lambda} V'_{1,l}(\vec{r}-\vec{r}_l).$$

Basis vectors of the ammonia molecules linked to the metal ion are denoted by $\vec{r}_l$, and the electron-ion interaction is described by a local pseudopotential $V_i(\vec{r})$. The prime on $V'_{1,l}$ indicates that because of molecular bonding and polarization the dipolar potential differs slightly from $V_1(\vec{r})$. Writing $\vec{K} = \vec{k}' - \vec{k}$ again, Eq. (8) becomes

$$\Big\langle \sum_{\vec{r}_2, \vec{r}_2'} e^{-i\vec{K}\cdot(\vec{r}_2 - \vec{r}_2')} \Big[V_i(\vec{K}) + \sum_{l=1}^{\lambda} e^{-i\vec{K}\cdot\vec{r}_l} V'_{1,l}(\vec{K}) \Big] \times \Big[V_i^*(\vec{K}) + \sum_{m=1}^{\lambda} e^{i\vec{K}\cdot\vec{r}_m} V'^{\,*}_{1,m}(\vec{K}) \Big] \Big\rangle_{S,A}, \quad (9)$$

where $V'_{1,m}(\vec{K})$ is the $\vec{K}$th Fourier component of the electron interaction with the dipole at site m on the ion at position $\vec{r}_2'$.

First consider terms with $\vec{r}_2 = \vec{r}_2'$. These give a contribution

$$N_i \langle | V_i(\vec{K}) + \sum_{l=1}^{\lambda} e^{-i\vec{K}\cdot\vec{r}_l} V'_{1,l}(\vec{K}) |^2 \rangle_A . \quad (10)$$

The remaining terms of (9) always involve ions at different positions ($\vec{r}_2 \neq \vec{r}_2'$), and now the angular averages (A) of any such two solvated ions are independent for rotations about each central ion:

$$\Big\langle \sum_{\vec{r}_2 \neq \vec{r}_2'} e^{-i\vec{K}\cdot(\vec{r}_2-\vec{r}_2')} \Big[V_i(\vec{K}) + \sum_{l=1}^{\lambda} e^{-i\vec{K}\cdot\vec{r}_l} V'_{1,l}(\vec{K}) \Big] \Big[V_i^*(\vec{K}) + \sum_{m=1}^{\lambda} e^{i\vec{K}\cdot\vec{r}_m} V'_{1,m}{}^*(\vec{K}) \Big] \Big\rangle_{S,A}$$

$$= \Big\langle \sum_{\vec{r}_2 \neq \vec{r}_2'} e^{-i\vec{K}\cdot(\vec{r}_2-\vec{r}_2')} V_i(\vec{K}) + \sum_{l=1}^{\lambda} e^{-i\vec{K}\cdot\vec{r}_l} V'_{1,l}(\vec{K}) \rangle_A \langle V_i^*(\vec{K}) + \sum_{m=1}^{\lambda} e^{i\vec{K}\cdot\vec{r}_m} V'_{1,m}{}^*(\vec{k}) \rangle_A \Big\rangle_S$$

$$= \Big\langle \sum_{\vec{r}_2, \vec{r}_2'} e^{-i\vec{K}\cdot(\vec{r}_2-\vec{r}_2')} |\langle V_i(\vec{K}) + \sum_{l=1}^{\lambda} e^{-i\vec{K}\cdot\vec{r}_l} V'_{1,l}(\vec{K}) \rangle_A|^2 \Big\rangle_S - \Big\langle \sum_{\vec{r}_2 = \vec{r}_2'} |\langle V_i(\vec{K}) + \sum_{l=1}^{\lambda} e^{-i\vec{K}\cdot\vec{r}_l} V'_{1,l}(\vec{K}) \rangle_A|^2 \Big\rangle_S$$

$$= N_i S_{ii}(\vec{K}) \Big| \Big\langle V_i(\vec{K}) + \sum_{l=1}^{\lambda} e^{-i\vec{K}\cdot\vec{r}_l} V'_{1,l}(\vec{K}) \Big\rangle_A \Big|^2 - N_i \Big| \Big\langle V_i(\vec{K}) + \sum_{l=1}^{\lambda} e^{-i\vec{K}\cdot\vec{r}_l} V'_{1,l}(\vec{K}) \Big\rangle_A \Big|^2 , \quad (11)$$

where N_i is the number of ions (and therefore of solvated atoms) and S_{ii} is the partial-structure factor for the solvated ions. Hence, combining (10) and (11), (9) becomes

$$N_i S_{ii}(\vec{K}) \Big| \Big\langle V_i(\vec{K}) + \sum_{l=1}^{\lambda} e^{-i\vec{K}\cdot\vec{r}_l} V'_{1,l}(\vec{K}) \Big\rangle_A \Big|^2 + N_i \Big[\Big\langle \Big| \sum_{l=1}^{\lambda} e^{-i\vec{K}\cdot\vec{r}_l} V'_{1,l}(\vec{K}) \Big|^2 \Big\rangle_A - \Big| \Big\langle \sum_{l=1}^{\lambda} e^{-i\vec{K}\cdot\vec{r}_l} V'_{1,l}(\vec{K}) \Big\rangle_A \Big|^2 \Big] . \quad (12)$$

[We have used the fact that $V_i(\vec{K})$ has no angular dependence.]

Finally, we consider the two cross terms (dipole, solvated-ion) corresponding to the second term of Eq. (2), one of which is

$$\Big\langle \sum_{\vec{r}_1, \vec{r}_2} e^{i\vec{K}\cdot(\vec{r}_1-\vec{r}_2)} [V_i(\vec{K}) + \sum_{l=1}^{\lambda} e^{-i\vec{K}\cdot\vec{r}_l} V'_{1,l}(\vec{K})][V^*_{1,\vec{\mu}_1}(\vec{K})] \Big\rangle_{S,A} ,$$

or
$$\left\langle \sum_{\vec{r}_1,\vec{r}_2} e^{i\vec{K}\cdot(\vec{r}_1-\vec{r}_2)} \langle V_i(\vec{K}) + \sum_{l=1}^{\lambda} e^{-i\vec{K}\cdot\vec{r}_l} V'_{1,l}(\vec{K})\rangle_A \cdot \langle V^*_{1,\vec{\mu}_1}(\vec{K})\rangle_A \right\rangle_S = 0 \quad ,$$

since $\langle V_1;\vec{\mu}_1(\vec{K})\rangle_A$ vanishes. We are thus left with (12) and (7) as the total scattering which corroborates our assertion of two independent classes of scatterers,[12] at least in the Born approximation.

We remark that the physical significance of performing independent angular averages (as we have done here) is familiar in another context: We refer to the incoherent terms in neutron scattering from liquids. These are independent of liquid structure and bear at least a formal similarity to (7). Contributions of this nature to electron scattering may also be referred to as incoherent. In the case of electron scattering from free dipoles, we have (for low-momentum transfers) completely incoherent scattering, in contrast to electron-solvated atom scattering where no simple division between coherent and incoherent is evident.

III. FORM OF RESISTIVITY FOR INDEPENDENT SCATTERERS

The dipoles and solvated atoms act as independent scattering groups; the scattering from the solvated ions still involves a knowledge of their correlations. In the Born approximation, the resistivity for this system may be written

$$\rho = (m/n_e e^2)(\tau_i^{-1} + \tau_d)^{-1} \quad ,$$

where n_e is the electron density

$$3\pi^2 n_e = (mv_F/\hbar)^3 = k_F^3 ,$$

and

$$\frac{1}{\tau_i} = 2\pi v_F n_e \int_0^\pi \left(\frac{d\sigma_i}{d\Omega}\right)(1-\cos\theta)\sin\theta\, d\theta , \qquad (13)$$

$$\frac{1}{\tau_d} = 2\pi v_F (n_d - \lambda n_e) \int_0^\pi \left(\frac{d\sigma_d}{d\Omega}\right)(1-\cos\theta)\sin\theta\, d\theta . \qquad (14)$$

By n_d and $n_d - \lambda n_e$ we mean, respectively, the total density of NH_3 molecules (whether bound or free) and the density of free NH_3 molecules. The quantities $(d\sigma_i/d\Omega)$ and $(d\sigma_d/d\Omega)$ denote the differential scattering cross sections per scatterer of solvated ions and of free NH_3 molecules.[13] These are easily calculated using the configuration averaged matrix element expressions of Sec. II:

$$\left(\frac{2\pi\hbar^2}{m}\right)^2\left(\frac{d\sigma_i}{d\Omega}\right) = S_{ii}(K)\left|\left\langle V_i(\vec{K}) + \sum_{l=1}^{\lambda} e^{-i\vec{K}\cdot\vec{r}_l} \times V'_{1,l}(\vec{K})\right\rangle_A\right|^2 + \left\langle\left|\sum_{l=1}^{\lambda} e^{-i\vec{K}\cdot\vec{r}_l} V'_{1,l}(\vec{K})\right|^2\right\rangle_A - \left|\left\langle\sum_{l=1}^{\lambda} e^{-i\vec{K}\cdot\vec{r}_l} V'_{1,l}(\vec{K})\right\rangle_A\right|^2 \qquad (15)$$

and

$$\left(\frac{2\pi\hbar^2}{m}\right)^2\left(\frac{d\sigma_d}{d\Omega}\right) = \langle|\langle\vec{k}'|V_{1,l}(\vec{r})|\vec{k}\rangle|^2\rangle_A . \qquad (16)$$

The (solvated-ion)–(solvated-ion) structure factor $S_{ii}(K)$ is calculated from the theory of Ashcroft and Langreth.[14] Expressed in units of mole % metal, its behavior at 9.8, at 17.4, and at saturation, 20, for the $Li(NH_3)_x$ system is shown in Fig. 1.

IV. DISCUSSION

We are now in a position to see why the resistivity drops so rapidly with increasing metallic con-

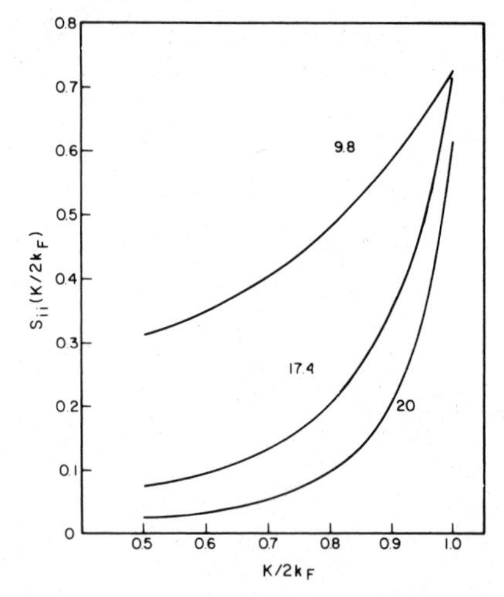

FIG. 1. Behavior of $S_{ii}(K)$ for the solvated ions in the $Li(NH_3)_x$ system (taken from Ref. 14). Expressed in units of mole % metal, the curves refer to concentrations of 9.8, 17.4, and 20 (the latter corresponding to saturation where the system is a single-component liquid). We draw attention to the substantial change which occurs in lowering the concentration from saturation to 17.4. The region of K that is plotted dominates the integrands in the formulas for ρ.

centration. As the concentration of the metallic component increases (n_e increases), the number of unbound NH_3 is speedily depleted ($n_d - \lambda n_e$ rapidly decreases). At the same time, the scattering from the solvated ions, measured by $d\sigma_i/d\Omega$, decreases rapidly and is, in fact, rather small in the first place. The reason for this can be seen by examining (15): The two terms involving only the dipoles are both numerically large, have approximately the same magnitude but clearly differ in sign. The subsequent cancellation is a reflection mainly of the characteristic size and geometry of solvated atoms, and although we demonstrate it here explicitly for the case of $Li(NH_3)_x$, it persists in other systems. The first term in (15) involves the angular average of the potential of the central ion plus the potentials of the bound dipoles. At large $\vec{K}$, where the most important scattering takes place, the ion potential is canceled by the dipole potentials. In other words, because of the internal structure of the solvated ion, the scattering (relative to that of a single ion) is greatly diminished. We also note that all of the bare potentials are divided (and hence reduced) by the Lindhard screening function[15] $\epsilon(K)$ which accounts for the response of the conduction electrons to the fields of dipoles and ions.

The combined rapid variations of τ_i^{-1} and τ_d^{-1} dominate the rather slow variation of n_e itself. In the case of lithium amine, the calculation results in the curve shown in Fig. 2. Following Sienko,[16] $\lambda = 4$ and the four NH_3 molecules are taken to have a tetrahedral arrangement about the central ion with dipole moments pointing radially outwards. Dipoles and ions are separated by a distance of 2.2 Å: The over-all size or hard-sphere diameters which are required in S_{ii} are taken from Ref. 5. For unbound NH_3 molecules, the dipole moment has a value of 1.47×10^{-18} esu.[17] Bound dipoles have a moment of value 1.86×10^{-18} esu.[18] The angular averages can all be calculated in closed form.[19]

In addition to the efficient cancellation of the second and third terms in (15), we draw attention to the behavior of the first term[20] as a function of n_e. As well as being important in the temperature dependence of ρ,[19] the presence of $S_{ii}(K)$ causes a sharp variation in its contribution. It can be seen in Fig. 1 that at concentrations near saturation almost the entire weight of the integral comes from around $2k_F$. With a decrease in metal concentration, contributions from other K values enter with

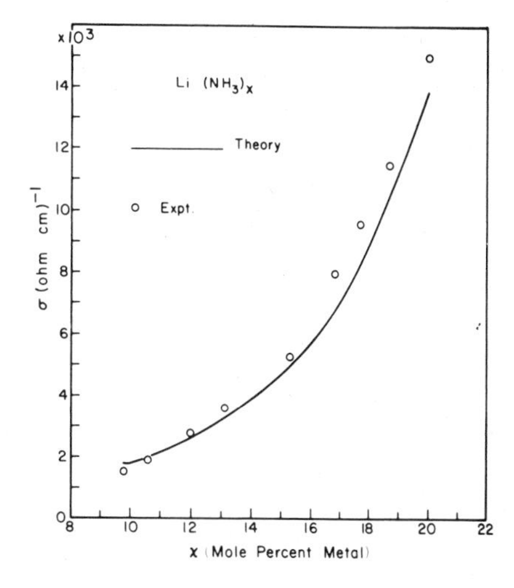

FIG. 2. Comparison of measured and calculated conductivities for the system $Li(NH_3)_x$ at 210° K. The experimental data is taken from J. A. Morgan, R. L. Schroeder, and J. C. Thompson, J. Chem. Phys. 43, 4494 (1965); and D. S. Kyser and J. C. Thompson, *ibid.*, 42, 3910 (1965).

rapidly increasing weight.

Finally we emphasize that although the trend of the theoretical curve given in Fig. 2 can now be understood, the rather exceptional numerical agreement must be regarded as partly fortuitous. In the first place, some of the parameters required in Eqs. (15) and (16) are not known with the best precision. In the second place, the calculation of ρ is performed only to lowest order. While it is true that because of the additional rotational degrees of freedom (and the associated phase averaging) some of the higher-order terms in ρ are bound to vanish, others – particularly those in the coherent scattering – will remain and the usual difficulties with this kind of calculation will prevail.

ACKNOWLEDGMENTS

We wish to thank Professor Wilkins, Professor Thompson, and Professor Sienko for helpful discussions. We are indebted to J. Perdew for assistance with the computations.

---

*Work supported by the Advanced Research Projects Agency through the Materials Science Center at Cornell University, MSC Report No. 1201.

[1]We will use the concentration notation adopted in the review article of M. H. Cohen and J. C. Thompson, Advan. Phys. 17, 857 (1968).

[2]See, for example, W. A. Harrison, Pseudopotentials in the Theory of Metals (W. A. Benjamin, Inc., New York, 1966).

[3]L. G. Christophorou and A. A. Christodoulides, J.

Phys. B2, 71 (1969).

[4]In some systems λ may be a function of concentration.

[5]R. L. Schroeder and J. C. Thompson, Phys. Rev. 179, 124 (1969); and (to be published).

[6]T. Faber and J. M. Ziman, Phil. Mag. 11, 153 (1965); N. W. Ashcroft and D. C. Langreth, Phys. Rev. 159, 500 (1967).

[7]The basic-wave number variable y is expressed in units of $2k_F$.

[8]As is common, we approximate the pseudowave function by a plane wave.

[9]We have already remarked that there will be finite size corrections to the dipole field. These are included in the definition of the molecular pseudopotential and they are important only for large $\vec{k}-\vec{k}'$.

[10]The orientation of $\vec{\mu}$ differs from molecule to molecule. As a notational simplification, we write $\vec{\mu}_1$ when, in fact, we mean $\vec{\mu}_1(\vec{r}_1)$.

[11]It is not *a priori* obvious that the orientational degrees of freedom of a dipole are independent of the spatial degrees. The fields tending to align the dipoles certainly relate to the configurations of the dipoles. While this argument cannot be disputed for a collection of pure dipoles, its efficiency is reduced substantially when electrons are present. Electron screening limits the dipole-dipole interactions to an exceedingly short range; the long-range spatial coupling of the dipoles is virtually eliminated and hence the angular orientations should become independent of the spatial configurations. It is precisely the long-range behavior which is of interest here.

[12]By independent we mean that the scattering from dipoles and solvated atoms suffers no interference. The solvated atoms, though independent of the free dipoles in the transport problem, give a contribution dependent on their own correlations.

[13]The momentum transfer $\vec{k}'-\vec{k}$ is related to θ by $|\vec{k}'-\vec{k}| = 2k_F \sin(\frac{1}{2}\theta)$.

[14]N. W. Ashcroft and D. C. Langreth, Phys. Rev. 156, 685 (1967).

[15]See, for example, D. Pines and P. Nozieres, The Theory of Quantum Liquids (W. A. Benjamin, Inc., New York, 1966), Vol. I. On physical grounds, we expect the density of the electron gas around the free dipoles to be slightly lower than average, while around the solvated ions it is slightly higher. By using Lindhard screening for both, there will be a slight overscreening of the dipoles and underscreening of the solvated ion. Correcting for this will lead to a somewhat sharper rise in the calculated σ.

[16]Metal Ammonia Solutions, edited by G. Lepoutre and M. J. Sienko (W. A. Benjamin, Inc., New York, 1964), p. 25.

[17]A. L. McClellan, Tables of Experimental Dipole Moments (W. H. Freeman and Co., San Francisco, 1963).

[18]Small changes either side of this value are found to cause minor increases in ρ. For example, increasing μ to 2.3×10^{-18} changes ρ by about 10%. The increase in μ is a reasonable estimate of the additional moment produced by the (slightly screened) field of the ion.

[19]G. Russakoff (to be published).

[20]The potential $V_i(K)$ for lithium is taken as an empty core pseudopotential [N. W. Ashcroft, J. Phys. C1, 232 (1968)]. Note, however, that the density of carriers is sufficiently low that core corrections to point ion scattering are small.

Errata: The $\vec{k}$ in equation 7 and in the first and second lines of equation 11, page397, should be $\vec{K}$. A $\langle$ bracket should precede the V_i in the second line of equation11, page 397. The first equation in section III, page 398, should read

$$\rho = (m/n_e e^2)\,(\tau_i^{-1} + \tau_d^{-1})$$

Reprinted from THE JOURNAL OF CHEMICAL PHYSICS, Vol. 53, No. 1, 109–116, 1 July 1970
Printed in U. S. A.

Hall Effect in Lithium–Ammonia Solutions*

60

R. D. NASBY†
The University of Texas at Austin, Austin, Texas 78712 and *Sandia Laboratory, Albuquerque, New Mexico 87115*
AND
J. C. THOMPSON
The University of Texas at Austin, Austin, Texas 78712

(Received 11 April 1969)

Measurements of the Hall coefficient and electrical conductivity are reported for solutions of lithium in liquid ammonia. The concentration range is from 1.2 to 14 mol % metal (MPM) and the temperature range is from near 200 to 240°K, except in the miscibility gap. Hall coefficients equal to the values expected from the metal-valence-electron density are found above 9 MPM. Below 9 MPM the measured Hall coefficient R_m increases to about twice the free-electron value R_0. The latter result does not agree with that previously reported by Kyser and Thompson. R_m is temperature independent at all concentrations. Data for the electrical conductivity σ show thermal activation of the mobility below 8 MPM. Changes in the concentration dependence of both R_m and σ are observed near 3, 5, and 8 MPM. The solutions behave as liquid metals above 9 MPM. As the concentration is decreased from 9 to 3 MPM, departure from metallic behavior occurs and various criteria for metallic conduction are no longer satisfied. We believe deep traps are present below 9 MPM. At concentrations near 3 MPH the influence of free electrons is no longer seen.

INTRODUCTION

Concentrated metal–ammonia solutions behave in most all respects as liquid metals. Above approximately 5 MPM[1] data on lithium solutions indicate metallic electrical conductivities[2,3] and mobilities,[4] metallic reflectivity,[5] approximately free-electron Lorenz numbers,[6] and free-electron Hall coefficients.[4] Sodium and potassium solutions exhibit similar behavior for those properties measured,[7] and in addition the thermoelectric power[8] and atomic susceptibility[9] are of the order observed in metals. As the concentration decreases through the range of the miscibility gap, Fig. 1, the behavior of the solutions departs from that of a metal, and the data indicate a transition to a nonmetallic state. Discussions of the transition region are found in several recent review articles.[10–13]

This paper presents a study of the transition in lithium–ammonia solutions through measurements of the Hall coefficient R_H and electrical conductivity σ as functions of solution concentration and temperature. For free electrons, the Hall coefficient R_0 equals $(nec)^{-1}$, where the electron density n may be calculated from mass density data,[14] upon assuming one free-electron per lithium atom. Kyser and Thompson[4] found agreement with free-electron theory for lithium–ammonia solutions with concentrations greater than 5 MPM at 212°K. Jaffe[15] found approximate agreement for the one liquid solution studied, 20 MPM at 193°K. Similar agreement was found for sodium solutions by Naiditch.[16]

Below 5 MPM the Hall data of Kyser and Thompson[4] increase to values 10^4 times R_0. Within free-electron theory this increase indicates the transition by a corresponding decrease in carrier density. This decrease was assumed to arise because of a rapidly increasing activition energy necessary for promotion of carriers into the conduction band.[4] However, little data on the temperature dependence were available to confirm the hypothesis.

We have measured R_H as a function of temperature and concentration in the transition region. *The present data disagree with the results of Kyser and Thompson*[4] at low concentrations. R_H was found to increase to a maximum of twice R_0 at 3 MPM and to be independent of temperature. The theoretical interpretation of the Hall coefficient data is unclear in this region.[17–19]

Previous conductivity data for concentrated solutions are reasonably plentiful and detailed.[2,3,20] Sparse data exist for solutions in the range 1–7 MPM. The data of this study confirm the metallic conductivities of concentrated solutions and yield detail on the concentration and temperature dependence in the transition region. Changes in the temperature dependence of σ with concentration correlate with criteria for a metal–nonmetal transition suggested by Mott and Allgaier.[21] Discussions of the metal–nonmental transition in these and other systems are given in Ref. 22.

EXPERIMENTAL

Solution Preparation

Solutions were prepared by distilling purified ammonia into a clean evacuated (10^{-3} torr) sample cell containing pure lithium metal. An all-glass system, with the exception of metal electrodes, was used for sample preparation and containment. Commercial ammonia, 99.95% pure, was further purified by preparing a sodium–ammonia solution and allowing it to stand for at least 12 h to remove traces of impurities, primarily water and oxygen. The amount of ammonia in a solution was determined by vaporizing ammonia from the sodium–ammonia solution into a bulb of known volume prior to condensing the gas into the sample cell. The number of moles was determined using

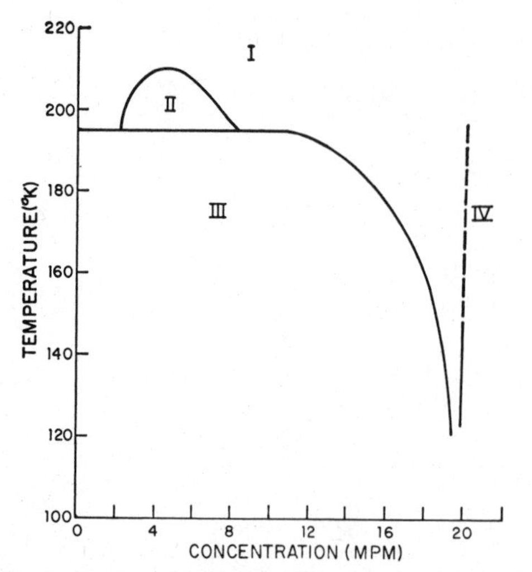

FIG. 1. The phase diagram for Li–NH_3 solutions. Region I is a homogenous solution; region II is the miscibility gap; homogenous solutions cannot be made with concentrations and temperatures in that range. In region III there is excess solid ammonia, and in region IV there is excess solid metal. See Ref. 12 for details and references.

the perfect gas law. Lithium metal, 99.8% pure, was prepared, cleaned, and weighed in a "dry box" containing helium gas and then introduced into a sample cell.

Solutions were stirred by magnetically raising and lowering a glass plunger containing an encapsulated bar magnet. The temperature was maintained by immersing the sample in a Dewar containing circulating cold methanol. A pressure of several pounds per square inch above the solution vapor pressure was established by introducing dried helium gas into the system. This overpressure was necessary to retard bubble formation in the sample cell. Sample preparation, containment, and the effects of various impurities on sample decomposition are discussed by several investigators.[2–9,23]

Sample Cell

The sample cell was of the conventional rectangular parallelepiped Hall bar geometry. Hall electrodes were positioned midway along the sample length and an additional electrode was positioned one-fourth the length from the end of the sample. This electrode and the adjacent Hall electrode provided detection probes for a conventional four-probe conductivity measurement.

The cells were constructed from two flat, rectangular Pyrex plates fused at the edges and with solution entrance and exit tubes sealed at the ends. The current electrodes were 25-mil tungsten wires sealed in the tubes. Hall and conductivity electrodes were sealed flush at the sample edges or in a small sidearm and were fashioned from 15-mil tungsten wire. The second construction yielded a larger electrode surface area. The interior cell dimensions were 4×1×0.065 cm.

Measurement Techniques

R_H was measured using a double-frequency technique similar to that first described by Russell and Wahlig[24] and to that used by Kyser.[4] The sample current I was operated at 570 Hz and the magnetic field B at 60 Hz. The linear Hall voltage, proportional to the product of I and B, occurred as equal components at 510 and 630 Hz.

The advantage of this technique is that most spurious and unwanted signals occur at frequencies different from those of the Hall signal or are reduced by the use of low audio frequencies. A primary advantage for these measurements is the ability to measure a small Hall voltage in the presence of a large and varying misalignment voltage.[4,16] A disadvantage arises from mixing of signals at the frequencies of I and B by non-Ohmic contacts or nonlinear components of the measuring system to yield false "Hall" signals. The double-frequency technique and the problems associated with Hall measurements on a liquid are described in several articles.[24–26]

The 60-Hz B field was generated by a magnet constructed from commercial wrapped transformer cores of 11-mil silicon steel. The air gap was 2.9 cm, and the pole face was a 10-cm square. A sinusoidal field of 4 kG (peak) was produced with a current of 10 A (rms). The total harmonic content was approximately 1% of the 60-Hz component.

The signal at the Hall electrodes was transformer coupled to a bucking supply and to an amplifier system tuned at 510 Hz. The bucking supply was used to reduce the misalignment voltage to allow detection of the Hall signal using the tuned amplifier system.[4] A commercial lock-in amplifier converted the Hall signal to a dc level which was recorded on a chart recorder.

The Hall voltage was determined using a "field reversal" technique (a 180° phase shift of B relative to the 60-Hz component used in generating the reference signal for the lock-in amplifier) to eliminate spurious 510-Hz signals which did not change phase with B. Hall voltages were measured as a function of B, and the data yielded a linear relationship from which R_H was determined. Hall voltages were also checked to confirm a linear dependence upon I.

The measurement system was prechecked by measuring R_H of a liquid-mercury sample. For this sample, Hall voltages were measured an order of magnitude lower than encountered in most lithium–ammonia solution measurements. R_H equaled -7.3×10^{-11} m^3/C. This value agrees well with the calculated free-electron value[26] of -7.7×10^{-11} m^3/C and the values of -7.6,

-7.5, -8.0, -7.3×10^{-11} m^3/C measured by Greenfield,[26] Cusack and Kendall,[27] Enderby,[28] and Tieche,[29] respectively.

The conductivity was measured using an ac four-probe technique with a current frequency of 570 Hz. The cell constant was determined by measuring the resistance of a liquid-mercury specimen and using mercury resistivity values from the *Handbook of Chemistry and Physics*, 50th edition. The Hall mobility $\mu_H = R_H\sigma$.

RESULTS

Measured values of R_H, denoted R_m, are shown in Fig. 2 as a function of concentration. Estimated maximum uncertainties are shown for each data point. The sign of R_m is taken as negative, as determined by Kyser,[4] since the sign is difficult to ascertain using a double-frequency technique. The dashed line represents the calculated R_0 values where the electron density is determined from the mass density data of Lo[14] assuming one free electron per lithium atom. The data are represented by straight lines to indicate probable trends with concentration even though this assumed power law dependence may not always be justified within experimental errors.

Above 6 MPM the values of R_m and R_0 are in agreement. However, trends indicated by the lines in Fig. 2 lead us to believe that *close* agreement is obtained only above $\sim$9 MPM. This agreement is extended to solution saturation by the data of Kyser.[4] Kyser's data are $\sim$10% higher than those of this study, but the discrepancies most likely are due to errors in measurement

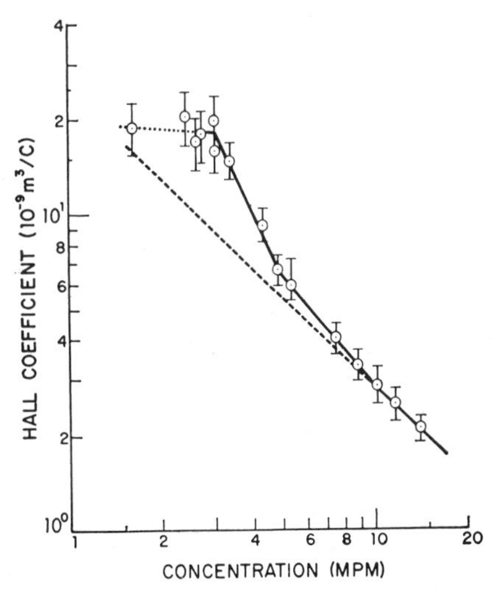

FIG. 2. The Hall coefficients as a function of concentration. The measured Hall coefficients R_m are shown as circles with error bars; free-electron Hall coefficients R_0 are shown as a dashed line. Changes in trends may be discerned near 9, 5, and 3 MPM. The Hall coefficient is negative (Ref. 4).

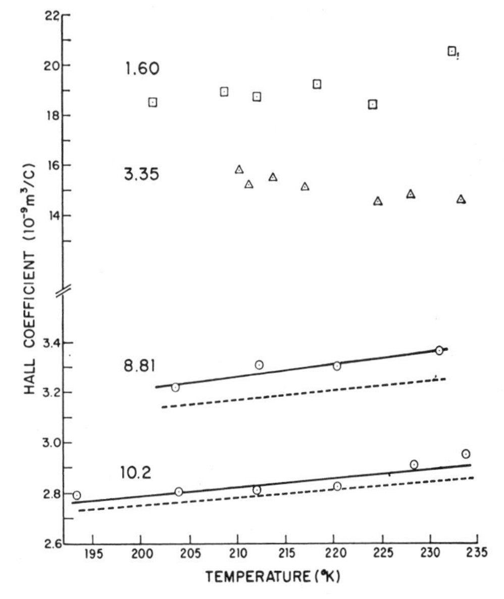

FIG. 3. The Hall coefficient as a function of temperature at four concentrations. Concentrations are listed at the left in mole percent metal. Free-electron Hall coefficients are shown as dashed lines for 8.81- and 10.2-MPM solutions. Note the break in the ordinate.

of the sample cell thickness t which were present in each case.

As the solution concentration is decreased, R_m increases to a maximum of twice R_0 at 3 MPM. These data are in strong disagreement with the results of Kyser and Thompson,[4] who reported R_H values 10^3 times R_0 at 2 MPM. During this study, R_m values $\sim$20 times R_0 were obtained at the lower concentrations. However, these latter data were found to be caused by spurious signals generated by non-Ohmic solution-to-electrode contacts,[30] as discussed below.

The increased scatter in the data at lower concentrations is due to a decreased signal-to-noise ratio resulting from decreased carrier mobilities and to the possible presence of spurious signals arising from non-Ohmic contacts. Problems with non-Ohmic contacts were particularly severe below $\sim$3 MPM, and for this reason the single data point at 1.6 MPM with no collaborating data must be viewed with caution (thus, the dotted line in Fig. 2).

Hall data as a function of temperature are shown in Fig. 3 for four representative solution concentrations. The precision of R_m with changes in temperature are estimated as 9%, 7%, and 5% for the 1.6-, 10.2-, and 8.81- and 3.35-MPM solutions, respectively. Changes or trends of R_m with temperature are less than the experimental error for all concentrations studied. For the 8.81- and 10.2-MPM solutions, trends in R_m agree with

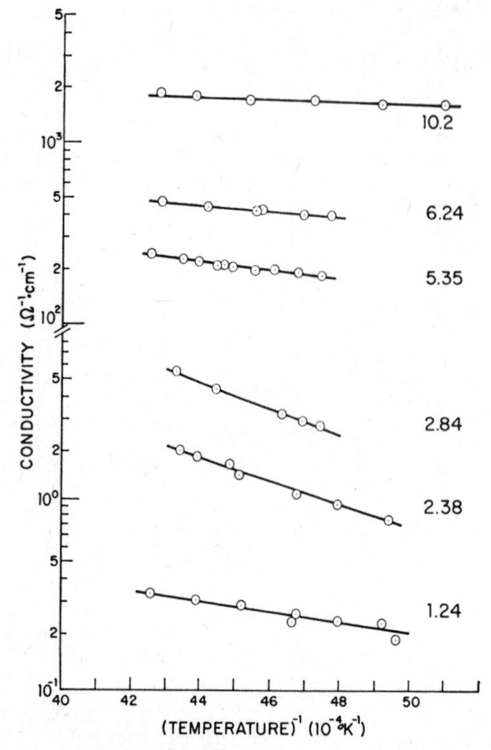

FIG. 4. The electrical conductivity as a function of reciprocal temperature at several concentrations. Above 8 MPM the data may be fitted by a linear power law *or* an exponential. Below 8 MPM only the exponential will suffice. Concentrations are given at the right in mole percent metal. Note the break in the ordinate.

free-electron theory wherein the change in R_0 with temperature arises from the temperature dependence of n as calculated from mass density data. Kyser[4] reported similar trends for 17.5- and 16.39-MPM lithium solutions. For concentrations less than 8.81 MPM, the variations of R_m with temperature are erratic but less than the experimental error. These data will hereinafter be described as temperature independent.

In Fig. 4, representative conductivity data for several concentrations are plotted as a function of temperature. Interpolated values of σ at 234 and 211°K for each concentration studied are plotted versus concentration in Fig. 5. The estimated maximum uncertainty in σ is 6%, and the estimated precision is 3%. The σ data of Morgan *et al.*[2] above 7 MPM are in agreement with those of this study. Kraus[20] measured σ at 240°K for three concentrations in the more dilute region. These data are adjusted to 234 and 211°K by using an extrapolation of the temperature dependence found in this study.

The temperature dependence of σ is depicted in Figs. 4 and 6. Over the concentration and temperature range of this study, the logarithm of σ is a linear function of the reciprocal of absolute temperature. Thus, it is assumed that the data can be represented by an activation energy ϵ defined by

$$\sigma = \sigma_0 \exp(-\epsilon/kT), \tag{1}$$

where T is the absolute temperature and σ_0 is a constant. The ϵ values of Fig. 6 were obtained by a least-squares fit to data such as those shown in Fig. 4.

It is necessary to point out that this simple model may not be justified over the entire range of this study. Above ~8 MPM the conductivity data may be represented by *either* a linear relationship in temperature or by the activation energy model. Near solution saturation the temperature dependence is definitely not exponential and even changes sign.[2,3] From Fig. 6 the conductivity temperature dependence changes in the region of ~8 MPM. The activation energy model appears appropriate for lower concentrations, while a dependence closer to linear in temperature is appropriate at higher concentrations. The activation energy plot indicates changes in the temperature

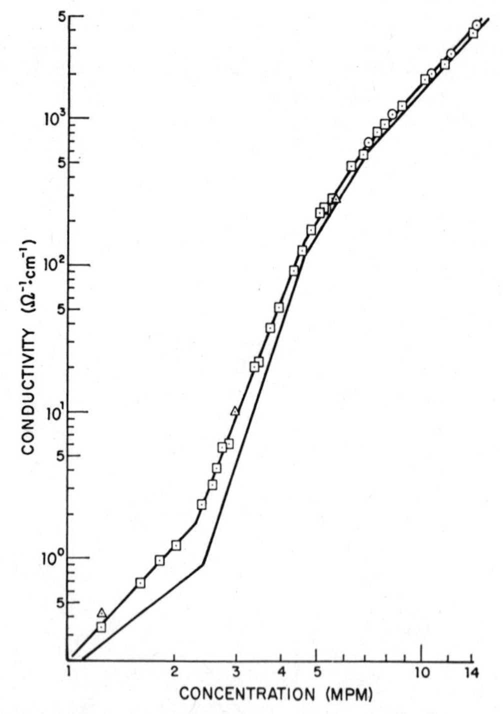

FIG. 5. The conductivity as a function of concentration at 211 and 234°K, data points are shown only at 234°K. The present data are represented by ⊡, data of Ref. 20 by △, and data of Ref. 2 by ⊙.

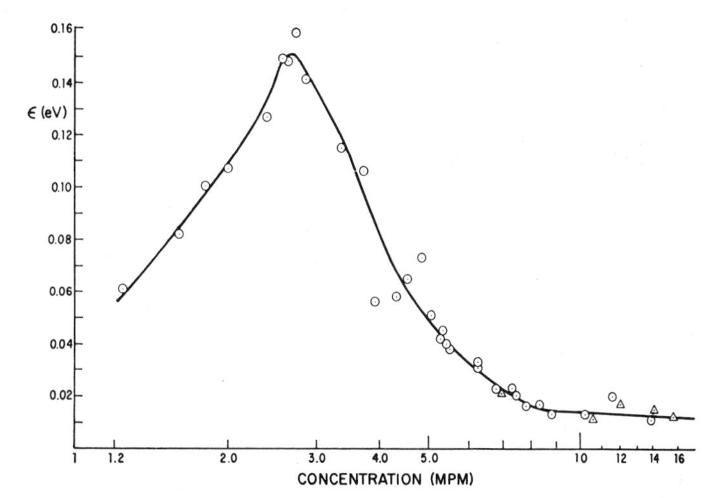

FIG. 6. The activation energy ϵ for conductivity or mobility. Above 8 MPM the data may be equally well represented by a power law (see Fig. 4), but ϵ is used at all concentrations to facilitate comparison. Changes in trends may be discerned near 8 and 2.7 MPM. Error bars are typically 0.02 eV in length. Present data are shown as ⊙; data of Ref. 2 by △.

dependence at approximately the same concentrations as those trends observed in R_m and σ.

The Hall mobility μ_H at 223°K is plotted versus concentration in Fig. 7. Since R_m was found essentially independent of temperature, μ_H exhibits the same temperature dependence as σ.

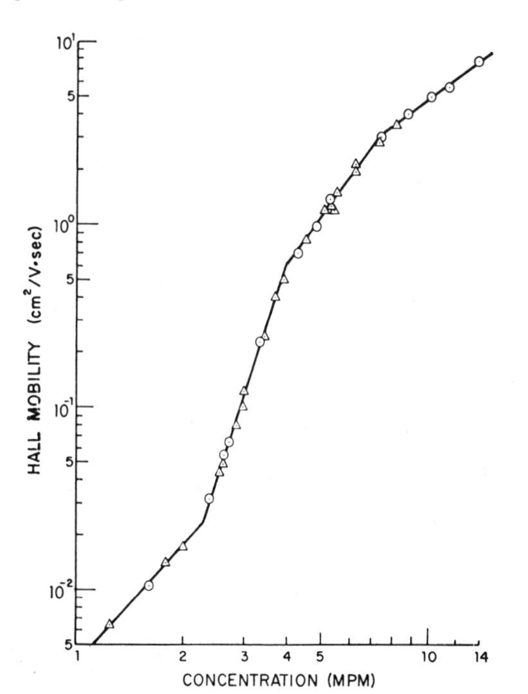

FIG. 7. The Hall mobility $\mu_H(=\sigma R_m)$ as a function of concentration at 223°K. The circles represent points where both σ and R_m were available; the triangles are based on interpolations of the R_m data. Note that μ_H falls below 1.0 cm²/V·sec near 5 MPM.

For concentrations greater than ~4 MPM, the mobility values from this study agree with those of Kyser and Thompson.[4] As the concentration is decreased, the mobility values of this investigation continue to decrease and a minimum in μ_H as reported by Kyser and Thompson[4] does not appear. An order-of-magnitude confirmation of the low mobility is provided by the electron diffusion coefficient data for electrons in potassium–ammonia solutions obtained by Gordon and Sundheim[31] using the method of chronopotentiometric waves. At the highest concentration studied, 0.35 MPM, the diffusion coefficient was 3×10^{-3} cm²/sec at 223°K. Through the use of the Nernst–Einstein equation a mobility of 2×10^{-3} cm²/V·sec is obtained. The extrapolated value of μ_H at 0.35 MPM is 1×10^{-3} cm²/V·sec. The close agreement is probably fortuitous, considering that this is a comparison between μ_H for lithium solutions and essentially a conductivity mobility in potassium solutions. However, the agreement offers a partial and an independent confirmation of the low μ_H values found in this study.

It is also noted that mobilities calculated from the diffusion coefficient decrease with increasing solution concentration as 0.35 MPM is approached. Conversely, μ_H decreases with decreasing concentration. A smooth connecting curve yields a minimum in the mobility versus concentration near the minimum in the equivalent conductance as reported by Kraus[20] for lithium solutions, 0.14 MPM.

False "Hall" Voltages

During this study large false "Hall" voltages were observed whenever one or both Hall electrodes exhibited non-Ohmic, rectifying contact with the solution. Near 3 MPM transverse potentials 10 times true Hall voltages were observed. Non-Ohmic contacts were ***not*** observed for the data presented above. The generation of the false signals can probably be understood as arising from mixing of 570- and 60-Hz signals present at the non-Ohmic contacts. However, as discussed below, neither the exact origin of rectifying contacts nor the mode of mixing was understood.

Rectifying contacts appeared erratically upon solution preparation and prolonged storage of solutions. Care in cleaning of electrodes greatly increased the possibility of Ohmic contacts, but did not assure Ohmic behavior. Some electrode contamination could not be avoided within the experimental techniques; in particular, a surface oxide would form in the 10^{-3} torr vacuum[32] and solution decomposition most likely occurred at the electrodes.[23,33,34] Non-Ohmic behavior was also found dependent on electrode surface area; larger electrodes exhibited less tendency to form rectifying contacts.[30] Non-Ohmic contact was noted only for solutions with concentrations less than ~9 MPM with particularly severe rectification occurring near and below ~3 MPM. The appearance of non-Ohmic behavior was found dependent on electrode contamination, electrode surface area, and solution concentration.[30]

If a constant dc current was passed through an electrode pair with one rectifying electrode, the reverse voltage increased with time in a manner similar to that of cathodic chronopotentiometric waveforms for ionic solutions.[31,35] Stirring of the solution decreased or eliminated the reverse voltage increase. These and further observations indicated that depletion of charged species near the electrode or electrode polarization coupled with reactions of electrode contaminates formed the basis for non-Ohmic behavior. The dependence of rectification on electrode contamination and surface area supports this supposition.[35] Furthermore, it may not then be coincidental that changes in carrier species, as previously indicated, coincide with the changes in rectification behavior with concentration.

Whenever rectification was observed, large and inconsistent "Hall" voltages were measured, and any increase in rectification produced increased voltages. In the absence of rectification, consistent data were obtained (Sec. III). The false "Hall" voltages were found to be linear in B and I and to change phase with B. In attempts to eliminate the spurious signals, all known magnetically induced 60-Hz voltages were altered with no change in the spurious signals. Further, all known 570-Hz signals were altered or reduced by bucking voltages with no change in the spurious signals. Direct discrimination between true and false Hall signals or elimination of spurious signals could not be accomplished. Further, the sources for the mixing hypothesis are in question.

One experiment yields a plausible, though admittedly incomplete, explanation of the mixing. If the experiment was readied for normal Hall measurements except $B=0$ and the solution was vibrated at 60 Hz with the mechanical stirrer, large false "Hall" signals were detected. In normal operation, magnetic forces and magnetohydrodynamic flow would produce solution movement; the 570-Hz source remains in question.

In summary, rectification at electrodes was observed and deemed responsible for spurious transverse voltages, false "Hall" voltages; however, a complete explanation of the behavior could not be made. Kyser[4] also reported rectification at electrodes and noted erratic changes in measured Hall voltages. Further, he reported that sample currents had to be reduced on several occasions to avoid signal distortion. It is possible that rectification effects led to the large Hall voltages he reported. Many of the false "Hall" signals measured in this study would lead to R_H values of the order reported by Kyser[4] in the concentration range 4–2 MPM; however, R_H values as high as 10^4 times R_0 near 1 MPM were not measured in this study. Mixing of signals to produce spurious voltages has been observed by other investigators on these and other materials.[16,36,37]

DISCUSSION

In discussing the results, we will first examine the metallic and the nonmetallic concentration regions as distinguished by the data of this study. Data from other studies will be discussed, and a qualitative model for intermediate concentrations will be proposed.

We begin with nearly free-electron theory as it provides us with a vantage point and helps locate those regions of concentration requiring a closer view. The equality of R_m and R_0 above ~9 MPM designates that concentration range as metallic in nature. This notion is supported by the magnitude of the conductivity, $\sigma > 10^3\ \Omega^{-1}\cdot\text{cm}^{-1}$, and by the smallness of the temperature dependence of the conductivity. Further, the latter is more appropriately represented by a simple power law in temperature above 8 MPM rather than by the activation energy model. The Hall mobility, like the conductivity, is large ($\mu_H > 4\ \text{cm}^2/\text{V}\cdot\text{sec}$) above 9 MPM. Other properties listed in the Introduction also indicate the solutions are metallic. That the solutions cannot be treated altogether as simple liquid metals is evidenced by the strong dependence of σ on concentration and by $d\sigma/dT > 0$.[2,3,38]

R_m departs from R_0 as the solutions are diluted below 9 MPM. If we adopt the naive view that R_m yields the number of free electrons, then we infer that up to half the metal valence electrons must be trapped, i.e., localized. The absence of a temperature dependence for R_m must be interpreted to mean that the traps are deep, so that the activation energy is much greater than kT. Further, we assume charge transport due to hopping of the trapped electron is insignificant when compared to free electrons. Optical data[5,10,39,40] indicate the appearance of an absorbing species as the concentration is decreased from 8 to 5 MPM with properties similar to those of the "solvated electron" found in dilute solutions. Beckman and Pitzer,[39] in particular, conclude that the

TABLE I. Simple metal parameters.[a]

Concentration (MPM)	n_H (10^{21} cm^{-3})	μ_H (cm^2/V·sec)	k_F (10^8 cm^{-1})	λ (10^{-8} cm)	r_0 (10^{-8} cm)
20	4.0	23	0.49	74	6.3
18	3.6	17	0.48	54	6.5
16	3.3	12	0.45	37	6.7
14	2.9	7.4	0.44	22	7.0
12	2.6	6.0	0.42	17	7.3
10	2.2	4.8	0.40	12	7.7
8	1.7	3.5	0.38	8.6	8.3
6	1.2	1.7	0.33	3.8	9.0
4	0.59	0.63	0.26	1.1	10.0
3	0.34	0.12	0.22	0.12	11
2	0.32	0.02	0.21	0.02	12

[a] Parameters were calculated from the present R_m and σ data below 14 MPM, using free-electron theory with $m^*=m_0$. All parameters above 14 MPM are based on the σ data of Ref. 2 and the mass density data of Ref. 14. The value of r_0 equals $N^{-1/3}$, where N is the density of lithium atoms.

absorbing species is present only below 5 MPM, while more recent data[40] indicate that it may persist to near 8 MPM. The concept of an increasing number of deep traps as the concentration decreases below 8 MPM will also promote the more rapid decrease in σ with concentration (Fig. 5) due to the simultaneous depletion of mobile carriers and increase of scattering centers. The decrease in free carrier density is also consistent with thermopower[8] and susceptibility[9] of sodium solutions. No ready explanation of the increased $d\sigma/dT$ (Fig. 6) nor of the exponential dependence of σ on T is provided thereby.

Qualitatively, at least, the properties of the solutions are consistent with there being a mixture of deep traps and free electrons between 3 and 9 MPM. This naive approach fails near 3 MPM. If one attempts to use free-electron densities obtained from R_m for calculation of the thermoelectric power[8] or susceptibility,[9] agreement with the data no longer is found. Furthermore, the optical data no longer show free-electron (Drude) character.[5] The above data indicate a drastic reduction in free carrier density below 3 MPM while the Hall data do not, at least in the naive view taken thus far. The maximum in the temperature dependence of σ (Fig. 6) and the maximum in R_m/R_0 which may be obtained from Fig. 2 provide other clues that competing conduction processes exist below 3 MPM. We next examine the departure from NFE theory below 9 MPM in more detail.

The conductivity (Fig. 5) and Hall mobility (Fig. 7) decrease rapidly throughout the 9–3-MPM range. Free-electron theory provides the basis for the numbers calculated from these data and displayed in Table I. We note that the electron mean free path λ is less than the interparticle distance r_0 for concentrations below 8 MPM, that $k_F\lambda<1.0$ below 6 MPM, where k_F is the Fermi wavenumber, and that μ_H becomes less than 1 cm^2/V·sec below 5 MPM. These inequalities have been suggested[21,41–43] as criteria for failure of the band model. Further, Mott and Allgaier[21] estimate that conductivities of order 300–1000 $\Omega^{-1}\cdot$cm^{-1} and less imply carrier localization and conduction by hopping. These arguments compel us to question the use of band theory throughout the 9–3-MPM range and to consider this as an intermediate region exhibiting aspects of both band and hopping conduction. The appearance of an exponential temperature dependence for σ below 8 MPM is consistent with a hopping process as is also the magnitude of σ.[18,21] On the other hand, the measured Hall coefficient below 9 MPM does not fit any of the currently available theories of hopping[17–19] as to either magnitude or temperature dependence. We must therefore question the agreement obtained with the naive interpretation of R_m above. Some non-NFE conduction process is necessary below 3 MPM, in any event.

In summary, the solutions behave as liquid metals above 9 MPM. As the concentration is decreased from 9 to 3 MPM, departure from metallic behavior occurs and various criteria for band conduction are no longer satisfied. The interpretation of the data is somewhat ambiguous; aspects of both free-electron band conduction and conduction by processes similar to small polaron hopping[18] are present. We believe deep traps are present below 9 MPM, although a complete description of all properties cannot be given. At concentrations near 3 MPM the influence of free electrons is no longer seen. The presence of competing conduction mechanisms is exemplified by the peak in ϵ (Fig. 6) at 2.7 MPM.

It is also noted that the intermediate range (3–9 MPM) discussed above spans most of the miscibility gap as seen in the phase diagram of Fig. 1. The onset of the transition region coincides approximately with the high concentration edge of the miscibility gap near the freezing line, rather than with the consolute point as has been previously asserted.[4] At the consolute point, 4–4.5 MPM, a somewhat more rapid change in R_m and σ with concentration is observed (Figs. 2 and 5).

Several models have been suggested for the concentration range in question. Arnold and Patterson[44] considered carriers thermally activated into a conduction band. Their model cannot be reconciled with the temperature independence of R_m and must be rejected. Duval *et al.*[45] suggest on the basis of Knight shift data that density-of-states effects are involved. Our results do not permit us to comment on that suggestion. Cohen and Thompson[11] suggest that the solutions are inhomogenous mixtures[46] of highly conductive clusters of ions and solvated electrons imbedded in a poorly conducting solution. Their assumption that solvated electrons exist at concentrations above 4 MPM is consistent with our assumption of deep traps in the 3–9-MPM range. That assumption is supported by Duval *et al.*,[45] who have pointed out that, in a cluster of solvated ions, the oriented solvent molecules surrounding a cluster of neighboring ions form an array much like that presumed to exist around an isolated solvated electron.[11,12] Cohen and Thompson place less emphasis on the role of hopping than we do.

Much attention[4,10,11,22,37] has been given to the origin of the metal–nonmetal transition observed in these solutions and we do not explore that point further.[47] It is interesting, however, to note the close similarity of these solutions to other systems exhibiting a metal–nonmetal transition, for example, doped semiconductors.[22,48]

* This work was supported in part by the U.S. National Science Foundation and the Robert A. Welch Foundation.

† Texas Instruments Fellow in Physics 1965–1966. Present address: Sandia Corp., Albuquerque, N.M.

[1] For this paper, the concentration will be expressed either in mole fractions x, where x = (moles metal)/(moles metal + moles NH_3), or as mole percent metal (MPM = 100x).

[2] J. A. Morgan, R. L. Schroeder, and J. C. Thompson, J. Chem. Phys. **43**, 4494 (1965).

[3] R. L. Schroeder, J. C. Thompson, and P. L. Oertel, Phys. Rev. **178**, 298 (1969).

[4] D. S. Kyser and J. C. Thompson, J. Chem. Phys. **42**, 3910 (1965).

[5] W. T. Cronenwett and J. C. Thompson, Advan. Phys. **16**, 439 (1967).

[6] P. G. Varlashkin and J. C. Thompson, J. Chem. Phys. **38**, 1974 (1963).

[7] J. C. Thompson, Advan. Chem. Ser. **50**, 96 (1965).

[8] J. F. Dewald and G. Lepoutre, J. Am. Chem. Soc. **76**, 3369 (1954); **78**, 2956 (1956).

[9] R. G. Suchannek, S. Naiditch, and O. J. Klejnot, J. Appl. Phys. **38**, 690 (1967).

[10] J. C. Thompson, Rev. Mod. Phys. **40**, 704 (1968).

[11] M. H. Cohen and J. C. Thompson, Advan. Phys. **17**, 857 (1968).

[12] *Metal-Ammonia Solutions*, edited by G. Lepoutre and M. J. Sienko (Benjamin, New York, 1964).

[13] T. P. Das, Advan. Chem. Phys. **4**, 303 (1962).

[14] R. E. Lo, Z. Anorg. Allgem. Chem. **344**, 230 (1966).

[15] H. Jaffe, Z. Physik **93**, 741 (1935).

[16] S. Naiditch, Final Tech. Rept., Contract NONR 3437(00) 1965 (unpublished) and (private communication).

[17] T. Holstein, Ann. Phys. (N.Y.) **8**, 325, 343 (1959); T. Holstein and L. Friedman, *ibid.* **21**, 494 (1963); Phys. Rev. **165**, 1019 (1968).

[18] J. Appel, Solid State Phys. **21**, 193 (1968).

[19] D. C. Langreth, Phys. Rev. **148**, 707 (1966).

[20] C. A. Kraus, J. Am. Chem. Soc. **43**, 749 (1921); C. A. Kraus and W. W. Lucasse, *ibid.* **43**, 2529 (1921); **44**, 1941 (1922).

[21] N. F. Mott and R. S. Allgaier, Phys. Status Solidi **21**, 343 (1967).

[22] See the Proceedings of the International Conference on Metal–Nonmetal Transitions, Rev. Mod. Phys. **40** (1968).

[23] G. W. Watt, G. D. Barnett, and L. Vask, Ind. Eng. Chem. **46**, 1023 (1954).

[24] B. R. Russell and C. Wahlig, Rev. Sci. Instr. **21**, 1028 (1950).

[25] N. E. Cusack, J. E. Enderby, P. W. Kendall, and Y. Tieche, J. Sci. Instr. **42**, 256 (1965).

[26] A. J. Greenfield, Phys. Rev. **135**, A1589 (1964).

[27] N. E. Cusack and P. W. Kendall, Phil. Mag. **6**, 419 (1961).

[28] J. E. Enderby, Proc. Phys. Soc. (London) **31**, 772 (1963).

[29] Y. Tieche, Helv. Phys. Acta **33**, 963 (1960).

[30] R. D. Nasby and J. C. Thompson, J. Chem. Phys. **49**, 969 (1968).

[31] R. P. Gordon and B. R. Sundheim, J. Phys. Chem. **68**, 3347 (1964).

[32] E. A. Gulbransen and W. S. Ysong, Trans. AIME **175**, 611 (1948).

[33] J. E. Wreede and S. Naiditch (private communications).

[34] B. Breyer and H. H. Bauer, *Chemical Analysis* (Wiley, New York, 1963), Vol. 13, p. 253.

[35] *Electroanalytical Chemistry*, edited by A. J. Bard (Marcel Dekker, New York, 1966), Vol. 1.

[36] R. G. Suchannek, Rev. Sci. Instr. **37**, 589 (1966).

[37] D. Parker and J. Yahia, Phys. Rev. **169**, 605 (1968).

[38] R. L. Schroeder and J. C. Thompson (unpublished); N. W. Ashcroft and G. Russakoff, Phys. Rev. A **1**, 39 (1970).

[39] T. A. Beckman and K. S. Pitzer, J. Phys. Chem. **65**, 1527 (1961).

[40] R. B. Somoano and J. C. Thompson, Phys. Rev. A **1**, 376 (1970).

[41] J. Kaplan and C. Kittel, J. Chem. Phys. **21**, 1429 (1953).

[42] A. F. Ioffe and A. R. Regel, Progr. Semicond. **4**, 237 (1960).

[43] N. F. Mott, Phil. Mag. **6**, 287 (1961); Advan. Phys. **16**, 49 (1967); Phil. Mag. **17**, 1259 (1968); N. F. Mott and W. D. Twose, Advan. Phys. **10**, 107 (1961).

[44] E. Arnold and A. Patterson, J. Chem. Phys. **41**, 3089 (1964); see also J. Poplelawski, Chem. Phys. Letters **2**, 71 (1968).

[45] E. Duval, P. Rigny, and G. Lepoutre, Chem. Phys. Letters **2**, 237 (1968).

[46] J. Volger, Phys. Rev. **79**, 1023 (1950); C. Herring, J. Appl. Phys. **31**, 1939 (1960).

[47] Suffice it to note that Mott's simple criterion for the Mott transition (Ref. 43) is satisfied at 4 MPM in these solutions.

[48] W. D. Straub, H. Roth, W. Bernard, S. Goldstein, and J. E. Mulhern, Phys. Rev. Letters **21**, 752 (1968).

A Theoretical Study of Solvated Electrons

Copeland, Kestner and Jortner have theoretically studied a microscopic model for electrons in polar solvents, with particular emphasis on its application to dilute metal–ammonia solutions. They quantitatively discuss the physical factors which determine the stability of a localized ground state, the physical properties of this state, and the excited electronic states. Unfortunately, space limitations have prevented us from reproducing their entire paper, and we have just reprinted the introduction and the discussion.

Reprinted from THE JOURNAL OF CHEMICAL PHYSICS, Vol. 53, No. 3, 1189–1216, 1 August 1970
Printed in U. S. A.

61

Excess Electrons in Polar Solvents*

DAVID A. COPELAND AND NEIL R. KESTNER†
Department of Chemistry, Louisiana State University, Baton Rouge, Louisiana 70803
AND
JOSHUA JORTNER
Institute of Chemistry, University of Tel-Aviv, Tel-Aviv, Israel and *Chemistry Department, James Franck Institute, University of Chicago, Chicago, Illinois 60637*

(Received 3 February 1970)

In this paper we consider a structural model for localized excess electron states in polar solvents with particular reference to dilute metal ammonia solutions and to the hydrated electron. The over-all energetic stability of these species was assessed by considering simultaneously the electronic energy and the medium rearrangement energy. The present model consists of a finite number of loosely packed molecules on the surface of the cavity which are subjected to thermal fluctuations and a polarizable continuum beyond. The electronic energy was computed utilizing an electrostatic microscopic short-range attraction potential, a Landau-type potential for long-range interactions, and a Wigner-Seitz potential for short-range repulsive interactions. The medium rearrangement energy includes the surface tension work, the dipole–dipole repulsion in the first solvation layer, and most importantly the short-range repulsive interactions between the hydrogen atoms of the molecules oriented by the enclosed charge. The gross features of localized electron states in different solvents can be rationalized in terms of different contributions to the medium rearrangement terms. The energetic stability of the localized state of excess electrons in polar solvents was established and the cavity size in the ground state of the solvated electron could be uniquely determined. Experimental energetic and structural data such as volume expansion, coordination numbers, heats of solution, and spectroscopic properties are in qualitative agreement with the predictions of the present model. Optical line shape data calculated from the theoretical model do not agree with experiment; this discrepancy suggests that more data are required in regard to the excited states of the solvated electron.

I. INTRODUCTION

There is currently a wealth of information available on the properties of chemically stable excess electron states in polar solvents (i.e., metal–ammonia solutions, solutions of metals in ammonia and ether)[1–7] and of metastable excess electron states (i.e., electrons in water, alcohols, acetonitrile).[8–14] The physical and chemical properties of dilute metal ammonia solutions and of the hydrated electron, which are assembled in Table I, provide a firm basis for the picture of the solvated electron in polar solvents. These features are independent of the nature of the positive ion, so that in the limit of low concentrations the electron–cation interaction is negligible. The excess electron is expected to be localized by a solvation mechanism which is reminiscent of that of an ordinary ion in an electrolyte solution. The electric field produced by the localized charge distribution of the excess electron polarizes the dipolar solvent molecule, whereupon the nearest-neighbor shell is presumably strongly oriented while the medium outside is subjected to a long-range polarization potential. Adopting this general picture, several theoretical models were proposed to account for the thermochemical and optical properties of the solvated electron. Following early primitive cavity models, the theoretical studies of Deigen,[15] Davydov,[16] and Jortner[17] applied Landau's[18] polaron self-trapping model for the excess electron. Some attempts were also made to account in greater detail for the nature of short-range attractive interaction, which is averaged out in a rather crude manner in the continuum. In an early treatment of this problem the potential was constructed as a superposition of the molecular field of a fixed number ($N=4$) of solvent molecules in the first coordination layer and a continuum contribution beyond it.[19] More recently, Land and O'Reilly[20,21] have constructed a potential due to a finite variable number of polar solvent molecules in the first layer.

Most of these studies were judged in terms of their success in producing "agreement" between theory and experiment. We feel that such criteria are not quite justified in view of the limitation of the crude one-electron scheme employed in all current treatments of the complex quantum mechanical problem of single-electron traps in solids and in polar liquids. At the present stage due to inherent limitations imposed on such one-electron calculations (which we do not hope to improve upon) and the complexity of the physical system, it may be more profitable to focus attention on some general questions pertinent to the physical factors which determine electron localization and the gross features of such a localized state.

The present study of excess electron states in polar solvents was motivated by the impressive progress which was recently achieved in the understanding of excess electron states in nonpolar fluids[22–42] such as He^3, He^4, Ar, Kr, Xe, H_2, and D_2. These studies focused attention on the general problem of the stability of the excess electron state in a liquid, the nature of electron solvent interactions, and the con-

TABLE I. Properties of solvated electrons in water and in liquid ammonia.

Property	System	
	e_{am} (240°K)	e_{aq} (300°K)
$h\nu_{max}$	0.80 eV[a]	1.72 eV[b]
ϵ_{max} Extinction coefficient	49 000$M^{-1}\cdot cm^{-1}$ [a]	15 800$M^{-1}\cdot cm^{-1}$
f Oscillator strength	0.77[a]	0.65[b]
$W_{1/2}$ Half-linewidth	0.46 eV[a]	0.92 eV[b]
$dh\nu_{max}/dT$	$-(1.5\pm0.2)\times10^{-3}$ eV/deg[a]	-2.9×10^{-3} eV/deg[b]
ΔH	1.7 ± 0.7 eV[c]	1.7 eV[d]
P Photoelectric threshold	1.6 eV[e]	Unknown
Electron mobility	1.08×10^{-2} cm²/V·sec[f]	2.5×10^{-3} cm /V·sec[g]

[a] R. K. Quinn and J. J. Lagowski, J. Phys. Chem. **73**, 2326 (1969).
[b] E. J. Hart and W. C. Gottschall, J. Am. Chem. Soc. **71**, 2102 (1967).
[c] J. Jortner, J. Chem. Phys. **30**, 839 (1959). The uncertainty in this value is due to the (unknown) absolute heat of solvation of the proton in liquid ammonia.
[d] J. H. Baxendale, Radiation Res. Suppl. **4**, 139 (1964). J. Jortner and R. M. Noyes, J. Phys. Chem. **70**, 770 (1966).
[e] G. V. Teal, Phys. Rev. **71**, 138 (1948).
[f] C. A. Kraus, J. Am. Chem. Soc. **43**, 749 (1921).
[g] K. H. Schmidt and W. L. Buck, Science **151**, 70 (1966).

figurational charges in the fluid which may be induced by the presence of an excess electron. Although the properties of excess electrons in nonpolar liquids are somewhat easier to handle by (approximate) theoretical methods, the basic problems involved are identical with those encountered in the study of excess electron states in polar solvents.

The question that should be asked in connection with excess electron states in liquids can be summarized as follows:

(a) *The nature of quasifree and localized excess electron states.* When an excess electron is introduced into a nonpolar or polar liquid, it is not immediately apparent what is the energetically stable state of the electron. The total ground-state energy E_t of the system can be always written in terms of two contributions: the electronic energy E_e and the medium rearrangement energy E_M, so that

$$E_t=E_e+E_M. \qquad (1)$$

The second term in Eq. (1) involves the structural modifications induced in the medium due to the presence of the excess electron, so that in general $E_M\geq0$. The electronic energy has to be computed in the spirit of the Born–Oppenheimer approximation for each nuclear configuration of the medium. In this context two limiting extreme cases should be distinguished:

(1) The quasifree electron state whereupon the excess electron can be described by a plane wave (e.g., a wave packet) which is scattered by the atoms or molecules constituting the dense fluid. Under these circumstances it is expected that the liquid structure is not perturbed by the presence of the excess electron, so that $E_M=0$. The electronic energy of the quasifree electron state, which we shall denote by V_0, is determined by a delicate balance between short-range repulsions and long-range polarization interactions, so that

$$E_t(\text{quasifree})=V_0. \qquad (2)$$

This energy obviously corresponds to the bottom of the conduction band in the liquid relative to the vacuum level. Theoretical studies based on the Wigner–Seitz model for quasifree excess electron states in nonpolar fluids have established that any dense fluid consisting of molecules which are light and saturated, and thus characterized by a low polarizability, is expected on the basis of pseudopotential theory to exhibit a positive ground-state energy for the quasifree excess electron state.[43] This situation prevails for liquid He, H_2, D_2, and Ne. On the other hand, for heavier rare gases such as Ar, Kr, and Xe the contribution of the attractive long-range polarization overwhelms the contribution of the short-range repulsive pseudopotential, so the $V_0<0$ in these systems. Nothing is currently known concerning the V_0 values for polar liquids.

(2) The localized excess electron state where the wavefunction for the excess electron tends to zero at large distances from the localization center. In this case the liquid structure has to be modified to form the localization center. Such a liquid rearrangement process requires the investment of energy. It will be convenient at this stage to specify the configuration of the solvent by a "configurational coordinate" R; the simplest choice is, of course, the mean cavity radius. The total energy of the system can be written

$$E_t(R)=(\text{localized})=E_e(R)+E_M(R). \qquad (3)$$

The most stable configuration of the localized state is obtained by minimizing the total energy [Eq. (3)] with respect to R,

$$\partial E_t(R)/\partial R=0 \quad \text{at } R=R_0. \tag{4}$$

The resulting energy $E_t(R_0)$ can be, of course, either positive or negative as the liquid rearrangement process requires the investment of energy [$E_M(R_0)>0$] which is (wholly or partially) compensated by the electronic energy term $E_e(R_0)$.

(b) *The stability of the localized excess electron state.* The absolute sign of the minimum electronic energy of the localized state does not determine whether this localized state will be energetically found. Cases are encountered when $E_t(R_0)>0$ for the localized state [e.g., liquid He where $E_t(R_0)\approx 0.25$ eV] and still the localized state is stable. To assess the energetic stability of the localized excess electron state in a liquid, one has to compare the energy of the localized state with that of the quasifree state. The general stability criterion for the localized state implies that

$$E_t(R_0)<V_0. \tag{5}$$

Following these general considerations, two specific problems have to be considered:

(c) *The nature of electron–solvent interactions.* This general quantum mechanical problem involves the elucidation of the various contributions to the electronic interaction energy of the excess electron in the quasifree and in the localized state.

(d) *Configurational changes in the medium.* In the case of a localized excess electron the liquid structure will be modified because of the following effects:

(1) Local conformational changes arising from short-range electron–solvent repulsion can lead to cavity formation. This effect operates both in nonpolar liquids (e.g., liquid helium) and in polar liquids (e.g., ammonia).

(2) The long-range polarization field induced by the excess electron may lead to marked structural changes in the fluid arising from electrostriction effects, which are operative in both polar and nonpolar liquids, and from the effects of rotational polarization, which are operative in a polar liquid such as water or ammonia.

In order to understand the properties of excess electron states the energy changes accompanying these structural modifications have to be estimated.

In this work we shall consider some of the problems related to the understanding of low-density localized excess electron states in polar liquids where no chemically stable bound states exist for the excess electron. We shall focus attention on the solvated electron in liquid ammonia and in liquid water which provide typical model systems for the localized state in polar solvents. The main theoretical points which will be handled in the present study can be summarized as follows:

(a) *The energy of the quasifree electron state in polar solvents* determines both the stability of the localized state and also the general thermodynamic and optical properties of the solvated electron. Although a reliable theoretical estimate of this energy cannot be provided, we shall present some model calculations which will lead to an "educated guess" of this energy term.

(b) *Energetic stability of the localized state.* Although experiment demonstrates that indeed relation (5) is satisfied in polar solvents, it will be useful to provide a more firm theoretical basis for the stability of the excess electron in polar solvents.

(c) *Configurational stability of the ground state of the solvated electron.* A complete description of the localized state has to involve the calculation of both E_e and E_M. Early studies of the electron cavity model in liquid ammonia introduced the medium rearrangement energy; however, the electronic energy was handled in a rather simplified fashion.[44–46] Later studies based on the polaron model[19] and on a molecular field model[20] were mainly centered on the calculation of the electronic energy at a fixed cavity radius, which was chosen to "fit" the volume expansion data (a "small" value for water and $R_0=3.2$ Å for ammonia). Some contributions of the medium rearrangement energy were introduced at this fixed R_0 to account for thermodynamic properties such as the heat of solution. However, to date no attempt was made to calculate the variation of the total energy accompanied by the change in structural parameters such as the cavity radius or the first-shell coordination number. Nevertheless, some considerations of these terms are to be found in work by O'Reilly[20,47] and, Iguchi.[48] In the present work we shall attempt to establish theoretically the stable configuration of the ground state of the solvated electron by the calculation of the configurational diagrams for the ground state.

(d) *The electronic energy for the cavity model.* We shall consider the following contributions within the framework of the one particle scheme: (1) electronic kinetic energy, (2) short-range repulsions, (3) short-range attractive interactions, (4) long-range attractive polarization interactions. In previous work contributions (3) and (4) were included within the framework of a continuum model. Later a molecular field model was proposed for the short-range attractive interactions. To account for the attractive electronic interactions we shall apply a microscopic molecular electrostatic potential for short-range attractive interactions (3) and a Landau-type polaron potential for the long-range polarization forces (4). The effect of short-range repulsions (2) which was not treated before will be handled by the Wigner–Seitz method which was successfully applied to localized electron states in nonpolar solvents.

(e) *The medium rearrangement energy.* In the case of nonpolar solvents one has to consider in this context just the contributions of the contractible surface work and the pressure volume work acting on the electron cavity. The situation is much more complicated for polar solvents where the following contributions to E_M have to be considered.

(1) The energy, E_v, required to form a void in the liquid. As in the case of the nonpolar solvents this term can be roughly estimated from the surface tension. Unlike that case, the contribution of this term for polar solvents (where the cavity radius is small) is rather small.

(2) The volume pressure work. This term which is again analogous to that encountered for the nonpolar solvents is important only at high pressures.

(3) The long-range polarization energy π of the medium required for the orientation of the permanent dipoles to form the potential well.

(4) The dipole–dipole repulsion term between E_{dd}, the oriented dipoles in the first coordination layer on the cavity boundary.

(5) Short-range repulsions between the reoriented solvent molecules on the cavity boundary.

(6) The energy required for the rupture of hydrogen bonds.

(f) *Configurational diagrams in excited electronic states.* In spite of the success of previous semiempirical theoretical work to obtain fair agreement with experiment concerning the location of the maximum in the optical absorption of the electron in metal ammonia solutions, many questions remain open. The half-line-width in the optical absorption amounts to about 60% of the optical excitation energy. The problem of line broadening in the optical spectrum was never properly resolved. These problems will be handled theoretically by calculating the configuration diagrams for the ground and excited electronic states. These configuration diagrams will be then applied for the understanding of the intensity distribution in absorption.

We hope that the present study will provide a better understanding of the structural, thermodynamic, and optical properties of the solvated electron.

X. DISCUSSION

In this paper we have attempted to present a microscopic model for excess electron states in polar solvents with a special emphasis on the physical properties of dilute metal ammonia solutions. The purpose of the exercise is not to reproduce experimental results, which are currently determined in a much more reliable manner in the laboratory, but rather to try to elucidate the pertinent physical factors which determine the stability of a localized ground state, the physical properties of this state and the nature of excited electronic states of excess electrons in polar solvents.

A detailed calculation of the electronic energy and of the medium rearrangement energy has been provided. Although these calculations are admittedly approximated and some of their features will be soon improved upon, we believe that the present results are useful to demonstrate the major theoretical ingredients which have to be introduced in such a study. In the calculation of the electronic energy we have considered the electronic kinetic energy and the short-range attractive interactions on the basis of a molecular model while short-range repulsions and long-range polarization effects were introduced by a "coarse grain" averaging method. The medium rearrangement energy was handled by the introduction of a molecular model for the first solvation layer, using a continuum approximation beyond it. The present treatment was rather successful in predicting and interpreting semiquantitatively a variety of structural, thermodynamic, and optical properties of metal ammonia solutions, and of the hydrated electron such as the size of the electron cavity, the heat of solution, optical excitation energies, line shapes in the optical spectrum, photoelectric thresholds, and photoionization onsets. The number of adjustable parameters entering into the theory is minimal. Apart from the unknown value of the short-range repulsion term V_0 (which is not expected to be far from $V_0=0$), the present treatment can provide a reasonable estimate of the ground-state configuration of the solvated electron in simple polar solvents.

The major general conclusions arising from the present study can be summarized as follows:

(a) In order to understand the properties of the solvated electron in polar solvents, one cannot get away with the calculation of only the electronic energy. The medium rearrangement energy plays a crucial role in determining the ground-state configuration and the physical properties of the solvated electron. Inclusion of the medium rearrangement energy is also crucial for the understanding of the different properties of the solvated electron in different solvents. It is hoped that people interested in the properties of the solvated electron now will stop performing calculations for an arbitrary cavity size which is chosen to fit some experimental data.

(b) The major physical reasons for the stability of a localized ground state of an excess electron in polar solvent can be now traced to the following physical factors:

(1) V_0 is not far from $V_0=0$ and the general stability conditions are thus satisfied, Eqs. (4) and (39). For a hypothetical solvent characterized by large negative V_0 (say $V_0 \approx -2$–3 eV), the quasifree electron state would have been energetically favored.

(2) The potential acting on the electron is characterized by a large negative potential within the cavity, which leads to large negative values of the electronic energy. This feature is not unique to the present microscopic model and can be considered as almost model independent since it also appears in the polaron model and can even be incorporated in the primitive electron in a box model.

(3) The present treatment of the electronic energy and of the total energy of the solvated electron involves a major extension of the old Landau self-trapping model, whereupon the role of short attractive interactions is now properly introduced.

(4) The present approach based on the configurational diagrams for the ground and excited states of the solvated electron has many features in common with the conventional picture of F center and impurity centers in solids. The major difference between electron trapping in a crystal anion vacancy and the solvated electron is that in the latter case the electron reorganizes the solvent ("digging its hole" in the old terminology) and thus the configuration diagrams have to be determined in a self-consistent manner. This is obviously also the reason that the configuration diagrams are temperature dependent.

(5) The interpretation optical spectrum of the solvated electron should not be limited to the energy of the band maximum, but rather the whole line shape should be considered. The broadening arising from thermal motion is responsible for a major portion of the linewidth of the bound-bond $1s \rightarrow 2p$ transition. Similar considerations should be applied for an interpretation of the energy dependence of an external photoemission cross sections and the intrinsic photoionization data, if these will become available.

(6) A unique feature of the extended self-trapping model which is strongly dependent on the physical picture of Landau involves the Coulombic form of the long-range attractive potential due to the permanent dipoles. This potential can sustain an infinite number of excited levels converging to the photoionization threshold which should be observed at about 2.5 eV in NH_3. Other models such as an electron in a finite box are unrealistic in this respect as they can sustain only a finite number of excited states. That is the reason why photoconductivity (or alternatively, laser flash photolysis) data are of such intrinsic importance for the understanding of the excited states of these systems and of their time-dependent behavior.

There are currently several interesting problems which deserve further theoretical study to gain a deeper insight into the nature of electrons in polar solvents:

(a) Theoretical calculation of V_0 in polar solvents.

(b) More elaborate calculations of the electronic energy of the ground and higher excited states using better trial functions. This task was recently carried out by Copeland and Kestner.[71]

(c) The location, intensity and width of higher excited states is of considerable interest for the understanding of the asymmetric broadening of the absorption band of the solvated electron. The nature of these states depends critically on the exact definition of V_0.[51]

(d) The nature of electronic nuclear coupling in the triply excited $2p$ state is of considerable interest both in the context of the general Jahn–Teller effect and in relation to the problem of the solvated electron. A general formulation of the line shape problem including nontotally symmetric distortions will be reported by one of us (J. J.).

(e) Relaxation phenomena of these excited electronic states are of considerable interest; these were recently handled by one of us (J. J.) using a general theory of nonradiative processes.[72] An additional complication in this case is that the dipole moments can also relax under the excited-state electronic distribution.

(f) A closely related problem involves the properties of the positron and of the positronium atom in polar solvents which can be studied by positronium annihilation methods.[73–75] It is amusing to notice that as the positron does not suffer from the exclusion principle, the energy of a quasifree positron in a polar solvent such as an ammonia is expected to consist just of the contribution of long-range polarization interactions. Thus adopting the notation of Sec. II for positron, $T=0$ and $V_0=U_p \approx -3$ eV, and this particle is not expected to be localized. On the other hand, for the neutral positronium atom the long-range polarization interaction are switched off, whereupon $U_p=0$ and $V_0=T=+3$ eV. Hence the positronium is expected to be localized in a polar solvent due to short-range repulsion. The effective potential will be $V(r)=0$ for $r<R$ and $V(r)=V_0$ for $r>R$. This picture is completely analogous to that applied for positronium bubbles in nonpolar liquids. This problem deserves a more serious further study, both theoretically and experimentally.[76]

* This work is partially supported by a National Science Foundation Grant to Neil R. Kestner and an AFOSR Grant to J. Jortner. Computer time was supported primarily by National Science Foundation Grant to Louisiana State University.

† Alfred P Sloan Fellow.

[1] L. Onsager, *Modern Quantum Chemistry-Istanbul Lectures*, edited by O. Sinanoğlu (Academic, New York, 1966), Vol II, p. 129.

[2] G. Lapoutre and M. J. Sienko, *Metal Ammonia Solutions* (Benjamin, New York, 1964).

[3] J. C. Thompson, *Chemistry of Nonaqueous Solvents*, edited by J. J. Lagowski (Academic, New York, 1967), p. 265.

[4] J. L. Dye, Acct. Chem. Res. **1**, 306 (1968).

[5] F. Cafasso and B. R. Sundeim, J. Chem. Phys. **31**, 809 (1959).

[6] J. Eloranta and H. Linschitz, J. Chem. Phys. 2214 (1963).

[7] R. Catteral, P. L. Stodulski, and M. C. R. Symmons. J. Chem.

Soc. A437 (1968).
[8] Advan. Chem. Ser. 50, (1964).
[9] J. W. Boag and E. J. Hart, J. Am. Chem. Soc. 84, 4090 (1962).
[10] L. Keens, Nature 188, 42 (1963).
[11] E. J. Hart, Science 146, 19 (1964).
[12] L. M. Dorfman and M. S. Matheson, Progr. Reaction Kinetics 3, 239 (1965).
[13] M. C. Sauer, S. Arai and L. M. Dorfman, J. Chem. Phys. 42, 708 (1965).
[14] S. Arai and M. C. Sauer, J. Chem. Phys. 44, 2297 (1966).
[15] M. F. Deigen, Zh. Eksp. Teor. Fiz. 26, 300 (1954).
[16] A. S. Davydov. Zh. Eksp. Teor. Fiz. 18, 913 (1948).
[17] J. Jortner, J. Chem. Phys. 30, 839 (1959).
[18] L. Landau, Physik. Z. Soviet Union 3, 664 (1933).
[19] J. Jortner, J. Chem. Phys. 27, 823 (1957).
[20] D. E. O'Reilly, J. Chem. Phys. 41, 3736 (1964).
[21] R. H. Land and D. E. O'Reilly, J. Chem. Phys. 46, 4496 (1967).
[22] G. Careri, F. Scaramuzi, and J. O. Thomson, Nuovo Cimento 13, 186 (1959).
[23] L. Meyer and F. Reif, Phys. Rev. 110, 279 (1958); 119, 1164 (1960); 123, 727 (1961).
[24] J. Levine and T. M. Saunders, Phys. Rev. Letters 8, 159 (1962).
[25] T. M. Sanders, Bull. Am. Phys. Soc. Ser. II 7, 606 (1962); J. Levine, Ph.D. thesis, University of Minnesota, 1964 (unpublished); T. M. Sanders and J. Levine (unpublished).
[26] W. T. Sommer, Phys. Rev. Letters 12, 271 (1964).
[27] M. A. Woolf and G. W. Rayfield, Phys. Rev. Letters 15, 235 (1965).
[28] K. R. Atkins, Phys. Rev. 116, 1399 (1959).
[29] G. Kuper, Phys. Rev. 122, 1007 (1961).
[30] J. Jortner, N. R. Kestner, S. A. Rice, and M. H. Cohen, J. Chem. Phys. 43, 2614 (1965); *Modern Quantum Chemistry—Istanbul Lectures*, edited by O. Sinanoğlu (Academic, New York, 1966), p. 129.
[31] K. Hiroike, N. R. Kestner, S. A. Rice, and J. Jortner, J. Chem. Phys. 43, 2625 (1965).
[32] B. Burdick, Phys. Rev. Letters 14, 11 (1965).
[33] B. E. Springett, M. H. Cohen, and J. Jortner, Phys. Rev. 159, 183 (1957).
[34] R. C. Clark, Phys. Letters 16, 42 (1965).
[35] L. Meyer, H. T. Davis, S. A. Rice, and R. J. Donnelly, Phys. Rev. 126, 1925 (1962).
[36] H. Schnyders, S. A. Rice, and L. Meyer, Phys. Rev. Letters 15, 187 (1965).
[37] H. Schnyders, S. A. Rice, and L. Meyer, Phys. Rev. 150, 127 (1967).
[38] D. W. Awan, Proc. Phys. Soc. (London) 83, 659 (1964).
[39] L. S. Miller, S. Howe, and W. E. Spear, Phys. Rev. 166, 861 (1968).
[40] B. Halpern, J. Lekner, S. A. Rice, and R. Gomer, Phys. Rev. 156, 351 (1967).
[41] J. Lekner, Phys. Rev. 158, 130 (1967).
[42] B. Halpern and R. Gomer, J. Chem. Phys. 43, 1069 (1968); 51, 1031, 1048, 3043, 5709 (1969).
[43] B. E. Springett, M. H. Cohen, and J. Jortner, J. Chem. Phys. 48, 2720 (1968).
[44] W. N. Lipscomb, J. Chem. Phys. 21, 52 (1953).
[45] R. A. Ogg, Phys. Rev. 69, 668 (1946).
[46] R. A. Stairs, J. Chem. Phys. 27, 1431 (1957).
[47] D. E. O'Reilly, J. Chem. Phys. 35, 1856 (1962).
[48] K. Iguchi, J. Chem. Phys. 48, 1735 (1968).
[49] M. H. Cohen and J. C. Thompson, Advan. Phys. 17, 857 (1968).
[50] J. Jortner and N. R. Kestner, in Metal-Ammonia Solns., Physicochem. Properties, Proc. Colloque Weyl II, Cornell Univ. 1969 (to be published).
[51] The reader should carefully note that our electronic energy in Eq. (26) is slightly inconsistent with our definition of V_0. In Eq. (27) we include contributions from the polarization of the medium but such effects were already included in the definition of V_0. This difference is minor since the charges are quite localized but in comparing V_0, which we use as an unknown parameter, with any direct measurement of it one must keep this slight difference in mind. The difference does not affect the ground state significantly but can affect the nature of the excited state, i.e., whether it is found or metastable. These points are being studied and will be reported in later publications.
[52] M. H. Cohen and J. Jortner, Phys. Rev. 180, 238 (1969).
[53] This last point was emphasized by Professor L. Onsager in his comments at the Colloque Weyl II, Cornell University, June 1969. We are grateful to Professor Onsager for discussions on this problem.
[54] A. D. Buckingham, Discussions Faraday Soc. 24, 151 (1967).
[55] See, for example, K. Iguchi, J. Chem. Phys. 48, 1735 (1968). Here the orientation is considered from the viewpoint of a continuum model.
[56] D. Eisenberg and W. Kauzmann, *The Structure and Properties of Water* (Oxford U. P., New York, 1969), p. 46.
[57] R. F. W. Bader and G. A. Jones, J. Chem. Phys. 38, 2791 (1963).
[58] U. Schindewolf, in Metal-Ammonia Solns., Physicochem. Properties, Proc. Colloque Weyl II, Cornell Univ., 1969 (to be published), and references discussed there.
[59] R. Catterall, in Metal-Ammonia Solns., Physicochem. Properties, Cornell University, 1969 (to be published), and references discussed there.
[60] See, for example, R. Catterall, in *Metal Ammonia Solutions*, edited by G. Lepoutre and M. J. Sienko (Benjamin, New York, 1964), pp. 41ff.
[61] This is based on preliminary unpublished calculations.
[62] H. Blades and J. W. Hodgins, Can. J. Chem. 83, 411 (1955).
[63] R. K. Quinn and J. J. Lagowski, J. Phys. Chem. 73, 2326 (1969).
[64] S. Golden and T. R. Tuttle, Jr., in Metal-Ammonia Solns., Physicochem. Properties, Colloque Weyl II, Cornell University, 1969 (to be published).
[65] W. R. Elliott, Science 157, 559 (1967).
[66] See, for example, M. S. Matheson, Advan. Chem. Ser. 50, 45 (1965).
[67] M. S. Matheson and L. M. Dorfman, *Pulse Radiolysis* (M. I. T. Press, Cambridge, Mass., 1969).
[68] M. C. Sauer, Jr., S. Arai, and L. M. Dorfman, J. Chem. Phys. 42, 708 (1965).
[69] L. R. Dalton, J. L. Dye, E. M. Fielden, and E. J. Hart, J. Phys. Chem. 70, 3358 (1966).
[70] R. R. Dewald and J. L. Dye, J. Phys. Chem. 68, 121 (1964).
[71] D. A. Copeland and N. R. Kestner, preliminary results in Metal-Ammonia Solns., Physicochem. Properties, Proc. Colloque Weyl II, Cornell University, 1969 (to be published); complete results will be published in J. Chem. Phys.
[72] See, for example, M. Bixon and J. Jortner, J. Chem. Phys. 48, 715 (1968); 50, 3284 (1969) or K. F. Freed and J. Jortner, *ibid.* 50, 2916 (1969).
[73] P. G. Varlashkin and A. T. Stewart, Phys. Rev. 148, 459 (1966).
[74] P. G. Varlashkin and A. T. Stewart, *Positron Annihilation*, edited by A. T. Stewart and L. O. Roelling (Academic, New York, 1967), p. 313.
[75] P. G. Varlashkin, J. Chem. Phys. 49, 3088 (1968).
[76] Such work is now in progress by P. G. Varlashkin, N. R. Kestner, and J. Jortner.
[77] M. Lax, J. Chem. Phys. 20, 1752 (1952).
[78] R. Kubo and Y. Toyozawa, Progr. Theoret. Phys. (Kyoto) 13, 161 (1955).

[On page 413, sections II–IX (original page numbers 1192–1212) were deleted.]

Solubilization of Alkali Metals with Cyclic Polyethers

The following paper by Dye, DeBacker, and Nicely marks an exciting new departure in the study and application of solutions of alkali metals. By stabilization of the cations by complexation with cyclic polyethers, it is possible to dissolve alkali metals in solvents in which they are ordinarily insoluble or very slightly soluble.

62

Solubilization of Alkali Metals in Tetrahydrofuran and Diethyl Ether by Use of a Cyclic Polyether

Sir:

We wish to report a new technique for dissolving alkali metals in solvents in which they are ordinarily either insoluble or only slightly soluble. This method may extend the range of solvents in which the properties of relatively stable solutions of solvated electrons and other species common to metal–amine solutions[1,2] can be studied. Of particular interest would be the ability to make extended comparisons with the properties of solvated electrons produced by pulse radiolysis.

The basis for this increased solubility is the ability of certain cyclic polyethers to complex alkali metal cations.[3,4] Noting that stabilization of the cations should increase the solubility of the metals, we studied the effect of adding cyclohexyl-18-crown-6,[5] **1**, to tetrahydrofuran (THF) and diethyl ether (Et_2O) in the presence of a mirror of potassium. With both solvents,

1

deep blue solutions were formed. Solutions of both potassium and cesium in THF formed readily at room temperatures and were stable for several hours even in the absence of excess metal. Metal concentrations of about 1×10^{-4} *M* were obtained by using 5×10^{-3} *M* solutions of **1**. Solutions stored at $-78°$ for several days showed no visible signs of decomposition. In order to form solutions of potassium in Et_2O it was necessary to first cool the system to $-78°$. Once formed, however, the blue solutions in this solvent were stable at room temperatures for 5–10 min and for hours at $-78°$. At this writing only the three metal–solvent pairs described above have been examined. It is probable that many other systems will behave in a similar fashion.

(1) R. R. Dewald and J. L. Dye, *J. Phys. Chem.*, **68,** 121 (1964).
(2) M. Ottolenghi, K. Bar-Eli, and H. Linschitz, *J. Chem. Phys.*, **43,** 206 (1965).
(3) C. J. Pedersen, *J. Amer. Chem. Soc.*, **89,** 7017 (1967).
(4) C. J. Pedersen, *ibid.*, **92,** 386 (1970).
(5) See ref 3 for a discussion of the nomenclature of this class of compounds.

It has been reported[6] that very dilute solutions of potassium in THF exhibit two epr signals, a four-line pattern characteristic of the potassium monomer and a single narrow line attributed to the solvated electron. In the presence of **1**, however, the much more concentrated potassium solutions in THF used in this work showed only a single epr line. In Et_2O at 25°, in addition to the single line ($C \simeq 10^{-7}$ M) a weak four-line pattern ($C \simeq 10^{-8}$ M, $A \simeq 11$ G) was observed, probably attributable to the potassium monomer. Both absorptions in Et_2O were absent at 0° and appeared upon warming to room temperature. The single line observed in THF solutions could be observed down to the freezing point of the solvent. In both solvents at low temperatures (−60° and below) a weak seven-line pattern was observed. The splitting value and relative intensities indicate that this absorption is probably from the benzenide anion.[7] This identification was strengthened by adding small amounts of benzene to a similar solution of potassium in THF containing **1**. The result was a marked increase in the intensity of the epr pattern but no change in the number of lines or the splitting value.

The epr results indicate that the relative concentration of monomer decreased when **1** is present. They also show, in comparison with the optical spectra, that only a small fraction of the total dissolved metal gives an epr pattern, a result which is consistent with the behavior of metal–amine solutions.[8]

The optical absorption spectra measured at room temperatures with a Beckman DK2 spectrophotometer are shown in Figure 1. These spectra have been cor-

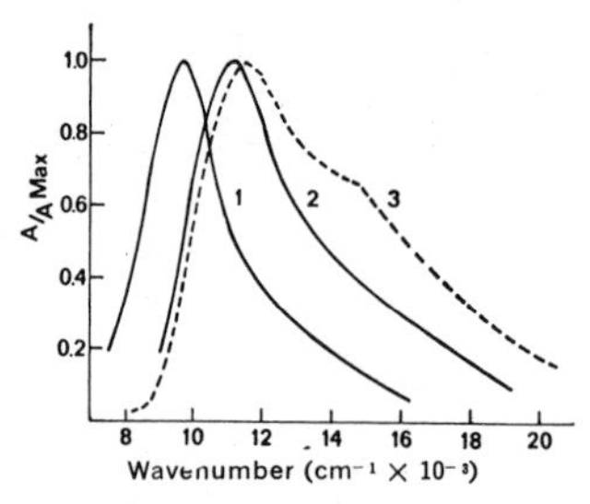

Figure 1. Absorption spectra in the presence of **1**: (1) cesium in THF, (2) potassium in THF, (3) potassium in Et_2O.

rected for decomposition by interpolating the results from successive scans. Potassium in THF in the presence of an excess of **1** showed only a single band with a maximum at 11,100 cm^{-1}, while the band maximum for cesium solutions occurred at 9700 cm^{-1}. With Et_2O the band maximum for potassium occurred at 11,400 cm^{-1} and a shoulder (probably caused by sodium from the Pyrex container[9]) was observed at $\sim$14,000 cm^{-1}. These peaks can be identified with the corresponding metal-dependent "R-bands" in metal–amine solutions.[1,2] Even when the absorbance at the peak was $\sim$3 in Et_2O (as judged by the positions of unit absorbance, 9500 and 16,000 cm^{-1}) there was no detectable absorption in the region from 8000 to 4000 cm^{-1}. A similar search was not possible for solutions in THF because of the strong solvent absorptions in this region. The absence of an optical band for the solvated electron is not necessarily at variance with the presence of an epr signal attributed to this species. The estimated spin concentration of 10^{-7} M is an order of magnitude below that which would have been detected optically.

Similar absorption bands in metal–amine solutions have been attributed to the alkali anion, M^-.[10] The absence of hyperfine splitting by the potassium nucleus in THF in the presence of **1** contrasts with the results in THF alone[6] and indicates that the equilibrium

$$M \rightleftarrows M^+ + e^- \quad (1)$$

is shifted to the right by the complexation equilibrium

$$M^+ + \mathbf{1} \rightleftarrows \text{complex} \quad (2)$$

Presumably the solubility equilibrium

$$2M(s) \rightleftarrows M^+ + M^- \quad (3)$$

is similarly shifted to the right. However, the absence of a solvated electron absorption band and the very low concentration of unpaired spin indicate that the equilibrium

$$M^- + \mathbf{1} \rightleftarrows \text{complex} + 2e^- \quad (4)$$

lies far to the left in these solvents.

Acknowledgment. Support of this research by the U. S. Atomic Energy Commission under Contract No. AT(11-1)-958 is gratefully acknowledged. We are indebted to Dr. H. K. Frensdorf of E. I. du Pont de Nemours Co. for providing us with a sample of the crown compound.

* Address correspondence to this author.

James L. Dye,* Marc G. DeBacker, Vincent A. Nicely
Department of Chemistry, Michigan State University
East Lansing, Michigan 48823
Received June 12, 1970

(6) R. Catterall, J. Slater, and M. C. R. Symons, *J. Chem. Phys.*, **52**, 1003 (1970).
(7) T. R. Tuttle, Jr., and S. I. Weissman, *J. Amer. Chem. Soc.*, **80**, 5342 (1958).
(8) L. R. Dalton, J. D. Rynbrandt, E. M. Hansen, and J. L. Dye, *J. Chem. Phys.*, **44**, 3969 (1966).
(9) I. Hurley, T. R. Tuttle, Jr., and S. Golden, *ibid.*, **48**, 2818 (1968).
(10) S. Matalon, S. Golden, and M. Ottolenghi, *J. Phys. Chem.*, **73**, 3098 (1969).

The Solvation Number of the Electron in Ammonia

Pinkowitz and Swift describe a proton magnetic resonance study of dilute potassium–ammonia solutions in the following paper. From line shape analyses, they calculated an average lifetime of $1–2 + 10^{-12}$ sec for the NH_3 molecules in electron–ammonia complexes and an electron solvation number of 20–40.

Reprinted from:

THE JOURNAL OF CHEMICAL PHYSICS VOLUME 54, NUMBER 7 1 APRIL 1971

Solvation Number of the Electron in Liquid Ammonia. A Nitrogen Magnetic Relaxation Study

R. A. PINKOWITZ AND T. J. SWIFT

Department of Chemistry, Case Western Reserve University, Cleveland, Ohio 44106

63

The high resolution proton magnetic resonance spectrum of a dilute solution of potassium in liquid ammonia shows a large contribution to the line shape from ^{14}N spin–lattice relaxation through magnetic dipolar interaction with unpaired electrons. The predominant ^{14}N-electron interaction occurs through Fermi contact. From total line shape analyses of the proton spectra recorded at 60 MHz and 100 MHz values were obtained for both $\tau(e^-)$, the average lifetime of a given ^{14}N in a given electron–ammonia complex, and for N, the average solvation number of the electron. The lifetime varies from approximately $1-2\times10^{-12}$ sec over the concentration and temperature range studied. The solvation number ranges from approximately 20 to 40. These results are most consistent with the lattice vacancy model of the solvated electron.

INTRODUCTION

In the century since the discovery that alkali metals dissolve reversibly in liquid ammonia much attention has been focused on metal solutions in a number of solvents. From these studies came the generally accepted conclusion that a metal such as potassium dissolves in liquid ammonia to produce as the primary species a solvated K^+ and a solvated electron.[1] Subsequently, it was discovered that high-energy pulsing can produce the solvated electron as a transient species in an aqueous system.[2]

A natural question of interest concerns the structural nature of this elemental electron in solution. The type of structure of the electron–solvent complex and the number of solvent molecules in a given complex are basic questions which must be answered. Further, one may inquire concerning kinetic parameters for the electron–solvent complex such as the average time that the solvent molecules remain in a given complex. This manuscript represents an attempt to give definitive answers to these questions through the application of high-resolution nuclear magnetic resonance techniques.

THEORY

In this section we will show how ^{14}N magnetic dipolar spin–lattice relaxation manifests itself in the high resolution proton magnetic resonance (PMR) spectra of metal–ammonia solutions. We will define a parameter associated with this relaxation and show how this parameter is obtained quantitatively from PMR spectra. In the subsequent section on results we will relate this parameter to the solvation number of the electron and to the average lifetime of the electron–solvent complex.

The PMR spectrum of pure liquid ammonia (Fig. 1) shows three distinctive features. The three relatively intense peaks arise from ^{14}N–H coupling, while the smaller doublet arises from ^{15}N–H coupling; in addition the center peak of the triplet is considerably taller and narrower than either of the outer peaks.

This last feature is a consequence[3] of ^{14}N nuclear spin–lattice relaxation through interaction of the ^{14}N nuclear quadrupole moment with the axial electric field gradient of the ammonia molecule. The three nondegenerate spin states of ^{14}N may be treated as supplying three distinct magnetic environments for protons coupled to them. The selection rules for nuclear transitions by the electric quadrupolar mechanism are such that we may write the following scheme of transition probabilities due to this mechanism.

$$+1 \underset{1/\tau}{\overset{1/\tau}{\rightleftharpoons}} 0 \underset{1/\tau}{\overset{1/\tau}{\rightleftharpoons}} -1, \qquad +1 \underset{2/\tau}{\overset{2/\tau}{\rightleftharpoons}} -1$$

Swift, Marks, and Sayre[4] showed that the parameters τ and J, the ^{14}N–H coupling constant, are all that are required to yield an excellent computed fit to any PMR spectrum of liquid ammonia recorded under low power and slow passage conditions. An example of such a spectral fit is given in Fig. 1.

There are other factors which could have affected the line shape of the PMR spectrum of pure liquid ammonia, but did not to any significant extent. Among these are proton spin–spin relaxation and ^{14}N spin–lattice relaxation through interaction between the ^{14}N magnetic dipole moment and the fluctuating magnetic surroundings of the lattice. The lack of importance of these two factors is perhaps best demonstrated by the sharpness of the doublet due to ^{15}N–H splitting.

As will be shown in the section on results ^{14}N magnetic dipolar spin–lattice relaxation is a very prominent factor in determining the PMR line shape for metal–ammonia solutions. Consequently we will consider it here in some detail.

The magnetic dipolar selection rules are such that we can write the following scheme of transition probabilities due to this mechanism:

$$+1 \underset{1/\tau'}{\overset{1/\tau'}{\rightleftharpoons}} 0 \underset{1/\tau'}{\overset{1/\tau'}{\rightleftharpoons}} -1.$$

Let us consider the low power, slow passage line

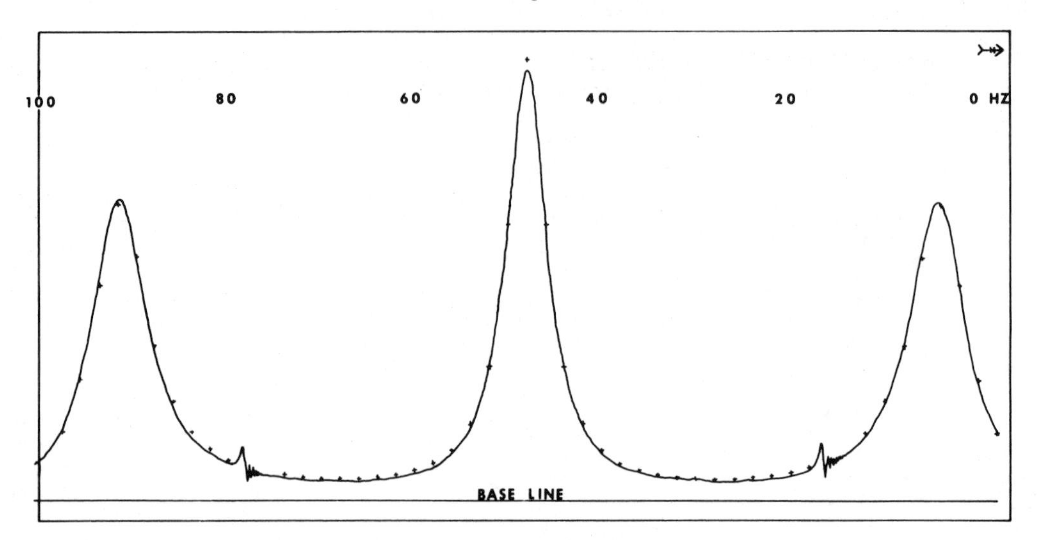

FIG. 1. PMR spectrum of pure liquid ammonia at +21.3 °C.

shape which would result for the PMR spectrum of liquid ammonia if the only parameters of importance were the ^{14}N–H coupling constant and the lifetime τ' above. For this situation we may employ the modified Bloch equations for steady state. We define transverse magnetizations G_{+1}, G_0 and G_{-1} for protons coupled to ^{14}N's in the three respective spin states and the equations are

$$0=i\gamma H_1M_0/3+G_{(1)}(-1/\tau'+i\Delta\omega_{(1)})+G_{(0)}/\tau',$$
$$0=i\gamma H_1M_0/3+G_{(0)}(-2/\tau'+i\Delta\omega_{(0)})+G_{(1)}/\tau'+G_{(-1)}/\tau',$$
$$0=i\gamma H_1M_0/3+G_{(-1)}(-1/\tau'+i\Delta\omega_{(-1)})+G_{(0)}/\tau', \quad (1)$$

where γ is the proton gyromagnetic ratio, H_1 is the radio frequency field strength, M_0 is the total longitudinal magnetization, and the $\Delta\omega$'s are defined as

$$\Delta\omega_{(1)}=\omega_{0(0)}+2\pi J-\omega,$$
$$\Delta\omega_{(0)}=\omega_{0(0)}-\omega,$$
$$\Delta\omega_{(-1)}=\omega_{0(0)}-2\pi J-\omega. \quad (2)$$

In Eqs. (2), $\omega_{0(0)}$ is the resonance frequency for those protons coupled to ^{14}N's in the state of $m_I=0$, ω is the applied radio frequency, and J is the ^{14}N–H coupling constant.

The absorption mode signal calculable from Eqs. (1) consists of three peaks; however the outer two peaks are approximately twice as high and half as broad as the center peak. Thus, the presence of a significant line shape contribution from ^{14}N spin–lattice relaxation is revealed by a distinct narrowing of the outer peaks, as compared to the center peak.

For solutions of potassium in ammonia the significant parameters are J, τ, and τ' and the appropriate modified Bloch equations are

$$0=i\gamma H_1M_0/3+G_{(1)}(-3/\tau-1/\tau'+i\Delta\omega_{(1)})$$
$$+G_{(0)}(1/\tau+1/\tau')+G_{(-1)}(2/\tau),$$
$$0=i\gamma H_1M_0/3+G_{(0)}(-2/\tau-2/\tau'+i\Delta\omega_{(0)})$$
$$+G_{(1)}(1/\tau+1/\tau')+G_{(-1)}(1/\tau+1/\tau'),$$
$$0=i\gamma H_1M_0/3+G_{(-1)}(-3/\tau-1/\tau'+i\Delta\omega_{(-1)})$$
$$+G_{(0)}(1/\tau+1/\tau')+G_{(1)}(2/\tau), \quad (3)$$

where we relate the $\Delta\omega$'s to a convenient variable X defined as, $X=(\omega-\omega_{0(0)})/2\pi J$, such that we have

$$\Delta\omega_{(1)}=-2\pi J(X-1),$$
$$\Delta\omega_{(0)}=-2\pi JX,$$
$$\Delta\omega_{(-1)}=-2\pi J(X+1). \quad (4)$$

From Eqs. (3) and (4) we can obtain the total complex magnetization $G=G_{(1)}+G_{(0)}+G_{(-1)}$ and then v the absorption mode signal as the imaginary component of G. The result is that v is proportional to A/B where

$$A=J^2[360\tau\tau'^4+792\tau^2\tau'^3+504\tau^3\tau'^2+72\tau^4\tau'$$
$$+32\pi^2J^2(5X^2+1)\tau^3\tau'^4+32\pi^2J^2(X^2+1)\tau^4\tau'^3],$$
$$B=J^2[(36X^2)\tau^4+(432X^2)\tau^3\tau'+(1656X^2)\tau^2\tau'^2$$
$$+(2160X^2)\tau\tau'^3+(900X^2)\tau'^4$$
$$+16\pi^2J^2(10X^4-10X^2+4)\tau^4\tau'^2$$
$$+32\pi^2J^2(14X^4-6X^2+4)\tau^3\tau'^3$$
$$+16\pi^2J^2(34X^4-2X^2+4)\tau^2\tau'^4$$
$$+64\pi^4J^4(X^6-2X^4+X^2)\tau^4\tau'^4]. \quad (5)$$

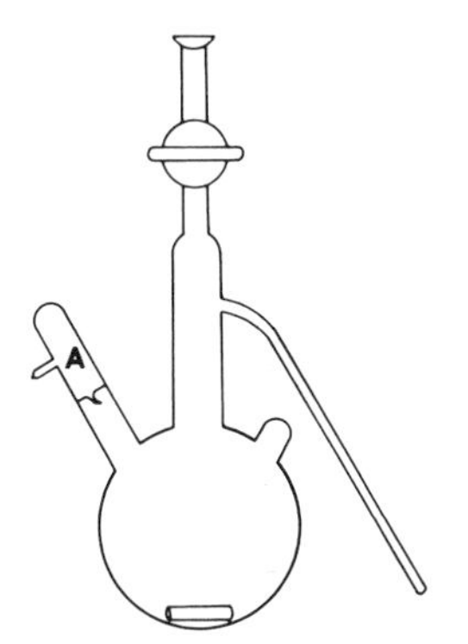

FIG. 2. Pyrex apparatus used in the preparation of potassium–ammonia solutions.

Thus the parameters J, τ, and τ' may be obtained by matching experimental spectra and spectra calculated from Eqs. (5). The matching procedure is discussed in some detail in the section on results.

A particularly troublesome factor in metal–ammonia solutions is the presence of traces of amide from solution decomposition. Amide ion catalyzes rapid ammonia–ammonia proton transfer, effectively producing ^{14}N–H decoupling and masking the ^{14}N magnetic dipolar spin–lattice relaxation effect. The procedure used by us for the effective suppression of amide is given in the experimental section.

EXPERIMENTAL

It has been recently shown by Dewald and Tsina[5] that the direct reaction between *t*-butanol and the solvated electron in liquid ammonia is quite slow. We can also conclude from the work of Birchall and Jolly[6] that (*t*-butanol)–ammonia proton exchange is slow and does not interfere with the ammonia proton spectrum. However, *t*-butanol is a strong enough acid to effectively suppress the amide concentration to an acceptable level.

Figure 2 shows the Pyrex apparatus used in the preparation of potassium–ammonia solutions. The breakseal tube A contains potassium metal which had been previously doubly distilled. The potassium was introduced into the breakseal tube by distillation under high vacuum and the breakseal was attached to the bulb. A trace amount of *t*-butanol was distilled into the apparatus under high vacuum. Ammonia which had been previously stored over potassium was then distilled into the bulb. By means of the stirring bar the breakseal was opened and the entire apparatus was washed with the solution. A sample of solution was poured into the attached NMR tube and the tube was sealed under vacuum.

The NMR sample tube was a coaxial cell unit purchased from Wilmad Glass Co. The inner tube contained the sample and the outer tube contained a sample of methylene chloride used to provide a lock signal for the Varian HA-100 spectrometer.

The PMR spectrum of the potassium–ammonia solution was then recorded as a function of temperature at both 60 MHz on a Varian A-60A spectrometer and 100 MHz on the HA-100. To insure that no significant decomposition occurred between runs at the two different frequencies the following procedure was employed. The spectrum was first obtained at 60 MHz and a given temperature. Then the spectrum was obtained at this temperature on the 100 MHz unit. Finally the 60 MHz spectrum was repeated to check that it agreed with the first run. The sample tube was stored in an isopropanol dry-ice bath when not in use to minimize decomposition.

After the NMR sample was prepared the ammonia from the bulb was distilled into a previously degassed and weighed bulb containing concentrated sulfuric acid. The amount of ammonia was determined from the weight increase of the bulb. The residual potassium in the bulb was reacted with water and titrated to a methyl red end point with standard HCl.

RESULTS

Figure 3 shows the PMR spectrum of a relatively dilute potassium–ammonia solution. From the fact that the outer peaks are significantly higher than the inner peak it is clear that the spectrum shows a large contribution from ^{14}N magnetic dipolar spin–lattice relaxation. This result is hardly surprising in view of the presence of a significant concentration[7] of unpaired electrons in the solution.

In the attempt to fit the spectrum J, τ, and τ' are known to be important parameters. Are there any others? One possible additional parameter is T_{2H}, the proton spin–spin relaxation time. Marks[8] has shown that this is not an important parameter by the following procedure. A relatively concentrated solution of amide in ammonia was prepared. The PMR spectrum of this sample was a sharp singlet due to rapid proton transfer. Addition of a sizeable concentration of potassium to this amide solution led to only a very slight increase in the PMR linewidth. Thus the effect of the solvated paramagnetic electrons on T_{2H} was insignificant.

The spectrum of Fig. 3 was fit to spectra calculated from Eq. (5) through use of a "damped least squares" procedure[9,10] on a Univac 1108 computer. The values obtained were for J, for τ, and for τ'. The values for J and τ are very nearly those for pure ammonia,[4] a result which was found for all the metal–ammonia solutions studied at all temperatures. This indicates that there is no large increase in the electric field gradient at the ^{14}N nuclei in the solvated electron species over that in bulk ammonia.

The large increase in $1/\tau'$ over bulk ammonia is due to the large magnetic field perturbation in the solvated unpaired electron species and it yields information about

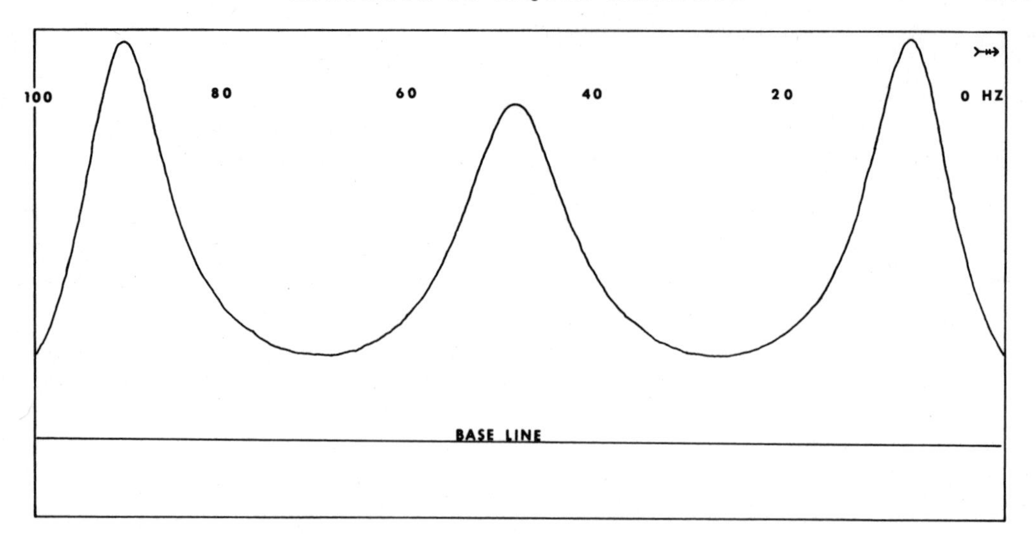

FIG. 3. PMR spectrum of a 10^{-2} M potassium–ammonia solution at +24.1°C.

either the ^{14}N magnetic dipolar spin–lattice relaxation in the solvation state or about the exchange between bulk and solvation states or both. From the narrowness of the width of the electron paramagnetic resonance signal[11] for ammoniated electrons it is known that the solvent exchange process is exceptionally rapid. The average lifetime of a given electron-^{14}N coupling has been estimated to be of the order of 10^{-12} sec. With an exchange this rapid the measured $1/\tau'$ represents an average of the contributions from bulk ammonia and the ammonias bound to unpaired electrons. Thus τ' is related to $T_1(e^-)$, the ^{14}N magnetic dipolar spin–lattice relaxation time in the solvated species through Eq. (6) below:

$$T_1^{-1}(e^-)=\{[NH_3](\tau')^{-1}/N[e^-]\}, \qquad (6)$$

where $[NH_3]$ is the bulk ammonia concentration, $[e^-]$ is the concentration of unpaired spins, and N is the average solvation number of the electron.

Magnetic dipolar spin–lattice relaxation arises from two sources, dipole–dipole coupling through space and Fermi contact coupling. Expressions for these two relaxation rates have been derived. That for nucleus–electron dipole–dipole coupling[12] is given in Eq. (7):

$$T_1^{-1}(e^-)=\frac{4S(S+1)\gamma_I^2g^2\beta^2}{30r^6}\left[\frac{\tau_c}{1+(\omega_I-\omega_s)^2\tau_c^2}+\frac{3\tau_c}{1+\omega_I^2\tau_c^2}+\frac{6\tau_c}{1+(\omega_I+\omega_s)^2\tau_c^2}\right]. \qquad (7)$$

In this equation, as it applies to the solvated electrons, S is the electron-spin quantum number, γ_I is the ^{14}N gyromagnetic ratio, $g\beta$ is the electron magnetic moment, r is the average electron-^{14}N separation, τ_c is the correlation time for the spatial coupling of the two magnetic moments, ω_I is the angular Larmor frequency of the ^{14}N nuclei at the magnetic field employed, and ω_s is the electron Larmor frequency.

The expression for spin–lattice relaxation through contact coupling[13] is given in Eq. (8):

$$T_1^{-1}(e^-)=[2S(S+1)A^2/3\hbar^2]\times\{\tau_e/[1+(\omega_I-\omega_s)^2\tau_e^2]\}. \qquad (8)$$

In this equation A is the electron-^{14}N coupling constant, and τ_e is the correlation time for the contact interaction.

To what extent is $T_1(e^-)$ in potassium–ammonia solutions determined by dipole–dipole coupling? To answer this question we compare the ^{14}N magnetic dipolar relaxation rates measured here with the proton T_1 data of Waugh[14] and co-workers. This comparison shows the $T_1(e^-)$ for ^{14}N is approximately two orders of magnitude smaller than $T_1(e^-)$ for protons in the same solution. Since γ_I for the proton is 13.65 that for ^{14}N and r for the proton should be smaller than r for ^{14}N (the ammonia molecule is expected to be oriented so that the protons are nearer the electron), the dipole–dipole contribution to $T_1(e^-)$ should be considerably less for ^{14}N than for the proton. Thus we may safely conclude that T_1 (e^-) is almost totally determined by the contact coupling mechanism. This conclusion is also consistent with the large Knight shift[15] observed for ^{14}N and the much smaller one observed for protons[16] in metal–ammonia solutions.

The correlation time τ_e in Eq. (8) is related to two well-defined lifetimes through Eq. (9):

$$\tau_e^{-1}=T_1^{-1}(\text{EPR})+\tau^{-1}(e^-), \qquad (9)$$

where T_1(EPR) is the electron spin–lattice relaxation

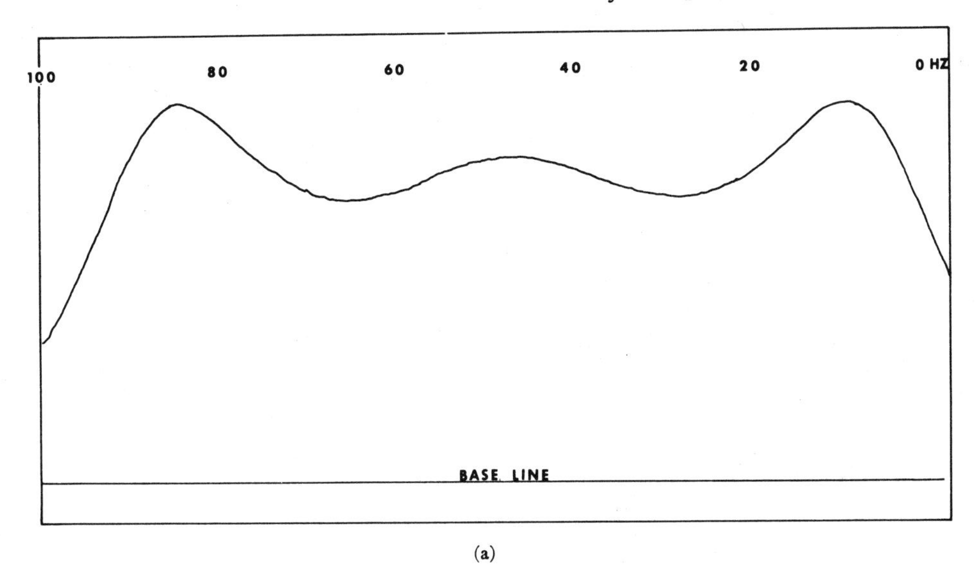

(a)

100 80 60 40 20 0 HZ

BASE LINE

(b)

FIG. 4. PMR spectra of a potassium–ammonia solution. (a) 60 MHz. (b) 100 MHz.

time obtainable from EPR data and $\tau(e^-)$ is the average lifetime of a given electron-^{14}N coupling (the average lifetime of a given ammonia molecule in a given solvated electron species). From the sharp singlet obtained as the EPR signal[11] of metal–ammonia solutions it is certain that $\tau(e^-) \ll T_1(\text{EPR})$. We combine this conclusion with Eqs. (6) and (8) to obtain

$$(\tau')^{-1} = \frac{2[e^-]S(S+1)(AN)^2}{3[\text{NH}_3]\hbar^2}\left[\frac{\tau(e^-)/N}{1+\omega_s^2\tau(e^-)^2}\right], \quad (10)$$

where we have also taken advantage of $\omega_s \gg \omega_I$. The term AN is grouped since it rather than A is obtainable from Knight shift data.

An examination of Eq. (10) reveals a most fortunate situation. As mentioned previously $\tau(e^-)$ has been estimated to be of the order of 10^{-12} sec. At the magnetic fields employed for the study of protons on both the 60 MHz and 100 MHz spectrometers ω_s is within an order of magnitude of 10^{12} sec^{-1} and hence τ' should be distinctly field dependent between these two fields. The field dependence permits an unequivocal determination of $\tau(e^-)$ and Eq. (10) may then be used with either 60 MHz or 100 MHz data to obtain N.

Figure 4 shows the spectra obtained for a typical metal–ammonia solution at 60 and 100 MHz. From the ratio of heights of outer peaks to inner peak it is clearly seen that the ^{14}N magnetic dipolar spin–lattice relaxation is significantly greater at 60 MHz than at 100 MHz in agreement with Eq. (10). Thus $\tau(e^-)$ could be obtained with a fair degree of precision.

In the calculation of N values, AN's were obtained from Knight shift data[15] for solutions of the same concentrations as those studied here. Unpaired electron concentrations were obtained from the EPR data of Hutchison and Pastor.[7]

Table I lists values of N and $\tau(e^-)$ for potassium–ammonia solutions at a number of temperatures and concentrations.

DISCUSSION

At the outset of this discussion it is important to indicate the error limits for $\tau(e^-)$ and N and to treat the sources of error. First $\tau(e^-)$ is known with a precision of approximately $\pm 10\%$ and the only appreciable error source comes in the spectral fits. The solvation number N is not known nearly as precisely as $\tau(e^-)$. One major source of uncertainty is the determination of (AN) based upon the conflicting Knight shift values contained in the literature.[15,17,18] This uncertainty resides principally in the determination of metal concentrations and is due to the combination of the inherent imprecision in the analysis procedures employed and of an indeterminate degree of solution decomposition. We have repeated these Knight shift measurements employing the double resonance technique with the Varian HA-100 instrument with a Monsanto frequency synthesizer for the production and systematic variation of the ^{14}N resonance frequency. We have obtained Knight shifts which are within the experimental uncertainty, discussed above, equal to those of O'Reilly and we have therefore employed his values. It should be added that the uncertainty in metal concentration will increase with increasing temperature due to increased decomposition. Thus interpolation of (AN) between the two temperatures presented by O'Reilly[15] also represents a source of uncertainty in these calculations. Thus we can only state that N is typically between 20 and 40 and we cannot say at this point if it varies systematically with either temperature or concentration. Nevertheless we can state with some conviction that N is quite large and the ammoniated electron is a highly delocalized species.

The numbers for $\tau(e^-)$ and N are quite consistent with a "cavity" model for the electron. In particular N agrees well with the number of 24 employed by O'Reilly[19] in his lattice vacancy model of the ammoniated electron. With use of the number of 24 for N, O'Reilly[11] has calculated $\tau(e^-)$ values from electron spin resonance linewidths which are within 20% of those given in Table I.

A problem of particular interest concerns the source of the time of $\approx 10^{-12}$ sec for $\tau(e^-)$. The orientational relaxation time obtained from dielectric relaxation measurements[20] is also approximately 10^{-12} sec for ammonia molecules in pure ammonia. Diffusion times for ammonia molecules from site to site in pure ammonia have been calculated[11] from diffusion coefficient data and they are typically between 10^{-12}–10^{-11} sec. Thus it is not possible to distinguish between diffusion and reorientation (or perhaps other rapid processes) on the basis of $\tau(e^-)$ values alone. However, the availability of a procedure for accurate $\tau(e^-)$ values together with such modifications as the study of metal–amine solutions may provide the answer to this question.

TABLE I. Calculated parameters for potassium ammonia solutions.

Temperature (°C)	Mole ratio ([K]/[NH_3])	(AN)* calc (ergs)	$\tau(e)$ (sec)	Solvation number
−37.0	1.49×10^{-3}	2.28×10^{-18}	1.85×10^{-12}	24.0
	1.72×10^{-3}	2.42×10^{-18}	1.03×10^{-12}	15.7
	1.87×10^{-3}	2.50×10^{-18}	2.10×10^{-12}	32.1
	3.88×10^{-3}	2.98×10^{-18}	1.70×10^{-12}	47.5
−24.0	1.49×10^{-3}	2.17×10^{-18}	1.70×10^{-12}	31.2
	1.72×10^{-3}	2.33×10^{-18}	1.32×10^{-12}	28.5
	1.87×10^{-3}	2.33×10^{-18}	1.89×10^{-12}	42.7
0.0	1.72×10^{-3}	2.14×10^{-18}	1.09×10^{-12}	31.2
	1.87×10^{-3}	2.06×10^{-18}	1.70×10^{-12}	41.9
	3.88×10^{-3}	2.34×10^{-18}	1.18×10^{-12}	38.2
+17.0	1.72×10^{-3}	2.03×10^{-18}	1.05×10^{-12}	32.1
	3.88×10^{-3}	2.03×10^{-18}	1.26×10^{-12}	24.8
+28.0	1.72×10^{-3}	1.95×10^{-18}	0.928×10^{-12}	29.8
	3.88×10^{-3}	1.85×10^{-18}	0.686×10^{-12}	25.1

* Calculated from Knight Shift data in Ref. 15.

ACKNOWLEDGMENT

We wish to acknowledge the financial support of the National Science Foundation in the form of research Grant GP-8562 to TJS.

[1] *Metal–Ammonia Solutions*, edited by G. Lepoutre and M. J. Sienko (Benjamin, New York, 1964).

[2] R. F. Gould (Ed.), Advan. Chem. Ser. **50**, (1965).

[3] J. Pople, Mol. Phys. **1**, 168 (1958).

[4] T. J. Swift, S. B. Marks, and W. G. Sayre, J. Chem. Phys. **44**, 2797 (1966).

[5] R. R. Dewald and R. V. Tsina, J. Phys. Chem. **72**, 4520 (1968).

[6] T. Birchall and J. Jolly, J. Am. Chem. Soc. **87**, 3007 (1965).

[7] C. Hutchison and R. Pastor, J. Chem. Phys. **21**, 1959 (1953).

[8] S. B. Marks, Ph.D. thesis, Case Institute of Technology, Cleveland, Ohio, 1966.

[9] K. Levenberg, Quart. Appl. Math. **2**, 164 (1944).

[10] D. Papousek and J. Pliva, Collection Czech. Chem. Commun. **30**, 3007 (1965).

[11] D. O'Reilly, J. Chem. Phys. **50**, 4743 (1969).

[12] I. Solomon, Phys. Rev. **99**, 559 (1955).

[13] I. Solomon and N. Bloembergen, J. Chem. Phys. **25,** 261 (1956).
[14] R. Newmark, J. Stephenson, and J. Waugh, J. Chem. Phys. **46,** 3514 (1967).
[15] D. O'Reilly, J. Chem. Phys. **41,** 3729 (1964).
[16] T. Hughes, J. Chem. Phys. **38,** 202 (1963).
[17] J. V. Acrivos and K. S. Pitzer, J. Phys. Chem. **66,** 1693 (1962).
[18] H. McConnell and C. Holm, J. Chem. Phys. **26,** 1517 (1957).
[19] D. E. O'Reilly, J. Chem. Phys. **50,** 5378 (1969).
[20] K. Breitschwerdt and H. Radscheit, Phys. Letters **29A,** 381 (1969).

Author Index

Subject Index